Springer INdAM Series

Volume 44

This series will publish textbooks, multi-authors books, thesis and monographs in English language resulting from workshops, conferences, courses, schools, seminars, doctoral thesis, and research activities carried out at INDAM - Istituto Nazionale di Alta Matematica, http://www.altamatematica.it/en. The books in the series will discuss recent results and analyze new trends in mathematics and its applications.
THE SERIES IS INDEXED IN SCOPUS

More information about this series at http://www.springer.com/series/10283

Onofrio Mario Di Vincenzo • Antonio Giambruno
Editors

Polynomial Identities in Algebras

Springer

Editors
Onofrio Mario Di Vincenzo
Dipartimento di Matematica, Informatica ed Economia
Università della Basilicata
Potenza, Italy

Antonio Giambruno
Dipartimento di Matematica e Informatica
Università di Palermo
Palermo, Italy

ISSN 2281-518X ISSN 2281-5198 (electronic)
Springer INdAM Series
ISBN 978-3-030-63110-9 ISBN 978-3-030-63111-6 (eBook)
https://doi.org/10.1007/978-3-030-63111-6

This Springer imprint is published by the registered company Springer Nature Switzerland AG.
The registered company address is: Gewerbestrasse 11, 6330 Cham, Switzerland

Preface

This volume contains the proceedings of the INDAM workshop on "Polynomial Identities in Algebras" held in Roma from September 16 to September 20, 2019. The purpose of the workshop was to present the current state of the art in the theory of PI-algebras.

The theory started with the discovery of special identities and with various structure theorems for primitive or prime rings satisfying a PI. Then, some deep results analyzing mainly the nil part of an algebra were proved leading to the theorem of Razmyslov on the nilpotency of the radical of a finitely generated PI-algebra over a field. A further major step was made by Kemer who developed a theory of varieties, leading to the solution of the Specht problem stating the finite generation of T-ideals in characteristic zero. The theory of Kemer introduced superalgebras and their superidentities as an essential tool. It turns out that the Grassmann algebra plays an important role and a basic result of Kemer states that a PI-algebra is PI equivalent to the Grassmann envelope of a finite-dimensional superalgebra.

Based on these grounds, the theory developed via two different methods: a geometric approach strongly related to invariants of matrices leading to the theory of trace identities and a combinatorial approach based on the representation theory of the symmetric group leading to the distinction of T-ideals through the analysis of some growth functions attached to them.

The workshop, inspired by the review of the classical results made in the last few years, revealed new perspectives and connections to other branches of mathematics suitable for the development of the theory.

The meeting brought together experts from different areas related to the theory of polynomial identities and focused on the computational and combinatorial aspects of the theory, its connection with invariant theory, representation theory, growth problems, and many other topics.

It was attended by experts from several countries, including Belgium, Brazil, Bulgaria, Canada, Israel, Poland, Russia, Ukraine, and the USA. The workshop featured 1-h lectures by E. Aljadeff, Y. Bahturin, A. Berele, V. Drensky, A. Giambruno, A. Kanel-Belov, P. Koshlukov, V. Petrogradsky, C. Polcino Milies, C.

Procesi, L. H. Rowen, and M. Zaicev and several other invited talks of shorter length.

The workshop was also an occasion for celebrating Antonio Giambruno's 70th birthday and his contribution to the theory of polynomial identities.

The papers of most of the principal speakers and of some of the invited speakers are included in the present volume. The contents span a broad range of themes in current active research areas.

The editors thank the Istituto Nazionale di Alta Matematica "Francesco Severi" for providing funding and logistical support for the workshop. They also wish to express their appreciation to the institutions that contributed financial support: Università della Basilicata, Università di Palermo, and Università di Roma "La Sapienza."

Potenza, Italy Onofrio Mario Di Vincenzo

Contents

About the Editors

Onofrio Mario Di Vincenzo has been a full professor of Algebra since 2000. He currently works at the University of Basilicata where he has been the head of the department for the last 8 years. His research interests focus on the polynomial identities of algebras, in particular in the presence of gradings or involutions acting on them.

Antonio Giambruno received his PhD in Mathematics from the University of Chicago in 1977 under the supervision of Israel N. Herstein. He became a professor at the department of mathematics of the University of Palermo in 1980. He has been a visiting researcher at numerous institutions, including University of Southern California, MSRI, University of Alberta, and University of San Paulo. His research focuses on noncommutative algebra, especially on polynomial identities and their connection with representation theory and asymptotic methods.

Some Thoughts on the Current State of the Theory of Identical Relations in Lie Algebras

Yuri Bahturin

Abstract This is an attempt to survey the progress made in the study of identical relations in Lie algebras during almost three decades since the publication of my book "Identical Relations in Lie Algebras", the only monograph devoted entirely to this area. Accordingly, we assume that that reader has access to either the Russian or the English version of this book.

Keywords Lie algebras · Identical relations

1 Introduction

Let us quickly recall that a free Lie algebra $L(X)$ with free generating set X over a field $\mathbb{F}$ is a Lie algebra generated by X and such that any map $vp : X \to M$, where M is another Lie algebra over $\mathbb{F}$, uniquely extends to a Lie algebra homomorphism $\overline{\varphi} : L(X) \to M$. This property is called the universal property of $L(X)$. It is easy that the free Lie algebra with the free generating set X is unique. The existence can be established in different ways, the two basic ways are to start with the associative algebras or with the groups.

Let $A(X)$ be an associative algebra on noncommutative polynomials in the variables X over $\mathbb{F}$. This algebras has the same universal property as $L(X)$ in the class of associative algebras, so $A(X)$ is a free associative algebra. Now $A(X)$ is a Lie algebra with respect to the commutator $[a, b] = ab - ba$. One can prove that the (Lie) subalgebra of $A(X)$ generated by X with respect to this new operation has the above universal property of $L(X)$, hence it is isomorphic to $L(X)$.

Another way to obtain $L(X)$ is to start with the free group $F = F(X)$ freely generated by X and consider its descending filtration by the lower central series

Y. Bahturin (✉)
Memorial University of Newfoundland, St. John's, NL, Canada
e-mail: bahturin@mun.ca

O. M. Di Vincenzo, A. Giambruno (eds.), *Polynomial Identities in Algebras*, Springer INdAM Series 44, https://doi.org/10.1007/978-3-030-63111-6_1

F_n, $n = 1, 2, \ldots$, where $F_1 = F$ and $F_{n+1} = [F_n, F]$, for all $n = 1, 2, \ldots$. One has $[F_p, F_q] \subset F_{p+q}$, for all $p, q = 1, 2, \ldots$. The associated graded abelian group

$$\mathrm{gr}F = \bigoplus_{n=1}^{\infty} F_n/F_{n+1}$$

can be endowed with the Lie commutator defined on the homogeneous elements $\xi = aF_{k+1}$, $\eta = bF_{\ell+1}$, where $a \in F_k$, $b \in F_\ell$, by the formula

$$[\xi, \eta] = [a, b]F_{k+\ell+1}.$$

One can prove that this operation can be extended to the whole of $\mathrm{gr}F$ and makes $\mathrm{gr}F$ to a Lie ring. If $\mathbb{F}$ is a field then $\mathbb{F} \otimes \mathrm{gr}F$ is a Lie algebra and the mapping $x \to xF_2$, for all $x \in X$ is an isomorphism of $L(X)$ and the Lie subalgebra generated by $\overline{X} = \{1 \otimes xF_2 | x \in X\}$.

Let M be a Lie algebra over a field $\mathbb{F}$. Let $x_1, \ldots, x_n \in X$ and $w(x_1, \ldots, x_n)$ a nonzero element in the subalgebra of $L(X)$ generated by $x_1, \ldots, x_n$. Given $a_1, \ldots, a_n \in M$ we denote by $w(a_1, \ldots, a_n)$ the image of $w(x_1, \ldots, x_n)$ under any homomorphism of $L(X)$ to M extending the map $x_1 \mapsto a_1, \ldots, x_n \mapsto a_n$. We call $w(a_1, \ldots, a_n)$ the value of $w(x_1, \ldots, x_n)$ when $x_1 = a_1, \ldots, x_n = a_n$. We say that $w(x_1, \ldots, x_n) = 0$ is an identical relation in M if $w(a_1, \ldots, a_n) = 0$, for all $a_1, \ldots, a_n \in M$.

Given a set (or even a class) $\mathfrak{V}$ of algebras over $\mathbb{F}$, the set V of all $w(x_1, \ldots, x_n) \in L(X)$ such that $w(x_1, \ldots, x_n) = 0$ is an identical relation in all algebras in $\mathfrak{V}$, is an ideal of $L(X)$ closed under any endomorphisms of $L(X)$. In other words, if $w(x_1, \ldots, x_n) = 0$ is an identity in a Lie algebra M and $u_1, \ldots, u_n \in L(X)$ then $w(u_1, \ldots, u_n) = 0$ is also an identity in M. Ideal with this property are called verbal ideals or ideals of identities or T-ideals.

For any subset $V \subset L(X)$ the class $\mathfrak{V}$ of all algebras satisfying $w = 0$, for all $w \in V$ is closed under subalgebras, factor-algebras and Cartesian products; it is called the variety of algebras defined by identical relations $w = 0$, for all $w \in V$. By Birkhoff's Theorem any class of algebras closed under taking subalgebras, factor-algebras and Cartesian products is a variety. The set of all $w \in L(X)$ such that $w = 0$ holds in all algebras in $\mathfrak{V}$ is the verbal ideal of $L(X)$ generated by V. If $X = \{x_1, x_2, \ldots\}$ is a countable set of free generators then there is a one-to-one Galois-type correspondence between varieties of Lie algebras and the verbal ideals of $L(X)$.

Two main questions in the theory of identical relations in any class of algebras—associative, Lie, Jordan, etc.—are the following:

1. Given an algebra A, describe all identities of A;
2. Describe all varieties of algebras in question.

Since no one seriously believes that these questions can be answered in full generality, one has to impose conditions under which the questions are answerable. It is always good to have reasons why these or those restrictions are imposed.

2 Finite Basis Problem

By Hilbert's Basis Theorem, every ideal in the polynomial ring in finitely many variables can be generated by a finite number of elements. In other words, any affine algebraic variety can be given by a finite number of equations. A distant analogy with polynomial identities and varieties of algebras in the case of Lie algebras lead to one of the main questions of the theory

Finite Basis Problem

> Is it true that any variety of Lie algebras can be given by a finite number of identical relations?
>
> Equivalently,
>
> Is it true that any verbal ideal of a free Lie algebra of countable rank can be generated, as a verbal ideal, by a finite number of elements?
>
> Keeping closer to Hilbert's Basis Theorem,
>
> Is it true that for any natural n, the verbal ideals of $L(x_1, \ldots, x_n)$ can be generated, as verbal ideals, by finitely many elements?

The Finite Basis Problem was solved in the negative in 1970 by **M. R. Vaughan-Lee** who provided an example of an infinite set of identical relations

$$w_s = [[x_1, x_2], x_3, \ldots, x_s, [x_1, x_2]] = 0, \ \ s = 3, 4, \ldots$$

In [3] we give the details of the generalization of this example by **V. Drensky** to the case of Lie algebras over arbitrary fields of positive characteristic $p > 0$. Drensky's system consists of

$$w_s = [[x_1, x_2]^{p-1}, x_3, \ldots, x_s, [x_1, x_2]] = 0, \ \ s = 3, 4, \ldots$$

Thanks to these authors we have examples of finite-dimensional dimensional Lie algebras over infinite fields of positive characteristic whose identical relations do not admit finite basis. In the case of a field $\mathbb{F}$ of characteristic $p > 0$ such algebra L belongs to the variety $\mathfrak{N}_p\mathfrak{A}$. In other words, $[L, L]^p = \{0\}$. One of the amazing facts is that the identities of the Lie algebra $\mathfrak{gl}_2(\mathbb{F})$ of 2-by-2 matrices over an infinite field of characteristic 2 are infinitely based! Also, these authors produced examples of Lie algebras of triangular matrices over infinite fields of characteristic $p > 0$ without finite bases for their identities.

The condition of the base field being infinite is essential because a theorem by Bahturin-Olshanskii states that the identities of any finite-dimensional Lie algebra over a finite field have finite basis. This theorem is described in detail in [3].

At the same time, Krasil'nikov [25] has proven that over any field of characteristic $p > 0$ identities of a Lie algebra L with $[L, L]^{p-1} = \{0\}$ do admit finite basis! In particular, a Lie algebra of triangular $n \times n$ matrices over a field of characteristic $p > n$ has a finite basis of identities.

In the case of characteristic zero, the Finite Basis Problem remains wide open. This means that there are no examples of infinitely based varieties of Lie algebras over the fields of characteristic zero. At the same time, there are many theorems where the authors prove the existence of a finite basis for identities of Lie algebras satisfying additional conditions. Often, people prove that a certain variety $\mathfrak{V}$ of Lie algebras is *Specht*. This name comes from PI-algebras and means that not only $\mathfrak{V}$ itself but also any subvariety of $\mathfrak{V}$ is finitely based. By Kemer's Theorem [20] PI-algebras in characteristic zero are finitely based. Most of these results are described in detail in [3, Chapter 4]. We only mention one more result by Krasilnilov [24].

Theorem 1 (A. N. Krasilnikov) *Let L be a Lie algebra over a field of characteristic zero. Assume that the commutator subalgebra $[L, L]$ is nilpotent. Then identical relations of L are finitely based.*

An important corollary is the following.

Corollary 1 *Identities of a finite-dimensional solvable Lie algebra over a field of characteristic zero are finitely based.*

One of the most general results concerning Finite Basis Problem is published in a paper [19] of A. Iltyakov. We define identities of representations of Lie algebras in Sect. 4.

Theorem 2 (A. Iltyakov) *Identities of finite-dimensional representations of any Lie algebra over a field of characteristic zero are finitely based.*

Corollary 2 *Identities of a finite-dimensional Lie algebra over a field of characteristic zero are finitely based.*

This paper has many interesting ideas and techniques, which should be carefully studied.

3 Engel Lie Algebras

A visible omission in [3] was a very important area of Engel Lie algebras. The main problem here is whether an analogue of the classical Engel's Theorem is true in the case of infinite-dimensional algebras. Let us call an element x of a Lie algebra L ad-nilpotent or Engel, or simply nil, if the inner derivation $\operatorname{ad} x : y \mapsto [x, y]$, for any

$y \in L$, is a nilpotent linear transformation of L. A Lie algebra in which all elements are nil is called a nil or Engel Lie algebra. The main question is the following.

Is it true that a finitely generated nil Lie algebra L is nilpotent?

The answer to this question is well-known to be in the negative. The first famous Golod's example dates back to 1963 [15].

After a long period of time, some new examples, based on different ideas, have appeared. An array of infinite-dimensional nil Lie algebras was built, starting 2006, in several papers authored by Petrogradsky, Shestakov and Zelmanov (see [35] and reference therein). These examples are based on self-similarity idea due to Grigorchuk and Gupta, and Sidki in Group Theory.

Another collection of examples was given in a 2007 paper by Bahturin-Olshanskii [8]. This time the techniques were similar to those people use while constructing infinite Burnside groups. One needs to add enough many relations to a free restricted Lie algebra to make it nil but not too many, so it remains infinite-dimensional. Note that if we deal with restricted Lie algebras over a filed of prime characteristic p (see the definition in [3]) then nil elements are those for which $x^{[p^n]} = 0$, for some n, which makes them closer to associative nil-algebras.

Also, in many important cases, a Lie algebra built by a central filtration of a periodic group is Engel. What is important is that if this Lie algebra is nilpotent, then also the original group is nilpotent. This made Lie algebras an important tool in the solution of the problems of Burnside type in the Group Theory.

Engel Lie algebras in which the nilpotency index of every element is bounded by a certain number n are called n-Engel Lie algebras. These algebras form a variety $\mathfrak{E}_n$ given by an identity $(\operatorname{ad} x)^n(y) = 0$. The main questions here are the following:

1. (Global nilpotence) *Is it true that for any natural n there is natural $f(n)$ such that $\mathfrak{E}_n \subset \mathfrak{N}_{f(n)}$?*
2. (Local nilpotence) *Is it true that for any natural n, r there is natural $g(n,r)$ such that any r-generator algebra in $\mathfrak{E}_n$ is nilpotent of class $g(n,r)$?*

3.1 Global Nilpotence

An example of P. M. Cohn (see [3]) tells us that if char $\mathbb{F} = p > 0$ then $\mathfrak{E}_{p+1}$ is not nilpotent. With much more work and ingenuity, Razmyslov [37] has shown that if $p > 3$ then $\mathfrak{E}_{p-2}$ is also not nilpotent. For $p = 5$ this was shown also in [2].

In 1987 Zelmanov proved the following major result [52]

Theorem 3 (E. I. Zelmanov) *n-Engel Lie algebra over a field of characteristic 0 is nilpotent.*

Corollary 3 *There is a function of natural argument $f(n)$ such that n-Engel algebra over a field of characteristic $p > f(n)$ is nilpotent.*

Zelmanov's result implies that there is a function $f(n)$ such that $\mathfrak{E}_n \subset \mathfrak{N}_{f(n)}$ (take as $f(n)$ the nilpotency class of the free algebra of countable rank in $\mathfrak{E}_n$), but it does not provide for such number. In a paper [45] Traustason proves the following.

Theorem 4 (G. Traustason) *If the characteristic of the ground field is zero or big, say, $p > \ell(n) = 1024 \cdot 10^{46} \cdot n^{40} \cdot 3^{[(n-5)/2]}$ then n-Engel Lie algebras are nilpotent of class $T(n, \ell(n))$ where $T(n, s)$ is given recursively by $T(n, 1) = n$, $T(n, s+1) = n^T(n, s)$, for $s > 1$*

One more mathematician was close to the solution of Engel's Problem. In 1984 paper [28] S. P. Mishchenko considered Engel Lie algebras in the varieties of exponential growth (see Sect. 5). His result is as follows.

Theorem 5 (S. P. Mishchenko) *Let $\mathfrak{V}$ be a variety of Lie algebras with exponential growth of codimensions, over a field of characteristic zero. If $L \in \mathfrak{V}$ satisfies an Engel identity then L is nilpotent.*

Although the growth of codimensions of varieties of Lie algebras does not need to be exponential, many varieties do have such growth. For examples, as shown in the same paper, varieties $\mathfrak{W}_n$ generated by Cartan Lie algebras W_n have exponential growth and any infinite-dimensional algebra simple algebra with a proper subalgebra of codimension n can be embedded in the Lie algebra $\widetilde{W}_n$ of derivations of the power series ring $\mathbb{F}[[x_1, \ldots, x_n]]$ over a proper extension of the base field of coefficients (which does not affect identical relations).

It is more or less obvious that the numbers $\ell(n)$ and $T(n, \ell(n))$ in Traustason's theorem are too big. For instance, Zelmanov's guess was that $\ell(n)$ could be replaced by just $2n$.

3.2 Restricted Burnside Problem: Local Nilpotence of Engel Lie algebras

In 1994 Zelmanov was awarded the Fields medal for his groundbreaking solution of Restricted Burnside Problem, which reads as follows:

> For what values of r and n is there an upper bound on the orders of finite r-generator groups of exponent n?

The answer turns out to be that such an upper bound exists for all r and n. From the classification of finite simple groups and the work of P. Hall and G. Higman, it follows that it is sufficient to consider n when n is a power of a prime. In 1959 Kostrikin [22, 23] proved that there is an upper bound if n is a prime. In 1989 Zelmanov [53] showed that an upper bound exists if n is a power of an odd prime. In 1991 Zelmanov [54] showed that an upper bound exists also if n is a power of 2. Thus he solved the Restricted Burnside problem in its full generality.

The theorems of Kostrikin and Zelmanov are in fact theorems about Lie algebras. The reason is that the following are equivalent.

1. There is a largest finite r-generator group of exponent p^m;
2. The associated Lie ring of $B(r, p^m)$ is nilpotent.

Here $B(r, n)$ is the (relatively) free r-generator group of exponent n. Now the associated Lie ring of $B(r, p)$ satisfies $(p - 1)$ Engel identity and has characteristic p, and so we can think of it as an Engel Lie algebra over the field $\mathbb{Z}/p\mathbb{Z}$. So it was sufficient for Kostrikin to prove the following.

Theorem 6 (A. I. Kostrikin) *Let L be a finitely generated Engel $(p-1)$-Engel Lie algebra over a field of characteristic p. Then L is nilpotent.*

A more general result, that the associated Lie-ring of $B(r, p^m)$ is nilpotent, follows from

Theorem 7 (E. I. Zelmanov) *Let L be a finitely generated Lie-algebra. Suppose that there exist positive integers s, t such that:*

$$\sum_{\sigma \in \mathrm{Sym}(s)} (\mathrm{ad}\, x_{\sigma(1)})(\mathrm{ad}\, x_{\sigma(2)}) \cdots (\mathrm{ad}\, x_{\sigma(s)})(x) = 0,$$

for all $x, x_1, x_2, ..., x_s \in L$, and

$$(\mathrm{ad}\, u)^t = 0$$

for all Lie monomials $u \in L$ in terms of generators of L Then L is nilpotent.

A somewhat more general Zelmanov's result reported to Kyoto ICM in 1991 and widely used in Group Theory was the following [55]. We call a subset $S \subset L$ is called a Lie set if, for arbitrary elements $a, \in S$, we have $[a, b] \in S$. For a subset $X \subset L$, the Lie set generated by X is the smallest Lie set $S\langle X\rangle$ containing X. It consists of X and of all iterated commutators in elements from X.

Theorem 8 (E. I. Zelmanov) *Let L be a Lie algebra satisfying a polynomial identity and generated by elements $a_1, \ldots, a_m$. If an arbitrary element $s \in S\langle a_1, \ldots, a_m\rangle$ is* ad*-nilpotent then the Lie algebra L is nilpotent.*

In conclusion, we mentioned one more book [46] on Restricted Burnside Problem.

4 Identities of Simple Lie Algebras

4.1 The Isomorphism Problem

In 1983 paper [26] the following important result was proven.

Theorem 9 *Over an algebraically closed field any simple finite-dimensional Lie algebra is completely determined, up to isomorphism, by its identical relations.*

Actually, this result was a consequence of previous deep results by Razmyslov [42] dealing with so called Ω-algebras and their representations. If $\Omega = \cup_{n=0}^{\infty}\Omega_n$ is a set, a vector space A is an Ω-algebra if with each $\omega \in \Omega_n$ one associates an n-ary operation on A, that is, an n-linear map $\omega : A^{\otimes n} \to A$. One can naturally define free Ω-algebra $F_\Omega(X)$ for a set of free generators X. Associative and Lie algebras are natural examples of Ω-algebras, with just one, binary, operation.

Let us fix in $F_\Omega(X)$ a subset Ω' of multilinear polynomials. Elements of Ω' have natural arity and can be viewed as operations on any subspace $\mathfrak{G}$ of an Ω-algebra B, which is closed under these operations. Razmyslov calls $(B, \mathfrak{G})$ an (Ω, Ω')-pair. Again, natural examples are associative Lie pairs, in particular, the pairs which arise when we consider a representation $\rho : \mathfrak{G} \to \operatorname{End} V$ of a Lie algebra $\mathfrak{G}$ by linear transformations of a vector space V and consider an associative algebra B generated by $\rho(\mathfrak{G})$.

The elements of $F_\Omega(X)$ are called Ω-polynomials. An Ω-polynomial $w(x_1, \ldots, x_n)$ is called an identity of an (Ω, Ω')-pair $(B, \mathfrak{G})$ if $w(g_1, \ldots, g_n) = 0$, for any $g_1, \ldots, g_n \in \mathfrak{G}$. Identities of a Lie algebra $\mathfrak{G}$ can be viewed as identities of the pair $(B, \mathfrak{G})$ where B is an associative subalgebra of $\operatorname{End} L$ generated by all $\operatorname{ad} g$, $g \in \mathfrak{G}$.

In [39] Razmyslov proves the following.

Theorem 10 *Let $(B_1, \mathfrak{G}_1)$ and $(B_2, \mathfrak{G}_2)$ be two (Ω, Ω')-algebras over an infinite field $\mathbb{F}$ such that B_1 and B_2 are centrally prime and $\mathfrak{G}_1$ and $\mathfrak{G}_2$ are finite-dimensional over $\mathbb{F}$. The identities of $(B_1, \mathfrak{G}_1)$ and $(B_2, \mathfrak{G}_2)$ are the same if and only if there exists an $\mathbb{F}$-linear automorphism σ of an algebraically closed extension $\mathbb{F}_1$ of infinite transcendental degree over $\mathbb{F}$ such that the pairs $(B_1, \mathfrak{G}_1)$ and $(B_2, \mathfrak{G}_2)$ are $\mathbb{F}$-isomorphic and this isomorphism is σ-semilinear.*

The above Theorem 9 is a corollary of Theorem 10. Another corollary is

Theorem 11 *Let $\mathbb{F}$ be an algebraically closed field. Let $\rho_i : \mathfrak{G}_i \to \operatorname{End} V_i$, $i = 1, 2$ are two faithful irreducible not necessarily finite-dimensional representations of finite-dimensional Lie algebras $\mathfrak{G}_1$ and $\mathfrak{G}_2$, respectively. Let us denote by B_i the associative subalgebra in End V_i generated by $\rho(\mathfrak{G}_i)$, $i = 1, 2$. If $(B_1, \mathfrak{G}_1)$ and $(B_2, \mathfrak{G}_2)$ have the same identities then they are isomorphic. If, moreover, V_1 and V_2 are finite-dimensional then there exists a Lie isomorphism $\varphi : \mathfrak{G}_1 \to \mathfrak{G}_2$ and a linear isomorphism $\psi : V_1 \to V_2$ such that $\psi(gv) = \varphi(g)\psi(v)$, for all $g \in \mathfrak{G}_1$ and all $v \in V_1$.*

Note that Theorem 9 is just one possible corollary of Theorem 10. It also follows that

> Any two (nonassociative) finite-dimensional simple algebras over an algebraically closed field have the same identities if and only if they are isomorphic.

Easy examples, say $\mathfrak{g}_1 = \mathfrak{su}_2(\mathbb{R})$ and $\mathfrak{g}_1 = \mathfrak{so}_3(\mathbb{R})$ show that the requirement of $\mathbb{F}$ being algebraically closed is essential. An example by Razmyslov [41] shows that the following conjecture is false:

Given two finite-dimensional simple Lie algebras $\mathfrak{G}_1$ and $\mathfrak{G}_2$ over an algebraically closed field $\mathbb{F}$, does it follow from var $\mathfrak{G}_1 \subset$ var $\mathfrak{G}_2$ that $\mathfrak{G}_1$ is isomorphic to a subalgebra in $\mathfrak{G}_2$?

4.2 Identities of Cartan Type Lie Algebras

This section mostly consists of the results of Razmyslov. Many of them can be found in his monograph [42]. In [40] he states the following conjecture, the answer to which is not known even now.

Razmyslov's Conjecture Any simple Lie algebra satisfying a nontrivial Lie identity is an algebra of Cartan type.

In the case of the fields of characteristic zero, Cartan type Lie algebra have been known in Geometry since more than by one hundred years ago and belong to one of the types W_n, S_n, H_n and K_n. Their analogues in the case of positive characteristic appear later in Kostrikin–Shafarevich conjecture for the classification of simple Lie algebras. However, these latter algebras are finite-dimensional. One of the basic properties of Cartan Lie algebras of rank n is the presence of a proper subalgebra of codimension n. Razmyslov chooses this property for the definition of simple Cartan type algebras. As a result, any finite-dimensional simple algebras are of Cartan type, which is at odds with the terminology accepted in the classification theory of modular simple Lie algebras.

Definition A simple Lie algebra $\mathfrak{g}$ over an arbitrary field $\mathbb{F}$ is called a Cartan type Lie algebra, if there is an extension $\overline{\mathbb{F}}$ of the centroid C of $\mathfrak{g}$ such that the extended $\overline{\mathbb{F}}$-Lie algebra $\overline{\mathfrak{g}} = \overline{\mathbb{F}} \otimes_C \mathfrak{g}$ has a proper $\overline{\mathbb{F}}$-subalgebra of finite codimension. It follows from the classical Cartan's result that, if $\mathbb{F}$ is a field of characteristic zero, any simple Lie algebra $\mathfrak{g}$ such that $\overline{\mathfrak{g}} = \overline{\mathbb{F}} \otimes \mathfrak{g}$ has a proper $\overline{\mathbb{F}}$-subalgebra of finite codimension n can be embedded in a Lie algebra $\widetilde{W}_n(\overline{\mathbb{F}})$ of all derivations of the power series $\overline{\mathbb{F}}[[t_1 \ldots, t_n]]$; in $\widetilde{W}_n(\overline{\mathbb{F}})$ there is a unique Lie subalgebra of codimension n.

Thus the algebras $\widetilde{W}_n(\overline{\mathbb{F}})$ are unique universal simple algebras containing arbitrary simple Cartan type Lie algebras. Similar universal simple algebras can be defined in the case of the fields of positive characteristic.

The following result is among the strongest aimed to the solution of Razmyslov's Conjecture [40].

Theorem 12 *Let $\mathfrak{V}$ be an arbitrary variety of Lie algebras over a field $\mathbb{F}$ of characteristic zero. Assume the growth of $\mathfrak{V}$ is at most exponential. Then any simple $\mathbb{F}$-algebra in $\mathfrak{V}$ is a Cartan type algebra*

As we mentioned in Sect. 3 earlier, S. P. Mishchenko proved that var $W_n(\mathbb{F})$, hence var $\widetilde{W}_n(\overline{\mathbb{F}})$ has exponential growth.

If in Theorem 12 we take $\mathfrak{V} =$ var $W_n(\mathbb{F})$ then Razmyslov proves that any simple algebra $\mathfrak{g}$ in $\mathfrak{V}$ such that $\dim \mathfrak{g} > n^2 + 2n$ is isomorphic to a subalgebra of $\widetilde{W}_n(\overline{\mathbb{F}})$.

An earlier result by Razmyslov, in the case of $n = 1$, gives an additional interesting piece of information:

Theorem 13 *Let $\mathfrak{g}$ be a simple Lie algebra over a field $\mathbb{F}$,* $\operatorname{char}\mathbb{F} \neq 2$. *Let C be the centroid of $\mathfrak{g}$. Then the following properties of $\mathfrak{g}$ are equivalent.*

1. *There is an extension $\mathbb{F}_1$ of C such the simple $\mathbb{F}_1$-algebra $\mathfrak{g}_1 = \mathbb{F}_1 \otimes_V \mathfrak{g}$ has an $\mathbb{F}_1$-subalgebra of codimension 1;*
2. $\mathfrak{g} \in \operatorname{var} W_1(\mathbb{F})$*;*
3. $\mathfrak{g}$ *satisfies standard Lie identity of degree 5:*

$$\sum_{\sigma\in S_4} (\operatorname{sgn}\sigma)[x_{\sigma(1)}, x_{\sigma(2)}x_{\sigma(3)}, x_{\sigma(4)}, x_5] = 0; \tag{1}$$

This brings us to another open question:

Open Problem Is it true that the standard Lie identity of degree 5 forms a basis for the identities of $W_1(\mathbb{F})$?

To answer this question in the positive, we would need to prove that for any algebra $\mathfrak{g}$ over $\mathbb{F}$, if $\mathfrak{g}$ satisfies (1), then $\mathfrak{g} \in \operatorname{var} W_1(\mathbb{F})$. Razmyslov proved this in the case where $\mathfrak{g}$ is simple.

M. V. Zaicev noted that the above problem has a "no" answer if $m \geq 2$, namely the identities of W_m do not follow from a standard identity.

We conclude with one more result aimed at the solution of Razmyslov's Conjecture. To state it, we recall that an algebra L is called locally finite-dimensional if every finite subset of L generates a finite-dimensional subalgebra. The following result is proven in [4].

Theorem 14 (Y. Bahturin and H. Strade) *Let L be a simple locally finite-dimensional Lie algebra over a field $\mathbb{F}$, satisfying a nontrivial identical relation. Assume* $\operatorname{char}\mathbb{F} = 0$ *or* $\operatorname{char}\mathbb{F} = p > 7$. *Then L is an algebra of Cartan type.*

The restriction $p > 7$ is related to the state of the classification theory of simple modular Lie algebras in 1994. Since then there was essential progress by Premet and Strade [36] in the case of $\operatorname{char}\mathbb{F} = 5, 7$, which raises the question about the validity of the above theorem in the case of these smaller characteristics.

4.3 Capelli Identities

For some time, people believed that infinite-dimensional Cartan type Lie algebras do not satisfy non-trivial identities. But in 1974 I. Sumenkov proved that W_1 satisfies standard identity of degree 5. Later Razmyslov found that standard identities of appropriate degrees are satisfied in all W_n, $n = 1, 2, \ldots$. Standard identities are particular case of so called Capelli identities which are defined, as follows.

Definition Let $w(x_1, \ldots, x_m; x_{m+1}, \ldots, x_n) \in L(x_1, \ldots, x_n)$ be a Lie polynomial which is multilinear and alternating with respect to the variables $x_1, \ldots, x_m$, that is,

$$w(x_1, \ldots, x_i, \ldots, x_j, \ldots, x_m; x_{m+1}, \ldots, x_n) \tag{2}$$

$$= -w(x_1, \ldots, x_j, \ldots, x_i, \ldots, x_m; x_{m+1}, \ldots, x_n) \tag{3}$$

for any $1 \leq i < j \leq m$.

Then we say that $w(x_1, \ldots, x_m; x_{m+1}, \ldots, x_n) = 0$ is a Capelli identity of order m.

Any finite-dimensional (Lie) algebra L satisfies all Capelli identities of order $n, n > \dim L$. In [38] Razmyslov proved several results for algebras and pairs of algebras far more general then (the representations of) Lie algebras. These results show that in a certain sense algebras satisfying all Capelli identities of order $m + 1$ can be viewed as m-dimensional over an extended domain of "scalars". One of the corollaries of the main theorems, closest to Lie algebras, says the following. Given a Lie algebra L over a field F, we denote by $\operatorname{Ad} L$ the associative subalgebra of $\operatorname{End}_{\mathbb{F}} L$ generated by all $\operatorname{ad} x$ where $x \in L$. Sometimes one calls $\operatorname{Ad} L$ the adjoint algebra of L.

Theorem 15 (Yu. P. Razmyslov) *Suppose that a Lie algebra L satisfies all Capelli identities of order m+1. Then in L there is an ideal J satisfying $J^{m-2} = \{0\}$ with the following property. The adjoint algebra $D = \operatorname{Ad}(L/J)$ is a PI-algebra, moreover, D satisfies all identities of a matrix algebra of certain order.*

Capelli identities enable one to define the rank $r(A, V)$ of a subspace V of an algebra A as follows. Suppose we have a Capelli polynomial $w(x_1, \ldots, x_m; x_{m+1}, \ldots, x_n)$, as before. A natural number k is called the rank of V in A if k is the smallest number m such that $w(v_1, \ldots, v_m; a_{m+1}, \ldots, a_n) = 0$ as soon as $v_1, \ldots, v_m \in V$ and $a_{m+1}, \ldots, a_n \in A$. The following Rank Theorem appears in the proof of the fact that simple algebras are determined by their identities (see Sect. 4.1). Given a semiprime algebra (could be an Ω-algebra!) A, one can view A as a module over the associative algebra $D(A)$ of multiplications of A ($\operatorname{Ad} A$ if A is a Lie algebra), take the injective hull P of this module, consider $E = \operatorname{End}_{D(A)} P$ and define the central closure $Q(A) = E * A$. If $p : E \to \operatorname{End}_{D(A)} Q(A)$ is the natural representation then $C(A) = \operatorname{Im} p = E/\operatorname{Ker} p$ is called the Martindale centroid of A. We have $Q(A) = C(A) * A$. For A simple, this is just the ordinary centroid of A. If A is prime then $C(A)$ is a field.

Theorem 16 (Yu. P. Razmyslov) (see [42]) *Let V be an $\mathbb{F}$-vector subspace of a prime algebra A. If rank$(A, V) < \infty$, then*

$$\dim_{C(A)} C(A)V = \mathit{rank}(A, V) - 1.$$

In [38] the author asks if for any m, the set of all Capelli identities of order m is finitely based. He mentions that in the case of weak identities of a pair (A, L), where A is associative and L is Lie, the answer is yes. No proof substantiated this claim. We conclude this section by a theorem of Stovba [44].

Theorem 17 (V. V. Stovba) *Let $\mathbb{F}$ be a field of characteristic zero. The following systems of identical relations in Lie and associative algebras are finitely based.*

1. *The set of Capelli identities of any given order m.*
2. *The set of symmetric identities of any given order m.*
3. *Any set of identities in finitely many variables such that the degrees of all variables but one are bounded.*

5 Codimension Growth

Let $\mathbb{F}$ be a field of characteristic zero and M a Lie algebra over $\mathbb{F}$. Let $L(X)$ be a free Lie algebra over $\mathbb{F}$, generated by a countable set $X = \{x_1, x_2, \ldots\}$, and Id (M) the verbal ideal of $L(X)$, consisting of all identities of L. Let V_n be the space of all multitlinear Lie polynomials in the variables $x_1, \ldots, x_n$ inside $L(X)$. Since char $\mathbb{F} = 0$, the ideal Id (M) is uniquely determined by its multilinear components, that is, $\mathrm{Id}(M) \cap V_n$, $n = 1, 2, \ldots$ Let us denote by $c_n(M)$ the so called nth codimension of identities of M, that is,

$$c_n = c_n(M) = \dim \frac{V_n}{V_n \cap \mathrm{Id}\,(M)}.$$

The asymptotic behavior of the sequence $\{c_n(M)\}$, $n = 1, 2, \ldots$, is an important numerical characteristic of L. In 1980 paper [10] the authors have described all the varieties with polynomial growth of codimensions in the language of Young diagrams. In [29], Mishchenko proved the following:

Theorem 18 (S. P. Mishchenko) *The codimensions $c_n(\mathfrak{V})$ have polynomial growth if and only $\mathfrak{N}_2\mathfrak{A} \not\subset \mathfrak{V} \subset \mathfrak{N}_c\mathfrak{A}$ for some c.*

A useful consequence of this result is the following.

Corollary 4 *If the codimensions of a variety are of subexponential growth, then their growth is polynomial.*

5.1 *Exponential Growth*

Many important classes of Lie algebras have exponentially bounded growth of the sequence $c_n(L)$.

Among them we mention finite-dimensional algebras (any, not necessarily associative or Lie, [7]), affine Kac-Moody Lie algebras [48], infinite-dimensional Cartan type simple Lie algebras [28], special Lie algebras and many others.

In the case where the growth of $c_n(L)$ is at most exponential, the sequence $\sqrt[n]{c_n(L)}$ is bounded and one can write its upper and lower limits

$$\overline{\text{EXP}}(L) = \overline{\lim_{n\to\infty}} \sqrt[n]{c_n(L)}, \qquad \underline{\text{EXP}}(L) = \underline{\lim_{n\to\infty}} \sqrt[n]{c_n(L)},$$

which are called *upper* and *lower exponents*, respectively. If they coincide, there exists the limit of the sequence

$$\text{EXP}(L) = \overline{\text{EXP}}(L) = \underline{\text{EXP}}(L),$$

called the *exponent of the growth of identities of* L or simply the *exponent of* L.

For an associative PI-algebra A its nth codimension, upper, lower and ordinary exponents are defined in a similar matter. Regev proved that $\{c_n(A)\}$ is always exponentially bounded [43]. Several decades ago, Amitsur conjectured that, given a PI-algebra A, its exponent $\text{EXP}(A)$ always exists and is an integral number. This conjecture was proven in the papers of Giambruno and Zaicev (see this and many other interesting results in their book [13]).

In the case of Lie algebras, there are many examples where $\text{EXP}(L)$ exists and equals an integer. For instance, if L is an algebra with nilpotent commutator subalgebra, $(L^2)^m = 0$, then $\text{EXP}(L)$ is an integer bounded by m [34].

At the same time, for arbitrary Lie algebras, even with exponentially bounded growth of codimensions, the answer to Amitsur's question is negative. It was shown in [50], that there exists an infinite-dimensional Lie algebra L, for which

$$3.1 < \underline{\text{EXP}}(L) \le \overline{\text{EXP}}(L) < 3.9.$$

In [31] the authors mention that actually for this algebra $\text{EXP}(L)$ exists and is a number close to 3.61. In fact, the integrality of the exponent does not hold for simple Lie algebras of Cartan type, as shown by an example of S. S. Mishchenko [30]): $13.1 < \text{EXP}(W_2) < 13.5$. In [31] the following is conjectured:

$$\text{EXP}(W_k) = k(k+1)\left(1 + \frac{1}{k}\right)^k.$$

In the case of finite-dimensional Lie algebras, Amitsur's problem was solved by Zaicev in 2001 [51].

Theorem 19 *Let L be a finite-dimensional Lie algebra over a field of characteristic zero. Then EXP(L) exists and is an integral number.*

It should be mentioned, that the techniques suggested in the proof allow one to explicitly compute $\text{EXP}(L)$, if one knows the structure of L.

5.2 Overexponential Growth

In distinction with the case of associative PI-algebras, the sequence of codimensions of a Lie algebra, satisfying a nontrivial identity has much more involved behavior. Volichenko in [47] showed that a Lie L algebra can have overexponential growth of codimensions already if L satisfies the identity $[[x_1, x_2, x_3], [x_1, x_2, x_3]] = 0$, that is, belongs to the product variety $\mathfrak{A}\mathfrak{N}_2$.

Still, there is "restriction" from above for the codimension growth of Lie algebras satisfying a nontrivial identity. This is given in the following theorem by Grishkov [17].

Theorem 20 (A. N. Grishkov) *For any number a the growth of a nontrivial varieties of Lie algebras is at most* $\frac{n!}{a^n}$.

Razmyslov [42] has associated with any nontrivial variety $\mathfrak{V}$ of (Lie) algebras a power series

$$C_{\mathfrak{V}}(z) = \sum_{n=1}^{\infty} \frac{c_n(\mathfrak{V})}{n!} z^n,$$

which he called the complexity function of $\mathfrak{V}$.

The following is a theorem of Razmyslov [42] that implies the above result of Grishkov:

Theorem 21 (Yu. P. Razmyslov) *For any nontrivial variety of Lie algebras* $\mathfrak{V}$ *the series* $C_{\mathfrak{V}}(z)$ *defines an entire function of complex variable.*

Starting from this, Petrogradsky in [33] exhibited a whole scale of overexponential functions in the process of describing the codimension growth of the polynilpotent Lie algebras, that is, Lie algebras in the product varieties $\mathfrak{N}_{c_1} \cdots \mathfrak{N}_{c_m}$.

Petrogradsky gave a better bound for the codimensions of any proper variety of Lie algebras. To state it, we define

$$\ln^{(1)} x = \ln x; \ln^{(s+1)} x = \ln(\ln^{(s)} x) \ s = 1, 2; \ldots; \tag{4}$$

Then the following is true.

Theorem 22 (V. M. Petrogradsky) *Let* $\mathfrak{V}$ *be a variety of Lie algebras satisfying a non-trivial identity of degree* $m > 3$. *Then there exists an infinitesimal* $o(1)$ *(depending only on* m*) such that*

$$c_n(\mathfrak{V}) \leq \frac{n!}{(\ln^{(m-3)} n)^n}(1 + o(1))^n.$$

For any variety $\mathfrak{V} = \mathfrak{N}_{c_q} \cdots \mathfrak{N}_{c_2}\mathfrak{N}_{c_1}$ of polynilpotent Lie algebras Petrogradsky determined the following:

$$c_n(\mathfrak{V}) = \frac{n!}{(\ln^{(q-2)} n)^{n/c_1}} \left(\frac{c_2 + o(1)}{c_1}\right)^{n/c_1} \quad \text{if } q \geq 3$$
$$c_n(\mathfrak{V}) = (n!)^{(c_1-1)/c_1} (c_2 + o(1))^{n/c_1} \text{ if } q = 2 \tag{5}$$

5.3 Colength

The study of the growth of the codimensions is done through the representation theory of the symmetric group S_n. One defines an S_n-action on P_n, where S_n is the symmetric group on n symbols, by setting

$$\sigma \circ f(x_1, \ldots, x_n) = f(x_{\sigma(1)}, \ldots, x_{\sigma(n)}),$$

where $\sigma \in S_n,\ f \in P_n$. Under this action $P_n \cap \mathrm{Id}\,(A)$ is an S_n-submodule of P_n and we consider the induced action on the quotient space $P_n(A) = P_n/(P_n \cap Id(A))$.

Let $\chi_n(A) = \chi(P_n(A))$ be the S_n-character of $P_n(A)$. The character $\chi_n(A)$ is called the nth cocharacter of A. It can be decomposed as the sum of irreducible characters

$$\chi_n(A) = \sum_{\lambda \vdash n} m_\lambda \chi_\lambda$$

where λ is a partition of n, χ_λ is the associated irreducible S_n-character and the integer m_λ is the corresponding multiplicity. Then if $d_\lambda = \deg \chi_\lambda$,

$$c_n(A) = \sum_{\lambda \vdash n} m_\lambda d_\lambda$$

Now the sequence of colengths is defined as follows. First of all, the nth colength $\ell_n(A)$ of A is defined as the sum of the multiplicities m_λ, as above. Hence

$$\ell_n(A) = \sum_{\lambda \vdash n} m_\lambda.$$

We can also speak about the sequence of colengths for any variety of nonassociative algebras: $\{\ell_n(\mathfrak{V})\}$. An easy fact is the folowing:

$$\ell_n(\mathfrak{V}) \geq \frac{c_n(\mathfrak{V})}{\sqrt{n!}}.$$

Now it follows easily from Petrogradsky's formula for codimensions (5) that the growth of the sequence of colengths for $\mathfrak{N}_b\mathfrak{N}_a$ is overexponential as soon as $a \geq 3$.

The following theorem is proven in [14].

Theorem 23 (A. Giambruno, S. P. Mischenko, and M. V. Zaicev) *If A is finite-dimensional algebra over a field of characteristic zero and* $\dim A = d$, *then* $\ell_n(A) \leq d(n+1)^{d^2+d}$.

Note that it is not required in this theorem for A to be a Lie algebra. Finally, in the same paper, the authors find the following

$$\ell_n(\mathfrak{A}\mathfrak{N}_2) \sim \exp\left(\sqrt{\frac{2}{3}n\pi}\right)$$

Thus the *growth of the sequence of colengths for a variety of Lie algebras can be intermediate between polynomial and exponential.*

6 Graded Lie Algebras and Identities

Let L be an algebra over a field $\mathbb{F}$ and G a group. We say that L is G-graded if $L = \bigoplus_{g\in G} L_g$ such that for any $g, h \in G$ one has $[L_g, L_h] \subset L_{gh}$. The subset $\mathrm{Supp}L = \{g \in G | L_g \neq 0\}$ is called the support of the grading. If L is a simple algebra, it is well-known that $gh = hg$ for any $g, h \in \mathrm{Supp}L$. The neutral component of the grading L_e is a Lie subalgebra and one of the natural questions is the following.

> Suppose that $L = \bigoplus_{g\in G} L_g$ is graded by a finite group G so that L_e satisfies a nontrivial identity. Is the same true for the whole of L?

The restriction on the finiteness of the number of homogeneous components is necessary: a free Lie algebra $L(X)$ is naturally graded by $G = \mathbb{Z}$, with $L(X)_0 = \{0\}$.

A positive answer to above question, asked by A. Zalessky, was given in 1996 in [6].

Theorem 24 (Y. Bahturin, M. Zaicev) *Let L be a Lie algebra over an arbitrary field $\mathbb{F}$, G a finite group. If the component L_e is a Lie algebra with a non-trivial identity, then so is L.*

Actually, the theorem deals with so called "Lie Type Algebras", which includes associative and Lie algebras and superalgebras as particular cases (see Sect. 7 below).

One of the consequences of the above theorem deals with fixed point of automorphisms in Group Theory. It was proved in [6] for solvable groups but later V. Linchenko [27] provided a proof without this restriction.

Theorem 25 (V. Linchenko) *Let G be a finite subgroup in the automorphism group of a Lie algebra L over a field $\mathbb{F}$ where* $\mathrm{char}\,\mathbb{F}$ *is not a divisor of $|G|$ and* *let the set of G-invariants L^G satisfy a non-trivial identity. Then also L satisfies a non-trivial identity*

If the answer to the above question is "yes", then one asks:

Is there a connection between the identities satisfied by L_e and L?

For example, a famous G. Higman's problem [18] asks:

What is the nilpotency class of a Lie algebra graded by a finite group so that the neutral component of the grading is trivial?

In connection with the study of groups admitting a fixed-points-free automorphism of prime-power order, Higman proved the following.

Theorem 26 (G. Higman) *To each prime p corresponds an integer $k(p)$ such that if a Lie ring L has an automorphism of order p which leaves fixed no element except zero, then L is nilpotent of class at most $k(p)$.*

It is easy to see that $k(2) = 1$, and that $k(3) = 2$; Higman proves that $k(5) = 6$, and that, for any odd p, $k(p) > \dfrac{p^2-1}{4}$. However, it was not possible to find any general upper bound for $k(p)$, and as Higman writes, "it appears to be quite difficult even to find its order of magnitude".

One of the consequences is the following

Theorem 27 (G. Higman) *If a finite solvable group has an automorphism of prime order which leaves no element fixed except the identity, it is nilpotent of class at most $k(p)$, where $k(p)$ is as in the previous theorem.*

7 Graded Identities of Lie Algebras and Generalizations

Let H be a Hopf algebra, say, a group algebra $\mathbb{F}G$ of a group G, or the universal enveloping algebra $U(\mathfrak{g})$ for a Lie algebra $\mathfrak{g}$ and so on (see [32]). We say that a Lie algebra L is an H-algebra, if L is a left H-module and $h * [a, b] = \sum_h [h_1 a, h_2 b]$. Here $\Delta h = \sum_h h_1 \otimes h_2$. For example, $\mathbb{F}G$ acts on L in such a way that the elements of G acts as automorphisms thanks to $\Delta g = g \otimes g$. Also, in $H = U(\mathfrak{g})$ one has $\Delta x = x \otimes 1 + 1 \otimes x$, for $x \in \mathfrak{g}$. As a result in an $U(\mathfrak{g})$-algebra L the elements of $\mathfrak{g}$ act as derivations. If $H = (\mathbb{F}G)^*$ is the dual to the group algebra of a group G, then the basis of H is formed by the projections $p_g, g \in G$, such that

$$\Delta p_g = \sum_{hk=g} p_h \otimes p_k.$$

One has $p_g[a, b] = [p_h a, p_k b]$. If we set $L_g = p_g L$ then $L = \oplus_{g \in G} L_g$ is a grading of L by the group G.

To define identities of H-algebras, we need to define the free H-algebra with the free set X of generators. For this, we consider the algebra $T(H)\otimes\mathbb{F}\langle X\rangle$, where $T(H)$ is the tensor algebra of a vector space H and $\mathbb{F}\langle X\rangle$ a free nonassociative algebra with free generating set X. First, we make $T(H)$ an H-algebra as follows. Using coassociativity of the coproduct in H we can correctly define $\Delta_1=\Delta$ and $\Delta_{m+1}=(\Delta_m\otimes\mathrm{id})\circ\Delta$, for $m\geq 1$. Then we write

$$\Delta_{n-1}(h)=\sum_h h_{(1)}\otimes h_{(2)}\otimes\cdots\otimes h_{(n)}$$

and define

$$h(a_1\otimes a_2\otimes\cdots\otimes a_n)=\sum_h (h_{(1)}a_1)\otimes(h_{(2)}a_2)\otimes\cdots\otimes(h_{(n)}a_n). \tag{6}$$

Now we choose inside $T(H)\otimes\mathbb{F}\langle X\rangle$ a subalgebra $F_H(X)$ spanned by $T^n(H)\otimes(\mathbb{F}\langle X\rangle)_n$, $n=1,2,\ldots$, where $(\mathbb{F}\langle X\rangle)_n$ is spanned by the monomials of degree n in $\mathbb{F}\langle X\rangle$. This will be an absolutely free H-algebra with free generators X. Any map $\varphi: X\to A$, A an H-algebra, uniquely extends to a homomorphism $\overline{\varphi}: F_H(X)\to A$ if one sets

$$\overline{\varphi}((h_1\otimes\cdots\otimes h_n)x_1\cdots x_n)=(h_1\varphi(x_1))\cdots(h_n\varphi(x_n)).$$

Finally, a free H-Lie algebra $L_H(X)$ is the factor-algebra of $F_H(X)$ by an H-invariant ideal generated by all $\{u(vw)+v(wu)+w(uv)|u,v,w\in F_H(X)\}$ and $\{u^2|u\in F_H(X)\}$.

If H is cocommutative and char $\mathbb{F}\neq 2$, then $L_H(X)$ is an ordinary free Lie algebra with free generators $h_\beta\otimes x$, where h_β runs through a basis of H and x runs through X. Otherwise, it does not need to be a Lie algebra. So when we use this approach to deal with identities of graded algebra, we must assume that the grading group is abelian, which is quite natural for dealing gradings on Lie algebras.

In the case of G-graded identities, G a group or even a semigroup, people normally choose an alphabet $X=\cup_{g\in G}X^g$, where $X^g=\{x_1^g,x_2^g,\ldots\}$, consider a free algebra $L(X)$. If $w(x_{i_1}^{g_1},\ldots,x_{i_n}^{g_n})$ is an element in $L(X)$ then $w(x_{i_1}^{g_1},\ldots,x_{i_n}^{g_n})=0$ is an identity in a G-graded algebra A if $w(a_1,\ldots,a_n)=0$ whenever $a_1\in A_{g_1},\ldots,a_n\in A_{g_n}$. In terms of H-identities one can rewrite $w(x_{i_1}^{g_1},\ldots,x_{i_n}^{g_n})$ as follows. The grading is given by the action of the dual Hopf algebra $H=(\mathbb{F}G)^*$. A natural basis of H consists of the projections p_g, $g\in G$. Graded monomial $x_{i_1}^{g_1},\ldots,x_{i_n}^{g_n}$ in $L(X)$ can now be associated with the element $(p_{g_1}\otimes\cdots\otimes p_{g_n})(x_{i_1}\cdots x_{i_n})$ in $L_H(X)$. This enables one to rewrite graded identities in terms of H-identities and *vice versa*.

The subalgebra $T(H)^H$ of invariants of the action of H on $T(H)$ via the left regular action (6) turns out to be important in the study of identities. In [5] the authors introduced a technical condition on H which, as they proved, was equivalent to the following

$$\dim T(H)/\mathrm{Ideal}(T(H)^H) < \infty. \tag{7}$$

Suppose that A is an H-algebra. We denote by $I(A)$ the subalgebra of H-invariants in A. Then the following is true.

Theorem 28 (Y. Bahturin, V. Linchenko) *Let H be a finite-dimensional Hopf algebra over a field $\mathbb{F}$. The following conditions are equivalent.*

1. *For any associative algebra A over $\mathbb{F}$ with an action of H it follows from $I(A)$ being a PI-algebra that also A is a PI-algebrai;*
2. *There exists a function $f : \mathbb{N} \to \mathbb{N}$ such that for any associative algebra A with an action of H if $I(A)$ satisfies a non-trivial identity of degree t; then A satisfies a non-trivial identity of degree $f(t)$*
3. *There exists a function $g(t)$ such that for any natural t and any H-algebra A with $I(A)^t = \{0\}$ one has $A^{g(t)} = \{0\}$;*
4. *There exists a number N such that any H-algebra A where $I(A)$ has zero multiplication satisfies $A^N = \{0\}$;*
5. *Condition (7) holds.*

Each of the above conditions implies that H is semisimple.

It is noted in [5] that (7) holds for the dual group algebras $(\mathbb{F}G)^*$, so that the result applies to algebras graded by a finite group. It also holds for semisimple group algebras and for the crossed product of algebras, satisfying (7). In particular, all semisimple Hopf algebras of dimension p^n, p a prime number, over a field of characteristic zero, satisfy (7). Thus from the above results one can recover well-known results on identities of graded associative algebras [9] and algebras with action of a group by automorphisms [21].

Condition (7) is crucial also in the case of Lie H-algebras. In the same way as in the case of Theorem 24, the result applies not only to Lie algebras with action of a Hopf algebras, but to a wider class of so called Lie type algebras, which incidentally includes associative algebras, Lie supearalgebras, and so on.

Let us call called a varuety of algebras a Lie type variety if the following conditions are satisfied:

1. There exist $\lambda, \mu \in \mathbb{F}$, $\lambda \neq 0$, such that in any algebra $\in \mathfrak{M}$ one has

$$a(bc) = \lambda(ab) + \mu(ac)b, \text{ where } \lambda, \mu \in \mathbb{F}, \lambda \neq 0.$$

2. The codimension growth $c_{\mathfrak{M}}(n)$ of M is faster than a function of the form $\frac{n!}{b^n}$, for some $b \in \mathbb{N}$.

An important theorem by V. Linchenko says the following.

Theorem 29 (V. Linchenko) *For a finite-dimensional Hopf algebra H the following conditions are equivalent*

(1) $\dim T(H)/Ideal(I(H)) < \infty$;
(2) *For any H-algebra L in a Lie type variety $\mathfrak{M}$, if L^H satisfies a non-trivial identity then so does L.*

As a corollary from Linchenko's theorem, one gets Bahturin and Zaicev's Theorem [6]. These theorems have been actively applied in Group Theory while studying, for example, groups with automorphisms, such that the subgroups of fixed points satisfy some finiteness conditions. As one of the latest references, please see [1].

7.1 Codimension Growth

Let L be an H-module Lie algebra. In complete analogy with the case of the sequence of ordinary codimensions $\{c_n(L)\}$ one can define the sequence of H-codimensions $\{c_n^H(L)\}$. The analog of Amitsur's conjecture for H-codimensions of L can be formulated as the question of the existence and integrality of the number $PI\exp^H(L) = \lim_{n\to\infty} c_n^H(L)$.

One calls $PI\exp^H(L)$ the Hopf PI-exponent of L.

A. Gordienko proved a number of results on the codimension growth of identical relations in finite-dimensional Lie algebras, with additional structure. The most general of them are the following. In the statement of them there is the notion of an H-nice Lie algebra L which is a finite-dimensional Lie algebra over an algebraically closed field $\mathbb{F}$ of characteristic zero. One says that L is H-nice if either L is semisimple or the following conditions hold:

1. The nilpotent radical N and the solvable radical R of L are H-invariant;
2. (Levi decomposition) There exists an H-invariant maximal semisimple subalgebra $B \subset L$ such that $L = B \oplus R$ (direct sum of H-modules);
3. (Wedderburn–Mal'cev decomposition) For any H-submodule $W \subset L$ and associative H-module subalgebra $A_1 \subset \mathrm{End}_{\mathbb{F}}(W)$, the Jacobson radical $J(A_1)$ is H-invariant and there exists an H-invariant maximal semisimple associative subalgebra $\widetilde{A}_1 \subset A_1$ such that $A_1 = \widetilde{A}_1 \oplus J(A_1)$ (direct sum of H-submodules);
4. For any H-invariant Lie subalgebra $L_0 \subset \mathfrak{gl}(L)$ such that L_0 is an H-module algebra and L is a completely reducible L_0-module disregarding H-action, L is a completely reducible (H, L_0)-module.

The following are Gordienko's main results on the codimension growth of H Lie algebras [16].

Theorem 30 (A. S. Gordienko) *Let L be a nonnilpotent H-nice Lie algebra over an algebraically closed field $\mathbb{F}$ of characteristic 0. Then there exist constants*

$C_1, C_2 > 0$, $r_1, r_2 \in \mathbb{R}$, $d \in \mathbb{N}$ *such that* $C_2 n^{r_1} d^n \leq c_n^H(L) \leq C_1 n^{r_2} d^n$ *for all* $n \in \mathbb{N}$.

Corollary 5 *The above analog of Amitsur's conjecture holds for such codimensions.*

Theorem 31 (A. S. Gordienko) *Let* $L = L_1 \oplus \cdots \oplus L_s$ *(direct sum of* H*-invariant ideals) be an* H*-module Lie algebra over an algebraically closed field* $\mathbb{F}$ *of characteristic 0 where* H *is a Hopf algebra. Suppose* L_i *are* H*-nice algebras. Then there exists* $PI \exp^H(L) = \max_{1 \leq i \leq s} PI \exp^H(L_i)$.

7.2 Isomorphism of H-Algebras

In the study of identities of H-algebras, such as graded algebras, algebras with fixed group of automorphisms or algebras with derivation, superalgebras one can use a corollary of the results of Razmyslov in Sect. 4.

We always assume that the signature of Ω-algebras contains at least one operation of arity at least 2.

Theorem 32 (Yu. P. Razmyslov) *Two simple finite-dimensional* Ω*-algebras over an algebraically closed field, satisfying the same polynomial identities, are isomorphic.*

An H algebra A can be turned to an Ω-algebra in the following way. Let μ be the original operation of A and for each $h \in H$ ρ_h is a unary operation given by $\rho_h(a) = h * a$. Set $\Omega_H = \{\mu\} \cup \{\rho_h | h \in H\}$. Now consider the relatively free Ω_H-algebra given by the identities

1. $\rho_h(x) + \rho_g(x) = \rho_{h+g}(x)$;
2. $\rho_{\alpha g}(x) = \alpha \rho_g(x)$;
3. $\rho_1(x) = x$;
4. $\rho_g(\rho_h(x)) = \rho_{gh}(x)$;
5. $\rho_h(\mu(x, y)) = \sum_h \mu(\rho_{h_{(1)}}(x), \rho_{h_{(2)}}(y))$.

Much in the same way as we did just above for the graded identities, one can rewrite H-identities of A to its identities, as Ω_H-algebra. Thus we obtain the following.

Theorem 33 (Y. Bahturin, F. Yasumura) *Two finite-dimensional* H*-algebras over an algebraically closed field, which are simple as* H*-algebras and satisfy the same* H*-identities are isomorphic as* H*-algebras.*

A particular case was settled in a paper [11]: under the restriction that G is an abelian group, the authors proved that any two finite-dimensional nonassociative G-graded simple algebras having the same G-graded identities must be isomorphic as G-graded algebras.

In particular, if we have two different (=non-isomorphic) G-gradings of a finite-dimensional simple Lie algebra then their graded identities are different. Thus it could be interesting to find identities that separate non-isomorphic gradings.

8 Special Lie Algebras

A Lie algebra L is called special, or SPI, if it is isomorphic to a Lie subalgebra of an associative PI-algebra. These algebras were introduced in 1963 by Latyshev. An excellent survey on the achievements in the area of special Lie algebras was written by Zaicev [49]. We refer the reader to this paper to learn about the theory of SPI-algebras, which seems to be in dormant condition right now.

To conclude, I will mention one open problem from [12] and a partial solution from the above survey.

Is it true that any variety of Lie algebras is locally solvable if it does not contain Sl_2?

In the case of special varieties, Zaicev proves the following.

Theorem 34 (M. V. Zaicev) *Let* $\mathfrak{V}$ *be a special variety of Lie algebras over an infinite field* $\mathbb{F}$. *If* $\mathfrak{V}$ *has no finite-dimensional semisimple Lie algebras then it is locally solvable. If* $\operatorname{char}\mathbb{F} = 0$ *then* $\mathfrak{V}$ *is solvable.*

References

1. Acciarri, C., Khukhro, E.I., Shumyatsky, P.: Profinite groups with an automorphism whose fixed points are right Engel. Proc. Am. Math. Soc. **147**(9), 3691–3703 (2019)
2. Bachmuth, S., Mochizuki, H.Y., Walkup, D.W.: A nonsolvable group of exponent 5. Bull. Am. Soc., **76**, 638–640 (1970)
3. Bahturin, Y.: Identical Relations in Lie Algebras, p. 309. VNU Press (1987)
4. Bahturin, Yu.A., Strade, H.: Locally finite-dimensional simple Lie algebras. Mat. Sb. **185**, 2–31 (1994)
5. Bahturin, Yu.A., Linchenko, V.V.: Identities of algebras with actions of Hopf algebras. J. Algebra **202**, 634–654 (1998)
6. Bahturin, Yu.A., Zaicev, M.V.: Identities of graded algebras. J. Algebra **205**, 1–12 (1998)
7. Bahturin, Yu.A., Drensky, V.: Identities of bilinear mappings and graded polynomial identities of matrices. Linear Algebra Appl. **369**, 95–112 (2003)
8. Bahturin, Yu.A., Olshanskii, A.Yu.: Large restricted Lie algebras. J. Algebra **310**, 413–427 (2007)
9. Bahturin, Yu.A., Giambruno, A., Riley, D.: Group-graded algebras with polynomial identity. Israel J. Math. **104**, 146–155 (1998)
10. Benediktovich, I.I., Zalesski, A.E.: T-ideals of free Lie algebras with polynomial growth of a sequence of codimensionalities. Vestsi Akad. Navuk BSSR Ser. Fiz.-Mat. Navuk (3), 5–10 (1980)
11. Bianchi, A., Diniz, D.: Identities and isomorphisms of finite-dimensional graded simple algebras. J. Algebra **526**, 333–344 (2019)

12. Dniester Notebook: Unsolved Problems of the Theory of Rings and Modules, 2nd edn. Novosibirsk (1976)
13. Giambruno, A., Zaicev, M.: Polynomial Identities and Asymptotic Methods, AMS Math. Surveys and Monographs, vol. 122, p. 352 (2005)
14. Giambruno, A., Mishchenko, S., Zaicev, M.: Algebras with intermediate growth of the codimensions. Adv. Appl. Math. **37**, 360–377 (2006)
15. Golod, E.S.: On nil-algebras and finitely approximable p-groups. Izv. Akad. Nauk SSSR Ser. Mat. **28**, 273–276 (1964)
16. Gordienko, A.S.: Amitsur's conjecture for polynomial H-identities of H-module Lie algebras. Trans. Am. Math. Soc. **367**, 313 –354 (2015)
17. Grishkov, A.N.: On the growth of varieties of Lie algebras. Mat. Zametki **44**, 51–54 (1988)
18. Higman, G.: Groups and rings having automorphisms without nontrivial fixed elements. J. Lond. Math. Soc. **32**, 321–334 (1958)
19. Iltyakov, A.V.: On finite basis of identities of Lie algebra representations. Nova J. Algebra Geom. **1**, 207–259 (1992)
20. Kemer, A.R.: Ideals of Identities of Associative Algebras. Amer. Math. Soc., Transl. of Monographs, vol. 87, p. 81 (1991)
21. Kharchenko, V.K.: Galois extensions and rings of fractions. Algebra Logika **13**, 460–484 (1974)
22. Kostrikin, A.I.: The Burnside problem. Izv. Akad. Nauk SSSR Ser. Mat. **23**, 3–34 (1959)
23. Kostrikin, A.I., Around Burnside, M.: Nauka, 1986, p. 232; Engl. Transl.: Springer, Berlin (1991), 216 pp
24. Krasilnikov, A.N.: Finite basis property for some varieties of Lie algebras. Vestnik Moscow Univ. Ser. 1 Mat. Mech. (2), 34–38 (1982)
25. Krasilnikov, A.N.: Identities of Lie algebras with nilpotent commutator ideal over a field of finite characteristic. Mat. Zametki **51**, 47–52 (1992)
26. Kushkulei, A.H., Razmyslov, Yu.P.: Varieties generated by irreducible representations of Lie algebras. Vestnik Moscow Univ. Ser. 1 Mat. Mech. (5), 4–7 (1983)
27. Linchenko, V.V.: Identities of Lie algebras with actions of Hopf algebras. Commun. Algebra **25**, 3179–3187 (1997)
28. Mishchenko, S.P.: Engel's problem. Mat. Sb. **124**, 56–67 (1984)
29. Mishchenko, S.P.: On varieties of polynomial growth of Lie algebras over a field of characteristic zero. Mat. Zametki **40**, 713–721, 828 (1986)
30. Mishchenko, S.S.: New example of a variety of Lie algebras with fractional exponent. Vestnik Moscow Univ., Ser. 1 Mat. Mech. 6, 44–47 (2011); Engl. Transl.: Moscow Univ. Math. Bull. **66**(6), 264–266 (2011)
31. Mishchenko, S.P., Zaicev, M.V.: Sequences of codimensions of identities and their asymptotic behavior. Fundam. Appl. Mat. **22**, 115–127 (2019)
32. Montgomery, S.: Hopf algebras and their actions on rings. In: CBMS Regional Conference Series in Mathematics, vol. 82, p. 238 (1993)
33. Petrogradskii, V.M.: Growth of polynilpotent varieties of Lie algebras and rapidly growing entire functions. Mat. Sb. **188**, 119–138 (1997)
34. Petrogradskii, V.M.: On numerical characteristics of subvarieties for three varieties of Lie algebras. Mat. Sb. **188**, 111–126 (1999); Engl. transl.: Sb. Math. **190**, 887–902 (1999)
35. Petrogradsky, V.M., Shestakov, I.P., Zelmanov, E.I.: Nil graded self-similar algebras. Groups Geom. Dyn. **4**, 873–900 (2010)
36. Premet, A., Strade, H.: Simple Lie algebras of small characteristic. VI. Completion of the classification. J. Algebra **320**, 3559–3604 (2008)
37. Razmyslov, Yu.P.: On Lie algebras satisfying the Engel condition. Algebra Logika **10**, 33–44 (1972)
38. Razmyslov, Yu.P.: Algebras satisfying Capelli type identities. Izv. AN SSSR, Ser. Mat. **45**, 143–166 (1981)
39. Razmyslov, Yu.P.: Varieties of representations of finite-dimensional algebras. Vestn. Mosc. Univ. Ser. 1 Mat. Mech. (6), 31–37 (1982)

40. Razmyslov, Yu.P.: Simple Lie algebras in the varieties generated by simple Lie algebras of Cartan type. Izvest. Akad. Nauk SSSR, Ser. Mat. **51**,1228–1264 (1987)
41. Razmyslov, Yu.P.: On inclusion of varieties generated by simple Lie algebras. Vestn. Mosc. Univ. Ser. 1 Mat. Mech. (2), 34–37 (1989)
42. Razmyslov, Yu.P.: Identities of Algebras and their Representations. AMS Transl. of Math. Mon. vol. 138, p. 318 (1994)
43. Regev, A.: Existence of identities in $A \otimes B$. Israel J. Math. **11**, 131–152 (1972)
44. Stovba, V.V.: Finite basis problem for certain varieties of Lie algebras and associative algebras. Vestn. Mosc. Univ. Ser. 1 Mat. Mech. (2), 54–58 1982
45. Traustason, G.: A constructive approach to Zel'manov's global nilpotency theorem for n-Engel Lie algebras over a field of characteristic zero. Internat. J. Algebra Comput. **8**, 317–326 (1998)
46. Vaughan-Lee, M.R.: The Restricted Burnside Problem, 2nd edn. London Math. Soc. Monographs (N.S.) vol. 5, p. 256. Oxford University Press, Oxford (1993)
47. Volichenko, I.B.: Varieties of Lie algebras with identity $[[X_1, X_2, X_3], [X_4, X_5, X_6]] = 0$ over a field of characteristic zero. Sibirsk. Mat. Zh. **25**, 40–54 (1984) (in Russian)
48. Zaicev, M.V.: Identities of affine Kac–Moody algebras. Vestn. Mosc. Univ. Ser. 1 Mat. Mech. (2), 34–38 (1982)
49. Zaicev, M.V.: Special Lie algebras. Usp. Mat. Nauk **48**, 103–140 (1993)
50. Zaicev, M.V., Mishchenko, S.P.: An example of a variety of Lie algebras with a fractional exponent. Algebra. J. Math. Sci. (New York) **93**, 977–982 (1999)
51. Zaicev, M.V.: Integrality of exponents of growth of identities of finite-dimensional Lie algebras. Izv. Ross. Akad. Nauk Ser. Mat. **66**, 23–48 (2002); Engl. Transl.: Izv. Math. **66**, 463–487 (2002)
52. Zel'manov, E.I.: Engel Lie-algebras. Dokl. Akad. Nauk SSSR **292**, 265–268 (1987)
53. Zel'manov, E.I.: The solution of the restricted Burnside problem for groups of odd exponent. Math. USSR Izvestia **36**, 41–60 (1991)
54. Zel'manov, E.I.: A solution of the restricted Burnside problem for 2-groups. Mat. Sb. **182**, 568–592 (1991); Engl. Transl.: Math. USSR-Sb. **72**, 543–565 (1992)
55. Zel'manov, E.I.: Lie algebras and torsion groups with identity. J. Comb. Algebra **1**, 289–340 (2017)

Minimal Degree of Identities of Matrix Algebras with Additional Structures

Dafne Bessades, Rafael Bezerra dos Santos, and Ana Cristina Vieira

Abstract In the ordinary context of the PI-theory, it is well known that $2n$ is the smallest degree of a standard polynomial identity of $M_n(F)$. Here we present some results about the minimal degree of polynomial identities of $M_n(F)$ in the graded and involution cases. Also we give some consequences in the graded involution case.

Keywords Matrix algebra · Standard identity · Involution · Superalgebra · Graded involution

1 Introduction

Throughout this paper, F will denote a field of characteristic zero and A is an associative algebra over F. Recall that an identity of A is a polynomial $f(x_1, \ldots, x_n)$ in the free associative algebra $F\langle X\rangle$ on a countable set X of noncommuting variables over F such that $f(a_1, \ldots, a_n) = 0$ for all $a_1, \ldots, a_n \in A$. In this case, we say that A satisfies the identity f and denote $f \equiv 0$ on A.

It is obvious that the null polynomial is an identity for any algebra and in the special case that A satisfies a non trivial identity, we say that A is a PI-algebra.

We denote by $Id(A) = \{f \in F\langle X\rangle : f \equiv 0 \text{ on } A\}$ the ideal of all identities satisfied by A. We have that $Id(A)$ is a T-ideal of $F\langle X\rangle$, i.e. an ideal invariant under all endomorphisms of $F\langle X\rangle$. In [15], Kemer proved that a T-ideal is generated by a finite set of multilinear polynomials. Recall that a multilinear polynomial is a polynomial which is linear in each of its variables.

D. Bessades · R. B. dos Santos · A. C. Vieira (✉)
Departamento de Matemática, Instituto de Ciências Exatas, Universidade Federal de Minas Gerais, Belo Horizonte, Brazil
e-mail: dafneclbessades@ufmg.br; rafaelsantos23@ufmg.br; anacris@ufmg.br

O. M. Di Vincenzo, A. Giambruno (eds.), *Polynomial Identities in Algebras*, Springer INdAM Series 44, https://doi.org/10.1007/978-3-030-63111-6_2

An important example of multilinear polynomial in $F\langle X\rangle$ is the standard polynomial of degree n defined by

$$St_n(x_1,\dots,x_n)=\sum_{\sigma\in S_n} sgn(\sigma)\, x_{\sigma(1)}\cdots x_{\sigma(n)},$$

where S_n denotes the symmetric group of degree n and $sgn(\sigma)$ is the sign of the permutation σ.

If A is a k-dimensional algebra, it is not difficult to prove that A satisfies $St_{k+1}(x_1,\dots,x_{k+1})$. The well known Amitsur–Levitzki theorem (see [2]) shows that the degree of a standard identity is lower in case A is the full algebra of matrices over F. In fact, $St_{2n}(x_1,\dots,x_{2n})$ is a polynomial identity of the algebra $M_n(F)$ and, in addition, $M_n(F)$ does not satisfy a polynomial identity of degree less than $2n$, that is, the smallest degree of an identity of $M_n(F)$ is $2n$.

An interesting question concerning the above situation can be setting when we consider the algebra $M_n(F)$ with additional structures, such as graded algebra or algebra with involution. We can ask if the minimal degree of a standard identity remains $2n$. More precisely, when only certain types of matrices are considered, what is the smallest degree of a standard identity satisfied by $M_n(F)$?

Our goal is to discuss this problem in the graded and involution cases, by presenting the results already established in order to respond the question about the smallest degree of a standard identity in each case. Furthermore, we consider the consequences of these results for the case in which $M_n(F)$ is endowed with a graded involution, giving our contribution to the theory in this situation.

More generally, we shall also present some results regarding the minimal degree of an identity satisfied by $M_n(F)$ with additional structures.

2 Involution Case

In this section we treat standard identities in symmetric and skew variables of an algebra with involution. An involution on an algebra A is a F-linear map $*: A\to A$ satisfying

$$(a^*)^*=a \text{ and } (ab)^*=b^*a^*, \text{for all } a,b\in A,$$

i.e. $*$ is an anti-automorphism of order at most 2 of A.

An algebra A endowed with an involution $*$ will be called an algebra with involution and will be denoted by $(A,*)$. In this case, we write $A=A^+\oplus A^-$ where $A^+=\{a\in A : a^*=a\}$ is the subspace of symmetric elements and $A^-=\{a\in A : a^*=-a\}$ is the subspace of skew elements of A.

A natural example of involution on the matrix algebra $M_n(F)$ is the transpose involution defined by $(a_{ij})^t=(a_{ji})$, where $(a_{ij})\in M_n(F)$.

If $n=2k$ is even, there exists another involution on $M_n(F)$ called the canonical symplectic involution s, which will be referred just as symplectic involution, and is defined by

$$a^s = Ta^tT^{-1}, \quad \text{for all } a \in M_{2k}(F),$$

where $T = \sum_{i=1}^{k}(e_{i,i+k} - e_{i+k,i})$. In other words, if $n = 2k$, a $2k \times 2k$ matrix is partitioned into four $k \times k$ matrices R, S, P, Q and s is defined as follows

$$\begin{pmatrix} R & S \\ P & Q \end{pmatrix}^s = \begin{pmatrix} Q^t & -S^t \\ -P^t & R^t \end{pmatrix}.$$

Recall that if $X = \{x_1, x_2, \dots\}$ is a countable set of noncommuting variables, we can consider $F\langle X, *\rangle = F\langle x_1, x_1^*, x_2, x_2^*, \dots\rangle$, the free algebra with involution on X over F. By setting $y_i = x_i + x_i^*$ and $z_i = x_i - x_i^*$, for every $i = 1, 2 \dots$, we consider $F\langle X, *\rangle = F\langle y_1, z_1, y_2, z_2, \dots\rangle$ as generated by symmetric and skew variables. A $*$-polynomial $f(y_1, \dots, y_n, z_1, \dots, z_m) \in F\langle X, *\rangle$ is a $*$-identity of A if $f(u_1, \dots, u_n, v_1, \dots v_m) = 0$ for all $u_1, \dots, u_n \in A^+$ and $v_1, \dots, v_n \in A^-$.

In the involution case, we will consider two types of standard polynomials of degree n. The first one is $St_n(y_1, \dots, y_n)$ in symmetric variables and the other is $St_n(z_1, \dots, z_n)$ in skew variables. Next, we will present the results concerning the smallest degree of standard identities of these types for $M_n(F)$ endowed with transpose and symplectic involutions.

In 1979, Slin'ko proved that the degree does not change when we consider symmetric matrices under the transpose involution.

Theorem 1 (Slin'ko, [21]) *The smallest degree of a standard identity in symmetric variables of* $(M_n(F), t)$ *is* $2n$.

In case that skew matrices under the transpose involution are considered, the first result was given in 1958 by Kostant as follows.

Theorem 2 (Kostant, [16]) *The standard polynomial* $St_{2n-2}(z_1, \dots, z_{2n-2})$ *is an identity of* $(M_n(F), t)$, *for all* $n > 1$ *even.*

Some years later, in 1974, Rowen extended the Kostant's result showing that $St_{2n-2}(z_1, \dots, z_{2n-2})$ is an identity of $(M_n(F), t)$, for all $n \geq 1$, and also proved the next theorem, which has independent proofs given by Owens [17] and Hutchinson [14].

Theorem 3 (Rowen, [19]) *The smallest degree of a standard identity in skew variables of* $(M_n(F), t)$ *is* $2n - 2$.

For the symplectic involution we consider the algebra $M_{2k}(F)$, where $k \geq 1$. In this case, when symmetric matrices are considered, Rowen proved the following.

Theorem 4 (Rowen, [20]) *The standard polynomial* $St_{4k-2}(y_1, \dots, y_{4k-2})$ *is an identity of* $(M_{2k}(F), s)$, *for all* $k \geq 1$.

Although this theorem does not give information about the smallest degree in this case, Rowen conjectured that the smallest degree of a standard identity in symmetric variables of $(M_{2k}(F), s)$ is $4k - 2$ and for $k = 2$, he showed that $(M_4(F), s)$ does not satisfy $St_5(y_1, \ldots, y_5)$.

Furthermore, in 1992, Adamsson (see [1]) considered the cases $k = 3, 4$ and proved that $(M_6(F), s)$ and $(M_8(F), s)$ do not satisfy $St_9(y_1, \ldots, y_9)$ and $St_{13}(y_1, \ldots, y_{13})$, respectively.

Recently, Bessades, Leal, dos Santos and Vieira treated the problem by considering that k is a power of 2. They proved the following theorem that confirms Rowen's conjecture in a particular case.

Theorem 5 (Bessades et al. [4]) *The minimal degree of a standard identity in symmetric variables of* $(M_{2^m}(F), s)$ *is* $2^{m+1} - 2$.

The cases discussed above suggest that the smallest degree of a standard identity in symmetric variables of $(M_{2k}(F), s)$, in fact, must be $4k - 2$. A general proof was not given until now.

On the other hand, for skew matrices under the symplectic involution, the smallest degree of a standard identity in skew variables of $(M_{2k}(F), s)$ has already been determined by Giambruno, Ioppolo and Martino in 2016. In this situation, the minimal degree is the same as in the ordinary case.

Theorem 6 (Giambruno et al. [10]) *The minimal degree of a standard identity in skew variables of* $(M_{2k}(F), s)$ *is* $4k$.

We remark that the general question concerning the minimal degree of identities of $(M_n(F), t)$ and $(M_{2k}(F), s)$ is still open.

For $(M_n(F), t)$ Giambruno proved in [8] that $n + 1$ is a lower bound for the degree of an identity of $(M_n(F), t)$. In the cases $n = 2, 3, 4$, D'Amour and Racine in [5] determined that the minimal degree are 2, 4 and 5, respectively.

When only symmetric variables are considered, Slin'ko [21] determined that the minimal degree of an identity is $2n$. Also, Wenxin and Racine described in [22] all identities in symmetric variables with this degree. On the other hand, when only skew variables are considered, Hill [13] proved the existence of an identity of degree $2n - 3$, for $n > 1$ even.

For $(M_{2k}(F), s)$, in the cases $k = 1, 2$, D'Amour and Racine [6] determined that the minimal degrees are 2 and 5, respectively. When only symmetric variables are considered, Rashkova [18] determined that for $k = 3$ the minimal degree is 9.

When $k \geq 3$, Drensky and Giambruno [7] proved that $2k + 2$ is a lower bound for the degree of an identity in symmetric variables of $(M_{2k}(F), s)$. Also, Hill [12] presented an upper bound by constructing a multilinear polynomial in symmetric variables of degree $4k - 3$ which is an identity of $(M_{2k}(F), s)$.

However, as was said above, a proof providing the precise minimal degree has not been presented in the general case for both cases $(M_n(F), t)$ and $(M_{2k}(F), s)$.

3 Graded Case

In the graded case, the smallest degree of an identity of the full matrix algebra with a non trivial $\mathbb{Z}_2$-grading has already been determined. In this section we shall present the results obtained by Antonov about the minimal degree of multilinear identities in even and odd variables.

We say that an algebra A is a superalgebra if there exists a vector space decomposition $A = A_0 \oplus A_1$, where A_0 and A_1 satisfy $A_0A_0 + A_1A_1 \subseteq A_0$ and $A_0A_1 + A_1A_0 \subseteq A_1$. Observe that the pair (A_0, A_1) is a $\mathbb{Z}_2$-grading on A. Notice that any algebra is a superalgebra with trivial grading $(A, \{0\})$.

The free associative algebra $F\langle X\rangle$ has a natural structure of superalgebra as follows. Write $X = Y \cup Z$, the disjoint union of two countable sets $Y = \{y_1, y_2, \ldots\}$ and $Z = \{z_1, z_2, \ldots\}$. If we denote by $\mathcal{F}_0$ the subspace of $F\langle Y \cup Z\rangle$ spanned by all monomials on X having an even number of variables from Z and by $\mathcal{F}_1$ the subspace spanned by all monomials with an odd number of variables from Z, then $F\langle Y \cup Z\rangle = \mathcal{F}_0 \oplus \mathcal{F}_1$ is a $\mathbb{Z}_2$-graded algebra called the free superalgebra on Y and Z over F. We say that the variables from Y have even degree, whereas the variables from Z have odd degree.

Given a superalgebra $A = (A_0, A_1)$, we say that a polynomial

$$f(y_1, \ldots, y_n, z_1, \ldots, z_m) \in F\langle Y \cup Z\rangle$$

is a graded identity of A if $f(a_1, \ldots, a_n, b_1, \ldots, b_m) = 0$ for all $a_1, \ldots, a_n \in A_0$ and $b_1, \ldots, b_m \in A_1$.

The ideal $Id^{gr}(A)$ of the graded identities satisfied by A is an ideal invariant under all endomorphisms of $F\langle Y \cup Z\rangle$ that preserve the grading and is completely determined by its multilinear polynomials.

Recall that, when F is an algebraically closed field, up to isomorphism, any $\mathbb{Z}_2$-grading on $M_n(F)$ is given by a pair $(M_{k,l}(F)_0, M_{k,l}(F)_1)$ defined by

$$M_{k,l}(F)_0 := \left\{ \begin{pmatrix} A & 0 \\ 0 & D \end{pmatrix} : A \in M_k(F);\ D \in M_l(F) \right\}$$

$$\text{and} \quad M_{k,l}(F)_1 := \left\{ \begin{pmatrix} 0 & B \\ C & 0 \end{pmatrix} : B \in M_{k\times l}(F);\ C \in M_{l\times k}(F) \right\},$$

where $k + l = n$ and $k \geq l \geq 0$.

Observe that if $l = 0$, then we have a trivial grading on $M_n(F)$. So, our interest is in the case $k \geq l > 0$. In this situation, $M_n(F)$ is a superalgebra endowed with a non trivial $\mathbb{Z}_2$-grading which will be denoted by $M_{k,l}(F)$.

The minimal degree of a polynomial identity in variables of even degree of $M_{k,l}(F)$ is an immediate consequence of Amitsur–Levitzki theorem.

Theorem 7 *The minimal degree of a polynomial identity in variables of even degree of $M_{k,l}(F)$ is $2k$, for all $k \geq l > 0$.*

In 2012, Antonov considered the case of polynomials containing only variables of odd degree.

Theorem 8 (Antonov, [3]) *The minimal degree of a polynomial identity in variables of odd degree of $M_{k,k}(F)$ is $4k - 1$, for all $k \geq 1$.*

Theorem 9 (Antonov, [3]) *The minimal degree of a polynomial identity in variables of odd degree of $M_{k,l}(F)$ is $4l + 1$, for all $k > l > 0$.*

Moreover, the author further provided identities in variables of odd degree having the minimal degrees established in Theorems 8 and 9. These identities are known as double Capelli polynomials and are defined as follows.

Definition 1 We define the so-called double Capelli polynomials of degree $2n - 1$ and of degree $2n$ as being, respectively,

$$\mathscr{C}_n^{n-1}(u_1, \ldots, u_n; v_1, \ldots, v_{n-1}) := \sum_{\substack{\sigma \in S_n, \\ \tau \in S_{n-1}}} sgn(\sigma\tau)\, u_{\sigma(1)} v_{\tau(1)} \ldots v_{\tau(n-1)} u_{\sigma(n)} \text{ and}$$

$$\mathscr{C}_n(u_1, \ldots, u_n; v_1, \ldots, v_n) := \sum_{\substack{\sigma \in S_n, \\ \tau \in S_n}} sgn(\sigma\tau)\, u_{\sigma(1)} v_{\tau(1)} \ldots u_{\sigma(n)} v_{\tau(n)},$$

where the indexes of $\mathscr{C}_n^{n-1}$ suggest that we have $n - 1$ variables v's and n variables u's, whereas the index of $\mathscr{C}_n$ suggest that we have n variables u's and n variables v's.

Assuming these definitions, Antonov showed the following theorems.

Theorem 10 (Antonov, [3]) *The double Capelli polynomial $\mathscr{C}_{2k}^{2k-1}$ in odd variables is an identity of minimal degree of $M_{k,k}(F)$, for all $k \geq 1$.*

Theorem 11 (Antonov, [3]) *The double Capelli polynomial $\mathscr{C}_{2l+1}^{2l}$ in odd variables is an identity of minimal degree of $M_{k,l}(F)$, for all $k > l > 0$.*

In the ordinary context, Giambruno and Sehgal [9] showed that $4n$ is the minimal degree of a double Capelli identity of $M_n(F)$. In particular, $8n$ is the minimal degree of a double Capelli identity of $M_{2n}(F)$. However, as we saw above, in the context of graded identities of $M_{k,k}(F)$, when only odd variables are considered, this degree drops drastically to $4k - 1$. In light of this, we can see that the answer to the question discussed in the introduction about the smallest degree of standard identities can be even more delicate when we consider gradings on matrix algebras.

4 Graded Involution Case

In this section we give our contribution to the graded involution case by presenting the minimal degrees of standard identities in symmetric and skew variables of even degree of the full matrix algebra with transpose and symplectic graded involutions.

Recall that an involution $*$ on a superalgebra $A = A_0 \oplus A_1$ that preserves the homogeneous components A_0 and A_1, i.e. $(A_0)^* = A_0$ and $(A_1)^* = A_1$, is called a graded involution. A superalgebra A endowed with a graded involution $*$ is called $*$-superalgebra.

The connection between the superstructure and the involution on A is given in the next lemma (see [11]).

Lemma 1 *Let A be a superalgebra over a field F of characteristic different from 2 endowed with an involution $*$ and φ the automorphism of order 2 determined by the superstructure. Then A is a $*$-superalgebra if and only if $* \circ \varphi = \varphi \circ *$.*

As a consequence of this lemma, if A is a superalgebra over a field F of characteristic different from 2 endowed with an involution $*$, then A is a $*$-superalgebra if and only the subspaces A^+ and A^- are graded subspaces. As a consequence, any $*$-superalgebra can be written as a sum of 4 subspaces

$$A = (A_0)^+ \oplus (A_1)^+ \oplus (A_0)^- \oplus (A_1)^-.$$

We can give a superstructure on the free algebra $F\langle X\rangle$ by writing the set X as the disjoint union of four countable sets $X = Y_0 \cup Y_1 \cup Z_0 \cup Z_1$, where $Y_0 = \{y_{1,0}, y_{2,0}, \ldots\}$, $Y_1 = \{y_{1,1}, y_{2,1}, \ldots\}$, $Z_0 = \{z_{1,0}, z_{2,0}, \ldots\}$ and $Z_1 = \{z_{1,1}, z_{2,1}, \ldots\}$. We define the free $*$-superalgebra $\mathcal{F} = F\langle X|\mathbb{Z}_2, *\rangle$ of countable rank on X by requiring that the variables from $Y_0 \cup Z_0$ are homogeneous of even degree and those from $Y_1 \cup Z_1$ are homogeneous of odd degree. We also define an involution on $\mathcal{F}$ by requiring that the variables from $Y_0 \cup Y_1$ are symmetric and those from $Z_0 \cup Z_1$ are skew.

Consider $\mathcal{F}^{(0)}$ to be the span of all monomials in the variables from X which have an even number of variables of odd degree and $\mathcal{F}^{(1)}$ to be the span of all monomials in the variables from X which have an odd number of variables of odd degree. Then $(\mathcal{F}^{(0)})^* = \mathcal{F}^{(0)}$ and $(\mathcal{F}^{(1)})^* = \mathcal{F}^{(1)}$ and so $\mathcal{F} = \mathcal{F}^{(0)} \oplus \mathcal{F}^{(1)}$ has a structure of $*$-superalgebra. The elements of $\mathcal{F}$ are called $(\mathbb{Z}_2, *)$-polynomials.

Let

$$f = f(y_{1,0}, \ldots, y_{m,0}, y_{1,1}, \ldots, y_{n,1}, z_{1,0}, \ldots, z_{p,0}, z_{1,1}, \ldots, z_{q,1}) \in \mathcal{F}.$$

We say that f is a $(\mathbb{Z}_2, *)$-identity for the $*$-superalgebra A, and we write $f \equiv 0$ on A, if

$$f(a_{1,0}^+, \ldots, a_{m,0}^+, a_{1,1}^+, \ldots, a_{n,1}^+, a_{1,0}^-, \ldots, a_{p,0}^-, a_{1,1}^-, \ldots, a_{q,1}^-) = 0,$$

for all $a_{1,0}^{+}, \dots, a_{m,0}^{+} \in (A_0)^{+}, a_{1,1}^{+}, \dots, a_{n,1}^{+} \in (A_1)^{+}, a_{1,0}^{-}, \dots, a_{p,0}^{-} \in (A_0)^{-}$ and $a_{1,1}^{-}, \dots, a_{q,1}^{-} \in (A_1)^{-}$.

It is clear that any algebra with involution $*$ endowed with trivial grading is a $*$-superalgebra. Also, notice that the identity map is a graded involution for any commutative superalgebra.

In [11], Giambruno, dos Santos and Vieira proved that, when F is an algebraically closed field, the only graded involutions defined on $M_{k,l}(F)$ are the transpose (t) and the symplectic (s), the latter case being allowed only when $k = l$ or $l = 0$ and k is even. We will consider only the case $l \neq 0$, since we are interested in non trivial gradings.

In the case of transpose graded involution (t), we have the following four subspaces of elements symmetric and skew of even and odd degree:

$$\left(M_{k,l}(F), t\right)_0^+ = \left\{\begin{pmatrix} S & 0 \\ 0 & S' \end{pmatrix} : S \in (M_k(F), t)^+ \text{and } S' \in (M_l(F), t)^+\right\},$$

$$\left(M_{k,l}(F), t\right)_0^- = \left\{\begin{pmatrix} K & 0 \\ 0 & K' \end{pmatrix} : K \in (M_k(F), t)^- \text{and } K' \in (M_l(F), t)^-\right\},$$

$$\left(M_{k,l}(F), t\right)_1^+ = \left\{\begin{pmatrix} 0 & A \\ A^t & 0 \end{pmatrix} : A \in M_{k\times l}(F)\right\},$$

$$\left(M_{k,l}(F), t\right)_1^- = \left\{\begin{pmatrix} 0 & A \\ -A^t & 0 \end{pmatrix} : A \in M_{k\times l}(F)\right\}.$$

Also, for the case of the symplectic graded involution (s) the four subspaces of elements symmetric and skew of even and odd degree are

$$\left(M_{k,k}(F), s\right)_0^+ = \left\{\begin{pmatrix} A & 0 \\ 0 & A^t \end{pmatrix} : A \in M_k(F)\right\},$$

$$\left(M_{k,k}(F), s\right)_0^- = \left\{\begin{pmatrix} A & 0 \\ 0 & -A^t \end{pmatrix} : A \in M_k(F)\right\},$$

$$\left(M_{k,k}(F), s\right)_1^+ = \left\{\begin{pmatrix} 0 & K \\ K' & 0 \end{pmatrix} : K \in (M_k(F), t)^- \text{and } K' \in (M_k(F), t)^-\right\},$$

$$\left(M_{k,k}(F), s\right)_1^- = \left\{\begin{pmatrix} 0 & S \\ S' & 0 \end{pmatrix} : S \in (M_k(F), t)^+ \text{and } S' \in (M_k(F), t)^+\right\}.$$

In what follows, we are interested in to determine the smallest degree of standard identities in symmetric and skew variables of even degree of the $*$-superalgebras $\left(M_{k,k}(F), s\right)$ and $\left(M_{k,l}(F), t\right)$. With this purpose, we begin with some remarks concerning well known properties of standard polynomials.

Remark 1 For all $n \geq 1$, if $A_1, \ldots, A_{n+1}$ are $k \times k$ matrices, then

(i) $St_{n+1}(A_1, \ldots, A_{n+1}) = \sum_{i=1}^{n+1}(-1)^{i+1} A_i St_n(A_1, \ldots, \hat{A}_i, \ldots, A_{n+1})$, where the symbol ^ means omission;

(ii) $St_n(A_1^t, \ldots, A_n^t) = (-1)^{\frac{n(n-1)}{2}} St_n(A_1, \ldots, A_n)^t$.

We shall use the properties described above to provide the smallest degree of a standard $(\mathbb{Z}_2, *)$-identity in the symmetric and skew variables of even degree of the $*$-superalgebra $\big(M_{k,k}(F), s\big)$.

Theorem 12 *The minimal degree of a standard identity in skew variables of even degree of* $(M_{k,k}(F), s)$ *is* $2k$.

Proof Let $K_{1,0}, \ldots, K_{r,0} \in \big(M_{k,k}(F), s\big)_0^-$. Then, we have that

$$K_{i,0} = \begin{pmatrix} A_i & 0 \\ 0 & -A_i^t \end{pmatrix}, \text{ where } A_i \in M_k(F), \text{ for all } 1 \leq i \leq r.$$

Now, by using the Remark 1, we have that

$$\begin{aligned}
St_r(K_{1,0}, \ldots, K_{r,0}) &= St_r\left(\begin{pmatrix} A_1 & 0 \\ 0 & -A_1^t \end{pmatrix}, \ldots, \begin{pmatrix} A_r & 0 \\ 0 & -A_r^t \end{pmatrix}\right) \\
&= \begin{pmatrix} St_r(A_1, \ldots, A_r) & 0 \\ 0 & St_r(-A_1^t, \ldots, -A_r^t) \end{pmatrix} \\
&= \begin{pmatrix} St_r(A_1, \ldots, A_r) & 0 \\ 0 & (-1)^r St_r(A_1^t, \ldots, A_r^t) \end{pmatrix} \\
&= \begin{pmatrix} St_r(A_1, \ldots, A_r) & 0 \\ 0 & (-1)^{r+\frac{r(r-1)}{2}} St_r(A_1, \ldots, A_r)^t \end{pmatrix}.
\end{aligned}$$

Therefore, we conclude that $St_r(z_{1,0}, \ldots, z_{r,0})$ is a $(\mathbb{Z}_2, *)$-identity of $(M_{k,k}(F), s)$ if and only if $St_r(x_1, \ldots, x_r)$ is an identity of $M_k(F)$. However, by Amitsur–Levitzki theorem, we know that this occurs if and only if $r \geq 2k$. In this way, we have that the minimal degree of a standard identity in skew variables of even degree of $(M_{k,k}(F), s)$ is $2k$. □

In a similar way it is also possible to prove the following.

Theorem 13 *The minimal degree of a standard identity in symmetric variables of even degree of* $(M_{k,k}(F), s)$ *is* $2k$.

Now, as a consequence of the Theorems 1 and 3, we find in the following theorems the smallest degree of a standard identity in the symmetric and skew variables of even degree of the $*$-superalgebra $\big(M_{k,l}(F), t\big)$.

Theorem 14 *The minimal degree of a standard identity in skew variables of even degree of* $(M_{k,l}(F), t)$ *is* $2k - 2$.

Proof Let $K_{1,0}, \ldots, K_{r,0} \in \big(M_{k,l}(F), t\big)_0^-$. We know that $K_{i,0} = \begin{pmatrix} A_i & 0 \\ 0 & B_i \end{pmatrix}$, $1 \leq i \leq r$, where $A_1, \ldots, A_r \in (M_k(F), t)^-$ and $B_1, \ldots, B_r \in (M_l(F), t)^-$.

Now, it is easily seen that

$$St_r(K_{1,0}, \ldots, K_{r,0}) = St_r\left(\begin{pmatrix} A_1 & 0 \\ 0 & B_1 \end{pmatrix}, \ldots, \begin{pmatrix} A_r & 0 \\ 0 & B_r \end{pmatrix}\right)$$

$$= \begin{pmatrix} St_r(A_1, \ldots, A_r) & 0 \\ 0 & St_r(B_1, \ldots, B_r) \end{pmatrix}.$$

Since $A_1, \ldots, A_r$ are $k \times k$ skew matrices and $B_1, \ldots, B_r$ are $l \times l$ skew matrices, with respect to the transpose involution, we conclude, by Theorem 3, that $St_r(A_1, \ldots, A_r) = 0$ and $St_r(B_1, \ldots, B_r) = 0$ if and only if $r \geq 2k - 2$ and $r \geq 2l - 2$. Since it was established that $k \geq l > 0$, we get that the minimal degree of a standard identity in skew variables of even degree of $(M_{k,l}(F), t)$ is in fact $2k - 2$. □

Theorem 15 *The minimal degree of a standard identity in symmetric variables of even degree of* $(M_{k,l}(F), t)$ *is* $2k$.

Proof Let $S_{1,0}, \ldots, S_{r,0} \in \big(M_{k,l}(F), t\big)_0^+$. Now, we have

$$S_{i,0} = \begin{pmatrix} A_i & 0 \\ 0 & B_i \end{pmatrix}, \ 1 \leq i \leq r,$$

where $A_1, \ldots, A_r \in (M_k(F), t)^+$ and $B_1, \ldots, B_r \in (M_l(F), t)^+$.

As in the previous theorem,

$$St_r(S_{1,0}, \ldots, S_{r,0}) = St_r\left(\begin{pmatrix} A_1 & 0 \\ 0 & B_1 \end{pmatrix}, \ldots, \begin{pmatrix} A_r & 0 \\ 0 & B_r \end{pmatrix}\right)$$

$$= \begin{pmatrix} St_r(A_1, \ldots, A_r) & 0 \\ 0 & St_r(B_1, \ldots, B_r) \end{pmatrix}.$$

Since $A_1, \ldots, A_r$ are $k \times k$ symmetric matrices and $B_1, \ldots, B_r$ are $l \times l$ symmetric matrices, with respect to the transpose involution, we now conclude, by Theorem 1, that $St_r(A_1, \ldots, A_r) = 0$ and $St_r(B_1, \ldots, B_r) = 0$ if and only if $r \geq 2k$ and $r \geq 2l$. Recalling that $k \geq l > 0$, we get the minimal degree of a standard identity in symmetric variables of even degree of $(M_{k,l}(F), t)$ is $2k$.

□

We can see from the results discussed above that there is a close relationship between the minimal degree of standard identities in symmetric and skew variables of $(M_k(F), t)$ in the involution case and the minimal degree of standard identities in symmetric and skew variables of even degree of $(M_{k,k}(F), s)$ and $(M_{k,l}(F), t)$ in the graded involution case.

In a future work, we will show this relationship is not only restricted to the symmetric and skew variables of even degree. In fact, it can also be extended in a some way to symmetric and skew variables of odd degree. In light of this, an interesting observation is that several approaches and tools used in the case of matrices with transpose involution can also be applied in the context of superalgebras with graded involution.

We finish by remarking that the minimal degree of identities of the $*$-superalgebras $(M_{k,l}(F), t)$ and $(M_{k,k}(F), s)$ seems to be out of reach at present.

Acknowledgement The first author was partially supported by CNPq—Conselho Nacional de Desenvolvimento Científico e Tecnológico.

References

1. Adamsson, J.M.H.: The standard polynomial as an identity on symplectic matrices. M.Sc. thesis. University of Ottawa (1992)
2. Amitsur, S.A., Levitzki, J.: Minimal identities for algebras. Proc. Am. Math. Soc. **1**, 449–463 (1950)
3. Antonov, S.Yu.: The least degree of identities in the subspace $M_1^{(m,k)}(F)$ of the matrix superalgebra $M^{(m,k)}(F)$. Russ. Math. (Iz. VUZ) **56**(11), 1–16 (2012)
4. Bessades, D., Leal, G., dos Santos, R., Vieira, A.C.: Matrix algebras with involution and standard polynomial identities in symmetric variables. Linear Algebra Appl. **582**, 467–481 (2019)
5. D'Amour, A., Racine, M.L.: $*$-Polynomial identities of matrices with the transpose involution: the low degrees. Trans. Am. Math. Soc. **12**, 5089–5106 (1999)
6. D'Amour, A., Racine, M.L.: $*$-Polynomial identities of matrices with the symplectic involution: The low degrees. Commun. Algebra **32**, 895–918 (2004)
7. Drensky, V., Giambruno, A.: On the $*$-polynomial identities of minimal degree for matrices with involution. Boll. Unione Mat. Ital. **7**, 471–482 (1995)
8. Giambruno, A.: On $*$-polynomial identities for $n \times n$ matrices. J. Algebra **133**(3), 433–438 (1990)
9. Giambruno, A., Sehgal, S.: On a polynomial identity for $n \times n$ matrices. J. Algebra **126**, 451–453 (1989)
10. Giambruno, A., Ioppolo, A., Martino, F.: Standard polynomial and matrices with superinvolutions. Linear Algebra Appl. **504**, 272–291 (2016)
11. Giambruno, A., dos Santos, R., Vieira, A.: Identities of $*$-superalgebras and almost polynomial growth. Linear and Multilinear Algebra **64**, 484–501 (2016)
12. Hill, J.D.: Polynomial identities for matrices symmetric with respect to the symplectic involution. J. Algebra **349**, 8–21 (2012)
13. Hill, J.D.: An identity of degree $2n - 3$ for the $n \times n$ skews, n even, and corollaries for standard identities . Linear Algebra Appl. **439**, 310–327 (2013)

14. Hutchinson, J.: Eulerian graphs and polynomial identities for sets of matrices. Proc. Nat. Acad. Sci. U.S.A. **71**, 1314–1316 (1974)
15. Kemer, A.R.: Finite basability of identities of associative algebras. Algebra i Logika **26**(5), 597–641, 650 (1987)
16. Kostant, B.: A theorem of Frobenius, a theorem of Amitsur-Levitski and cohomology theory. J. Math. Mech. **7**, 237–264 (1958)
17. Owens, F.W.: Graph theory and identities on matrices. Not. Am. Math. Soc. **14**, 367–373 (1973)
18. Rashkova, T.G.: $*$-Identities of minimal degree in matrix algebras of low order. Period. Math. Hung. **34**, 229–233 (1998)
19. Rowen, L.H.: Standard polynomials in matrix algebras. Trans. Am. Math. Soc. **190**, 253–284 (1974)
20. Rowen, L.H.: A simple proof of Kostant's theorem and an analogue for the symplectic involution. Contemp. Math. Am. Math. Soc. **13**, 207–215 (1982)
21. Slin'ko, A.M.: Special varieties of Jordan algebras. Mat. Zametki. **26**, 337–344 (1979)
22. Wenxin, M., Racine M. Minimal identities of symmetric matrices. Trans Am. Math. Soc. **320**, 171–192 (1990)

On the Asymptotics of Capelli Polynomials

Francesca Saviella Benanti and Angela Valenti

To Antonio Giambruno on the occasion of his 70th birthday

Abstract We present old and new results about Capelli polynomials, $\mathbb{Z}_2$-graded Capelli polynomials, Capelli polynomials with involution and their asymptotics.

Let $Cap_m = \sum_{\sigma \in S_m} (\mathrm{sgn}\sigma) t_{\sigma(1)} x_1 t_{\sigma(2)} \cdots t_{\sigma(m-1)} x_{m-1} t_{\sigma(m)}$ be the m-th Capelli polynomial of rank m. In the ordinary case (see Giambruno and Zaicev, Israel J Math 135:125–145, 2003) it was proved the asymptotic equality between the codimensions of the T-ideal generated by the Capelli polynomial Cap_{k^2+1} and the codimensions of the matrix algebra $M_k(F)$. In (Benanti, Algebr Represent Theory 18:221–233, 2015) this result was extended to superalgebras proving that the $\mathbb{Z}_2$-graded codimensions of the T_2-ideal generated by the $\mathbb{Z}_2$-graded Capelli polynomials Cap^0_{M+1} and Cap^1_{L+1} for some fixed M, L, are asymptotically equal to the $\mathbb{Z}_2$-graded codimensions of a simple finite dimensional superalgebra. Recently, the authors proved that the $*$-codimensions of a $*$-simple finite dimensional algebra are asymptotically equal to the $*$-codimensions of the T-$*$-ideal generated by the $*$-Capelli polynomials Cap^+_{M+1} and Cap^-_{L+1}, for some fixed natural numbers M and L.

Keywords Algebras with involution · Capelli polynomials · Codimension · Growth · Superalgebras

F. S. Benanti (✉)
Dipartimento di Matematica e Informatica, Universitá di Palermo via Archirafi, Palermo, Italy
e-mail: francescasaviella.benanti@unipa.it

A. Valenti
Dipartimento di Ingegneria, Università di Palermo Viale delle Scienze, Palermo, Italy
e-mail: angela.valenti@unipa.it

O. M. Di Vincenzo, A. Giambruno (eds.), *Polynomial Identities in Algebras*,
Springer INdAM Series 44, https://doi.org/10.1007/978-3-030-63111-6_3

1 Introduction

From Kemer's theory (see [35]), the polynomial identities of the matrix algebra $M_k(F)$ over a field F of characteristic zero are among the most intriguing topics in the PI-theory. There are a lot of open problems and conjectures concerning the bases of polynomial identities of $M_k(F)$, the minimal degree of identities which do not follow from the standard polynomial, the numerical invariants of polynomial identities, etc. Similar problems are also to consider for matrix algebras with additional structure as $\mathbb{Z}_2$-gradings, group gradings or involution. The Capelli polynomial plays a central role in the combinatorial PI-theory and in particular in the study of polynomial identities of the matrix algebra $M_k(F)$ in fact it was determinated a precise relation between the growth of the corresponding T-ideal and the growth of the T-ideal of the matrix algebra. Moreover the Capelli polynomials characterize the algebras having the cocharacter contained in a given strip (see [41]). Let us recall that, for any positive integer m, the m-th Capelli polynomial is the element of the free algebra $F\langle X\rangle$ defined as

$$Cap_m = Cap_m(t_1, \dots, t_m; x_1, \dots, x_{m-1}) =$$

$$= \sum_{\sigma \in S_m} (\mathrm{sgn}\, \sigma) t_{\sigma(1)} x_1 t_{\sigma(2)} \cdots t_{\sigma(m-1)} x_{m-1} t_{\sigma(m)}$$

where S_m is the symmetric group on $\{1, \dots, m\}$. It is an alternating polynomial and every polynomial which is alternating on $t_1, \dots, t_m$ can be written as a linear combination of Capelli polynomials obtained by specializing the x_i's. These polynomials were first introduced by Razmyslov (see [39]) in his construction of central polynomials for $k \times k$ matrices. It is easy to show that if A is a finite dimensional algebra A and dim $A = m - 1$ then A satisfies Cap_m. Moreover, any finitely generated PI-algebra A satisfies Cap_m for some m (see, for example, Theorem 2.2 in [35]). Then the matrix algebra $M_k(F)$ satisfies Cap_{k^2+1} and $k^2 + 1$ is actually the minimal degree of a Capelli polynomial satisfied by $M_k(F)$.

The main purpose of this paper is to present a survey on old and new results concerning the Capelli polynomials. In particular, in Sect. 2 we recall the results about the T-ideal generated by the m-th Capelli polynomial Cap_m and in Sect. 3 the results concerning the T_2-ideal generated by the $\mathbb{Z}_2$-graded Capelli polynomials Cap^0_{M+1} and Cap^1_{L+1}. We show their relations with the T-ideal of the polynomial identities of $M_k(F)$ and, respectively, with the T_2-ideals of the $\mathbb{Z}_2$-graded identities of the simple finite dimensional superalgebra $M_k(F)$, $M_{k,l}(F)$ and $M_s(F \oplus tF)$. In Sect. 4 we present the recent results obtained by the authors about the study of the $*$-codimensions of the T-$*$-ideal generated by the $*$-Capelli polynomials Cap^+_{M+1} and Cap^-_{L+1}. These results has been announced in a complete version at the preprint server of Cornell University (https://arxiv.org/pdf/1911.04193.pdf) and has been submitted elsewhere.

2 Ordinary Case

Let F be a field of characteristic zero and let $F\langle X\rangle = F\langle x_1, x_2, \ldots\rangle$ be the free associative algebra on a countable set X over F. Recall that an ideal I of $F\langle X\rangle$ is a T-ideal if it is invariant under all endomorphisms of $F\langle X\rangle$. Let A be an associative algebra over F, then an element $f = f(x_1, \ldots, x_n) \in F\langle X\rangle$ is a polynomial identity for A if $f(a_1, \ldots, a_n) = 0$ for any $a_1, \ldots, a_n \in A$. If f is a polynomial identity for A we usually write $f \equiv 0$ in A. Let $Id(A) = \{f \in F\langle X\rangle \mid f \equiv 0 \text{ in A}\}$ be the ideal of polynomial identities of A. When A satisfies a non trivial identity (i.e. $Id(A) \neq (0)$), we say that A is a PI- algebra. The connection between T-ideals of $F\langle X\rangle$ and PI-algebras is well understood: for any F-algebra A, $Id(A)$ is a T-ideal of $F\langle X\rangle$ and every T- ideal I of $F\langle X\rangle$ is the ideal of identities of some F-algebra A. For $I = Id(A)$ we denote by $\text{var}(I) = \text{var}(A)$ the variety of all associative algebras having the elements of I as polynomial identities. The language of varieties is effective for investigations of PI-algebras.

An important class of T-ideals is given by the so-called verbally prime T-ideals. They were introduced by Kemer (see [35]) in his solution of the Specht problem as basic blocks for the study of arbitrary T-ideals. Recall that a T-ideal $I \subseteq F\langle X\rangle$ is verbally prime if for any T-ideals I_1, I_2 such $I_1 I_2 \subseteq I$ we must have $I_1 \subseteq I$ or $I_2 \subseteq I$. A PI-algebra A is called verbally prime if its T-ideal of identities $I = Id(A)$ is verbally prime. Also, the corresponding variety of associative algebras $\text{var}(A)$ is called verbally prime. By the structure theory of T-ideals developed by Kemer (see [35]) and his classification of verbally prime T-ideals in characteristic zero, the study of an arbitrary T-ideal can be reduced to the study of the T-ideals of identities of the following verbally prime algebras

$$F,\ F\langle X\rangle,\ M_k(F),\ M_k(G),\ M_{k,l}(G)$$

where G is the infinite dimensional Grassmann algebra, $M_k(F)$, $M_k(G)$ are the algebras of $k \times k$ matrices over F and G, respectively, and

$$M_{k,l}(G) = \begin{array}{c} \\ k \\ l \end{array}\!\!\begin{array}{c} \begin{matrix} k & l \end{matrix} \\ \begin{pmatrix} G_0 & G_1 \\ G_1 & G_0 \end{pmatrix} \end{array}.$$

Recall that G is the algebra generated by a countable set $\{e_1, e_2, \ldots\}$ subject to the conditions $e_i e_j = -e_j e_i$ for all $i, j = 1, 2, \ldots$, and $G = G_0 \oplus G_1$ is the natural $\mathbb{Z}_2$-grading on G, where G_0 and G_1 are the spaces generated by all monomials in the generators e_i's of even and odd length, respectively.

It is well known that in characteristic zero every T-ideal is completely determined by its multilinear elements. Hence, if P_n is the space of multilinear polynomials of degree n in the variables $x_1, \ldots, x_n$, the relatively free algebra $F\langle X\rangle/Id(A)$ is determined by the sequence of subspaces $\{P_n/(P_n \cap Id(A))\}_{n\geq 1}$.

The integer $c_n(A) = \dim P_n/(P_n \cap Id(A))$ is called the n-th codimension of A and gives a quantitative estimate of the polynomial identities satisfied by A.

Thus to each T-ideal $I = Id(A)$ one can associate the numerical sequence of codimensions $\{c_n(I)\}_{n\geq 1} = \{c_n(A)\}_{n\geq 1}$ of I, or A, that plays an important role in the study of $Id(A)$. It is well known that A is a PI-algebra if and only if $c_n(A) < n!$ for some $n \geq 1$. Regev in [40] showed that if A is an associative PI-algebra, then $c_n(A)$ is exponentially bounded i.e., there exist constants α, β which depend on A such that $c_n(A) \leq \alpha\beta^n$ for any $n \geq 1$ (see also [36] and [42] for the best known estimates). Giambruno and Zaicev improved this result and, in [23] and [24], proved that for a PI-algebra A

$$\exp(A) = \lim_{n\to\infty} \sqrt[n]{c_n(A)}$$

exists and is an integer; $\exp(A)$ is called the PI-exponent of the algebra A. For the verbally prime algebras we have (see [14, 43, 44] and [24])

$$\exp(M_k(F)) = k^2, \ \exp(M_k(G)) = 2k^2, \ \exp(M_{k,l}(G)) = (k+l)^2.$$

In [43] Regev obtained the precise asymptotic behavior of the codimensions of the verbally prime algebra $M_k(F)$. It turns out that

$$c_n(M_k(F)) \simeq C(\frac{1}{n})^{(k^2-1)/2} k^{2n},$$

where C is a certain constant explicitly computed. For the other verbally prime algebras $M_k(G)$, $M_{k,l}(G)$ there are only some partially results (see [14] and [16]). More precisely,

$$c_n(M_{k,l}(G)) \simeq an^g\alpha^n, \ c_n(M_k(G)) \simeq bn^h\beta^n,$$

with $\alpha = (k+l)^2$, $g = -\frac{1}{2}(k^2+l^2-1)$, $\beta = 2k^2$, $h = -\frac{1}{2}(k^2-1)$, and a and b are undetermined constants. It turns out that it is in general a very hard problem to determine the precise asymptotic behavior of such sequences.

In [29] and in [10] it was found a relation among the asymptotics of codimensions of the verbally prime T-ideals and the T-ideals generated by Capelli polynomials or Amitsur's Capelli-type polynomials.

Now, if $f \in F\langle X\rangle$ we denote by $\langle f\rangle_T$ the T-ideal generated by f. Also for $V \subset F\langle X\rangle$ we write $\langle V\rangle_T$ to indicate the T-ideal generated by V. Let C_m be the set of 2^m polynomials obtained from the m-th Capelli polynomial Cap_m by deleting any subset of variables x_i (by evaluating the variables x_i to 1 in all possible ways) and let $\langle C_m\rangle_T$ denotes the T-ideal generated by C_m. If $U_m = \text{var}(C_m)$ is the variety corresponding to $\langle C_m\rangle_T$ then $\exp(C_m) = \exp(U_m)$. In case $m = k^2$, it follows from [43] that

$$\exp(C_{k^2+1}) = k^2 = \exp(M_k(F)).$$

Mishchenko, Regev and Zaicev in [37] computed the $\exp(C_m)$, for an arbitrary m, and in particular they proved (see also [30, Theorem 9.1.5])

Theorem 1 ([37, Theorem])

(1) $m-3 \leq \exp(C_{m+1}) \leq m$.
(2) $\exp(C_{m+1}) = max\{a_1, a_2, a_3, a_4\}$ *where*
$a_j = max\{d_1^2 + \cdots + d_j^2 \mid d_1, \ldots, d_j \in \mathbb{Z}, d_1, \ldots, d_j > 0, d_1^2 + \cdots + d_j^2 + j \leq m+1\}$.
(3) $\exp(C_{m+1}) \leq m \Leftrightarrow m = q^2$, for some q.

The proof applies, in an essential way, the classical Lagrange's four square theorem.

In [29] Giambruno and Zaicev proved that the codimensions of U_{k^2+1} are asymptotically equal to the codimensions of the verbally prime algebra $M_k(F)$

Theorem 2 ([29, Theorem 3, Corollary 4]) *Let* $m = k^2$. *Then* $\mathrm{var}(C_{m+1}) = \mathrm{var}(M_k(F) \oplus B)$ *for some finite dimensional algebra* B *such that* $\exp(B) < k^2$. *In particular*

$$c_n(C_{k^2+1}) \simeq c_n(M_k(F)).$$

This result has been extended to the others verbally prime algebras by the so called Amitsur's Capelli-type polynomials.

Let L and M be two natural numbers, let $\hat{n} = (L+1)(M+1)$ and let μ be a partition of $\hat{n}$ with associated rectangular Young diagram, $\mu = ((L+1)^{M+1}) \vdash \hat{n}$. In [6] the following polynomials, denoted Amitsur's Capelli-type polynomials, were introduced

$$e^*_{M,L} = e^*_{M,L}(t_1, \ldots, t_{\hat{n}}; x_1, \ldots, x_{\hat{n}-1}) = \sum_{\sigma \in S_{\hat{n}}} \chi_\mu(\sigma) t_{\sigma(1)} x_1 t_{\sigma(2)} \cdots x_{\hat{n}-1} t_{\sigma(\hat{n})},$$

where $\chi_\mu(\sigma)$ is the value of the irreducible character χ_μ corresponding to the partition $\mu \vdash \hat{n}$ on the permutation σ. We note that for $L = 0$ we have $\mu = (1^{\hat{n}})$ and $e^*_{M,L} = Cap_{\hat{n}}$ is the $\hat{n}$-th Capelli polynomial. The Amitsur's Capelli-type polynomials generalize the Capelli polynomials in the sense that the Capelli polynomials characterize the algebras having the cocharacter contained in a given strip (see [41]) and the Amitsur's polynomials characterize the algebras having a cocharacter contained in a given hook (see [6, Theorem B]).

Let $E^*_{M,L}$ denote the set of $2^{\hat{n}-1}$ polynomials obtained from $e^*_{M,L}$ by evaluating the variables x_i to 1 in all possible ways. Also we denote by $\Gamma_{M,L} = \langle E^*_{M,L} \rangle_T$ the T-ideal generated by $E^*_{M,L}$. Moreover we write $\mathcal{V}_{M,L} = \mathrm{var}(E^*_{M,L}) = \mathrm{var}(\Gamma_{M,L})$, $c_n(E^*_{M,L}) = c_n(\Gamma_{M,L})$ and $\exp(E^*_{M,L}) = \exp(\Gamma_{M,L})$. The following relations between the exponent of the Capelli-type polynomials and the exponent of the verbally prime algebras are well known (see [15])

$$\exp(E^*_{k^2,k^2}) = 2k^2 = \exp(M_k(G)), \quad \exp(E^*_{k^2+l^2,2kl}) = (k+l)^2 = \exp(M_{k,l}(G)).$$

In [15] (see also [30]) Berele and Regev, by using the generalized-six-square theorem [17], proved that

Theorem 3 ([15, Proposition 4.4.]) *Let* $l \le k$. *Then* $k+l-3 \le \exp(E^*_{k,l}) \le k+l$.

Finally, in [10] it was shown the following asymptotical equalities

Theorem 4 ([10, Theorem 5]) *Let* $k, l \in \mathbb{N}$. *Then* $\mathrm{var}(E^*_{k^2+l^2,2kl}) = \mathrm{var}(M_{k,l}(G) \oplus G(D'))$, *where* D' *is a finite dimensional superalgebra such that* $exp(D') < (k+l)^2$. *In particular*

$$c_n(E^*_{k^2+l^2,2kl}) \simeq c_n(M_{k,l}(G)).$$

Theorem 5 ([10, Theorem 10]) *Let* $k \in \mathbb{N}$, $k > 0$. *Then* $\mathrm{var}(E^*_{k^2,k^2}) = \mathrm{var}(M_k(G) \oplus G(D'))$, *where* D' *is a finite dimensional superalgebra such that* $\exp(D') < 2k^2$. *In particular*

$$c_n(E^*_{k^2,k^2}) \simeq c_n(M_k(G)).$$

3 $\mathbb{Z}_2$-Graded Case

Recall that an algebra A is a superalgebra (or $\mathbb{Z}_2$-graded algebra) with grading $(A^{(0)}, A^{(1)})$ if $A = A^{(0)} \oplus A^{(1)}$, where $A^{(0)}$, $A^{(1)}$ are subspaces of A satisfying:

$$A^{(0)}A^{(0)} + A^{(1)}A^{(1)} \subseteq A^{(0)} \text{ and } A^{(0)}A^{(1)} + A^{(1)}A^{(0)} \subseteq A^{(1)}.$$

The elements of $A^{(0)}$ and of $A^{(1)}$ are called homogeneous of degree zero (or even elements) and of degree one (or odd elements), respectively. If we write $X = Y \cup Z$ as the disjoint union of two countable sets, then the free associative algebra $F\langle X\rangle = F\langle Y \cup Z\rangle = \mathcal{F}^{(0)} \oplus \mathcal{F}^{(1)}$ has a natural structure of free superalgebra with grading $(\mathcal{F}^{(0)}, \mathcal{F}^{(1)})$, where $\mathcal{F}^{(0)}$ is the subspace generated by the monomials of even degree with respect to Z and $\mathcal{F}^{(1)}$ is the subspace generated by the monomials having odd degree in Z.

Recall that an element $f(y_1, \ldots, y_n, z_1, \ldots, z_m)$ of $F\langle Y \cup Z\rangle$ is a $\mathbb{Z}_2$-graded identity or a superidentity for A if $f(a_1, \ldots, a_n, b_1, \ldots, b_m) = 0$, for all $a_1, \ldots, a_n \in A^{(0)}$ and $b_1, \ldots, b_m \in A^{(1)}$. The set $Id^{sup}(A)$ of all $\mathbb{Z}_2$-graded identities of A is a T_2-ideal of $F\langle Y \cup Z\rangle$ i.e., an ideal invariant under all endomorphisms of $F\langle Y \cup Z\rangle$ preserving the grading. Moreover, every T_2-ideal Γ of $F\langle Y \cup Z\rangle$ is the ideal of $\mathbb{Z}_2$-graded identities of some superalgebra $A = A^{(0)} \oplus A^{(1)}$, $\Gamma = Id^{sup}(A)$. For $\Gamma = Id^{sup}(A)$ a T_2-ideal of $F\langle Y \cup Z\rangle$, we denote by $\mathrm{supvar}(\Gamma)$ or $\mathrm{supvar}(A)$ the supervariety of superalgebras having the elements of Γ as $\mathbb{Z}_2$-graded identities.

As it was shown by Kemer (see [34, 35]), superalgebras and their $\mathbb{Z}_2$-graded identities play a basic role in the study of the structure of varieties of associative algebras over a field of characteristic zero. More precisely Kemer showed that any variety is generated by the Grassmann envelope of a suitable finite dimensional superalgebra (see Theorem 3.7.8 [30]) and moreover he established that an associative variety is a prime variety if and only if it is generated by the Grassmann envelope of a simple finite dimensional superalgebra.

Recall that, if F is an algebraically closed field of characteristic zero, then a simple finite dimensional superalgebra over F is isomorphic to one of the following algebras (see [30, 35]):

1. $M_k(F)$ with trivial grading $(M_k(F), 0)$;
2. $M_{k,l}(F)$ with grading $\left(\begin{pmatrix} F_{11} & 0 \\ 0 & F_{22} \end{pmatrix}, \begin{pmatrix} 0 & F_{12} \\ F_{21} & 0 \end{pmatrix}\right)$, where F_{11}, F_{12}, F_{21}, F_{22} are $k \times k$, $k \times l$, $l \times k$ and $l \times l$ matrices respectively, $k \geq 1$ and $l \geq 1$;
3. $M_s(F \oplus tF)$ with grading $(M_s(F), tM_s(F))$, where $t^2 = 1$.

Thus an important problem in the theory of PI-algebras is to describe the T_2-ideals of $\mathbb{Z}_2$-graded identities of finite dimensional simple superalgebra: $Id^{sup}(M_k(F))$, $Id^{sup}(M_{k,l}(F))$, $Id^{sup}(M_s(F \oplus tF))$.

In case $\mathrm{char} F = 0$, it is well known that $Id^{sup}(A)$ is completely determined by its multilinear polynomials and an approach to the description of the $\mathbb{Z}_2$-graded identities of A is based on the study of the $\mathbb{Z}_2$-graded codimensions sequence of this superalgebra. If P_n^{sup} denotes the space of multilinear polynomials of degree n in the variables $y_1, z_1, \ldots, y_n, z_n$ (i.e., y_i or z_i appears in each monomial at degree 1), then the sequence of spaces $\{P_n^{sup} \cap Id^{sup}(A)\}_{n \geq 1}$ determines $Id^{sup}(A)$ and

$$c_n^{sup}(A) = \dim_F \left(\frac{P_n^{sup}}{P_n^{sup} \cap Id^{sup}(A)} \right)$$

is called the n-th $\mathbb{Z}_2$-graded codimension of A. The asymptotic behaviour of the $\mathbb{Z}_2$-graded codimensions plays an important role in the PI-theory of superalgebras. In 1985, Giambruno e Regev (see [22]) proved that the sequence $\{c_n^{sup}(A)\}_{n \geq 1}$ is exponentially bounded if and only if A satisfies an ordinary polynomial identity. In [12] it was proved that if A is a finitely generated superalgebra satisfying a polynomial identity, then $\lim_{n \to \infty} \sqrt[n]{c_n^{sup}(A)}$ exists and is a non negative integer. It is called superexponent (or $\mathbb{Z}_2$-exponent) of A and it is denoted by

$$\mathrm{supexp}(A) = \lim_{n \to \infty} \sqrt[n]{c_n^{sup}(A)}.$$

We remark that in [21] the existence of the G-exponent has been proved when G is a group of prime order and, in general, in [2, 31] and [1] for an arbitrary PI-algebras graded by a finite abelian group G.

Now, if $f \in F\langle Y \cup Z \rangle$ we denote by $\langle f \rangle_{T_2}$ the T_2-ideal generated by f. Also for a set of polynomials $V \subset F\langle Y \cup Z \rangle$ we write $\langle V \rangle_{T_2}$ to indicate the T_2-ideal generated by V. Let denote by $Cap_m[Y, X] = Cap_m(y_1, \ldots, y_m; x_1, \ldots, x_{m-1})$

and $Cap_m[Z, X] = Cap_m(z_1, \dots, z_m; x_1, \dots, x_{m-1})$ the m-th $\mathbb{Z}_2$-graded Capelli polynomial in the alternating variables of homogeneous degree zero $y_1, \dots, y_m$ and of homogeneous degree one $z_1, \dots, z_m$, respectively. Then Cap_m^0 indicates the set of 2^{m-1} polynomials obtained from $Cap_m[Y, X]$ by deleting any subset of variables x_i (by evaluating the variables x_i to 1 in all possible way). Similarly, we define by Cap_m^1 the set of 2^{m-1} polynomials obtained from $Cap_m[Z, X]$ by deleting any subset of variables x_i.

If L and M are two natural numbers, let $\Gamma_{M+1,L+1}$ be the T_2-ideal generated by the polynomials Cap^0_{M+1}, Cap^1_{L+1}, $\Gamma_{M+1,L+1} = \langle Cap^0_{M+1}, Cap^1_{L+1}\rangle_{T_2}$. We also write $\mathcal{U}^{sup}_{M+1,L+1} = \operatorname{supvar}(\Gamma_{M+1,L+1})$.

In [8] it was calculated the $\operatorname{supexp}(\mathcal{U}^{\text{sup}}_{M+1,L+1})$. We recall the following

Definition 1 (see [8]) Let M and L be fixed. Then, for any integers $s, t \geq 0, r \geq 1$ such that $r - 1 = r_0 + r_1$ for some non-negative integers r_0, r_1, we define the set

$$\overline{A}_{r,s,t;r_0,r_1} = \{a_1, \dots, a_r, k_1, l_1, \dots, k_s, l_s, b_1, \dots, b_t \in \mathbb{Z}^+ \mid$$

$$a_1^2 + \cdots + a_r^2 + (k_1^2 + l_1^2) + \cdots (k_s^2 + l_s^2) + b_1^2 + \cdots + b_t^2 + r_0 + s + t \leq M,$$

$$\text{and } 2k_1l_1 + \cdots + 2k_sl_s + b_1^2 + \cdots + b_t^2 + r_1 + s + t \leq L\}.$$

Also, given integers $s, t \geq 0$ $(r = 0)$, we define the set

$$\widetilde{A}_{s,t} = \{k_1, l_1, \dots, k_s, l_s, b_1, \dots, b_t \in \mathbb{Z}^+ \mid$$

$$(k_1^2 + l_1^2) + \cdots (k_s^2 + l_s^2) + b_1^2 + \cdots + b_t^2 + s + t \leq M + 1,$$

$$\text{and } 2k_1l_1 + \cdots + 2k_sl_s + b_1^2 + \cdots + b_t^2 + s + t \leq L + 1\}.$$

Moreover, let

$$\overline{a}_{r,s,t;r_0,r_1} = \max_{a_i,k_i,l_i,b_i \in \overline{A}_{r,s,t;r_0,r_1}} \{a_1^2 + \cdots + a_r^2 + (k_1 + l_1)^2 + \cdots (k_s + l_s)^2 + 2b_1^2 + \cdots + 2b_t^2\}$$

and

$$\widetilde{a}_{s,t} = \max_{k_i,l_i,b_i \in \widetilde{A}_{s,t}} \{(k_1 + l_1)^2 + \cdots (k_s + l_s)^2 + 2b_1^2 + \cdots + 2b_t^2\},$$

then we define

$$a_0 = \max\{\overline{a}_{r,s,t;r_0,r_1}, \widetilde{a}_{s,t} \mid r + s + t \leq 11\}.$$

Theorem 6 ([8, Theorem 4]) *If $M \geq L \geq 0$, then*

(1) $\operatorname{supexp}(\mathcal{U}^{\text{sup}}_{M+1,L+1}) = a_0$;
(2) $(M + L) - 10 \leq \operatorname{supexp}(\mathcal{U}^{\text{sup}}_{M+1,L+1}) \leq (M + L)$.

This result was inspirated by the ordinary case. Moreover, we should mention that in the previous theorem an essential tool is the generalized-six-square theorem proved in [17] (see also Appendix A of [30]).

The following relations between the superexponent of the $\mathbb{Z}_2$-graded Capelli polynomials and the superexponent of the simple finite dimensional superalgebras are well known (see [8, 12, 28])

$$\text{supexp}(\mathcal{U}^{\text{sup}}_{k^2+1,1}) = k^2 = supexp(M_k(F))$$

$$\text{supexp}(\mathcal{U}^{\text{sup}}_{k^2+l^2+1,2kl+1}) = (k+l)^2 = supexp(M_{k,l}(F))$$

$$\text{supexp}(\mathcal{U}^{\text{sup}}_{s^2+1,s^2+1}) = 2s^2 = supexp(M_s(F \oplus tF)).$$

In [9] it was found a close relation among the asymptotics of $\mathcal{U}^{sup}_{k^2+l^2+1,2kl+1}$ and $M_{k,l}(F)$ and the asymptotics of $\mathcal{U}^{sup}_{s^2+1,s^2+1}$ and $M_s(F \oplus tF)$. More precisely it was showed that

Theorem 7 ([9, Theorem 9]) *Let $M = k^2 + l^2$ and $L = 2kl$ with $k, l \in \mathbb{N}$, $k > l > 0$. Then $\mathcal{U}^{sup}_{M+1,L+1} = supvar(\Gamma_{M+1,L+1}) = supvar(M_{k,l}(F) \oplus D')$, where D' is a finite dimensional superalgebra such that $supexp(D') < M + L$. In particular*

$$c_n^{sup}(\Gamma_{M+1,L+1}) \simeq c_n^{sup}(M_{k,l}(F)).$$

Theorem 8 ([9, Theorem 14]) *Let $M = L = s^2$ with $s \in \mathbb{N}$, $s > 0$. Then $\mathcal{U}^{sup}_{M+1,L+1} = supvar(\Gamma_{M+1,L+1}) = supvar(M_s(F \oplus tF) \oplus D'')$, where D'' is a finite dimensional superalgebra such that $supexp(D'') < M + L$. In particular*

$$c_n^{sup}(\Gamma_{M+1,L+1}) \simeq c_n^{sup}(M_s(F \oplus tF)).$$

In [29] Giambruno and Zaicev proved that $c_n^{sup}(\Gamma_{k^2+1,1}) \simeq c_n^{sup}(M_k(F))$.

4 Involution Case

Let $F\langle X, *\rangle = F\langle x_1, x_1^*, x_2, x_2^*, \ldots\rangle$ denote the free associative algebra with involution $*$ generated by the countable set of variables $X = \{x_1, x_1^*, x_2, x_2^*, \ldots\}$ over a field F of characteristic zero. Let $(A, *)$ be an algebra with involution $*$ over F, recall that an element $f(x_1, x_1^*, \cdots, x_n, x_n^*)$ of $F\langle X, *\rangle$ is a $*$-polynomial identity (or $*$-identity) for A if $f(a_1, a_1^*, \cdots, a_n, a_n^*) = 0$, for all $a_1, \ldots, a_n \in A$. We denote by $Id^*(A)$ the set of all $*$-polynomial identities satisfied by A. $Id^*(A)$ is a T-$*$-ideal of $F\langle X, *\rangle$ i.e., an ideal invariant under all endomorphisms of $F\langle X, *\rangle$ commuting with the involution of the free algebra. Similar to the case of ordinary identities any T-$*$-ideal Γ of $F\langle X, *\rangle$ is the ideal of $*$-identities of some algebra A

with involution $*$, $\Gamma = Id^*(A)$. For $\Gamma = Id^*(A)$ we denote by $\mathrm{var}^*(\Gamma) = \mathrm{var}^*(A)$ the variety of $*$-algebras having the elements of Γ as $*$-identities.

It is well known that in characteristic zero $Id^*(A)$ is completely determinated by the multilinear $*$-polynomials it contains. To the T-$*$-ideal $\Gamma = Id^*(A)$ one associates a numerical sequence called the sequence of $*$-codimensions $c_n^*(\Gamma) = c_n^*(A)$ which is the main tool for the quantitative investigation of the $*$-polynomial identities of A. Recall that $c_n^*(A)$, $n = 1, 2, \ldots$, is the dimension of the space of multilinear polynomial in n-th variables in the corresponding relatively free algebra with involution of countable rank. Thus, if we denote by P_n^* the space of all multilinear polynomials of degree n in $x_1, x_1^*, \cdots, x_n, x_n^*$ then

$$c_n^*(A) = \dim P_n^*(A) = \dim \frac{P_n^*}{P_n^* \cap Id^*(A)}.$$

It is clear that the ordinary free associative algebra $F\langle X\rangle$ (without involution) can be considered as a subalgebra of $F\langle X, *\rangle$ and, in particular, an ordinary polynomial identity (without involution) can be considered as an identity with involution. Hence if A is a $*$-algebra, then $Id(A) \subseteq Id^*(A)$. Moreover, a celebrated theorem of Amitsur ([4, 5], see also [30]) states that if an algebra with involution satisfies a $*$-polynomial identity then it satisfies an ordinary polynomial identity. At the light of this result in [22] it was proved that, as in the ordinary case, if A satisfies a non trivial $*$-polynomial identity then $c_n^*(A)$ is exponentially bounded, i.e. there exist constants a and b such that $c_n^*(A) \leq ab^n$, for all $n \geq 1$. Later (see [7]) an explicit exponential bound for $c_n^*(A)$ was exhibited and in [28] a characterization of finite dimensional algebras with involution whose sequence of $*$-codimensions is polynomial bounded was given. This result was extended to non-finite dimensional algebras (see [27]) and $*$-varieties with almost polynomial growth were classified in [26] and [38]. The asymptotic behavior of the $*$-codimensions was determined in [13] in case of matrices with involution.

Recently (see [33]), for any algebra with involution, it was studied the exponential behavior of $c_n^*(A)$, and it was showed that the $*$-exponent of A

$$\exp^*(A) = \lim_{n\to\infty} \sqrt[n]{c_n^*(A)}$$

exists and is a non negative integer. It should be mentioned that the existence of the $*$-exponent was proved in [25] for finite dimensional algebra with involution.

An interesting problem in the theory of PI-algebras with involution $*$ is to describe the T-$*$-ideals of $*$-polynomial identities of $*$-simple finite dimensional algebras. Recall that, if F is an algebraically closed field of characteristic zero, then, up to isomorphisms, all finite dimensional $*$-simple are the following ones (see [30, 45]):

- $(M_k(F), t)$ the algebra of $k \times k$ matrices with the transpose involution;
- $(M_{2m}(F), s)$ the algebra of $2m \times 2m$ matrices with the symplectic involution;

- $(M_h(F) \oplus M_h(F)^{op}, exc)$ the direct sum of the algebra of $h \times h$ matrices and the opposite algebra with the exchange involution.

Let G be the infinite dimensional Grassmann algebra over F. G is generated by the elements $e_1, e_2, \ldots$ subject to the following condition $e_i e_j = -e_j e_i$, for all $i, j \geq 1$. Recall that G has a natural Z_2-grading $G = G_0 \oplus G_1$ where G_0 (resp. G_1) is the span of the monomials in the e_i's of even length (resp. odd length). If $B = B_0 \oplus B_1$ is a superalgebra, then the Grassmann envelope of B is defined as $G(B) = (G_0 \otimes B_0) \oplus (G_1 \otimes B_1)$. The relevance of $G(A)$ relies in a result of Kemer ([35, Theorem 2.3]) stating that if B is any PI-algebra, then its T-ideal of polynomial identities coincides with the T-ideal of identities of the Grassmann envelope of a suitable finite dimensional superalgebra. This result has been extended to algebras with involution in fact in [3] it was proved that, if A is a PI-algebra with involution over a field F of characteristic zero, then there exists a finite dimensional superalgebra with superinvolution B such that $Id^*(A) = Id^*(G(B))$.

Recall that a superinvolution $*$ of B is a linear map of B of order two such that $(ab)^* = (-1)^{|a||b|} b^* a^*$, for any homogeneous elements $a, b \in B$, where $|a|$ denotes the homogeneous degree of a. It is well known that in this case $B_0^* \subseteq B_0$, $B_1^* \subseteq B_1$ and we decompose $B = B_0^+ \bigoplus B_0^- \bigoplus B_1^+ \bigoplus B_1^-$. We can define a superinvolution $*$ on G by requiring that $e_i^* = -e_i$, for any $i \geq 1$. Then it is easily checked that $G_0 = G^+$ and $G_1 = G^-$. Now, if B is a superalgebra one can perform its Grassmann envelope $G(B)$ and in [3] it was shown that if B has a superinvolution $*$ we can regard $G(B)$ as an algebra with involution by setting $(g \otimes a)^* = g^* \otimes a^*$, for homogeneous elements $g \in G, a \in B$. By making use of the previous theorem, in [33] it was proved the existence of the $*$-exponent of a PI-algebra with involution A and also an explicit way of computing $\exp^*(A)$ was given. The $*$-exponent is computed as follows: if B is a finite dimensional algebra with superinvolution over F, then by Giambruno et al. [32] we write $B = \bar{B} + J$ where $\bar{B}$ is a maximal semisimple superalgebra with induced superinvolution and $J = J(B) = J^*$. Also we can write $\bar{B} = B_1 \oplus \cdots \oplus B_k$, where $B_1, \cdots, B_k$ are simple superalgebras with induced superinvolution. We say that a subalgebra $B_{i_1} \oplus \cdots \oplus B_{i_t}$, where $B_{i_1}, \ldots, B_{i_t}$ are distinct simple components, is admissible if for some permutation $(l_1, \ldots, l_t)$ of $(i_1, \ldots, i_t)$ we have that $B_{l_1} J B_{l_2} J \cdots J B_{l_t} \neq 0$. Moreover if $B_{i_1} \oplus \cdots \oplus B_{i_t}$ is an admissible subalgebra of B then $B' = B_{i_1} \oplus \cdots \oplus B_{i_t} + J$ is called a reduced algebra. In [33] it was proved that $\exp^*(A) = \exp^*(G(B)) = d$ where d is the maximal dimension of an admissible subalgebra of B. It follows immediately that if A is a $*$-simple algebra then $exp^*(A) = \dim_F A$. If $\mathcal{V} = \mathrm{var}^*(A)$ is the variety of $*$-algebras generated by A we write $Id^*(\mathcal{V}) = Id^*(A)$, $c_n^*(\mathcal{V}) = c_n^*(A)$ and $\exp^*(\mathcal{V}) = \exp^*(A)$.

The reduced algebras are basic elements of any $*$-variety in fact we have the following (see [11])

Theorem 9 *Let $\mathcal{V}$ be a proper variety of $*$-algebras. Then there exists a finite number of reduced superalgebras with superinvolution $B_1, \ldots, B_t$ and a finite* *dimensional superalgebra with superinvolution D such that*

$$\mathcal{V} = \text{var}(G(B_1) \oplus \cdots \oplus G(B_t) \oplus G(D))$$

with $\exp^*(\mathcal{V}) = \exp^*(G(B_1)) = \cdots = \exp^* G((B_t))$ *and* $\exp^*(G(D)) < \exp^*(\mathcal{V})$.

In terms of $*$-codimensions we obtain

Corollary 1 *Let* $\mathcal{V} = \text{var}^*(A)$ *be a proper variety of* $*$*-algebras. Then there exists a finite number of reduced superalgebras with superinvolution* $B_1, \ldots, B_t$ *and a finite dimensional superalgebra with superinvolution D such that*

$$c_n^*(A) \simeq c_n^*(G(B_1) \oplus \cdots \oplus G(B_t)).$$

If A is a finite dimensional $*$-algebra, then we have the following

Corollary 2 *Let A be a finite dimensional* $*$*-algebra. Then there exists a finite number of reduced* $*$*-algebras* $B_1, \ldots, B_t$ *and a finite dimensional* $*$*-algebra D such that*

$$\text{var}^*(A) = \text{var}^*(B_1 \oplus \cdots \oplus B_t \oplus D)$$

$$c_n^*(A) \simeq c_n^*(B_1 \oplus \cdots \oplus B_t)$$

and

$$\exp^*(A) = \exp^*(B_1) = \cdots = \exp^*(B_t), \ \exp^*(D) < \exp^*(A).$$

4.1 $*$-Capelli Polynomials and the $*$-Algebra $UT^*(A_1, \ldots, A_n)$

In this paragraph we shall recall the relation among the asymptotics of the $*$-codimensions of the $*$-simple finite dimensional algebras and the T-$*$-ideals generated by the $*$-Capelli polynomials recently proved by the authors. If $(A, *)$ is any algebra with involution $*$, let $A^+ = \{a \in A \mid a^* = a\}$ and $A^- = \{a \in A \mid a^* = -a\}$ denote the subspaces of symmetric and skew elements of A, respectively. Since $\text{char} F = 0$, we can regard the free associative algebra with involution $F\langle X, *\rangle$ as generated by symmetric and skew variables. In particular, for $i = 1, 2, \ldots$, we let $y_i = x_i + x_i^*$ and $z_i = x_i - x_i^*$, then we write $X = Y \cup Z$ as the disjoint union of the set Y of symmetric variables and the set Z of skew variables and $F\langle X, *\rangle = F\langle Y \cup Z\rangle$. Hence a polynomial $f = f(y_1, \ldots, y_m, z_1, \ldots, z_n) \in F\langle Y \cup Z\rangle$ is a $*$-polynomial identity of A if and only if $f(a_1, \ldots, a_m, b_1, \ldots, b_n) = 0$ for all $a_i \in A^+$, $b_i \in A^-$. Let $Cap_m^*[Y, X] = Cap_m(y_1, \ldots, y_m; x_1, \ldots, x_{m-1})$ denote the m-th $*$-Capelli polynomial in the alternating symmetric variables $y_1, \ldots, y_m$

and let $Cap^*_m[Z, X] = Cap_m(z_1, \dots, z_m; x_1, \dots, x_{m-1})$ be the m-th $*$-Capelli polynomial in the skew variables $z_1, \dots, z_m$. Then we denote by Cap^+_m the set of 2^{m-1} polynomials obtained from $Cap^*_m[Y, X]$ by deleting any subset of variables x_i (by evaluating the variables x_i to 1 in all possible way). Similarly, we define by Cap^-_m the set of 2^{m-1} polynomials obtained from $Cap^*_m[Z, X]$ by deleting any subset of variables x_i. If L and M are two natural numbers, we denote by $\Gamma^*_{M+1,L+1} = \langle Cap^+_{M+1}, Cap^-_{L+1}\rangle$ the T-$*$-ideal generated by the polynomials Cap^+_{M+1}, Cap^-_{L+1}. We also write $\mathcal{U}^*_{M+1,L+1} = \mathrm{var}^*(\Gamma_{M+1,L+1})$ for the $*$-variety generated by $\Gamma^*_{M+1,L+1}$.

The following results give us a characterization of the $*$-varieties satisfying a Capelli identity. The proof of the next result follows closely the proof given in [30, Theorem 11.4.3]

Theorem 10 *Let $\mathcal{V}$ be a variety of $*$-algebras. If $\mathcal{V}$ satisfies the Capelli identity of some rank then $\mathcal{V} = \mathrm{var}^*(A)$, for some finitely generated $*$-algebra A.*

Let M, L be two natural numbers. Let $A = A^+ \oplus A^-$ be a generating $*$-algebra of $\mathcal{U}^*_{M+1,L+1}$. It is easy to show that A satisfies a Capelli identity. Hence by the previous theorem, we may assume that A is a finitely generated $*$-algebra. Moreover by Sviridova [46, Theorem 1] we may consider A as a finite dimensional $*$-algebra. Since any polynomial alternating on $M+1$ symmetric variables vanishes in A (see [30, Proposition 1.5.5]), we get that $\dim A^+ \le M$. Similarly we get that $\dim A^- \le L$ and $\exp^*(A) \le \dim A \le M + L$. Thus we have the following

Lemma 1 $\exp^*(\mathcal{U}^*_{M+1,L+1}) \le M + L$.

Now, we recall the construction of the $*$-algebra $UT^*(A_1, \dots, A_n)$ given in Section 2 of [18]. Let $A_1, \dots, A_n$ be a n-tuple of finite dimensional $*$-simple algebras, then $A_i = (M_{d_i}(F), \mu_i)$, where μ_i is the transpose or the symplectic involution, or $A_i = (M_{d_i}(F) \oplus M_{d_i}(F)^{op}, exc)$, where exc is the exchange involution.

Let γ_d be the orthogonal involution defined on the matrix algebra $M_d(F)$ by putting, for all $a \in M_d(F)$, $a^{\gamma_d} = g^{-1}a^t g = g a^t g$, where a^t is the transposed of the matrix a and

$$g = \begin{pmatrix} 0 & \dots & 1 \\ & & \cdot \\ & \cdot & \\ \cdot & & \\ 1 & \dots & 0 \end{pmatrix}.$$

If $d = \sum_{i=1}^n \dim_F A_i$, then we can consider an embedding of $*$-algebras

$$\Delta : \bigoplus_{i=1}^n A_i \to (M_{2d}(F), \gamma_{2d})$$

defined by

$$(a_1, \ldots, a_n) \to \begin{pmatrix} \bar{a}_1 & & & & & \\ & \ddots & & & & \\ & & \bar{a}_n & & & \\ & & & \bar{b}_n & & \\ & & & & \ddots & \\ & & & & & \bar{b}_1 \end{pmatrix}$$

where, if $a_i \in A_i = (M_{d_i}(F), \mu_i)$, then $\bar{a}_i = a_i$ and $\bar{b}_i = a_i^{\mu_i \gamma_{d_i}}$, and if $a_i = (\tilde{a}_i, \tilde{b}_i) \in A_i = (M_{d_i}(F) \oplus M_{d_i}(F)^{op}, exc)$, then $\bar{a}_i = \tilde{a}_i$ and $\bar{b}_i = \tilde{b}_i$. Let $D = D(A_1, \ldots, A_n) \subseteq M_{2d}(F)$ be the $*$-algebra image of $\bigoplus_{i=1}^n A_i$ by Δ and let U be the subspace of $M_{2d}(F)$ so defined:

$$\begin{pmatrix} 0 & U_{12} & \cdots & U_{1t} & & & & \\ & \ddots & \ddots & \vdots & & & & \\ & & 0 & U_{t-1t} & & & & \\ & & & 0 & & & & \\ & & & & 0 & U_{tt-1} & \cdots & U_{t1} \\ & & & & & \ddots & \ddots & \vdots \\ & & & & & & 0 & U_{21} \\ & & & & & & & 0 \end{pmatrix}$$

where, for $1 \le i, j \le n, i \ne j$, U_{ij} denote the vector space of the rectangular matrices of dimensions $d_i \times d_j$. Let define (see section 2 of [18]) $UT^*(A_1, \ldots, A_n) = D \oplus U \subseteq M_{2d}(F)$. It is easy to show that $UT^*(A_1, \ldots, A_n)$ is a subalgebra with involution of $(M_{2d}(F), \gamma_{2d})$ in which the algebras A_i are embedded as $*$-algebras and whose $*$-exponent is given by

$$\exp^*(UT^*(A_1, \ldots, A_n)) = \sum_{i=1}^n \dim_F A_i.$$

In [20] and [19] the link between the degrees of $*$-Capelli polynomials and the $*$-polynomial identities of $UT^*(A_1, \ldots, A_n)$ was investigated. If we set $d^+ := \sum_{i=1}^n \dim_F A_i^+$ and $d^- := \sum_{i=1}^n \dim_F A_i^-$, then the following result applies

Lemma 2 *Let $R = UT^*(A_1, \ldots, A_n)$. Then $Cap^*_M[Y, X]$ and $Cap^*_L[Z, X]$ are in $Id^*(R)$ if and only if $M \ge d^+ + n$ and $L \ge d^- + n$.*

4.2 Asymptotics for $*$-Capelli Polynomials

In this section we shall state our main results about the $*$-Capelli polynomials and their asymptotics (see [11]).

The following two key lemmas hold for any $*$-simple finite dimensional algebra.

Lemma 3 *Let $A = \bar{A} \oplus J$ where $\bar{A}$ is a $*$-simple finite dimensional algebra and $J = J(A)$ is its Jacobson radical. Then J can be decomposed into the direct sum of four $\bar{A}$-bimodules*

$$J = J_{00} \oplus J_{01} \oplus J_{10} \oplus J_{11}$$

where, for $p, q \in \{0, 1\}$, J_{pq} is a left faithful module or a 0-left module according to $p = 1$, or $p = 0$, respectively. Similarly, J_{pq} is a right faithful module or a 0-right module according to $q = 1$ or $q = 0$, respectively. Moreover, for $p, q, i, l \in \{0, 1\}$, $J_{pq}J_{ql} \subseteq J_{pl}$, $J_{pq}J_{il} = 0$ for $q \neq i$ and there exists a finite dimensional nilpotent $*$-algebra N such that $J_{11} \cong \bar{A} \otimes_F N$ (isomorphism of $\bar{A}$-bimodules and of $*$-algebras).

Proof It follows from the proof of Lemma 2 in [29].

Lemma 4 *Let $\bar{A}$ be a $*$-simple finite dimensional algebra. Let $M = \dim_F \bar{A}^+$ and $L = \dim_F \bar{A}^-$. Then $\bar{A}$ does not satisfy $Cap^*_M[Y, X]$ and $Cap^*_L[Z, X]$.*

Proof The result follows immediately from [21, Lemma 3.1].

Lemma 5

(1) *Let $M_1 = k(k + 1)/2$ and $L_1 = k(k - 1)/2$ with $k \in \mathbb{N}$, $k > 0$ and let $J_{11} \cong M_k(F) \otimes_F N$, as in Lemma 3. If $\Gamma_{M_1+1,L_1+1} \subseteq Id^*(M_k(F) + J)$, then $J_{10} = J_{01} = (0)$ and N is commutative.*
(2) *Let $M_2 = m(2m - 1)$ and $L_2 = m(2m + 1)$ with $m \in \mathbb{N}$, $m > 0$ and let $J_{11} \cong M_{2m}(F) \otimes_F N$, as in Lemma 3. If $\Gamma_{M_2+1,L_2+1} \subseteq Id^*(M_{2m}(F) + J)$, then $J_{10} = J_{01} = (0)$ and N is commutative.*
(3) *Let $M_3 = L_3 = h^2$ with $h \in \mathbb{N}$, $h > 0$ and let $J_{11} \cong (M_h(F) \oplus M_h(F)^{op}) \otimes_F N$, as in Lemma 3. If $\Gamma_{M_3+1,L_3+1} \subseteq Id^*((M_h(F) \oplus M_h(F)^{op}) + J)$, then $J_{10} = J_{01} = (0)$ and N is commutative.*

Lemma 6

(1) *Let $M_1 = k(k + 1)/2$ and $L_1 = k(k - 1)/2$ with $k \in \mathbb{N}$, $k > 0$. Then*

$$\exp^*(\mathcal{U}^*_{M_1+1,L_1+1}) = M_1 + L_1 = k^2 = \exp^*((M_k(F), t)).$$

(2) *Let* $M_2 = m(2m-1)$ *and* $L_2 = m(2m+1)$ *with* $m \in \mathbb{N}$, $m > 0$. *Then*

$$\exp^*(\mathcal{U}^*_{M_2+1,L_2+1}) = M_2 + L_2 = 4m^2 = \exp^*((M_{2m}(F), s)).$$

(3) *Let* $M_3 = L_3 = h^2$ *with* $h \in \mathbb{N}$, $h > 0$. *Then*

$$\exp^*(\mathcal{U}^*_{M_3+1,L_3+1}) = M_3 + L_3 = 2h^2 = \exp^*((M_h(F) \oplus M_h(F)^{op}, exc)).$$

Proof (1) The exponent of $\mathcal{U}^*_{M_1+1,L_1+1}$ is equal to the exponent of some minimal variety lying in $\mathcal{U}^*_{M_1+1,L_1+1}$ (for the definition of minimal variety see [30]). Let $d^+ := \sum_{i=1}^n \dim_F A_i^+$ and $d^- := \sum_{i=1}^n \dim_F A_i^-$, then, by Di Vincenzo and Spinelli [20, Theorem 2.1] and Lemma 2, we have that

$$\exp^*(\mathcal{U}^*_{M_1+1,L_1+1}) = \max\{\exp^*(UT^*(A_1, \ldots, A_n)) \mid d^+ + n \leq M_1 + 1 \text{ and } d^- + n \leq L_1 + 1\}.$$

Then

$$\exp^*(\mathcal{U}^*_{M_1+1,L_1+1}) \geq M_1 + L_1 = k^2 = \exp^*(UT^*(M_k(F))).$$

Since by Lemma 1, $\exp^*(\mathcal{U}^*_{M_1+1,L_1+1}) \leq M_1 + L_1$ then the proof is completed.
(2), (3) The proof is the same of that of point (1).

Now we are able to prove the main results.

Theorem 11 *Let* $M_1 = k(k+1)/2$ *and* $L_1 = k(k-1)/2$ *with* $k \in \mathbb{N}$, $k > 0$. *Then*

$$\mathcal{U}^*_{M_1+1,L_1+1} = \operatorname{var}^*(\Gamma^*_{M_1+1,L_1+1}) = \operatorname{var}^*(M_k(F) \oplus D'),$$

where D' *is a finite dimensional* $*$*-algebra such that* $\exp^*(D') < M_1 + L_1$. *In particular*

$$c_n^*(\Gamma^*_{M_1+1,L_1+1}) \simeq c_n^*(M_k(F)).$$

Sketch of the Proof By the previous Lemma we have that $\exp^*(\mathcal{U}^*_{M_1+1,L_1+1}) = M_1 + L_1$.

Let $A = A^+ \oplus A^-$ be a generating finite dimensional $*$-algebra of $\mathcal{U}^*_{M_1+1,L_1+1}$. By Corollary 2, there exist a finite number of reduced $*$-algebras $B_1, \ldots, B_s$ and a finite dimensional $*$-algebra D' such that $\mathcal{U}^*_{M_1+1,L_1+1} = \operatorname{var}^*(A) = \operatorname{var}^*(B_1 \oplus \cdots \oplus B_s \oplus D')$, with $\exp^*(B_1) = \cdots = \exp^*(B_s) = \exp^*(\mathcal{U}^*_{M_1+1,L_1+1}) = M_1 + L_1$ and $\exp^*(D') < \exp^*(\mathcal{U}^*_{M_1+1,L_1+1}) = M_1 + L_1$. Then, it is enough to analyze the structure of a finite dimensional reduced $*$-algebra R such that $\exp^*(R) = M_1 + L_1 = \exp^*(\mathcal{U}^*_{M_1+1,L_1+1})$ and $\Gamma^*_{M_1+1,L_1+1} \subseteq Id^*(R)$. Let write $R = R_1 \oplus \cdots \oplus R_q + J$, where $J = J(R)$, $R_1 J \cdots J R_q \neq 0$ and R_i is isomorphic to one of the following algebras :$(M_{k_i}(F), t)$ or $(M_{2m_i}(F), s)$ or $(M_{h_i}(F) \oplus M_{h_i}(F)^{op}, exc)$.

Let t_1 be the number of $*$-algebras R_i of the first type, t_2 the number of $*$-algebras R_i of the second type and t_3 the number of R_i of the third type, with $t_1 + t_2 + t_3 = q$. By [18, Theorem 4.5] and [18, Proposition 4.7] there exists a $*$-algebra $\overline{R}$ isomorphic to the $*$-algebra $UT^*(R_1, \ldots, R_q)$ such that $\exp^*(R) = \exp^*(\overline{R}) = \exp^*(UT^*(R_1, \ldots, R_q))$. Let observe that

$$k^2 = M_1 + L_1 = \exp^*(R) = \exp^*(\overline{R}) = \exp^*(UT^*(R_1, \ldots, R_q)) =$$

$$\dim_F R_1 + \cdots + \dim_F R_q = k_1^2 + \cdots + k_{t_1}^2 + (2m_1)^2 + \cdots + (2m_{t_2})^2 + 2h_1^2 + \cdots + 2h_{t_3}^2.$$

Let $d^{\pm} = \dim_F (R_1 \oplus \cdots \oplus R_q)^{\pm}$ then

$$d^+ + d^- = d = \dim_F (R_1 \oplus \cdots \oplus R_q) = \exp^*(\overline{R}) = M_1 + L_1.$$

By [20, Lemma 3.2] $\overline{R}$ does not satisfy the $*$-Capelli polynomials $Cap^*_{d^+ + q-1}[Y; X]$ and $Cap^*_{d^- + q-1}[Z; X]$, but $\overline{R}$ satisfies $Cap^*_{M_1+1}[Y; X]$ and $Cap^*_{L_1+1}[Z; X]$. Thus $d^+ + q - 1 \leq M_1$ and $d^- + q - 1 \leq L_1$. Hence $d^+ + d^- + 2q - 2 \leq M_1 + L_1$. Since $d^+ + d^- = M_1 + L_1$ we obtain that $q = t_1 + t_2 + t_3 = 1$. Since t_1, t_2 and t_3 are nonnegative integers by considering all possible cases we get that $t_1 = 1$ and $R \cong M_k(F) + J$. From Lemmas 3 and 5 we obtain

$$R \cong (M_k(F) + J_{11}) \oplus J_{00} \cong (M_k(F) \otimes N^{\sharp}) \oplus J_{00}$$

where $N^{\sharp}$ is the algebra obtained from N by adjoining a unit element.

Thus var*(R) $=$ $\mathrm{var}^*(M_k(F) \oplus J_{00})$ with J_{00} a finite dimensional nilpotent $*$-algebra. Hence, recalling the decomposition given above, we get

$$\mathcal{U}^*_{M_1+1, L_1+1} = \mathrm{var}^*(\Gamma_{M_1+1, L_1+1}) = \mathrm{var}^*(M_k(F) \oplus D'),$$

where D' is a finite dimensional $*$-algebra with $\exp^*(D') < M_1 + L_1$. Then

$$c_n^*(\Gamma_{M_1+1, L_1+1}) \simeq c_n^*(M_k(F))$$

and the theorem is proved.

In a similar way we can prove the next two theorems.

Theorem 12 *Let* $M_2 = m(2m - 1)$ *and* $L_2 = m(2m + 1)$ *with* $m \in \mathbb{N}$, $m > 0$. *Then*

$$\mathcal{U}^*_{M_2+1, L_2+1} = \mathrm{var}^*(\Gamma_{M_2+1, L_2+1}) = \mathrm{var}^*(M_{2m}(F) \oplus D''),$$

where D'' *is a finite dimensional* $*$*-algebra such that* $\exp^*(D'') < M_2 + L_2$. *In particular*

$$c_n^*(\Gamma_{M_2+1,L_2+1}) \simeq c_n^*(M_{2m}(F)).$$

Theorem 13 *Let* $M_3 = L_3 = h^2$ *with* $h \in \mathbb{N}$, $h > 0$. *Then*

$$\mathcal{U}^*_{M_3+1,L_3+1} = \text{var}^*(\Gamma_{M_3+1,L_3+1}) = \text{var}^*((M_h(F) \oplus M_h(F)^{op}) \oplus D'''),$$

where D''' *is a finite dimensional* $*$*-algebra such that* $\exp^*(D''') < M_3 + L_3$. *In particular*

$$c_n^*(\Gamma_{M_3+1,L_3+1}) \simeq c_n^*(M_h(F) \oplus M_h(F)^{op}).$$

Acknowledgement The authors were partially supported by INDAM-GNSAGA of Italy.

References

1. Aljadeff, E., Giambruno, A.: Multialternanting graded polynomials and growth of polynomial identities. Proc. Am. Math. Soc. **141**, 3055–3065 (2013)
2. Aljadeff, E., Giambruno, A., La Mattina, D.: Graded polynomial identities and exponential growth. J. Reine Angew. Math. **650**, 83–100 (2011)
3. Aljadeff, E., Giambruno, A., Karasik, Y.: Polynomial identities with involution, superinvolutions and Grassmann envelope. Proc. Am. Math. Soc. **145**, 1843–1857 (2017)
4. Amitsur, S.A.: Rings with involution. Israel J. Algebra **6**, 99–106 (1968)
5. Amitsur, S.A.: Identities in rings with involution. Israel J. Algebra **7**, 63–68 (1968)
6. Amitsur, S.A., Regev, A.: PI-algebras and their cocharacters. J. Algebra **78**, 248–254 (1982)
7. Bahturin, Y., Giambruno, A., Zaicev, M.: G-identities on associative algebras. Proc. Am. Math. Soc. **127**, 63–69 (1999)
8. Benanti, F.: On the exponential growth of graded Capelli polynomials. Israel J. Math. **196**, 51–65 (2013)
9. Benanti, F.: Asymptotics for graded Capelli polynomials. Algebr. Represent. Theory **18**, 221–233 (2015)
10. Benanti, F., Sviridova, I.: Asymptotics for Amitsur's Capelli-type polynomials and verbally prime PI-algebras. Israel J. Math. **156**, 73–91 (2006)
11. Benanti, F., Valenti, A.: Asymptotics for Capelli polynomials with involution (2019). arXiv preprint arXiv:1911.04193
12. Benanti, F., Giambruno, A., Pipitone, M.: Polynomial identities on superalgebras and exponential growth. J. Algebra **269**, 422–438 (2003)
13. Berele, A., Giambruno, A., Regev, A.: Involution codimensions and trace codimensions of matrices are asymptotically equal. Israel J. Math. **96**, 49–62 (1996)
14. Berele, A., Regev, A.: On the codimensions of verbally prime P.I.-algebras. Israel J. Math. **91**, 239–247 (1995)
15. Berele, A., Regev, A.: Exponential growth for codimensions of some P.I. algebras. J. Algebra **241**, 118–145 (2001)
16. Berele, A., Regev, A.: Asymptotic codimensions of $M_k(E)$. Adv. Math. **363**, 106979 (2020)
17. Cohen, P.B., Regev, A.: A six generalized squares theorem, with application to polynomial identity algebras. J. Algebra **239**, 174–190 (2001)

18. Di Vincenzo, O.M., La Scala, R.: Minimal algebras with respect to their $*$-exponent. J. Algebra **317**, 642–657 (2007)
19. Di Vincenzo, O.M., Spinelli, E.: Some results of $*$-minimal algebras with involution. In: Giambruno, A., et al. (eds.) Group, Rings and Group Rings 2008. International Conference, Ubatuba, Brazil, July 27 August 2, 2008, Contemporary Mathematics, vol. 499, pp. 75–88. American Mathematical Society, Providence (2009)
20. Di Vincenzo, O.M., Spinelli, E.: A characterization of $*$-minimal algebras with involution. Israel J. Math. **186**, 381–400 (2011)
21. Di Vincenzo, O.M., Nardozza, V.: On the existence of the graded exponent for finite dimensional Z p-graded algebras. Can. Math. Bull. **55**, 271–284 (2012)
22. Giambruno, A., Regev, A.: Wreath products and P.I. algebras. J. Pure Appl. Algebra **35**, 133–149 (1985)
23. Giambruno, A., Zaicev, M.: On codimension growth of finitely generated associative algebras. Adv. Math. **140**, 145–155 (1998)
24. Giambruno, A., Zaicev, M.: Exponential codimension growth of P.I. algebras: an exact estimate. Adv. Math. **142**, 221–243 (1999)
25. Giambruno, A., Zaicev, M.: Involution codimensions of finite dimensional algebras and exponential growth. J. Algebra **222**, 471–484 (1999)
26. Giambruno, A., Mishchenko, S.: On star-varieties with almost polynomial growth. Algebra Colloq. **8**, 33–42 (2001)
27. Giambruno, A., Mishchenko, S.: Polynomial growth of the $*$-codimensions and Young diagrams. Commun. Algebra **29**, 277–284 (2001)
28. Giambruno, A., Zaicev, M.: A characterization of algebras with polynomial growth of the codimensions. Proc. Am. Math. Soc. **129**, 59–67 (2001)
29. Giambruno, A., Zaicev, M.: Asymptotics for the standard and the Capelli identities. Israel J. Math. **135**, 125–145 (2003)
30. Giambruno, A., Zaicev, M.: Polynomial Identities and Asymptotics Methods, Surveys, vol. 122. American Mathematical Society, Providence (2005)
31. Giambruno, A., La Mattina, D.: Graded polynomial identities and codimensions: computing the exponential growth. Adv. Math. **259**(2), 859–881 (2010)
32. Giambruno, A., Ioppolo, A., La Mattina, D.: Varieties of algebras with superinvolution of almost polynomial growth. Algebr. Represent. Theory **19**, 599–611 (2016)
33. Giambruno, A., Polcino Milies, C., Valenti, A.: Star-polynomial identities: computing the exponential growth of the codimensions. J. Algebra **469**, 302–322 (2017)
34. Kemer, A.R.: Varieties of $\mathbb{Z}_2$-graded algebras (Russian). Izv. Akad. Nauk SSSR Ser. Math. **48**(5), 1042–1059 (1984)
35. Kemer, A.R.: Ideals of identities of associative algebra. Translations of Math. Monographs, vol. 87. American Mathematical Society, Providence (1991)
36. Latyshev, V.H.: On the theorem of Regev about identities in the tensor product of PI-algebras. Uspekhi Mat. Nauk **27**, 213–214 (1972)
37. Mishchenko, S.P., Regev, A., Zaicev, M.: The exponential growth of codimension for Capelli identities. Israel J. Math. **115**, 333–342 (2000)
38. Mishchenko, S.P., Valenti, A.: A star-variety with almost polynomial growth. J. Algebra **223**, 66–84 (2000)
39. Razmyslov, Yu.P.: On the Kaplansky problem (Russian). Izv. Akad. Nauk SSSR (Ser. Mat.) **37**, 483–501 (1973)
40. Regev, A.: Existence of identities in $A \otimes B$. Israel J. Math. **11**, 131–152 (1972)
41. Regev, A.: Algebras satisfying Capelli identity. Israel J. Math. **33**, 148–154 (1979)
42. Regev, A.: Asymptotic values for degrees associated with strips of Young diagrams. Adv. Math. **41**, 115–136 (1981)
43. Regev, A.: Codimensions and trace codimensions of matrices are asymtotically equal. Israel J. Math. **47**, 246–250 (1984)

44. Regev, A.: On the identities of subalgebras of matrices over the Grassmann algebra. Israel J. Math. **58**, 351–369 (1987)
45. Rowen, L.H.: Polynomial Identities in Ring Theory. Academic Press, New York (1980)
46. Sviridova, I.: Finitely generated algebras with involution and their identities. J. Algebra **383**, 144–167 (2013)

Regev's Conjecture for Algebras with Hopf Actions

Allan Berele

Abstract Let A be a p.i. algebra in characteristic zero with action from the finite dimensional Hopf algebra H, and let $\{c_n^H(A)\}$ be the H-codimension sequence. If $1 \in A$ then we prove the H-analogue of Regev's conjecture, namely, $c_n^H(A) \simeq \alpha n^t e^n$ for some $\alpha > 0$, $2t \in \mathbb{Z}$ and $e \in \mathbb{N}$. We also prove weaker results under weaker hypotheses.

Keywords Codimensions · Regev's conjecture · Hopf actions

1 Introduction

Throughout this paper A will be a non-nilpotent p.i. algebra in characteristic zero. If $\{c_n(A)\}_{n=0}^{\infty}$ is the codimension sequence, then Giambruno and Zaicev in [8] and [9] proved the following theorem conjectured by Amitsur:

Theorem (Amitsur's Conjecture) *The limit* $\lim_{n\to\infty} (c_n(A))^{1/n}$ *exists and is a non-negative integer.*

The limit in this theorem is denoted $exp(A)$ or simply e. Giambruno and Zaicev's papers prove the stronger statement that

$$\alpha_1 n^{t_1} e^n \le c_n(A) \le \alpha_2 n^{t_2} e^n \tag{1}$$

for some positive α_1, α_2 and some t_1, t_2

Giambruno and Zaicev's theorem has been generalized to other types of codimensions, including *-codimension of p.i. algebras with involution in [12] and H-codimensions of p.i. algebras with action from a finite dimensional semisimple Hopf algebra in [13].

A. Berele (✉)
Department of Mathematical Science, De Paul University, Chicago, IL, USA
e-mail: aberele@depaul.edu

O. M. Di Vincenzo, A. Giambruno (eds.), *Polynomial Identities in Algebras*, Springer INdAM Series 44, https://doi.org/10.1007/978-3-030-63111-6_4

Regev conjectured a stronger estimate of the codimensions, namely,

Conjecture (Regev's Conjecture) For any non-nilpotent characteristic zero p.i. algebra A the codimensions satisfy $c_n(A) \simeq \alpha n^t e^n$, for some $e \in \mathbb{N},\ 2t \in \mathbb{Z}$ and $\alpha > 0$.

Regev also conjectured that the constant α should belong to a field extension of $\mathbb{Q}$ of the form $\mathbb{Q}[\sqrt{2\pi}, \sqrt{a_1}, \ldots, \sqrt{a_n}]$, where the a_i are positive integers. Although this part of Regev's conjecture is interesting and important, we have nothing to say about it in this paper and so shall ignore it.

It will be convenient to refer to the following as Regev's Weak Conjecture:

Conjecture (Regev's Weak Conjecture) For any non-nilpotent characteristic zero p.i. algebra A the codimensions satisfy

$$\alpha_1 n^t e^n \leq c_n(A) \leq \alpha_2 n^t e^n,$$

where $e \in \mathbb{N},\ 2t \in \mathbb{Z}$ and $0 < \alpha_1 \leq \alpha_2$.

Comparing Regev's Weak Conjecture to (1), the former is stronger in that it states that the two powers of t are equal integers or half-integers. Regev's Weak Conjecture is now known for all p.i. algebras: In [6] and [4] Berele and Regev proved it under the assumption that the codimension sequence is non-decreasing, and Giambruno and Zaicev proved that the codimension sequence is always non-decreasing in [11]. The former two papers also proved Regev's Conjecture for algebras with 1. Our main goal in this paper is to generalize these results from to H-codimensions of p.i. algebras with action from finite dimensional Hopf algebras.

Theorem 1.1 *Let A be a p.i. algebra with action from the finite dimensional Hopf algebra H, with H-codimensions $c_n^H(A)$. If the codimensions are eventually non-decreasing then Regev's Weak Conjecture holds, and if $1 \in A$, then Regev's Conjecture holds.*

The proof of this theorem is based on the proofs from [4], and for the reader not familiar with that work we now turn to a brief summary.

2 Regev's Conjectures for Algebras Without Actions

2.1 Background on Magnums

We start with these two fundamental theorems of Kemer, see [14].

Theorem 2.1 (The Specht Conjecture) *The set of T-ideals in characteristic zero satisfies the ascending chain condition.*

Let E be the infinite dimensional Grassmann algebra with its natural $\mathbb{Z}_2$-grading. Given any $\mathbb{Z}_2$-graded algebra A the Grassmann envelope $G(A)$ is defined to be $E_0 \otimes A_0 \oplus E_1 \otimes A_1 \subseteq E \otimes A$.

Theorem 2.2 (Kemer's Finite Representability Theorem) *If A is any characteristic zero p.i. algebra, there exists a finite dimensional, $\mathbb{Z}_2$-graded algebra B such that A and $G(B)$ satisfy the same identities.*

At this point we need to talk about cocharacters, not just codimensions, and it is helpful to use hooks. For the reader not familiar with the theory we recommend [10].

Definition 2.3 The hook $H(k, \ell; n)$ is the set of all partitions of n with at most k parts greater than ℓ, and $H(k, \ell)$ is the union $\cup_n H(k, \ell; n)$.

Henceforth, let $m_\lambda = m_\lambda(A)$ be the multiplicity of the irreducible character corresponding to the partition λ in the cocharacter sequence of A. Of course, the codimensions can be computed from the multiplicities via

$$c_n(A) = \sum_{\lambda \vdash n} m_\lambda f^\lambda, \tag{2}$$

where f^λ is the degree of the S_n-character on λ.

The following theorem is due to Amitsur and Regev from [2]. It is at least implicit in the work of Kemer, see Proposition 1.3 in [14].

Theorem 2.4 (Amitsur-Regev) *Given any p.i. algebra A there exists k, ℓ such that $m_\lambda = 0$ for all $\lambda \notin H(k, \ell)$.*

In the special case of $A = G(B)$ where B is finite dimensional with graded dimension (d_0, d_1) we may take $(k, \ell) = (d_0, d_1)$.

Definition 2.5 Let $A = G(B)$ and let $U^{k,\ell} = U^{k,\ell}(A)$ be the universal algebra for B, i.e., the free $\mathbb{Z}_2$-graded algebra on k degree zero variables and ℓ degree one variables modulo the graded identities of B. The algebra $U^{k,\ell}(A)$ is called a magnum of A.

The connection to cocharacters is because of this theorem of Berele and Regev from [5].

Theorem 2.6 *The magnum $U^{k,\ell}(A)$ has a natural $(k+\ell)$-fold grading with respect to which it has a Poincaré series $P_{k,\ell}(A)$. This series can be expanded as a series in the hook Schur functions*

$$P_{k,\ell}(x_1, \ldots, x_k, y_1, \ldots, y_\ell) = \sum_{n=0}^{\infty} \sum_{\lambda \in H(k,\ell;n)} m_\lambda HS_\lambda(x; y)$$

where the $m_\lambda = m_\lambda(A)$ are the multiplicities of the irreducible characters in the cocharacter sequence of A.

2.2 Codimension Sequences

Definition 2.7 A sequence $\{a_n\}$ is asymptotically almost polynomial times exponential (asymptotically APE for short) if there exists a modulus d such that for each $0 \le i \le d-1$ there exists $e \in \mathbb{N}$, $\alpha_i \ge 0$, $2t_i \in \mathbb{Z}$ such that $a_n \simeq \alpha_i n^{t_i} e^n$ where $n = qd + i, q \to \infty$.

Our main theorem in [4] is that $c_n(A)$ is always asymptotically APE. The proof follows combinatorially from two properties of codimension sequences. The first is due to [9], as follows. A p.i. algebra is said to be varietally irreducible if its T-ideal of identities cannot be written as the intersection of strictly larger T-ideals. Equivalently, A is not p.i. equivalent to a direct sum of algebras each satisfying more identities than A. Every algebra is equivalent to a direct sum of varietally irreducible ones and Giambruno and Zaicev proved the following theorem about their cocharacters

Theorem 2.8 *Let m_λ be the multiplicities in the cocharacter of the varietally irreducible algebra A. Then there exists k, ℓ, K such that*

(a) $m_\lambda = 0$ if the Young diagram of λ has more than K boxes outside of the hook $H(k,\ell)$, i.e., if $\sum_{i>k} \max\{\lambda_i - \ell, 0\}$ is greater than K.
(b) For all large t there exists λ with $m_\lambda \neq 0$ and λ within K of a partition of the form $((t+\ell)^k, \ell^t)$, i.e., such that

$$\sum_{i\le k} |\lambda_i - (t+\ell)| + \sum_{i=k+1}^{k+t} |\lambda_i - \ell| + \sum_{i>k+t} \lambda_i$$

is at most K.

In general, if a sequence of S_n characters satisfies (a) we will say that it almost lies in the $k \times \ell$ hook and write the set of such as $H'(k,\ell)$, surpressing the K; and if it satisfies both conditions we will say it satisfies the almost square hook conditions or the (k,ℓ) almost square hook conditions, ASH for short. If it is a finite sum of characters each satisfying some ASH, not necessarily with the same k, ℓ, we say it satisfies the multiple almost square hook condition, or MASH for short.

The next ingrediant we need is Theorem 3.3 of [4] which is based on Belov's theorem from [3].

Theorem 2.9 *For any characteristic zero p.i. algebra A and for any k, ℓ the series $P_{k,\ell}(A)(x_1, \ldots, x_k; y_1, \ldots, y_\ell)$ is the Taylor series of a rational function which can* *be written as a fraction with denominator a product of terms* $1-u$ *where* u *is a monic monomial.*

We call rational functions whose denominators can be taken to have this form nice rational functions.

It turns out that Theorems 2.8 and 2.9 are just what we need to prove that the codimension sequence is asymptotically almost polynomial times exponential (almost APE). Note that this next theorem taken from [4] is purely combinatorial and makes no reference to p.i. algebras.

Theorem 2.10 *Let* $\chi_n = \sum_{\lambda \in H(k,\ell;n)} m_\lambda \chi^\lambda$ *be a sequence of* S_n*-characters supported by some hook* $H(k,\ell)$*, and let* $c(n) = \sum_{\lambda \vdash n} m_\lambda f^\lambda$ *where* f^λ *is the degree of* χ^λ.

Assume that the characters satisfy MASH and that $\sum m_\lambda HS_\lambda(x; y)$ *is a nice rational function. Then* $c(n)$ *is APE.*

It is not hard to see that if $c(n)$ is also assumed to be eventually non-decreasing then the exponents t_i must all be equal and $c(n)$ will satisfy the Weak Regev Conjecture. This condition will be satisfied if whenever $f(x_1, \ldots, x_n)$ is not an identity for A neither is $f(x)x_{n+1}$, which in turn will be satisfied if A has a non-zero divisor. However, this is unnecessary as Giambruno and Zaicev proved in [11] that codimensions of p.i. algebras are eventually non-decreasing proving the Weak Regev Conjecture in general. As for the Regev Conjecture, if $1 \in A$ then we need the concept of Young derived characters.

Definition 2.11 A sequence $\{\chi_n\}$ of S_n-characters is said to be Young derived if there exists a sequnce $\{\phi_n\}$ of S_n-characters such that for each n

$$\chi_n = \sum_{i=0}^{n} \phi_i \uparrow,$$

where the arrow indicates inducing the S_i character up to S_n.

Drensky proved in [7] that cocharacter sequences of p.i. algebras with 1 are always Young derived. Together with the following theorem from [4] we see that Regev's Conjecture holds for all such algebras.

Theorem 2.12 *Let* χ_n *be as in the previous theorem and assume in addition that it is Young derived. Then* $c(n)$ *is asymptotic to* $\alpha n^t e^n$*, where* $\alpha > 0$*,* $2t \in \mathbb{Z}$ *and* $e \in \mathbb{N}$*, as in Regev's Conjecture.*

Explicitely, here are the theorems we intend to generalize in the next section:

Theorem 2.13 *Let A is any characteristic zero p.i. algebra.*

1. *The codimensions* $c_n(A)$ *are asymptotically APE.*
2. *If in addition they are known to be increasing, then they satisfy Regev's Weak Conjecture.*
3. *If* $1 \in A$ *then the codimensions satisfy Regev's Conjecture*

3 H-Cocharacters

Henceforth we assume that H is a finite dimensional semisimple Hopf algebra, and that A is a H-module algebra.

For the reader's convenience we define the H-cocharacter. Let H have basis $h_1, \ldots, h_m$. It is convenient to take $h_1 = 1$ so $h_1(x) = x$ for all x. Let V_n be the vector space with basis equal to the (finite) set of symbols

$$h_{i_1}(x_{\sigma(1)}) \cdots h_{i_n}(x_{\sigma(n)}),\ 1 \leq i_1, \ldots, i_n \leq m,\ \sigma \in S_n.$$

V_n is an H module in a natural way and elements of V_n can be identified with multilinear H-polynomials. If $f(x_1, \ldots, x_n)$ is such a polynomial and $\sigma \in S_n$ we may define $\sigma f(x_1, \ldots, x_n) = f(x_{\sigma(1)}, \ldots, x_{\sigma(n)})$. Finally, if I_n is the identities of the algebra A in V_n, then I_n is an S_n submodule. We let $c_n^H(A)$ be the dimension of the quotient, $\chi_n^H(A)$ the S_n-character and m_λ^H the multiplicities of the irreducible components.

Happily, most of what we need to generalize Theorem 2.13 was already done by Karasik in [13]. Most importantly he generalized Kemer's two main theorems:

Theorem 3.1 (H-Specht Conjecture) *The set of H-T-ideals in characteristic zero satisfies the ACC*

Theorem 3.2 (H-Finite Representability Theorem) *If A is any characteristic zero p.i. algebra with H action over the field F, and assume that F contains the complex numbers, then there exists a field extension L of F and a finite dimensional $\mathbb{Z}_2$-graded L-algebra B, with H action compatible with the grading, such that A and the Grassmann envelope $G(B)$ satisfy the same H-identities over F. Moreover, if A satisfies a Capelli identity, then there is an extension L of the base field F and a finite dimensional L-algebra B, with H action compatible, such that A and B satisfy the same H-identities over F.*

Two remarks: If the base field F does not contain $\mathbb{C}$, let K be a field containing both F and $\mathbb{C}$. Then the F-algebra A and the K-algebra $A \otimes K$ have the same codimension sequences, so for the study of codimensions we may assume without loss of generality that F contains $\mathbb{C}$. Secondly, that the algebra B has compatible H-action and $\mathbb{Z}_2$-grading is equivalent to B having an action from the finite dimensional semisimple Hopf algebra $H_2 = H \otimes (\mathbb{Z}_2)^*$.

As in the case of algebras without actions, we let $U^{k,\ell}(A)$ be the universal algebra for B as an F-algebra with H_2 action, in k degree 0 generators and ℓ degree one generators, and we let $P_{k,\ell}^H(x; y)$ be the corresponding Poincaré series. Then we have the exact analogue of Theorem 2.6.

Theorem 3.3 *The magnum $U^{k,\ell}(A)$ has a natural $(k+\ell)$-fold grading with respect to which it has a Poincaré series $P^H_{k,\ell}(A)$. This series can be expanded as a series in the hook Schur functions*

$$P^H_{k,\ell}(x_1,\ldots,x_k,y_1,\ldots,y_\ell)=\sum_{n=0}^{\infty}\sum_{\lambda\in H(k,\ell;n)} m^H_\lambda(A)HS_\lambda(x;y)$$

where the $m^H_\lambda(A)$ are the multiplicities of the irreducible characters in the H-cocharacter sequence of A.

Karasik also proved the analogues of Amitsur and Regev's theorem Theorem 2.4, and Giambruno and Zaicev's theorem, Theorem 2.8. Here is his theorem.

Theorem 3.4 *Every H-p.i. algebra is H-p.i. equivalent to a finite direct sum of basic ones, and the cocharacter sequence of each basic H-p.i. algebras is ASH.*

In light of Theorem 2.10 all we need to prove in order to obtain that $c^H_n(A)$ is asymptotically APE is that $P_{k,\ell}(x;y)$ is a nice rational function, which we do in the next section; and in light of Theorem 2.12 in order to prove Regev's Conjecture for algebras with 1 we need to prove that the cocharacter sequence is Young derived, which we do in the last section.

4 Rationality of Poincaré Series

Our main goal in this section is to prove that if B is any H_2 algebra (as above, $H_2 = H\otimes(\mathbb{Z}_2)^*$) satisfying a Capelli identity, then the corresponding generic algebra has nice rational Poincaré series. As remarked above, we may assume with out loss that F contains the complex numbers. If there were H_2-algebras satisfying Capelli identities whose generic algebras were not nice rational functions, then by the Karasik-Specht theorem we may assume that the ideal J of H-identities of B is maximal with this property. We could also assume that J is varietally irreducible, because if J were the intersection of two strictly larger ideals, $J = J_1\cap J_2$, and if $F_{k,\ell}$ is the free H algebra then the Poincaré series satisfy

$$P(F_{k,\ell}/J_1\cap J_2)=P(F_{k,\ell}/J_1)+P(F_{k,\ell}/J_2)-P(F_{k,\ell}/J_1+J_2).$$

But since each of J_1, J_2 and J_1+J_2 is bigger than J, each of the series on the right hand side would be nice rational and so $P(F_{k,\ell}/J_1\cap J_2)$ would be also.

By the finite representability theorem, we may assume that B is finite dimensional over a field $L\supset F$. Moreover, Karasik proved that every finite dimensional H-p.i is equivalent to a direct sum of H-p.i. basic ones, and so we may assume that B is H-p.i. basic as an L-H-algebra.

Let B have graded L basis $a_1, \ldots, a_k, b_1, \ldots, b_\ell$ and we consider B as a subalgebra of the matrices $M_{k+\ell}(L)$. Define generic graded elements via

$$x_i = t_{i1}a_1 + \ldots + t_{ik}a_k \text{ and } y_i = s_{i1}b_1 + \ldots + s_{i\ell}b_\ell$$

where the t_{ij} and s_{ij} are central indeterminants. Let K be the polynomial ring $L[t_{ij}, s_{ij}]_{ij}$

For each $h \in H_2$ we define $h(x_i)$ and $h(y_i)$ in the natural way.

We now define four algebras:

- U is the L-algebra generated by the $h(x_i)$ and the $h(y_i)$.
- R is the F-algebra generated by the $h(x_i)$ and $h(y_i)$.
- Let K be the polynomial algebra $L[t_{ij}, s_{ij}]_{ij}$. Since R is a subalgebra of the matrix algebra $M_{k+\ell}(K)$ there is a trace function $R \to L$. Let $\bar{C}$ be the F-algebra generated by all traces of elements of R.
- Let $\bar{R}$ be the F-subalgebra of $M_{k+\ell}(K)$ generated by R and $\bar{C}$.

It is not hard to see that U is the universal algebra for B as an L-algebra with H_2 action. This means that U satisfies all of the identities of B as an H_2-L-algebra and given any $a'_1, \ldots, a'_k \in B_0$ and any $b'_1, \ldots, b'_\ell \in B_1$ there is an L-H-homomorphism $T : U \to B$ that takes each x_i to a'_i and each y_i to b'_i. Since T restricted to R is a F-linear, it follows that R is the generic algebra for B as an F-H-algebra. In particular, B and R satisfy the same F-H_2-identities. Referring to Theorems 2.10 and 3.3, it is important to prove that the Poincaré series of R is a nice rational function.

A key step in the proof is that $\bar{C}$ must be a Noetherian ring and $\bar{R}$ must be a Noetherian module over it. This follows from two theorems:

Theorem 4.1 (Shirshov's Theorem) *There exist a finite number of words $u_1, \ldots, u_N$ in the $h_i(x_j)$ and $h_i(y_j)$ and an integer α such that R is spanned by the Shirshov words $u_{i_1}^{n_1} \cdots u_{i_\beta}^{n_\beta}$, where $\beta \leq \alpha$.*

and

Theorem 4.2 (Cayley-Hamilton Theorem) *$M_{k+\ell}(L)$ satisfies the mixed trace identity*

$$x^n + c_1(x)x^{n-1} + \cdots + c_n(x)1 = 0$$

where each $c_i(x)$ is a degree i trace polynomial in x (with coefficients in $\mathbb{Q}$).

It follows that $\bar{R}$ is generated over $\bar{C}$ by Shirshov words in which each exponent is at most $n-1$, and there are finitely many of these.

Also, multiplying the Cayley-Hamilton theorem by x and taking trace implies the pure trace identity

$$tr(x^{n+1}) + c_1(x)tr(x^n) + \cdots = 0$$

in which all but the first term (including inside the c_j) involves only powers x^i with $i \leq n$. This implies that $\bar{C}$ is generated by traces of Shirshov words in which each exponent is at most n, and there are finitely many of these. Summarizing:

Theorem 4.3 *$\bar{C}$ is commutative graded Noetherian, and $\bar{R}$ is a finite graded module over it.*

We now need to pass from $\bar{C}$ and $\bar{R}$ to R.

Because B is H_2-p.i. basic it has an H_2-Kemer polynomial, $h(x, y)$. Among the properties of Kemer polynomials is that it is not an identity for B and every evaluation $h(r)$, $r \in \bar{R}$ and every product $h(r)s$ is in R, where $r, s \in \bar{R}$, see Lemma 8.1 and Corollary 8.2 of [13]. Moreover, since $h(x, y)$ can be constructed from the Capelli identity by substitutions and alternations, it can be taken to have all its coefficients in $\mathbb{Q}$, and hence in F. Let J be the ideal generated by all evaluations of h on R. Then J is an ideal of both R and $\bar{R}$. Moreover, R/J is the universal algebra for the H_2-T-ideal generated by I and h. Consider

$$P(R) = P(R/J) + P(J).$$

The first term is the Poincaré series of a generic H_2-algebra satisfying more identities than B, so it is nice and rational. The second term is the Poincaré series of J, which is a module for the Noetherian ring $\bar{C}$ and contained in the Noetherian module $\bar{R}$. So it is Noetherian and has nice rational Poincaré series. Therefore R does also and we have now proven the following:

Theorem 4.4 *If B is any finite dimensional H_2-p.i. algebra and $U^{k,\ell}$ is the $\mathbb{Z}_2$-graded generic algebra in k degree zero and ℓ degree one variables, then the Poincaré series of $U^{k,\ell}$ is a nice rational function.*

We note that Theorem 4.4 was proven by Aljadeff and Kanel-Belov in [1] in the case of group gradings.

For our purposes, the main interest is in this consequence.

Theorem 4.5 *If A is any H-p.i. algebra, then the codimension sequence $c_n^H(A)$ is asymptotically APE. In particular, if $c_n^H(A)$ is known to be eventually non-decreasing then the Weak Regev Conjecture holds.*

Proof Referring to Theorem 2.10 we need to show that the cocharacters are MASH and that the Poincaré series of $U^{k,\ell}(A)$ is a nice rational function for all k, ℓ. The former statement is Theorem 3.4. As for the latter, $U^{k,\ell}(A)$ is the algebra R constructed above and the nice rationality of the Poincaré series is Theorem 4.4. □

5 Young Derived Sequences

By way of background, we start this section by describing the results for p.i. algebras without actions.

The space of multilinear, degree n polynomials in $x_1, \dots, x_n$ has $n!$ elements and so a basis can be indexed by elements of S_n. One well-known way to do this is due to Regev, but Specht, see [15], had an earlier one which we illustrate with a few examples:

$$(1)(23)(456) \leftrightarrow x_1[x_2, x_3][x_4, [x_5, x_6]]$$
$$(123)(4)(67) \leftrightarrow x_4[x_6, x_7][x_1, [x_2, x_3]]$$
$$(12)(3)(456)(78)(9) \leftrightarrow x_3x_9[x_1, x_2][x_7, x_8][x_4, [x_5, x_6]]$$

The exact rules are not important for us. What is important is that every $f \in V_n$ can be written as a sum

$$f = \sum_{1 \le i_1 < \cdots < x_k \le n} x_{i_1} \cdots x_{i_k} f_{i_1,\dots,i_k},$$

where $f_{i_1,\dots,i_k}$ is a commutator polynomial in the remaining variables. Drensky proved in [7] that if f is an identity for an algebra A with 1, then so is every $f_{i_1,\dots,i_k}$, and that this implies that the cocharacter sequence of A is Young derived.

We now turn to the case of Hopf algebra actions. Each semisimple Hopf algebra has a special element t called an integral. For example, in the special case that $H = FG$ is group algebra of a finite group, we let $t = \frac{1}{|G|}\sum_{g\in G} g$, noting that t is central, $t^2 = t$ and $tg = t$ for all $g \in G$. In a different special case, that of group gradings $H = (FG)^*$, we let $t = \pi_1$, the projection onto the identity component. Again, t is central and $t^2 = t$ and, in this case, $t\pi_i = \delta_{1i}t$. In the general case we have this theorem of Larson and Sweedler.

Theorem 5.1 (Larson-Sweedler) *If H is a finite dimensional semisimple Hopf algebra with count $\epsilon : H \to F$, then H contains a central element t such that for all h, $ht = \epsilon(h)t$ and $\epsilon(t) = 1$*

Let V_n be the set of all multilinear, homogeneous degree n H-polynomials in $x_1, \dots, x_n$. For each i we write x_i as $\tilde{x}_i + t(x_i)$ and note that if we substitute 1 for x_i then $t(x_i)$ becomes 1 and $\tilde{x}_i$ becomes 0.

By Specht's argument each f in V_n can be written as

$$f = t(x_1)g(x_2, \dots, x_n) + h(x_1, \dots, x_n)$$

where $h = h_1 + h_2$; in h_1 $t(x_1)$ occurs in commutators only; and h_2 is degree one in $\tilde{x}_1$ and does not involve $t(x_1)$. In short: h becomes zero if $x_1 = 1$.

If we substitute $x_1 + 1$ for x_1 then f becomes $f + g$. This means that if f is an identity for an algebra A, so is g and therefore so is h.

Now repeat the argument with $x_2, \ldots, x_n$. In the end we get that each $f \in V_n$ can be written as a sum of terms of the form

$$t(x_{i_1}) \cdots t(x_{i_a}) f_{i_1,\ldots,i_a}$$

summed over $i_1 < \cdots < i_a$, and where $f_{i_1,\ldots,i_a}$ is a multilinear polynomial in the remaining variables with the property that it becomes zero if any of those variables is 1. And, most importantly, f is an identity for A if and only in each $f_{i_1,\ldots,i_a}$ is.

Let $W_n \subseteq V_n$ be all of the polynomials which become zero if any variable is substituted by 1. Then it follows that as an S_n-module

$$\frac{V_n}{V_n \cap Id^H(A)} \cong \sum_{i=0}^{n} \frac{W_i}{W_i \cap Id^H(A)} \uparrow$$

This is the definition of Young derived.

It follows that the H-cocharacter sequence of A is Young derived and so Regev's conjecture is true for the codimensions.

References

1. Aljadeff, E., Kanel-Belov, A.: Hilbert series of PI relatively free G-graded algebras are rational functions. Bull. Lond. Math. Soc. **44**(3), 520–532 (2012)
2. Amitsur, S., Regev, A.: P.I. algebras and their cocharacters. J. Algebra **78**, 248–254 (1982)
3. Belov, A.: Rationality of Hilbert series with respect to free algebras. Russian Math. Surveys **52**, 394–395 (1997)
4. Berele, A.: Properties of hook Schur functions with applications to p.i. algebras. Adv. Appl. Math. **41**, 52–75 (2008); with Corrigendum in **43** (2009)
5. Berele, A., Regev, A.: Applications of hook Young diagrams to P.I. algebras. J. Algebra **82**, 559–567 (1983)
6. Berele, A., Regev, A.: Asymptotic behaviour of codimensions of P. I. algebras satisfying Capelli identities. Trans. A. M.S. **360**, 5155–5172 (2008)
7. Drensky, V.: Codimensions of T-ideals and Hilbert series of relatively free algebras. J. Algebra **91**, 1–17 (1984)
8. Giambruno, A., Zaicev, M.: On codimension growth of finitely generated associative algebras. Adv. Math. **140**, 145–155 (1988)
9. Giambruno, A., Zaicev, M.: Exponential codimension growth of P.I. algebras: an exact estimate. Adv. Math. **142**, 221–243 (1999)
10. Giambruno, A., Zaicev, M.: Polynomial Identities and Asymptotic Methods. Mathematical Surveys and Monographs, vol. 122. American Mathematical Society, Providence (2005)
11. Giambruno, A., Zaicev, M.: Growth of polynomial identities: is the sequence of codimensions eventually non-decreasing? Bull. Lond. Math. Soc. **46**, 771–778 (2014)
12. Giambruno, A., Polcino Milies, C., Valenti, A.: Star-polynomial identities: computing the exponential growth of the codimensions. J. Algebra **469**, 302–322 (2017)

13. Karasik, Y.: Kemer's theory for H-module algebras with application to the PI exponent. J. Algebra **457**, 194–227 (2016)
14. Kemer, A.R.: Ideals of Identities of Associative Algebras. AMS Translations of Mathematical Monograph, vol. 87. American Mathematical Society, Providence (1988)
15. Specht, W.: Gesetze in Ringen I. Math. Z. **52**, 557–589 (1950)

ℓ-Weak Identities and Central Polynomials for Matrices

Guy Blachar, Eli Matzri, Louis Rowen, and Uzi Vishne

Abstract We develop the theory of ℓ-weak identities in order to provide a feasible way of studying the central polynomials of matrix algebras. We describe the weak identities of minimal degree of matrix algebras in any dimension.

Keywords Weak identities · Central polynomials · Identities of matrix algebras

1 Introduction

One basic question in PI-theory is to determine the polynomial identities (PI's) of the matrix algebra $\mathrm{M}_n(\mathbb{Q})$. Specht's celebrated problem is whether every set of polynomial identities of an algebra is **finitely based**, i.e., is a consequence of a finite number of identities, solved affirmatively by Kemer in 1988 and 1990, cf. [10]. However, his solution is difficult to implement to obtain a finite (PI) base for the identities of $\mathrm{M}_n(\mathbb{Q})$, in the sense that every PI of the algebra is a consequence of the base identities. Indeed, a base is known only for $\mathbb{Q}$ and $\mathrm{M}_2(\mathbb{Q})$. Our overriding goal here is to obtain partial information about bases, mostly in terms of weak identities and weak central polynomials. A multilinear polynomial $f(x_1, \ldots, x_m)$ is an ℓ-**weak identity** of $\mathrm{M}_n(\mathbb{Q})$ if substitution of matrices for x_i sends f to zero whenever $\mathrm{tr}(x_1) = \cdots = \mathrm{tr}(x_\ell) = 0$, and an ℓ-**weak central polynomial** if such substitution sends f to a central element.

Section 2 provides a brief overview of polynomial identities. We define and discuss ℓ-weak identities in Sect. 3, developing an inductive procedure to compute spaces of ℓ-weak identities (see Remark 3.4). Aided by computer computations, we obtain the following results.

(1) Explicit generators for the ℓ-weak identities of $\mathrm{M}_2(F)$ in degrees 3 and 4, for any ℓ (Sect. 6).

G. Blachar · E. Matzri · L. Rowen · U. Vishne (✉)
Department of Mathematics, Bar-Ilan University, Ramat Gan, Israel
e-mail: blachag@biu.ac.il; eli.matzri@biu.ac.il; rowen@math.biu.ac.il; vishne@math.biu.ac.il

O. M. Di Vincenzo, A. Giambruno (eds.), *Polynomial Identities in Algebras*, Springer INdAM Series 44, https://doi.org/10.1007/978-3-030-63111-6_5

(2) When char$F \neq 3$ there are no weak identities of degree 5 for $\mathrm{M}_3(F)$ (Sect. 7.1).
(3) However, s_4 is a weak central polynomial of $\mathrm{M}_3(F)$ over a field of characteristic 3, so $[s_4, x_5]$ is a 4-weak polynomial identity of degree 5 (Sect. 7.2).
(4) We present dimensions and module decomposition for the ℓ-weak identity spaces in degree 6 for $\mathrm{M}_3(F)$, correcting a minor omission in [6] (Sect. 8.1).
(5) We obtain a trace identity of degree 4 for $\mathrm{M}_3(F)$ from the Okubo composition algebra, and deduce Halpin's 4-weak identity of degree 6 from it (Sect. 8.3).
(6) For $n \geq 4$, there are no weak identities of $\mathrm{M}_n(F)$ in degree $2n$ other than the standard identity (Sect. 9).

2 Preliminaries

Let F be a field. The free (associative) F-algebra generated by noncommuting variables $x_1, \ldots, x_m$ is denoted $F\{x_1, \ldots, x_m\}$; we refer to the elements of $F\{x_1, \ldots, x_m\}$ as **polynomials**.

Definition 2.1 A polynomial $p \in F\{x_1, \ldots, x_m\}$ is called a **polynomial identity** (PI) of the F-algebra A if $p(a_1, \ldots, a_m) = 0$ for all $a_1, \ldots, a_m \in A$. We write $\mathrm{id}(A)$ for the set of identities of A.

2.1 *Identities, Central Polynomials and Examples*

The free algebra has no nonzero identities, almost by definition. An algebra A is PI if $\mathrm{id}(A) \neq 0$. The most basic examples of PI-algebras are the matrix algebra $\mathrm{M}_n(F)$ for arbitrary n, f.d. algebras over a field, and the Grassmann algebra G, cf. [2, Definition 1.35].

Here is a notion closely related to PI.

Definition 2.2 (Central Polynomials) A polynomial $f(x_1, \ldots, x_n)$ is A-**central** if $f(A) \subseteq \mathrm{Cent}(A)$. A central polynomial $f(x_1, \ldots, x_n)$ is **strictly** A-**central** if $f \notin \mathrm{id}(A)$; in other words, $0 \neq f(A) \subseteq \mathrm{Cent}(A)$.

A polynomial $p(x_1, \ldots, x_n)$ is k-**multilinear** if each of the variables $x_1, \ldots, x_k$ appears exactly once in each of the monomials of p. We omit the preamble if p is multilinear in all of its variables. Let P_m be the subspace of multilinear polynomials in $F\{x_1, \ldots, x_m\}$, for $m \geq 1$. Any PI f can be transformed into a multilinear PI through the multilinearization process (see [2]), and the process is reversible in characteristic 0; likewise any central polynomial f can be transformed into a multilinear central polynomial through the multilinearization process, which is reversible in characteristic 0. Thus in what follows we consider polynomials in P_m.

Example 2.3

(i) The polynomial x_1 is central for any commutative algebra.
(ii) The polynomial $[x_1, x_2]$ is central for the Grassmann algebra.
(iii) Let $\mathrm{UT}(n)$ denote the algebra of upper triangular matrices over a given commutative base ring C. Any product of n strictly upper triangular $n \times n$ matrices is 0. Since $[a, b]$ is strictly upper triangular, for any upper triangular matrices a, b, we conclude that the algebra $\mathrm{UT}(n)$ satisfies the identity

$$[x_1, x_2][x_3, x_4] \cdots [x_{2n-1}, x_{2n}].$$

(iv) (Wagner's identity) The matrix algebra $\mathrm{M}_2(F)$ satisfies the identity $g_2 := [[x, y]^2, z]$ or, equivalently, the central polynomial $[x, y]^2$ and its multilinearization. (This is because the square of a trace-zero 2×2 matrix is scalar.)
(v) Fermat's Little Theorem translates to the fact that any field F of q elements satisfies the identity $x^q - x$. Its multilinearization is the symmetric polynomial, but in going back we only get qx^q which is identically zero.
(vi) The **standard polynomial**

$$s_m := \sum_{\pi \in S_m} \mathrm{sgn}(\pi) x_{\pi(1)} \cdots x_{\pi(m)}$$

is a PI of $\mathrm{M}_n(\mathbb{Q})$ precisely when $m \geq 2n$.
(vii) By Razmyslov [12] and Drensky [4] $\{s_4, g_2\}$ is a PI base for $\mathrm{M}_2(F)$. A base for $\mathrm{M}_3(\mathbb{Q})$ remains unknown.

The **PI degree** of an algebra A, denoted $\mathrm{PIdeg}\, A$, is the minimal degree of an identity of this algebra. Thus $\mathrm{PIdeg}\, \mathrm{M}_n(F) = 2n$, and $\mathrm{PIdeg}\, G = 3$. By [13], $\mathrm{PIdeg}\, \mathrm{M}_2(G) = 8$

2.2 *Spechtian Polynomials*

A multilinear polynomial is i-**Spechtian** if it vanishes when 1 is substituted for x_i. We write Sp_m^i for the subset of i-Spechtian polynomials in P_m, and Sp_m^I for the subset $\bigcap_{i \in I} \mathrm{Sp}_m^i$ of polynomials that vanishes when 1 is substituted for x_i, for any $i \in I$. In particular $\mathrm{Sp}_m^{\emptyset} = P_m$. We write $\mathrm{Sp}_m = \mathrm{Sp}_m^{\{1,\dots,m\}}$ for the set of **Spechtian** polynomials (also called **proper** in the literature). The standard polynomial s_m is Spechtian.

Definition 2.4 Define **higher commutator** inductively, as a commutator $[f, g]$ of either letters or higher commutators.

In the proof of [2, Proposition 6.2.1], by specializing x_i to 1, we see that a polynomial f can be written as $f_1 + f_2$ where x_i does not appear in f_1 and f_2 is i-Spechtian. It follows that f is Spechtian if and only if it is a sum of products of higher commutators.

We write $\mathrm{id}_{\mathrm{Sp}}(A)$ for the subset of Spechtian identities of A and $\mathrm{id}_{m,\mathrm{Sp}}(A)$ for $\mathrm{Sp}_m \cap \mathrm{id}(A)$.

In [2, Corollary 6.2.2] it is shown that any base of identities can be comprised of Spechtian identities.

3 Weak Identities

3.1 Weak and Strong Variables

We refine Definition 2.1 with respect to the matrix algebra $A = \mathrm{M}_n(F)$.

Definition 3.1 Let $p(x_1, \dots, x_m)$ be an ℓ-multilinear polynomial. We say that p is an ℓ-**weak identity** of A if it vanishes under every substitution of matrices of trace 0 in $x_1, \dots, x_\ell$ and arbitrary matrices in the other variables.

More generally, for $I \subseteq \{1, \dots, m\}$, we say that p is an I-**weak identity** of A if it vanishes under every substitution of matrices of trace 0 in $\{x_i : i \in I\}$ and arbitrary matrices in the other variables (in this context we say that x_i, $i \in I$ are **weak** variables in p, while x_i, $i \notin I$ are **strong**).

We write $\mathrm{id}_m^I = \mathrm{id}_m^I(A)$ for the set of I-weak multilinear identities of degree m. In particular, a 0-weak identity is simply an identity, namely $\mathrm{id}_m^{\emptyset} = \mathrm{id}_m$. On the other extreme, if p is m-weak we omit the prefix and say that p is a weak identity. For $I \subseteq J$ we have that $\mathrm{id}_m^I \subseteq \mathrm{id}_m^J$ and $\mathrm{Sp}_m^I \supseteq \mathrm{Sp}_m^J$.

Lemma 3.2 *Assume* char F *does not divide* n.

(1) $\mathrm{id}_m^I \cap \mathrm{Sp}_m^J \subseteq \mathrm{id}_m^{I \setminus J}$ *for every* $I, J \subseteq X$.
(2) $\mathrm{id}_m^I(A) \cap \mathrm{Sp}_m \subseteq \mathrm{id}_m$ *for every* I.
(3) *A weak identity which is a Specht polynomial is in fact an identity.*

Proof

(1) Let $\mathrm{M}_n(F)_0 = \{a \in \mathrm{M}_n(F) \mid \mathrm{tr}(a) = 0\}$. Since $\mathrm{M}_n(F) = F \cdot 1 \oplus \mathrm{M}_n(F)_0$, the condition for an I-weak identity $f \in \mathrm{id}_m^I$ to be in $\mathrm{id}_m^{I \setminus J}$ is that for every $j \in I \cap J$, substitution $x_j \mapsto 1$ sends f to an identity.
(2) Take $J = \{1, \dots, m\}$ in (1).
(3) Take $I = \{1, \dots, m\}$ in (2) to obtain $\mathrm{id}_m^m(A) \cap \mathrm{Sp}_m = \mathrm{id}_m$. □

3.2 Modules of Weak Identities

Write id_m^ℓ for $\mathrm{id}_m^{\{1,\dots,\ell\}}$, the set of ℓ-weak identities. We clearly have

$$\mathrm{id}_m(A) = \mathrm{id}_m^0(A) \subseteq \mathrm{id}_m^1(A) \subseteq \cdots \subseteq \mathrm{id}_m^m(A). \quad (1)$$

Following the Amitsur–Levitzki theorem [1], it is known that the minimal identities appear in $\mathrm{id}_m(\mathrm{M}_n(F))$ for $m = 2n$, where this space is 1-dimensional. As a refinement, it is desirable to describe the chain (1), at least for the minimal m for which it is nonzero.

Note that $\mathrm{id}_m^\ell(A)$ is not a submodule of P_m for $0 < \ell < m$, since a permutation could send a weak indeterminate to a strong indeterminate.

Remark 3.3 The space of ℓ-weak identities is a module through the natural action on weak and strong variables over the ring $F[S_\ell \times S_{m-\ell}] \cong F[S_\ell] \otimes F[S_{m-\ell}]$, which is semisimple when char $F = 0$, being a direct sum of matrix rings over F.

In particular $\mathrm{id}_m^m(A)$ and $\mathrm{id}_m^0(A)$ are S_m-modules, which can be described through their irreducible decompositions.

The level of details in a description of $\mathrm{id}_m^\ell(A)$ is a matter of taste. In increasing level of details, such a description might include:

(1) An indication that the space is nonempty (for m minimal).
(2) The dimension of the space, possibly given by a computer program.
(3) Better still would be explicit identities, preferably ones that can be understood and demonstrated to be identities (and not just computer verified).
(4) Computations in the module $\mathrm{id}_m^\ell(A)$ can be facilitated by generators and relations. Or, more generally, the module can be endowed with a resolution of permutation modules (defined through the action on indices in a generating set).
(5) A decomposition into irreducible submodules is not hard to obtain for small m, although our experience [13] and [14] show that by itself it is not very illuminating.
(6) Finally, it is desirable to explicitly exhibit the embedding $\pi_m^{\ell-1}(A) \hookrightarrow \pi_m^\ell(A)$.

In order to study the chain of weak identity spaces (1), we compare two consecutive chains.

Remark 3.4 The substitution map $x_\ell \mapsto 1$ defines a projection $\pi_\ell : P_m \to P_{m-1}$ (reducing the indices $\ell' > \ell$ by one), which induces the maps

$$\begin{array}{ccccccccccccc}
\mathrm{id}_m^0(A) & \subseteq & \mathrm{id}_m^1(A) & \subseteq \cdots \subseteq & \mathrm{id}_m^{\ell-1}(A) & \subseteq & \mathrm{id}_m^\ell(A) & \subseteq \cdots \subseteq & \mathrm{id}_m^m(A) & \subseteq & P_m \\
\downarrow & & \downarrow & & \downarrow & & \downarrow & & \downarrow & & \downarrow{\scriptstyle \pi_\ell} \\
\mathrm{id}_{m-1}^0(A) & \subseteq & \mathrm{id}_{m-1}^1(A) & \subseteq \cdots \subseteq & \mathrm{id}_{m-1}^{\ell-1}(A) & \subseteq & P_{m-1} & = \cdots = & P_{m-1} & = & P_{m-1}
\end{array}$$

Indeed, for every $k < \ell$, if $p \in \mathrm{id}_m^k(A)$ then $p(x_1, \ldots, x_k, \ldots, 1, \ldots, x_m)$ is a k-weak identity of degree $m - 1$, so the downwards arrows are defined.

Even more is true:

Remark 3.5 Assume char F is prime to m. For $\ell \leq m$,

$$\mathrm{id}_m^{\ell-1}(A) = \mathrm{id}_m^{\ell}(A) \cap \pi_\ell^{-1}(\mathrm{id}_{m-1}^{\ell-1}(A)).$$

Indeed, if $p \in \mathrm{id}_m^{\ell}(A)$ and $\pi_\ell(p) \in \mathrm{id}_{m-1}^{\ell-1}(A)$, then as long as $x_1, \dots, x_{\ell-1}$ are weak variables in p, x_ℓ is weak by the former assumption, and becomes strong by the latter.

We thus have an inductive procedure to compute the chain (1): once the chain was computed in degree $m-1$, the chain in degree m can be computed from $\mathrm{id}_m^m(A)$ by reverse induction on ℓ. In order to apply the condition $\pi_\ell(p) \in \mathrm{id}_{m-1}^{\ell-1}(A)$, we will need a hold on $\pi_\ell(\mathrm{id}_m^{\ell}(A)) \subseteq P_{m-1}$, whose elements in general are not even weak identities. For example, π_ℓ induces an embedding $\pi_\ell : \mathrm{id}_m^{\ell}(A)/\mathrm{id}_m^{\ell-1}(A) \hookrightarrow P_{m-1}/\mathrm{id}_{m-1}^{\ell-1}(A)$ which bounds the dimension of $\mathrm{id}_m^{\ell-1}(A)$ from below in terms of previously known quantities:

$$\dim(\mathrm{id}_m^{\ell-1}(A)) \geq \dim(\mathrm{id}_m^{\ell}(A)) - [(m-1)! - \dim(\mathrm{id}_{m-1}^{\ell-1}(A))].$$

For the minimal degree we can state this procedure more explicitly:

Remark 3.6 Assume char F is prime to n. Assume m is the minimal degree of a weak identity for A. Then for every $\ell < m$,

$$\mathrm{id}_m^{\ell}(A) = \left\{f \in \mathrm{id}_m^m(A) \mid \pi_{\ell+1}(f) = \cdots = \pi_m(f) = 0\right\}.$$

4 Central Polynomials for Matrices

The polynomials comprising a base of the T-ideal are hard to ascertain, unknown even for $\mathrm{M}_3(\mathbb{Q})$. So we look for minimal identities (e.g., s_{2n} for $\mathrm{M}_n(\mathbb{Q})$) and central polynomials. Surprisingly, even the minimal possible degree of a nonidentity which is a 1-weak identity (and thus provides a strict central polynomial, see Theorem 4.3 below) for $\mathrm{M}_n(F)$ is not known in general.

Halpin found an example of a central polynomial:

Lemma 4.1 ([2, Lemma 1.4.14]) *The multilinearization of*

$$s_{n-1}([x,y],[x^2,y],\dots,[x^{n-2},y],[x^n,y])$$

is an $\frac{n^2-n+2}{2}$*-weak identity of* $\mathrm{M}_n(F)$*, of degree*

$$\frac{n^2-n+2}{2} + n - 1 = \frac{n^2+n}{2} = \frac{n(n+1)}{2}.$$

As explained in [2, p. 37], this yields a 1-weak identity of total degree n^2:

Remark 4.2 For $0 \leq \ell' < \ell$, every ℓ-weak identity of degree m can be viewed as an ℓ'-weak identity of degree $m + (\ell - \ell')$, by substituting $x_i \mapsto [x_i', x_i'']$ for $i = \ell' + 1, \dots, \ell$. In particular every ℓ-weak identity of degree m can be viewed as an identity of degree $m + \ell$.

However, the 1-weak identity resulting from Halpin's polynomial is not an identity of $\mathrm{M}_n(\mathbb{Q})$. We thus have the existence of strict central polynomials. Formanek's polynomial [7] also has degree n^2, and for some time this was thought the lowest possible, but in 1983, 1985, Drensky and Kasparian [5] discovered by a computer search a strict central polynomial for $M_3(\mathbb{Q})$ of degree 8, further explained in terms of weak identities by Drensky and Kasparian in 1993. Drensky showed 8 is optimal for $n = 3$. The space of central polynomials of degree 8 is described in [13]: the rank of $\mathrm{id}_8(\mathrm{M}_3(F))$ is 43; the Drensky-Kasparian identity adds 2 to the rank; and the full rank of $\text{c-id}_8(\mathrm{M}_3(F))$ is 47.

In 1994 Drensky and Piacentini found a strict central polynomial for $\mathrm{M}_4(\mathbb{Q})$ of degree 13, also obtainable via weak identities. In 1995 Drensky [3] discovered a strict central polynomial for arbitrary $\mathrm{M}_n(\mathbb{Q})$ of degree $(n-1)^2+4$, which is minimal for $n = 3$ and $n = 4$, but its uniqueness is still open for $n = 4$, and minimality of degree is open for $n > 4$. We treat $n = 3$ in Sect. 8.

4.1 ℓ-Weak Central Polynomials

Similarly to Definition 3.1, a polynomial p of degree m is an ℓ**-weak central polynomial** of $\mathrm{M}_n(F)$ if it takes central values under the substitutions of $x_1, \dots, x_\ell$ to matrices of trace zero and $x_{\ell+1}, \dots, x_m$ to arbitrary matrices. More generally, p is an I**-weak central polynomial**, for $I \subseteq \{1, \dots, m\}$, if it takes central values under substitution of matrices provided that x_i maps to a zero trace matrix for all $i \in I$.

In particular, a 0-weak central polynomial is simply a central polynomial. On the other extreme, if p is m-weak we omit the prefix and say that p is a weak central polynomial.

Also let $\text{c-id}_m^\ell(A)$ be the space of ℓ-weak central polynomials of A, so that

$$\text{c-id}_m(A) = \text{c-id}_m^0(A) \subseteq \text{c-id}_m^1(A) \subseteq \cdots \subseteq \text{c-id}_m^m(A) \tag{2}$$

contains (1) component-wise. A natural question is to ask what is the minimal m for which $\mathrm{id}_m(A) \subset \text{c-id}_m(A)$.

By Razmyslov (cf. [2, Lemma 1.4.16]), central polynomials can be obtained from 1-weak identities, trading a weak variable in an identity for a strong variable in a central identity. We can copy the proof to get a more general result.

Let $p(x) = \sum a_i x b_i$ be a polynomial which is multilinear in x, where a_i, b_i are monomials over F in some variables other than x. We denote $p^*(x) = \sum b_i x a_i$, which defines an involution. For new variables y, z, consider $q(y,z) = p([y,z]) = \sum(a_i yzb_i - a_i zyb_i)$. Conjugating $q(y,z)$ with respect to y, we have that $q^*(y,z) = \sum(zb_i ya_i - b_i ya_i z) = \sum[z, b_i ya_i] = [z, p^*(y)]$. Therefore $p(x)$ is a weak identity in terms of x if and only if $q(y,z)$ is identically zero, if and only if $q^*(y,z) = [z, p^*(y)]$ is identically zero, if and only if all values of $p^*(y)$ are central. This procedure respects restrictions, such as zero trace, on any other variable involved. We thus proved a major result:

Theorem 4.3 (Razmyslov) *For $\ell \geq 1$, there is a degree-preserving one-to-one correspondence* $\mathrm{id}_m^{\ell}(A) \to \text{c-id}_m^{\ell-1}(A)$ *between ℓ-weak identities and $(\ell-1)$-weak central polynomials, given by $f \mapsto f^*$ (pivoting around x_ℓ).*

Consequently, we have a chain of isomorphisms between the components of the chains (1) and (2), albeit with non-commuting squares:

$$\begin{array}{ccccccccc}
\mathrm{id}_m^0(A) \subseteq \mathrm{id}_m^1(A) & \subseteq \cdots \subseteq & \mathrm{id}_m^{\ell}(A) & \subseteq & \mathrm{id}_m^{\ell+1}(A) & \subseteq \cdots \subseteq & \mathrm{id}_m^{m}(A) & & \\
\downarrow\cong & & \downarrow\cong & & \downarrow\cong & & \downarrow\cong & & \\
\text{c-id}_m^0(A) & \subseteq \cdots \subseteq & \text{c-id}_m^{\ell-1}(A) & \subseteq & \text{c-id}_m^{\ell}(A) & \subseteq \cdots \subseteq & \text{c-id}_{m-1}^{m}(A) & \subseteq & \text{c-id}_m^{m}(A)
\end{array}$$

Moreover, $\mathrm{id}_m^{\ell}(A)$ is an $(S_\ell \times S_{m-\ell})$-module, and $\text{c-id}_m^{\ell-1}(A)$ is an $(S_{\ell-1} \times S_{m-(\ell-1)})$-module. The groups intersect in the common stabilizer of the pivot variable x_ℓ, which is $S_{\ell-1} \times S_1 \times S_{m-\ell}$, and the isomorphism of Theorem 4.3 is of modules over this group.

5 The Connection to the Representation Theory

We view $\mathrm{id}_m(\mathrm{M}_n(F))$ as a module over S_m, and apply the representation theory of the group to obtain symmetrical identities (the same considerations holds for $\mathrm{id}_m^{\ell}(\mathrm{M}_n(F))$ over $S_\ell \times S_{m-\ell}$).

5.1 *Identities and the Group Algebra*

Given a multilinear polynomial

$$\sum_{\sigma \in S_m} a_\sigma x_{\sigma(1)} \ldots x_{\sigma(m)} \in P_m,$$

we may associate it with the element

$$\sum_{\sigma \in S_m} a_\sigma \sigma$$

of the group algebra $F[S_m]$.

The action of S_m on P_m translates to the usual multiplication in the group algebra. A natural left action of S_m on $F\{x_1, \ldots, x_m\}$ is defined by $\sigma(x_i) = x_{\sigma(i)}$, which induces an action of S_m on P_m by

$$(\sigma \cdot f)(x_1, \ldots, x_m) = f(x_{\sigma(1)}, \ldots, x_{\sigma(m)})$$

for all $\sigma \in S_m$ and $f \in F\{x_1, \ldots, x_m\}$, making P_m a cyclic faithful S_m-module. But $F[S_m]$ is semisimple by Maschke's Theorem (assuming char $F = 0$ or char $F > m$), so the module P_m is semisimple, and decomposes as a direct sum of simple submodules, some of which are generated by PIs of $M_n(F)$.

Each irreducible component of $F[S_m]$ corresponds to a partition λ of m. We denote the matrix subring corresponding to λ by Type_λ. We also denote the irreducible module corresponding to λ by Irr_λ. Notice that while Type_λ is a uniquely defined subset of $F[S_m]$ (and by identification, of P_m), Irr_λ is only defined up to isomorphism, as the decomposition of Type_λ into $\dim(\text{Irr}_\lambda)$ copies of Irr_λ is not unique.

Remark 5.1 The set Sp_m of Spechtian polynomials of degree m is a submodule of P_m.

Proof It is closed under the action. □

Being submodules of P_m, $\text{id}_{m,\text{Sp}}(A) \subseteq \text{id}_m(A)$ both are direct sums of minimal left ideals.

Given a submodule $L \leq P_m$, the corresponding subspace $\hat{L}$ of $F[S_m]$ is a left ideal. Since $F[S_m]$ is semisimple, $\hat{L}$ may be written as

$$\hat{L} = \bigoplus_{\lambda \vdash m} (\hat{L} \cap \text{Type}_\lambda).$$

We call each $\hat{L} \cap \text{Type}_\lambda$ the projection of L to λ.

5.2 *Identities and Representations*

While we may be able to decompose the weak identities ideal quite nicely using representation theory, it is not obvious that each projection has an "elegant" representative. The following proposition proves the existence of a relatively simple one.

Proposition 5.2 *Let L be a submodule of P_m. Suppose the projection of L on a partition $\lambda = (\lambda_1, \dots, \lambda_r) \vdash m$ is nonzero. Then there exists a nonzero multilinear polynomial $f(x_1, \dots, x_m) \in L$ which is fixed under the action of*

$$H = S_{\{1,\dots,\lambda_1\}} \times S_{\{\lambda_1+1,\dots,\lambda_1+\lambda_2\}} \times \cdots \times S_{\{\lambda_1+\cdots+\lambda_{r-1}+1,\dots,m\}}.$$

In other words, f is a multilinearization of a polynomial in r (noncommuting) variables $y_1, \dots, y_r$, where the degree of y_i in each monomial is λ_i.

Proof Recall that $P_m \cong F[S_m]$. Let $\hat{L}$ be the left ideal of $F[S_m]$ corresponding to L, and let $\hat{L}_\lambda = \hat{L} \cap \mathrm{Type}_\lambda$ be the projection of L on λ, which is a left ideal of Type_λ.

Following the notation of [9, Section 3.3], associate to λ the subgroups P_λ and Q_λ of S_m, fixing the rows and columns respectively in the standard tableau corresponding to λ. We also set

$$a_\lambda = \sum_{\sigma \in P_\lambda} \sigma, \quad b_\lambda = \sum_{\sigma \in Q_\lambda} (-1)^\sigma \cdot \sigma, \quad \text{and } c_\lambda = a_\lambda b_\lambda.$$

Then $c_\lambda F[S_m]$ is an irreducible module V_λ of $F[S_m]$, contained in the representation type Type_λ. In particular, $c_\lambda \in \mathrm{Type}_\lambda$. The elements fixed under the action of the above subgroup H of S_m are precisely the elements t such that $a_\lambda t = |H|t$. Since $a_\lambda^2 = |H|a_\lambda$, we conclude that $a_\lambda c_\lambda = |H|c_\lambda$, and thus every element of the right ideal $c_\lambda \mathrm{Type}_\lambda$ of Type_λ is fixed under H. Take any $0 \neq f \in \hat{L}_\lambda \cap c_\lambda \mathrm{Type}_\lambda$, which exists because left and right ideals in the prime ring Type_λ intersect nontrivially. □

6 Weak Identities and the Case $n = 2$

Our goal in this section is to describe the minimal (and next to minimal) ℓ-weak identities for the matrix algebra $\mathrm{M}_2(F)$, exemplifying the approach described in Remark 3.5.

6.1 Polynomials of Degree $m = 2$

Write $a \circ b = ab + ba$. Although the PI-degree of $\mathrm{M}_2(F)$ is 4, the Wagner identity provides a weak central polynomial of degree 2, namely $x_1 \circ x_2$. Nevertheless, the space of 1-weak central polynomials of degree 2 is trivial.

6.2 Weak Identities of Degree $m = 3$

The first nonzero instance of the chain (1) occurs for $m = 3$. Let

$$\psi_i = [x_i, x_j \circ x_k],$$

where $\{i, j, k\}$ is a permutation of the index set $\{1, 2, 3\}$. All the ψ_i are 3-weak identities, and ψ_3 is in fact 2-weak. We also observe that

$$\psi_1 + \psi_2 + \psi_3 = 0. \tag{3}$$

Therefore

$$0 = \mathrm{id}_3^0(\mathrm{M}_2(F)) \quad = \quad \mathrm{id}_3^1(\mathrm{M}_2(F)) \;\subset\; \mathrm{id}_3^2(\mathrm{M}_2(F)) \quad \subset \quad \mathrm{id}_3^3(\mathrm{M}_2(F)) \;, \tag{4}$$

where $\mathrm{id}_3^3(\mathrm{M}_2(F)) = \langle \psi_1, \psi_2, \psi_3 \rangle$ is 2-dimensional ($\cong \mathrm{Irr}_{(2,1)}$), and $\mathrm{id}_3^2(\mathrm{M}_2(F)) = \langle \psi_3 \rangle$ is 1-dimensional.

Anticipating the computation of $\mathrm{id}_4^\ell(\mathrm{M}_2(F))$ through Remark 3.5, let us further point out specific submodules of P_3. For an even permutation i, j, k of $1, 2, 3$, let

$$g_i = x_i[x_j, x_k], \qquad g_i' = [x_i, x_j]x_k,$$

and $G = \langle g_1, g_2, g_3 \rangle$, $G' = \langle g_1', g_2', g_3' \rangle$ the generated submodules. Observing that $g_1 + g_2 + g_3 = s_3 = g_1' + g_2' + g_3'$ generates the intersection $G \cap G'$, we conclude that

$$G \cong G' \cong \mathrm{Irr}_{(2,1)} \oplus \mathrm{Irr}_{(1,1,1)}$$

(the latter component is the sign representation). It follows that $G + G' = \mathrm{Type}_{(2,1)} \oplus \mathrm{Type}_{(1,1,1)}$ is the complement of $\langle \sum x_{\sigma 1} x_{\sigma 2} x_{\sigma 3} \rangle = \mathrm{Type}_{(3)}$ in P_3.

6.3 Weak Identities of Degree $m = 4$

We now consider the chain

$$\mathrm{id}_4^0(\mathrm{M}_2(F)) \subset \mathrm{id}_4^1(\mathrm{M}_2(F)) \subset \mathrm{id}_4^2(\mathrm{M}_2(F)) \subset \mathrm{id}_4^3(\mathrm{M}_2(F)) \subset \mathrm{id}_4^4(\mathrm{M}_2(F)). \tag{5}$$

For a permutation i, j, a, b of $1, 2, 3, 4$, let

$$h_{ij} = x_i[x_a \circ x_b, x_j], \qquad h'_{ij} = [x_j, x_a \circ x_b]x_i,$$

on which S_4 acts by the natural action on the indices. Both are weak identities, immediate consequences of the Wagner identity ψ_j. Let $H = \langle h_{ij} \mid i \neq j \rangle$ and $H' = \langle h'_{ij} \mid i \neq j \rangle$ be the generated submodules of P_4.

Proposition 6.1 *The space of weak identities* $\mathrm{id}_4^4(\mathrm{M}_2(F))$ *has dimension* 15, *isomorphic to* $2\mathrm{Irr}_{[2,1,1]} \oplus 2\mathrm{Irr}_{[3,1]} \oplus \mathrm{Irr}_{[2,2]} \oplus \mathrm{Irr}_{[1,1,1,1]}$. *It is generated as a module by* s_4, $h_{34} = x_3[x_1 \circ x_2, x_4]$, *and* $h'_{34} = [x_4, x_1 \circ x_2]x_3$.

Proof We apply a computer program to find the dimension as described in [13], which is indeed 15. We then guess and verify easy-to-describe identities in this space; and analyze the submodule they generate to the extent that its dimension becomes apparent, until we obtain a set of generators.

For every i, it follows from (3) that $\sum_{j \neq i} h_{ij} = \sum_{j \neq i} h'_{ij} = 0$. There are no other relations, so $\dim H = \dim H' = 8$. But since $8 + 8 > 15$, the spaces must intersect. The intersection is most easily computed by passing to the dual space. Elements $\sum \alpha_\sigma \sigma \in H$ are characterized by the "right transposition condition" $\alpha_{ijk\ell} + \alpha_{i\ell kj} = 0$ and the condition $\alpha_{ij_0j_1j_2} + \alpha_{ij_1j_2j_0} + \alpha_{ij_2j_0j_1} = 0$. Likewise H' is characterized by the "left transposition condition" $\alpha_{ijk\ell} + \alpha_{jki\ell} = 0$ and the condition $\alpha_{i_0i_1i_2j} + \alpha_{i_1i_2i_0j} + \alpha_{i_2i_0i_1j} = 0$. So $H \cap H'$ is characterized by the transposition conditions, as well as $\alpha_{ijk\ell} = \alpha_{ji\ell k}$ and $\alpha_{1234} + \alpha_{2314} + \alpha_{3124} = 0$; computation then indicates that $\dim(H \cap H') = 2$. Indeed, acting with $\sum_{\sigma \in K_4} \sigma$, where K_4 is the Klein 4-group, we find the equality $h_{ij} + h_{ji} + h_{k\ell} + h_{\ell k} = h'_{ij} + h'_{ji} + h'_{k\ell} + h'_{\ell k}$ for any partition $ij|k\ell$ of the index set. These are three equalities, each defining an element of $H \cap H'$, whose sum is zero. Thus $H \cap H' \cong \mathrm{Irr}_{[2,2]}$. The characters of H, H' can be computed from the action on the basis, and knowing the characters of S_4 we conclude that $H \cong H' \cong \mathrm{Irr}_{[2,2]} \oplus \mathrm{Irr}_{[2,1,1]} \oplus \mathrm{Irr}_{[3,1]}$ (of dimensions $2 + 3 + 3$). It follows that $\langle s_4 \rangle \cong \mathrm{Irr}_{[1,1,1,1]}$ cannot intersect $H + H'$, so that $H + H' + \langle s_4 \rangle$ is of dimension 15, and thus equal to the full space of identities. □

Remark 6.2 The dimensions in the chain (5) are $1 < 3 < 8 < 12 < 15$. The ℓ-weak identity spaces are given as follows.

(3) The space $\mathrm{id}_4^3(\mathrm{M}_2(F))$ of 3-weak identities has dimension 12, spanned as an $S_{\{1,2,3\}}$-module by $\{[s_3, x_4], h_{43}, h_{34}, t\}$, where

$$t = [x_1 \circ [x_2, x_4], x_3].$$

We have a direct sum decomposition, $\langle [s_3, x_4] \rangle \oplus \langle h_{43} \rangle \oplus \langle h_{34} \rangle \oplus \langle t \rangle$, with the components isomorphic to $\mathrm{Irr}_{[1,1,1]}$, $\mathrm{Irr}_{[2,1]}$ (as $h_{43} + h_{42} + h_{41} = 0$), $\mathrm{Irr}_{[3]} \oplus \mathrm{Irr}_{[2,1]}$, and the regular representation, respectively. Namely, $\mathrm{id}_4^3(\mathrm{M}_2(F))$ is twice the regular representation. We also note that $[s_3, x_4] = \frac{1}{2}(1 + (123) + (132))(34)t$.

(2) The space $\mathrm{id}_4^2(\mathrm{M}_2(F))$ of 2-weak identities has dimension 8, spanned as an $S_{\{1,2\}}S_{\{3,4\}}$-module by $\{s_4, t, h_{34}, q\}$, where $q = [x_1 \circ x_3, x_2 \circ x_4] + [x_2 \circ x_3, x_1 \circ x_4]$. In fact, $\mathrm{id}_4^2 = \langle s_4\rangle \oplus \langle t\rangle \oplus \langle h_{34}\rangle \oplus \langle q\rangle$, of dimensions 1+4+2+1 respectively.

(1) $\mathrm{id}_4^1(\mathrm{M}_2(F))$ is the 3-dimensional space spanned as an $S_{\{2,3,4\}}$-module by $(34)t = [x_1 \circ [x_2, x_3], x_4]$. This is a 1-weak identity, $x_1 \circ [x_2, x_3]$ being central when $\mathrm{tr}(x_1) = 0$. In fact, $(34)t + (24)t + (23)t = s_4$, explaining how $\mathrm{id}_4^0 \subset \mathrm{id}_4^1$.

(0) $\mathrm{id}_4^0(\mathrm{M}_2(F)) = F \cdot s_4$ is the well-known 1-dimensional space of degree 4 identities.

Remark 6.3 The spaces of ℓ-weak central polynomials of $\mathrm{M}_2(F)$ in degree 4, for $\ell = 0, 1, 2, 3, 4$, have dimensions 3, 8, 12, 15 and 18, respectively.

(The dimensions $3 < 8 < 12 < 15$ follow from Remark 6.2 by Theorem 4.3; and the dimension 18 for the space of weak central polynomials was found, once more, by a computer program).

7 The Weak PI-Degree of $\mathrm{M}_3(F)$

This section is concerned with weak identities of degree 5 for $\mathrm{M}_3(F)$. We show that there are none if $\mathrm{char}\, F \neq 3$, and describe the weak identities in degree 5 when $\mathrm{char}\, F = 3$.

7.1 Fields of Characteristic Not 3

Proposition 7.1 *The algebra* $\mathrm{M}_3(F)$ *has no weak identities of degree* 5 *when* $\mathrm{char}\, F \neq 3$.

Proof Suppose that

$$f(x_1, \ldots, x_5) = \sum_{\sigma \in S_5} a_\sigma x_{\sigma(1)} \cdots x_{\sigma(5)}$$

is a weak identity for $\mathrm{M}_3(F)$. Note that for all $\pi \in S_5$,

$$f(x_{\pi(1)}, \ldots, x_{\pi(5)}) = \sum_{\sigma \in S_5} a_\sigma x_{\pi(\sigma(1))} \cdots x_{\pi(\sigma(5))} = \sum_{\tau \in S_5} a_{\pi^{-1}\tau} x_{\tau(1)} \cdots x_{\tau(5)},$$

so permutation of the variables acts on the coefficients from the right by $a_\sigma \cdot \pi = a_{\pi^{-1}\sigma}$. We write permutations by the cycle decomposition.

Substituting $x_1, \dots, x_5 = e_{12}, e_{23}, e_{32}, e_{23}, e_{31}$, the resulting matrix satisfies

$$f(e_{12}, e_{23}, e_{32}, e_{23}, e_{31})_{1,1} = a_1 + a_{(2,4)}.$$

Hence $a_{(2,4)} = -a_1$. Applying a permutation $\pi \in S_5$ yields

$$a_{\pi(2,4)} = -a_\pi \tag{6}$$

for every $\pi \in S_5$.

Next, we substitute $x_1, \dots, x_5 = e_{13}, e_{31}, e_{12}, e_{23}, e_{32}$, and the $(1, 2)$ entry of the resulting matrix is

$$a_1 + a_{(2,5,3,4)} + a_{(1,3,2,4)} = 0.$$

Using (6) and acting with an arbitrary $\pi \in S_5$, we get

$$a_\pi - a_{\pi(3,4,5)} - a_{\pi(1,3,2)} = 0. \tag{7}$$

Tracing this equation over $(3, 4, 5)$ (that is applying $(3, 4, 5)$ and $(3, 5, 4)$, then summing the three equations) and applying $(1, 2, 3)$ yields the equation

$$a_1 + a_{(1,4,5)} + a_{(1,5,4)} = 0. \tag{8}$$

We now substitute $x_1, \dots, x_5 = e_{13}, e_{32}, e_{23}, e_{22} - e_{33}, e_{31}$. The $(1, 1)$ entry of the resulting matrix is

$$-a_1 + a_{(3,4)} - a_{(2,4,3)} = 0.$$

Using (6), we see that

$$a_1 - a_{(3,4)} - a_{(2,3)} = 0.$$

Applying $(1, 3)$ yields the equation $a_{(1,3,2)} = a_{(1,3)(2,3)} = a_{(1,3)} - a_{(1,3,4)}$. We substitute this expression in (7) (with $\pi = \mathrm{Id}$) to achieve

$$a_1 - a_{(3,4,5)} - a_{(1,3)} + a_{(1,3,4)} = 0.$$

By applying $(1, 3)$ on the last equation, we get

$$a_{(1,3)} - a_{(1,3,4,5)} - a_1 + a_{(3,4)} = 0.$$

Summing up the last two equations, we get

$$-a_{(3,4,5)} + a_{(1,3,4)} - a_{(1,3,4,5)} + a_{(3,4)} = 0.$$

Applying $(3, 4)$ means

$$a_1 + a_{(1,4)} - a_{(4,5)} - a_{(1,4,5)} = 0.$$

Applying $(1, 5)$ yields the equation

$$a_{(1,5)} + a_{(1,4,5)} - a_{(1,5,4)} - a_{(1,4)} = 0.$$

Subtracting the second equation from the first, we see that

$$a_1 - 2a_{(1,4,5)} + a_{(1,5,4)} = -2a_{(1,4)} + a_{(1,5)} + a_{(4,5)}.$$

So, using (8),

$$3a_{(1,4,5)} = 3a_{(1,4)},$$

and $a_1 = a_{(4,5)}$ since we assume char $F \neq 3$. We may again apply $\pi \in S_5$ to get

$$a_{\pi(4,5)} = a_{\pi}. \tag{9}$$

We now see that using (6) and (9),

$$a_{\pi(2,5)} = a_{\pi(2,4)(4,5)(2,4)} = a_{\pi},$$

but also

$$a_{\pi(2,5)} = a_{\pi(4,5)(2,4)(4,5)} = -a_{\pi},$$

implying that $a_{\pi} = 0$ for all $\pi \in S_5$. Hence $f = 0$, as required. □

Since there are identities of degree 6, we conclude that the "weak PI degree" of $M_3(F)$ is 6:

Corollary 7.2 *The minimal degree of a weak identity of* $M_3(F)$ *is* 6.

In Sect. 8 we indicate that in degree 6 there are weak identities other than the standard identity, so the "strict weak PI degree" of $M_3(F)$ is 6 as well.

7.2 *The Case* char $F = 3$

Proposition 7.1 holds when char $F \neq 3$. Interestingly, the situation is quite different in characteristic 3.

Proposition 7.3 *Assume* char $F = 3$. *The standard identity* s_4 *is a weak central identity of* $\mathrm{M}_3(F)$. *In particular* $\mathrm{M}_3(F)$ *has* 4*-weak identity of degree* 5*, namely*

$$[s_4(x_1, \ldots, x_4), x_5].$$

Proof The value of $s_4(x_1, \ldots, x_4)$ under substitution of matrix units e_{ij} $(i \neq j)$ or matrices of the form $e_{ii} - e_{jj}$, results in either $\pm 3e_{ij}$ $(i \neq j)$ or $\pm(1 - 3e_{ii})$. Over a field of characteristic 3, this implies all values of s_4 under weak substitutions are central. Hence $[s_4(x_1, \ldots, x_4), x_5]$ is a 4-weak identity.

(Incidentally, if even one variable is strong, the $\mathbb{Z}$-span of $s_4(x_1, \ldots, x_4)$ is the zero-trace part of $\mathrm{M}_3(\mathbb{Z})$; so $[s_4(x_1, \ldots, x_4), x_5]$ is not 3-weak). □

For any m, let $\psi_m = [s_{m-1}(x_1, \ldots, x_{m-1}), x_m]$. Let $F \oplus N_0$ be the natural representation of S_m, decomposed into the trivial module and its irreducible complement.

Proposition 7.4 *The* S_m*-module generated by* ψ_m *is:*

(1) $F[S_m]\psi_m \cong N_0 \otimes \mathrm{sgn}$ *when* m *is odd.*
(2) $F[S_m]\psi_m \cong (F \oplus N_0) \otimes \mathrm{sgn}$ *when* m *is even.*

Proof Fix $\sigma = (1\,2\,3\,\ldots m)$. Since S_{m-1} alternates ψ_m, the module is generated by the cyclic permutations $\sigma^j \psi_m$.

Every monomial appears in exactly two of the polynomials $\sigma^j \psi_m$. When m is odd, the signs are opposite. Therefore $\sum \sigma^j \psi_m = 0$ and there are no other relations, so the module is $N_0 \otimes$ sgn. When m is even, the signs are equal (opposite) when the difference of the indices of the first and last variables is even (odd); so the $\sigma^j \psi_m$ are linearly independent, and the module is $N \otimes$ sgn. □

Going back to the case $m = 5$ when char $F = 3$,

$$U = F[S_5] \cdot [s_4(x_1, x_2, x_3, x_4), x_5] \tag{10}$$

is 4-dimensional, isomorphic as an S_5-module to the nontrivial irreducible component of the natural representation, tensored with the sign character.

Proposition 7.5 *Assume* char $F = 3$. *The space* $\mathrm{id}_5^5(\mathrm{M}_3(F))$ *of weak identities of degree* 5 *has dimension* 5. *As an* S_5*-module, the representation space is uniquely an extension*

$$0 \longrightarrow U \longrightarrow \mathrm{id}_5^5(\mathrm{M}_3(F)) \longrightarrow F \longrightarrow 0$$

where U *is given in (10) and* F *denotes the trivial module.*

Proof The dimension is based on a Sage program. We find the 4-weak identity

$$\varphi = [x_1[x_2, x_3 \circ x_4] + x_2[x_1, x_3 \circ x_4] - x_3[x_4, x_1 \circ x_2] - x_4[x_3, x_1 \circ x_2],\ x_5] + \\ + \sum_{\sigma \in S_4} x_{\sigma(1)}[x_5,\ x_{\sigma(2)} x_{\sigma(3)}] x_{\sigma(4)},$$

generating $\mathrm{id}_5^5(\mathrm{M}_3(F))$ as a module; indeed, $\psi_5 = (1-(23))\varphi$. Notice that $(12)\varphi = (34)\varphi = \varphi$, showing that $\mathrm{id}_5^5(\mathrm{M}_3(F))/U$ is the trivial (and not the sign) module. □

A Sage computation also shows that (when $\mathrm{char}\, F = 3$) $\mathrm{id}_5^3(\mathrm{M}_3(F)) = 0$, and $\mathrm{id}_5^4(\mathrm{M}_3(F))$ is 2-dimensional, spanned by φ and ψ_5. Again $F\psi_5$ is the unique irreducible S_4-submodule, and $(F\varphi + F\psi_5)/(F\psi_5)$ is the trivial S_4-module.

8 Weak Identities for $\mathrm{M}_3(F)$ in Degree 6

Assuming $\mathrm{char}\, F = 0$, in this section we describe the sets $\mathrm{id}_6^\ell(\mathrm{M}_3(F))$ of ℓ-weak identities of $\mathrm{M}_3(F)$ in degree 6, which by Corollary 7.2 is the minimal degree of weak identities.

In [6] the authors study weak identities (when all variables are weak, namely the case $\ell = 6$) of $\mathrm{M}_3(F)$. Decomposing the S_6-module $\mathrm{id}_6^6(\mathrm{M}_3(F))$ into the representation components, their computations indicate that there are four nonzero summands, whose Young diagrams are 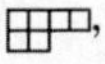, 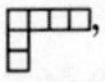, 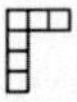 and .

We correct a minor omission in the literature by observing the following:

Proposition 8.1 *The space* $\mathrm{id}_6^6(\mathrm{M}_3(F))$ *has five nonzero components, namely the above four, as well as* .

In the first subsection we supply complete details on the dimensions of the spaces of weak identities, and in the second subsection we present explicit 4-weak identities and use the Okubo algebra to prove that they indeed have this property.

8.1 Weak Identities of $\mathrm{M}_3(F)$

We used a Sage program to find an F-basis for each weak identity space $\mathrm{id}_6^\ell(\mathrm{M}_3(F))$, and compute the intersection with each representation ideal Type_λ. The dimensions of the intersections $\mathrm{id}_6^\ell(\mathrm{M}_3(F)) \cap \mathrm{Type}_\lambda$ (for the partitions λ with nonzero intersection) are listed in the table below. In all participating representations, $\mathrm{id}_6^6(\mathrm{M}_3(F)) \cap \mathrm{Type}_\lambda$ happens to have rank 1, so the dimension of the representation is equal to the dimension of the intersection at the bottom line.

ℓ	$\dim \mathrm{id}_6^{\ell}(\mathrm{M}_3(F))$					
0	1	0	0	0	0	1
1	1	0	0	0	0	1
2	1	0	0	0	0	1
3	2	0	0	1	0	1
4	6	1	1	3	0	1
5	15	4	4	6	0	1
6	35	9	10	10	5	1

It follows that there are no 2-weak identities except for the standard identity; and there is a unique 3-weak identity modulo the standard identity (whose explicit description, in an appealing form, remains a challenge). The bottom line proves Proposition 8.1.

8.2 *Halpin's Identity and Its Projections*

For $n = 3$, Halpin's identity from Lemma 4.1 is

$$f(x, z) = [[x, z], [x^3, z]], \tag{11}$$

which (when multilinearized) is a 4-weak identity of $\mathrm{M}_3(F)$, namely we restrict x to have zero trace.

Proposition 8.2 *The (multilinearization of the) polynomials*

$$f'(x, z_1, z_2) = [[x, z_1], [x^3, z_2]] + [[x, z_2], [x^3, z_1]], \tag{12}$$

and

$$f''(x, z_1, z_2) = [x, z_1] \circ [x^3, z_2] - [x, z_2] \circ [x^3, z_1], \tag{13}$$

are the unique (up to scalar) 4-weak identities of degree 6 *of* $\mathrm{M}_3(F)$ *corresponding to the components*  *and* *, respectively.*

Proof The representation type follows from symmetries, so uniqueness follows from the line $\ell = 4$ in the table above. It remains to show that these are indeed 4-weak identities.

Linearizing z in (11), we get the 4-weak identity f' defined in (12), which can be decomposed as $f' = f_1 + f_2$ where $f_1(x, z_1, x_2) = [x, z_1][x^3, z_2] - [x^3, z_1][x, z_2]$ is the sum of monomials in which z_1 precedes z_2, and $f_2(x, z_1, z_2) = f_1(x, z_2, z_1)$ is the sum of monomials in which z_2 precedes z_1. By Drensky and Rashkova [6,

Theorem 1.3(ii)], both f_1 and f_2 are 4-weak identities for $M_3(F)$. It is easy to verify that $f'' = f_1 - f_2$ is the is the polynomial f'' defined in (13). □

8.3 Identities from the Okubo Algebra

Some surprising identities of $M_3(F)$ arise from the Okubo algebra, which we now describe. A nonassociative F-algebra $(A, \star)$ is a **composition algebra** if it is endowed with a nondegenerate quadratic form $N : A \to F$ such that $N(x \star y) = N(x)N(y)$. The algebra is **symmetric** if it further satisfies

$$y \star (x \star y) = (y \star x) \star y = N(y)x. \tag{14}$$

A major example of a symmetric composition algebra is the **Okubo algebra** [11], whose underlying vector space is the space $M_3(F)_0$ of zero-trace matrices. Assuming F has a cubic root of unity which we denote ρ, the multiplication is defined by

$$x \star y = \frac{1-\rho}{3} xy + \frac{1-\rho^2}{3} yx - \frac{1}{3} \operatorname{tr} a(xy).$$

(There is an analogous description for the case $\rho \notin F$, which does not concern us here). The norm form is $N(x) = -\frac{1}{3} s_2(x)$, where $s_2(x)$ is the second coefficient of the characteristic polynomial of x.

We can now prove the following trace identity:

Proposition 8.3 *Assume* $x, y \in M_3(F)_0$. *Then*

$$[x^2, y^2] - [y, xyx] = \operatorname{tr}(xy)[x, y].$$

Proof Write $\alpha = \frac{1-\rho}{3}$ and $\alpha' = \frac{1-\rho^2}{3}$, so that $\alpha + \alpha' = 1$ and $\alpha^2 = \alpha - \frac{1}{3}$, and therefore $\alpha^2 + \alpha'^2 = \alpha\alpha' = \frac{1}{3}$. By assumption,

$$x \star y = \alpha xy + \alpha' yx - \frac{1}{3} \operatorname{tr} a(xy).$$

Multiplying by y from left, we have

$$\begin{aligned} y \star (x \star y) &= y \star (\alpha xy + \alpha' yx - \frac{1}{3} \operatorname{tr} a(xy)) = \\ &= \alpha y(\alpha xy + \alpha' yx - \frac{1}{3} \operatorname{tr} a(xy)) \\ &\quad + \alpha'(\alpha xy + \alpha' yx - \frac{1}{3} \operatorname{tr} a(xy))y - \frac{1}{3} \operatorname{tr} a(y(\alpha xy + \alpha' yx - \frac{1}{3} \operatorname{tr} a(xy))) = \end{aligned}$$

$$= (\alpha^2 + \alpha'^2)yxy + \alpha\alpha'(y^2x + xy^2) - (\alpha + \alpha')\frac{1}{3}\operatorname{tr}(xy)y - \frac{1}{3}\operatorname{tr} a(y(\alpha xy + \alpha' yx))$$
$$= \frac{1}{3}yxy + \frac{1}{3}(y^2x + xy^2) - \frac{1}{3}\operatorname{tr}(xy)y - \frac{1}{3}\operatorname{tr} a(\alpha yxy + \alpha' y^2 x).$$

Since $y \star (x \star y) = N(y)x$, the above expression commutes with x. Hence

$$0 = [x, yxy + y^2x + xy^2 - \operatorname{tr}(xy)y] =$$
$$= xyxy - yxyx + x^2y^2 - y^2x^2 - \operatorname{tr}(xy)[x, y]$$
$$= -[y, xyx] + [x^2, y^2] - \operatorname{tr}(xy)[x, y].$$
□

Taking $y = [z, x]$ we get $y \in M_3(F)_0$ and $\operatorname{tr}(xy) = \operatorname{tr}(x[z, x]) = \operatorname{tr}([xz, x]) = 0$ so Proposition 8.3 gives the 4-weak identity

$$[[z, x], x[z, x]x] - [x^2, [z, x]^2] = 0;$$

but we already know the 4-weak identities, and this is indeed Halpin's identity (11):

Remark 8.4 We have the tautological identity

$$[[z, x], x[z, x]x] - [x^2, [z, x]^2] = [[x, z], [x^3, z]]. \tag{15}$$

Indeed, let $y = [x, z]$. Then $xy + yx = [x^2, z]$, and the left hand side is equal to

$$[y, xyx] - [x^2, y^2] = y(xy + yx)x - x(yx + xy)y$$
$$= y[x^2, z]x - x[x^2, z]y$$
$$= zx^3zx + xz^2x^3 - zxzx^3 + x^3zxz - x^3z^2x - xzx^3z$$
$$= [zx^3, zx] - [zx^3, xz] + [x^3z, xz] - [x^3z, zx]$$
$$= [[x, z], [x^3, z]].$$

9 Matrices of Size $n \geq 4$

In Sects. 6 and 8 we have seen that $M_n(F)$ has properly weak identities of degree $2n$ when $n = 2, 3$. Here we show that for $n \geq 4$ the only weak identity of $M_n(F)$ in degree $2n$ is the standard identity, slightly improving on Amitsur–Levizki [1] who proved that s_{2n} is the only identity of $M_n(F)$ in this degree.

An easy argument, similar to that of [8, Lemma 1.10.7], rules out identities of degree $2n - 2$:

Proposition 9.1 *The minimal degree of a weak identity of* $M_n(F)$ *is* $\geq 2n - 1$.

Proof There is a vector space embedding $M_{n-1}(F) \subseteq M_n(F)_0$ by sending $a \mapsto (a, -\operatorname{tr}(a))$, which preserves multiplication in the first component. It follows that s_{2n-2} is the only possible identity of degree $< 2n - 1$. But the standard identity s_{2n-2} is ruled out as a weak identity for $M_n(F)$ by the path $1 \to 2 \to \cdots \to n \to \cdots \to 2 \to 1$. □

9.1 Shadows of Identities

We begin by developing a simple decomposition technique for multilinear identities.

Definition 9.2 Let $f \in P_m$ be a multilinear polynomial. Writing

$$f = \sum_{i \neq j} x_i f_{i,j}(x_1, \ldots, \widehat{x_i}, \ldots, \widehat{x_j}, \ldots, x_m) x_j,$$

for strong variables x_i, x_j, we call each $f_{i,j}$ a **shadow** of f.

As usual, $\widehat{x_i}$ denotes omission of x_i from the list. Each $f_{i,j}$ is an $(m-2)$-multilinear polynomial (on the variables $\{x_1, \ldots, \widehat{x_i}, \ldots, \widehat{x_j}, \ldots, x_m\}$). The action of S_m on P_m induces an action on the shadows by

$$(\sigma f)_{\sigma(i),\sigma(j)} = f_{i,j}. \tag{16}$$

Proposition 9.3 *Suppose $f \in P_m$ is an I-weak identity for $M_n(F)$. Then the shadow $f_{i,j}$ is an $(I \setminus \{i, j\})$-weak identity for $M_{n-1}(F)$.*

In particular, if $f \in P_m$ is a (weak) identity for $M_n(F)$, then each $f_{i,j}$ is a (resp. weak) identity for $M_{n-1}(F)$.

Proof The latter statement follows from the former by taking $I = \emptyset$ (resp. $\ell = \{1, \ldots, m\}$). We view $M_{n-1}(F) \subseteq M_n(F)$ in the natural way, embedded in the upper-left corner. Fix $u, v = 1, \ldots, n-1$, and substitute $x_i \mapsto e_{nu}$ and $x_j \mapsto e_{vn}$. By substituting matrices from $M_{n-1}(F)$ into the other variables, we see that

$$f(x_1, \ldots, e_{nu}, \ldots, e_{vn}, \ldots, x_m)_{nn} = f_{i,j}(x_1, \ldots, \widehat{x_i}, \ldots, \widehat{x_j}, \ldots, x_m)_{uv},$$

since any monomial is zero unless e_{nu} appears first and e_{vn} last in the product.

By assumption we are forced to assume the variables whose indices are in I are weak, and this condition for the variables other than x_i, x_j remains on the substitution in $f_{i,j}$. □

For distinct $i, j = 1, \ldots, m$, let $[i, j]\ell$ denote the quantity $|\{1, \ldots, \ell\} - \{i, j\}|$. Thus $[i, j]\ell \in \{\ell - 2, \ell - 1, \ell\}$. By Proposition 9.3, if $f \in P_m$ is an ℓ-weak identity for $M_n(F)$, then $f_{i,j}$ is an $[i, j]\ell$-weak identity for $M_{n-1}(F)$.

Corollary 9.4 *For every ℓ there is an injective map*

$$\mathrm{id}_m^{\ell}(\mathrm{M}_n(F)) \hookrightarrow \bigoplus_{u \neq v} \mathrm{id}_{m-2}^{[u,v]\ell}(\mathrm{M}_{n-1}(F)).$$

In particular there are injective maps for identities,

$$\mathrm{id}_m^0(\mathrm{M}_n(F)) \hookrightarrow \mathrm{id}_{m-2}^0(\mathrm{M}_{n-1}(F))^{m(m-1)},$$

and for weak identities,

$$\mathrm{id}_m^m(\mathrm{M}_n(F)) \hookrightarrow \mathrm{id}_{m-2}^{m-2}(\mathrm{M}_{n-1}(F))^{m(m-1)}. \tag{17}$$

Corollary 9.5 $\mathrm{PIdeg}^{\infty}(\mathrm{M}_n(F)) \geq 2 + \mathrm{PIdeg}^{\infty}(\mathrm{M}_n(F))$. *Indeed, if we have* $\mathrm{id}_{m-2}^{m-2}(\mathrm{M}_{n-1}(F)) = 0$ *then* $\mathrm{id}_m^m(\mathrm{M}_n(F)) = 0$ *by (17).*

Proposition 9.6 *The matrix algebra* $\mathrm{M}_4(F)$ *has no weak identities of degree* 7.

Proof For fields of characteristic different than 3, $\mathrm{M}_3(F)$ has no weak identities of degree 5 by Corollary 7.2, so $\mathrm{M}_4(F)$ has no weak identities of degree 7 by Corollary 9.5. For the remaining case of fields of characteristic 3, the claim was verified by a Sage program (computing over $\mathbb{F}_3$). □

Corollary 9.7 *The weak PI degree of* $\mathrm{M}_n(F)$ *is* $2n$ *for all* $n \geq 3$.

Proof We have that $\mathrm{PIdeg}^{\infty}(\mathrm{M}_n(F)) \leq \mathrm{PIdeg}(\mathrm{M}_n(F)) = 2n$ by Amitsur–Levizki. The lower bound $2n \leq \mathrm{PIdeg}^{\infty}(\mathrm{M}_n(F))$ is given for $n = 4$ in Proposition 9.6, and follows for $n > 4$ by induction applying Corollary 9.5. □

9.2 Weak Identities Degree 2n

We will now strengthen this result, and show that in the minimal degree $2n$, the standard identity is the only weak identity, namely $\mathrm{id}_{2n}^{2n}(\mathrm{M}_n(F))$ is one dimensional for all $n \geq 4$.

Theorem 9.8 *Let F be a field of characteristic zero. For $n \geq 4$,*

$$\mathrm{id}_{2n}^{2n}(\mathrm{M}_n(F)) = F s_{2n},$$

where s_{2n} is the standard identity.

Proof We prove this theorem by induction. The case $n = 4$ was verified using a Sage program (computing over $\mathbb{Q}$).

Suppose the proposition is true for some $n \geq 4$. We consider a weak identity $f \in \mathrm{id}_{2n+2}^{2n+2}(\mathrm{M}_{n+1}(F))$. Since this is an S_{2n+2}-module, we may assume f lies in the λ-component of $\mathrm{id}_{2n+2}^{2n+2}(\mathrm{M}_{n+1}(F))$, for some partition $\lambda = (\lambda_1, \dots, \lambda_r) \vdash 2n+2$.

By (17) we have an embedding $\mathrm{id}_{2n+2}^{2n+2}(\mathrm{M}_{n+1}(F)) \hookrightarrow \mathrm{id}_{2n}^{2n}(\mathrm{M}_n(F))^{(2n+2)(2n+1)}$. Let us denote the right-hand side by M. As an S_{2n+2}-module, M is isomorphic to the induced representation $\mathrm{Ind}_{S_{2n}}^{S_{2n+2}}(\mathrm{sgn})$. The irreducible subrepresentations of M are, by Frobenius reciprocity, those whose restriction from S_{2n+2} to S_{2n} is the sign representation of degree $2n$, namely, by the Branching Theorem [8, Theorem 2.3.1], the representations $[3^1 1^{2n-1}]$, $[2^2 1^{2n-2}]$, $[2^1 1^{2n}]$ and the sign representation $[1^{2n+2}]$.

By Proposition 5.2, we may assume that f is fixed under the action of

$$H = S_{\{1,\dots,\lambda_1\}} \times S_{\{\lambda_1+1,\dots,\lambda_1+\lambda_2\}} \times \cdots \times S_{\{\lambda_1+\cdots+\lambda_{r-1}+1,\dots,2n+2\}}.$$

In particular, each shadow $f_{i,j}$ is symmetric under the stabilizer of i, j in H, namely under $H_{ij} = \{\sigma \in H \mid \sigma(i) = i,\ \sigma(j) = j\}$.

On the other hand, by Proposition 9.3, each shadow $f_{i,j}$ is a weak identity for $\mathrm{M}_n(F)$ of degree $2n$. According to the induction hypothesis, this is only possible if

$$f_{i,j} = \alpha_{i,j} \cdot s_{2n}(x_1, \dots, \widehat{x_i}, \dots, \widehat{x_j}, \dots, x_{2n+2})$$

for some $\alpha_{i,j} \in F$, and so the shadow is antisymmetric. We conclude that if H_{ij} contains odd permutations, then necessarily $f_{ij} = 0$. In other words for $f_{ij} \neq 0$ it is necessary that removing i and j will leave no more than a single point in each part of λ (reaffirming the list of possible partitions).

CASE I. $\lambda = [31^{2n-1}]$. Here the only nonzero shadows $f_{i,j}$ of f must be those where $1 \leq i, j \leq 3$. Since f must be symmetric with respect to x_1, x_2, x_3, their coefficients $\alpha_{i,j}$ must also be equal to each other, so up to multiplication by a scalar, f has to be

$$f = \sum_{1 \leq i,j \leq 3} x_i \cdot s_{2n}(x_1, \dots, \widehat{x_i}, \dots, \widehat{x_j}, \dots, x_{2n+2}) \cdot x_j.$$

In other words, f is the multilinearization of

$$\hat{f}(x, x_4, \dots, x_{2n+2}) = x \cdot s_{2n}(x, x_4, \dots, x_{2n+2}) \cdot x.$$

Substitute $x \mapsto e_{11} - e_{22}$ and for the variables $x_4, x_5, \dots, x_{2n+2}$ take the "ladder" matrix units $e_{12}, e_{23}, \dots, e_{n,n+1}, e_{n+1,n}, \dots, e_{32}$. By direct computation, one can verify that

$$\hat{f}(x, x_4, \dots, x_{2n+2})_{1,2} = s_{2n}(x, x_4, \dots, x_{2n+2})_{1,2} = 3,$$

which proves that $\hat{f}$ is not a weak identity for $\mathrm{M}_{n+1}(F)$.

CASE II. $\lambda = (2, 2, 1^{2n-2})$. In this case, f is symmetric with respect to x_1 and x_2 and with respect to x_3 and x_4. The possible nonzero shadows are $f_{i,j}$ where $i \in \{1, 2\}$ and $j \in \{3, 4\}$, or vice versa. A similar explanation shows that f is the multilinearization of an identity of the form

$$\hat{f} = \alpha \cdot x \cdot s_{2n}(x, y, x_5, \ldots, x_{2n+2}) \cdot y + \beta \cdot y \cdot s_{2n}(x, y, x_5, \ldots, x_{2n+2}) \cdot x$$

for some $\alpha, \beta \in F$. Set

$$x, y, x_5, \ldots, x_{2n+2} = e_{12}, e_{21}, e_{13}, e_{31}, \ldots, e_{1,n+1}, e_{n+1,1}.$$

A simple calculation shows that $s_{2n}(e_{12}, e_{21}, e_{13}, e_{31}, \ldots, e_{1,n+1}, e_{n+1,1}) = n!e_{11} - \sum_{k=2}^{n+1}(n-1)!e_{kk}$. Hence $\hat{f}(e_{12}, e_{21}, e_{13}, e_{31}, \ldots, e_{1,n+1}, e_{n+1,1}) = -\alpha(n-1)!e_{11} + \beta n!e_{22}$, showing that $\alpha = \beta = 0$.

CASE III. $\lambda = (2, 1^{2n})$. In a similar manner, one may see that f must be a multilinearization of a weak identity of the form

$$\hat{f}(x, x_1, \ldots, x_{2n}) = \sum_{i=1}^{2n} \alpha_i\, x\, s_{2n}(x, x_1, \ldots, \widehat{x_i}, \ldots, x_{2n})\, x_i + \\ + \sum_{i=1}^{2n} \beta_i\, x_i\, s_{2n}(x, x_1, \ldots, \widehat{x_i}, \ldots, x_{2n})\, x + \\ + \gamma\, x\, s_{2n}(x_1, \ldots, x_{2n})\, x.$$

Fixing $1 \leq j < 2n$, we substitute x_j in place of x_{j+1} and keep all the other variables in place. Most summands vanish, and the resulting polynomial is

$$(\alpha_j + \alpha_{j+1})\, x\, s_{2n}(x, x_1, \ldots, \widehat{x_{j+1}}, \ldots, x_{2n})\, x_j + \\ +(\beta_j + \beta_{j+1})\, x_j\, s_{2n}(x, x_1, \ldots, \widehat{x_{j+1}}, \ldots, x_{2n})\, x.$$

This must be a weak identity for $M_{n+1}(F)$. Since its multilinearization is symmetric with respect to two pairs of variables, it lies in the component of $(2, 2, 1^{2n-2})$, hence must be zero by CASE II. This shows that $\alpha_{j+1} = -\alpha_j$ and $\beta_{j+1} = -\beta_j$. But the argument holds for all j, so $\alpha_i = (-1)^{i-1}\alpha_1$ and $\beta_i = (-1)^{i-1}\beta_1$.

Next we substitute $x_1 = x$. Again, most terms become zero, and the result is

$$(\alpha_1 + \beta_1 + \gamma)\, x\, s_{2n}(x, x_2, \ldots, x_{2n})\, x.$$

This should be a weak identity for $M_{n+1}(F)$ lying in the component of $(3, 1^{2n-1})$, and by CASE I must be zero. This proves that $\alpha_1 + \beta_1 + \gamma = 0$.

We have therefore shown that our weak identity has the form

$$\hat{f} = \alpha \sum_{i=1}^{2n} (-1)^{i-1} x\, s_{2n}(x, x_1, \ldots, \widehat{x_i}, \ldots, x_{2n})\, x_i +$$

$$+ \beta \sum_{i=1}^{2n} (-1)^{i-1} x_i\, s_{2n}(x, x_1, \ldots, \widehat{x_i}, \ldots, x_{2n})\, x -$$

$$- (\alpha + \beta)\, x\, s_{2n}(x_1, \ldots, x_{2n})\, x =$$

$$= \alpha \sum_{i=1}^{2n} x\, s_{2n}(x_1, \ldots, x_{i-1}, x, x_{i+1}, \ldots, x_{2n})\, x_i +$$

$$+ \beta \sum_{i=1}^{2n} x_i\, s_{2n}(x_1, \ldots, x_{i-1}, x, x_{i+1}, \ldots, x_{2n})\, x -$$

$$- (\alpha + \beta)\, x\, s_{2n}(x_1, \ldots, x_{2n})\, x$$

for appropriate $\alpha, \beta \in F$.

We substitute

$$x, x_1, x_2, \ldots, x_{2n} = e_{12} + e_{23}, e_{12}, e_{21}, \ldots, e_{1,n+1}, e_{n+1,1}.$$

We know that $s_{2n}(x_1, \ldots, x_{2n}) = n! e_{11} - (n-1)! \sum_{k=2}^{n+1} e_{kk}$, so $x\, s_{2n}(x_1, \ldots, x_{2n})\, x = -(n-1)! e_{13}$. We next compute $s_{2n}(x_1, \ldots, x_{i-1}, e_{23}, x_{i+1}, \ldots, x_{2n})$. Consider the directed graph G_i on the vertices $1, 2, \ldots, n+1$, with an edge $j \to j'$ if and only if $e_{j,j'}$ appears in the list $x_1, \ldots, \widehat{x_i}, \ldots, x_{2n}, e_{23}$ after the substitution above. Any nonzero summand in the expression $s_{2n}(x_1, \ldots, x_{i-1}, e_{23}, x_{i+1}, \ldots, x_{2n})$ corresponds to an Eulerian path in G_i. We consider the following cases:

- $i = 2\ell - 1$ is odd, in which case $x_i = e_{1,\ell+1}$. Then $\deg^-(1) - \deg^+(1) = 1$, so any Hamiltonian path must end at 1. But if $\ell \neq 2$, we also have $\deg^-(\ell + 1) - \deg^+(\ell + 1) = 1$, so G_i has no hamiltonian path. There are two types of Hamiltonian paths in G_3: those that begin with $2 \to 3 \to 1$, and those that begin with $2 \to 1$. One can see that each path of the first type contributes $+1$ to the sum, and each path of the second type contributes -1 to the sum. Since their number is identical, the result is 0.
- $i = 3$. We want to compute $s_{2n}(e_{12}, e_{21}, e_{23}, e_{31}, e_{14}, \ldots, e_{n+1,1})$. Using the same considerations, every Hamiltonian path must start at 2 and end at 1.
- $i = 2\ell$ is even, in which case $x_i = e_{\ell+1,1}$. But then $\deg^-(1) - \deg^+(1) = -1$, and also $\deg^-(3) - \deg^+(3) = -1$ (or -2 if $i = 4$), which again shows that G_i has no Hamiltonian path.

To conclude, we know that $s_{2n}(x_1, \ldots, x_{i-1}, e_{23}, x_{i+1}, \ldots, x_{2n}) = 0$ for all i. Hence, for $i > 1$ we have

$$s_{2n}(x_1, \ldots, x_{i-1}, x, x_{i+1}, \ldots, x_{2n}) = s_{2n}(e_{12}, x_2, \ldots, x_{i-1}, e_{12}, x_{i+1}, \ldots, x_{2n}) + \\ + s_{2n}(x_1, \ldots, x_{i-1}, e_{23}, x_{i+1}, \ldots, x_{2n}) = 0,$$

and for $i = 1$ we have

$$s_{2n}(x, x_2, \ldots, x_{2n}) = s_{2n}(e_{12}, x_2, \ldots, x_{2n}) + s_{2n}(e_{23}, x_2, \ldots, x_{2n}) = \\ = s_{2n}(x_1, x_2, \ldots, x_{2n}) = n!e_{11} - (n-1)! \sum_{k=2}^{n+1} e_{kk}.$$

The appropriate summands are thus

$$x\, s_{2n}(x, x_2, \ldots, x_{2n})\, x_1 = (e_{12} + e_{23})\, s_{2n}(x, x_2, \ldots, x_{2n})\, e_{12} = 0$$
$$x_1\, s_{2n}(x, x_2, \ldots, x_{2n})\, x = e_{12}\, s_{2n}(x, x_2, \ldots, x_{2n})\, (e_{12} + e_{23}) = -(n-1)!e_{13}.$$

Therefore, the substitution above in $\hat{f}$ yields a matrix whose $(1, 3)$ component is $(n-1)!\alpha$, hence $\alpha = 0$. Similarly, one may show that $\beta = 0$, so $\hat{f} = 0$ as required.

In conclusion, we are left with the case where $\lambda = (1^{2n+2})$, which indeed corresponds to the standard identity s_{2n+2}. □

Acknowledgments This work was supported by the Israel science foundation grants #1623/16 (LR and UV) and #630/17 (EM).

References

1. Amitsur, S., Levitzki, J.: Minimal identites for algebras. Proc. Amer. Math. Soc. **1**, 449–463 (1950)
2. Belov-Kanel, A., Karasik, Y., Rowen, L.H.: Computational Aspects of Polynomial Identities: Volume I, Kemer's Theorems. Chapman & Hall/CRC Monographs and Research Notes in Mathematics, vol. 15, 2nd edn. CRC Press, Boca Raton (2016)
3. Drensky, V.S.: New central polynomials for the matrix algebra. Israel J. Math. **92**, 235–248 (1995)
4. Drensky, V.S.: Free Algebras and PI-algebras: Graduate Course in Algebra. Springer, Berlin (2000)
5. Drensky, V., Kasparian, A.: A new central polynomial for the 3 × 3 matrices. Comm. Algebra **13**, 745–752 (1985)
6. Drensky, V., Rashkova, T.: Weak polynomial identities for the matrix algebras (English summary) Comm. Algebra **21**(10), 3779–3795 (1993)
7. Formanek, E.: Central polynomials for matrix rings. J. Algebra **23**, 129–132 (1972)
8. Giambruno, A., Zaicev, M.: Polynomial Identities and Asymptotic Methods. Math Surveys and Monographs, vol. 122. American Mathematical Society, Providence (2005)
9. Huang, J.S.: Lectures on Representation Theory. World scientific, Singapore (1999)

10. Kemer, A.R.: Identities of Associative Algebras. Translations of Mathematical Monographs, vol. 87. American Mathematical Society, Providence (1991)
11. Matzri, E., Vishne, U.: Isotropic subspaces in symmetric composition algebras and Kummer subspaces in central simple algebras of degree 3. Manuscripta Math. **137**(3), 497–523 (2012)
12. Razmyslov, Yu.: Finite basing of the identities of a matrix algebra of second order over a field of characteristic zero. Algebra i Logika **12**(1), 83–113 (1973); English Translation, Algebra and Logic **12**(1), 47–63 (1973)
13. Vishne, U.: Polynomial Identities of 2×2 matrices over the Grassmannian. Commun. Algebra **30**(1), 443–454 (2002)
14. Vishne, U.: Polynomial Identities of $M_{2,1}(G)$. Comm. Algebra **39**(6), 2044–2050 (2011)

Computing Multiplicities in the Sign Trace Cocharacters of $M_{2,1}(F)$

Luisa Carini

Abstract In Regev (Linear Multilinear Algebra 21:1–28, 1987), Regev applied the representation theory of the general Lie superalgebra to generalize the theory of trace identities as developed by Procesi and Razmyslov. Regev showed that certain cocharacters arising from sign trace identities were given by

$$\sum_{\lambda \in H(k,l;n)} \chi_\lambda \otimes \chi_\lambda$$

where $\chi_\lambda \otimes \chi_\lambda$ denotes the Kronecker product of the irreducible character of the symmetric group associated with the partition λ with itself and $H(k, l; n)$ denotes the set of partitions of n $\lambda = (\lambda_1 \geq \lambda_2 \geq \ldots \geq \lambda_n)$ such that $\lambda_{k+1} \leq l$. In case of $k = 2, l = 1$, we show how to compute some multiplicities which occur in the expansion of the cocharacter in terms of irreducible characters by using the reduced notation Scharf et al. (J Phys A Math Gen 26:7461–7478, 1993).

Keywords Trace identity · Invariant theory · Kronecker product · Schur functions

1 Introduction

The theory of trace identities, developed independently by Procesi [14] and Razmyslov [15] has proved to be a powerful tool in the study of identities of the algebra $M_k(F)$ of $k \times k$ matrices over a field F of characteristic 0. In [17] Regev has given a "hook" generalization of the theory of trace identities, which has applications to the study of certain P.I. algebras. Briefly in the usual theory of trace identities, the group algebra $C(S_n)$ of the symmetric group is identified with

L. Carini (✉)
MIFT, University of Messina, Messina, Italy
e-mail: lcarini@unime.it

O. M. Di Vincenzo, A. Giambruno (eds.), *Polynomial Identities in Algebras*, Springer INdAM Series 44, https://doi.org/10.1007/978-3-030-63111-6_6

multilinear trace polynomials. Then one can use the classical work of Schur and Weyl [20] on the polynomial representations of the general linear algebra $gl(k, C)$ and show that the trace cocharacter of the $M_k(F)$ equals $\sum_{\lambda \in \Lambda_k(n)} \chi_\lambda \otimes \chi_\lambda$, where $\otimes$ denotes the Kronecker or inner product of the irreducible S_n−character χ_λ with itself and $\Lambda_k(n)$ denotes the set of partitions of n with k or fewer parts. It follows from the basic properties of the Kronecker products that:

$$\sum_{\lambda \in \Lambda_k(n)} \chi_\lambda \otimes \chi_\lambda = \sum_{\mu \in \Lambda_{k^2}(n)} m_\mu(M_k(F))\chi_\mu.$$

The multiplicities $m_\mu(M_k(F))$ are non negative integers and are explicitly known only for $k = 2$ and partially for $k = 3, 4$ (see [1–3, 5–7]) and they are not yet well understood. In [17] Regev generalizes the notion of trace polynomials to that of signed trace polynomials by multiplying their coefficients by the sign function ϵ. Now $C(S_n)$ can be identified with certain classes of signed trace polynomials in such a way that one can apply the representation theory of the general Lie superalgebra $pl(k, l)$ which is a generalization of the representation theory of $gl(k, C)$. In [17] Regev defines a certain quasi Z_2-grading on $M_{k+l}(F)$ depending on k and l and he denotes the resulting (k, l) quasi-structure by $M_{k,l}(F)$. One can then easily define by analogy the n-th sign trace cocharacter of $M_{k,l}(F)$, denoted by $\chi_n^{ST}(M_{k,l}(F))$, and as one of the major results of [17], Regev proves that $\chi_n^{ST}(M_{k,l}(F))$ equals

$$\sum_{\lambda \in H(k,l;n)} \chi_\lambda \otimes \chi_\lambda \tag{1}$$

where $H(k, l; n)$ denotes the set of partitions $\lambda = (\lambda_1 \geq \lambda_2 \geq \ldots \geq \lambda_n \geq 0)$ where $\lambda_{k+1} \leq l$. This character is associated with several objects in PI theory: with sign trace identities, with the PI's of the identities of the 3×3 matrices with the (2,1) superalgebra structure (see [16]) and it is also related to the cocharacters of the ordinary 3×3 matrices (see [6]).

In [18] Remmel gives an explicit formula for (1) in the case where $k = l = 1$ and he proves the following theorem:

Theorem 1 *Let*

$$\sum_{\lambda \in H(1,1;n)} \chi_\lambda \otimes \chi_\lambda = \sum_{r=1}^{n} \chi_{(r,1^{n-r})} \otimes \chi_{(r,1^{n-r})} = \sum_{\mu} c_\mu \chi_\mu.$$

Then

1. $c_\mu = 0$ *if* μ *is not a hook or a double hook shape;*
2. $c_{(r,1^{n-r})} = \begin{cases} r & \text{if } n\text{-}r \text{ is even} \\ r-1 & \text{if } n\text{-}r \text{ is odd} \end{cases}$
3. $c_{(q,p,2^b,1^a)} = 2(q - p + 1)$ *if* $q \geq p \geq 2$ *and* $q + p + 2b + a = n$.

In the case $k = 2, l = 1$, some partial new results about the decomposition of $\chi_n^{ST}(M_{2,1})$ are contained in [2] and in [4]. More specifically, let

$$\chi_n^{ST}(M_{2,1}) = \sum_{\lambda \in H(2,1;n)} \chi_\lambda \otimes \chi_\lambda = \sum_\mu c_\mu \chi_\mu$$

where $H(2, 1; n) = \{\lambda \vdash n : \lambda_3 \leq 1\}$. In [4], it has been conjectured that:
if n is even:

- $c_{(n-1,1)} = \frac{(n-2)^2}{2}$
- $c_{(n-2,2)} = \frac{3n^2-19n+34}{2}$
- $c_{(n-1,1^2)} = n^2 - 6n + 10;$

if n is odd:

- $c_{(n-1,1)} = \frac{n^2-4n+5}{2}$
- $c_{(n-2,2)} = \frac{3n^2-19n+32}{2}$
- $c_{(n-1,1^2)} = n^2 - 6n + 10.$

The detailed computation of the coefficient $c_{(n-1,1)}$ can be found in [4]. Here we will show how to compute $c_{(n-2,2)}$ and $c_{(n-2,1^2)}$ by using the reduced notation method which we believe might lead to further results.

The outline of this paper is as follows. In Sect. 2 we shall state some preliminaries on reduced notation and Littlewood's modification rules. Then in Sect. 3, we shall apply this method to carry out our main computation.

A Remark About Notation

In this paper, we shall freely mix the traditional notation of Littlewood with that of Macdonald [12], which is more convenient for algebraic manipulations. So, the Schur function corresponding to a partition λ will be indifferently denoted by $\{\lambda\}$ or s_λ.

2 Reduced Notation

The concept of reduced notation for the symmetric group was introduced by Murnaghan in [13] and later used by Littlewood [9–11] for the calculation of inner plethysm and Kronecker products for the symmetric group S_n.

The irreducible representation $\{\lambda\}$ of S_n may be labelled by ordered partitions (λ) of integers where $\lambda \vdash n$. In reduced notation the label $\{\lambda\} = \{\lambda_1, \lambda_2, \ldots, \lambda_p\}$ for S_n is replaced by $< \lambda > = < \lambda_2, .., \lambda_p >$. Given any irreducible representation $< \mu >$ in reduced notation it can be converted back into a standard irreducible representation of S_n by prefixing $< \mu >$ with the integer $(n - |\mu|)$.

For example, an irreducible representation $< 2, 1 >$ in reduced notation corresponds to $\{3, 2, 1\}$ in S_6 or $\{9, 2, 1\}$ in S_{12}. It is just this feature that leads to an n-independent notation for S_n. If $n - |\mu| \geq \mu_1$, then the resulting irreducible representation $\{n - |\mu|, \mu\}$ is assuredly a standard irreducible representation of S_n. However, if $n - |\mu| < \mu_1$, then the irreducible representation $\{n - |\mu|, \mu\}$ is non standard and must be converted into standard form using the following s-function modification rules [8]:

(i) In any s-function two consecutive parts may be interchanged provided that the preceding part is decreased by unity and the succeeding part increased by unity, the resulting s-function being thereby changed in sign, i.e.

$$\{\lambda_1, .., \lambda_i, \lambda_{i+1}, .., \lambda_k\} = -\{\lambda_1, .., \lambda_{i+1} - 1, \lambda_i + 1, \ldots, \lambda_k\}.$$

(ii) In any s-function if any part exceeds by unity the preceding part, the value of s-function is zero, i.e. if $\lambda_{i+1} = \lambda_i + 1$ then

$$\{\lambda_1, .., \lambda_i, \lambda_{i+1}, .., \lambda_k\} = 0.$$

(iii) The value of any s-function is zero if the last part is negative.

Example 1 Consider in reduced notation $< \mu >=< 2, 1 >$; in S_3 $\mu =< 2, 1 >$ becomes $\{n - |\mu|, \mu\} = \{0, 2, 1\}$ which is not standard and must be made standard using the above Littlewood's modification rules. Therefore by (i) we get $\{0, 2, 1\} = -\{1, 1, 1\} = -\{1^3\}$,
Instead in S_4, $\mu =< 2, 1 >$ becomes $\{n - |\mu|, 2, 1\} = \{1, 2, 1\}$ and, by (ii), we get $\{1, 2, 1\} = 0$.

Example 2 $< 4, 2 >$ in S_8 becomes $\{2, 4, 2\} = -\{3, 3, 2\}$ while in S_9 we get $\{3, 4, 2\}$ which is zero by (ii); in S_4, $\mu =< 4, 2 >$ becomes $\{n - |\mu|, 4, 2\} = \{-2, 4, 2\}$ and by applying (i) twice we get:

$$\{-2, 4, 2\} = -\{3, -1, 2\} = \{3, 1, 0\} = \{3, 1\}.$$

A reduced Kronecker product $< \lambda > \circ < \mu >$ may be evaluated by the recursive relation (see [11, 19])

$$< \lambda > \circ < \mu >= \sum_{\alpha,\beta,\gamma} < \{\lambda/\alpha\beta\} \cdot \{\mu/\alpha\gamma\} \cdot \{\beta \circ \gamma\} >$$

where "/" indicates an s-function skew, i.e. $\{\lambda/\mu\} = D_{s_\mu} s_\lambda$, (see [12]), a dot is for Littlewood-Richardson s-function multiplication and "o" is the ordinary inner (Kronecker) product. By the notation $\lambda/\alpha\beta$ and $\mu/\alpha\gamma$ we mean the Schur functions corresponding to all those partitions obtained by removing all possible β and γ with the same weight, (i.e. same number of cells in their corresponding diagrams), from the skew diagrams λ/α and μ/α for all possible partitions α contained in λ.

3 The Sign Trace Cocharacter of $M_{2,1}(F)$

Let $M_n(F)$, the algebra of $n \times n$ matrices over a field F of characteristic zero. In [17] Regev defines a certain quasi Z_2-grading on $M_{k+l}(F)$ depending on k and l and he denotes the resulting (k, l) quasi-structure by $M_{k,l}(F)$. As one of the major results of [17], Regev proves that the sign trace cocharacter of $M_{k,l}(F)$, $\chi_n^{ST}(M_{k,l}(F))$ equals

$$\sum_{\lambda \in H(k,l,n)} \chi_\lambda \otimes \chi_\lambda = \sum_\mu c_\mu \chi_\mu$$

where $H(k, l; n)$ denotes the set of partition $\lambda = (\lambda_1 \geq \lambda_2 \geq \ldots \geq \lambda_n \geq 0)$, where $\lambda_{k+1} \leq l$.

By Berele [2, Theorem 5.2] and Carini [4, Section 4] it follows that $c_{(n)}=|H(k, l; n)|=$ the number of partitions in $H(k, l; n)$ and $c_{(1^n)}$ equals the number of self conjugate partitions in $H(k, l; n)$.

In this section we consider the case $k = 2, l = 1$ and compute some multiplicities which occur in the expansion of the n-th sign trace cocharacter of $M_{2,1}(F)$ in terms of irreducible characters. In symbols:

$$\chi_n^{ST}(M_{2,1}(F)) = \sum_{\lambda \in H(2,1;n)} \chi_\lambda \otimes \chi_\lambda = \sum_\mu c_\mu \chi_\mu.$$

Denote by $\lfloor x \rfloor$ the greatest integer less than or equal to x and $\lceil x \rceil$ the least integer greater than or equal to x.

An easy computation shows that

$$c_{(n)} = |H(2, 1; n)| = \left\lfloor \frac{3n-2}{2} \right\rfloor + \sum_{i=3}^{n-2} \left\lfloor \frac{n-i}{2} \right\rfloor$$

.

Also, $c_{(1^n)}$ is always equal to 1 except the case $n = 2$, when is zero. In fact, the only self conjugate partitions contained in $H(2, 1, n)$ are the hooks $(\lceil \frac{n}{2} \rceil, 1^{\lfloor \frac{n}{2} \rfloor})$ for n odd and the partition $(a, 2, 1^{a-2})$ for n even with $2a = n$.

From now on we will use the reduced notation.

3.1 Computation of the Coefficient for $\{n - 2, 2\}$

We want to establish the coefficient $\{n - 2, 2\}$ as a polynomial in n. Essentially we are interested in the sums of the inner squares of all the partitions of n in which the third part is 1 or 0. In reduced notation they are of the form $< k1^x >$. Thus for $n = 8$ the partitions of interest are

$$\begin{array}{lllll}
\{8\} & +\{71\} & +\{62\} & +\{61^2\} & +\{53\} \\
+\{521\} & +\{51^3\} & +\{4^2\} & +\{431\} & +\{421^2\} \\
+\{41^4\} & +\{3^21^2\} & +\{321^3\} & +\{31^5\} & +\{2^21^4\} \\
+\{21^6\} & +\{1^8\} & & &
\end{array}$$

In reduced notation, the single hooks are (not in the same order as above)

$$\begin{array}{lllll}
< 4 > & + < 31^2 > & + < 31 > & + < 3 > & + < 21^4 > \\
+ < 21^3 > & + < 21^2 > & + < 21 > & + < 2 > & + < 1^7 > \\
+ < 1^6 > & + < 1^5 > & + < 1^4 > & + < 1^3 > & + < 1^2 > \\
+ < 1 > & + < 0 > & & &
\end{array}$$

Notice that the above hooks can be arranged in groups as follows

$$\begin{array}{llllll}
< 4 > & < 3 > & < 2 > & < 1 > & < 0 > & \\
< 31 > & < 21 > & & & & \\
< 31^2 > & & & & & \\
< 21^4 > & < 21^3 > & < 21^2 > & & & \\
< 1^7 > & < 1^6 > & < 1^5 > & < 1^4 > & < 1^3 > & < 1^2 >
\end{array}$$

We do this to show that the various reduced inner squares can be divided into classes of hooks and each class can be enumerated and treated separately. The irreducible representations considered can, in the reduced notation, be divided into three classes:

$$< 1^x >; \qquad < x >; \qquad < k1^x > \tag{2}$$

The values of x and k must satisfy certain constraints which depend on the coefficient of interest and the particular value on n. We then endeavour to extract from the reduced Kronecker squares the desired coefficient. Throughout we use the basic reduced Kronecker product result

$$< \lambda > \circ < \mu > = \sum_{\alpha,\beta,\gamma} < \lambda/\alpha\beta \cdot \mu/\alpha\gamma \cdot \beta \circ \gamma > \tag{3}$$

In going from the reduced notation to standard notation one must, where necessary, apply the usual s-function modification rules. The modification rules give $\{1, n - 1\} = -\{n - 2, 2\}$ and hence in computing the coefficient of $\{n - 2, 2\}$ we need to know not only the multiplicity of $< 2 >$ but also the multiplicity of $< n - 1 >$. The coefficient of $\{n - 2, 2\}$ will involve the difference of these two multiplicities. In what follows we shall consider the contribution of each class in (2) to the coefficient for $\{n-2, 2\}$ and then put together the pieces to yield the coefficient as a polynomial in n.

Contribution from the Class $< 1^x >$
Since in reduced notation $\{1^n\} \equiv < 1^{n-1} >$ and $\{1^n\} \circ \{1^n\} = \{n\}$ we can limit x to

$$n - 2 \geq x \geq 1 \tag{4}$$

Noting from (3) that

$$< 1^x > \circ < 1^x > = \sum_{\alpha,\beta,\gamma} < 1^x/\alpha\beta \cdot 1^x/\alpha\gamma \cdot \beta \circ \gamma > \tag{5}$$

If $x = 1$ the only choice for α, β, γ is $\alpha, \beta, \gamma = \{0\}$. For $n-3 \geq x \geq 2$ we have two choices $\alpha, \beta, \gamma = \{1^{x-1}\}, \{0\}, \{0\}$ and $\alpha, \beta, \gamma = \{1^{x-2}\}, \{1^2\}, \{1^2\}$. For $x = n - 2$ we still have the preceding two choices but the product also gives rise to $< n - 1 >$ which upon application of the s-function modification rules cancels one of the two choices and hence we end with the counting algorithm

1. Count 1 for $x = 1$
2. Count 2 for $n - 3 \geq x \geq 2$
3. Count 1 for $x = n - 2$
4. Giving a total count of

$$2(n - 3) \tag{6}$$

Note that for this class the result does not depend on the parity of n.

Contribution from the Class $< x >$
Here we must treat the odd and even values of n separately but the results derive in a very similar fashion to the above to give the two counting algorithms:

n even

1. Count 2 for $\frac{n-2}{2} \geq x \geq 2$
2. Count 1 for $x = \frac{n}{2}$
3. Giving a total count of

$$n - 3 \tag{7a}$$

n odd

1. Count 2 for $\frac{n-3}{2} \geq x \geq 2$
2. Count 1 for $x = \frac{n-1}{2}$
3. Giving a total count of

$$n - 4 \tag{7b}$$

Contribution from the Class $< k1^x >$

This is the most complex part of the derivation and we do it in three steps. We first compute the multiplicity of $< n-1 >$ in $< k1^x > \circ < k1^x >$ and then the multiplicity of $< 2 >$ in $< k1^x > \circ < k1^x >$ and then subtract the two sets of multiplicities. In general we will have to consider those cases where $k > 1$ with

$$n - 2k \geq x \geq 1 \tag{8}$$

Step 1: The Coefficient of $< n-1 >$ in $< k1^x > \circ < k1^x >$

Two distinct cases arise: (i) $x = n - 2k - 1$ and (ii) $x = n - 2k$. For all other values of x the coefficient is null. Throughout we assume $k > 1$.

(i) $x = n - 2k - 1$. It follows from (3) that the only possible choice for α, β, γ is $\alpha = \{0\}$, $\beta = \gamma = \{1^{n-2k-1}\}$ and we note that $\{k \cdot k \cdot (n-2k-1)\} \supset \{n-1\}$. Thus we conclude that in this case the coefficient of $< n-1 >$ in $< k1^{n-2k-1} > \circ < k1^{n-2k-1} >$ is 1.

(ii) $x = n - 2k$. It follows from (3) that there are three choices for α, β, γ

$$\alpha = \{0\}, \quad \beta = \gamma = \{1^{n-2k+1}\} \tag{9a}$$

$$\alpha = \{0\}, \quad \beta = \gamma = \{21^{n-2k-1}\} \tag{9b}$$

$$\alpha = \{1\}, \quad \beta = \gamma = \{1^{n-2k-1}\} \tag{9c}$$

Each case yields $< n-1 >$ just once and hence for this case the total coefficient of $< n-1 >$ in $< k1^{n-2k} > \circ < k1^{n-2k} >$ is 3.

Step 2: The Coefficient of $< 2 >$ in $< k1^x > \circ < k1^x >$

There are six choices of α, β, γ that can yield the coefficient $< 2 >$:

$$\alpha = \{k1^{x-1}\}, \quad \beta = \gamma = \{0\} \tag{10a}$$

$$\alpha = \{k-1, 1^x\}, \quad \beta = \gamma = \{0\} \tag{10b}$$

$$\alpha = \{k, 1^{x-2}\}, \quad \beta = \gamma = \{1^2\} \tag{10c}$$

$$\alpha = \{k-2, 1^x\}, \quad \beta = \gamma = \{2\} \tag{10d}$$

$$\alpha = \{k-1, 1^{x-1}\}, \quad \beta = \gamma = \{2\} \tag{10e}$$

$$\alpha = \{k-1, 1^{x-1}\}, \quad \beta = \gamma = \{1^2\} \tag{10f}$$

Each of the above give a count of 1 or 0 depending on the values of k and x. Specifically we have the following counting rules

$$k = 2, x = 1 \quad \text{count } 4 \tag{11a}$$

$$k = 2, n - 4 \geq x > 1 \quad \text{count } 5 \tag{11b}$$

$$k > 2, x = 1 \quad \text{count } 5 \tag{11c}$$

$$k > 2, n - 2k \geq x > 1 \quad \text{count } 6 \tag{11d}$$

Step 3: Combining the Coefficients for the Class $< k1^x >$
Steps 1 and 2 can be carried out for any member of the class $< k1^x >$ for any value of n. Thus for $n = 8$ we have the classes involving just $k = 2, 3$ and can readily deduce from Steps 1 and 2 that for $k = 2$ we obtain a contribution of 15 and for $k = 3$ a contribution of 7 which combined with the contributions of the classes $< 1^x >$ and $< x >$ yields a total contribution of 37 and hence the coefficient for $\{62\}$ is 37.

It is illuminating to arrange the contributions for each value of n even

n	$< 21^x >$	$< 31^x >$	$< 41^x >$	$< 51^x >$	$< 61^x >$	Total
6	5					5
8	15	7				22
10	25	19	7			51
12	35	31	19	7		92
14	45	43	31	19	7	145

(12)

It is readily seen how the pattern continues and that for a given n even the total contribution is

$$\frac{3n^2 - 25n + 52}{2} \tag{13}$$

Likewise for n odd one obtains the pattern

n	$< 21^x >$	$< 31^x >$	$< 41^x >$	$< 51^x >$	$< 61^x >$	$< 71^x >$	Total
7	10	2					12
9	20	13	2				35
11	30	25	13	2			70
13	40	37	25	13	2		117
15	50	49	37	25	13	2	176

and again the total contribution is

$$\frac{3n^2 - 25n + 52}{2} \tag{14}$$

and the result is independent of the parity of n.

4 The Final Result

The coefficient $c_{(\{n-2,2\})}$ comes from simply summing the contributions for each class to give the coefficient as

$$c_{(\{n-2,2\})} = \begin{cases} \frac{3n^2-19n+34}{2} & \text{if n is even} \\ \frac{3n^2-19n+32}{2} & \text{if n is odd} \end{cases}$$

The Coefficient of $\{n-2, 1^2\}$
Here one has to look at the three generic classes of partitions (1^x), (x), $(k1^x)$ and determines the bounds of x in each case and establishes a counting rule for each class starting with the Eq. (3):

$$< \lambda > \circ < \mu >= \sum_{\alpha,\beta,\gamma} < \lambda/\alpha\beta \cdot \mu/\alpha\gamma \cdot \beta \circ \gamma >$$

As before we need to consider the n even and odd cases separately even though the final result does not depend on the parity of n. In this particular derivation, the modification rules

$$\{-1, n-1, 2\} > \{n-2, 1^2\} \quad \text{and} \quad \{0, n-1, 1\} > \{n-2, 1^2\} \tag{15}$$

are required. The class (1^x) and (x) are almost the same as in the earlier derivation. Care is needed for the class $(k1^x)$.

From (3)

$$< k1^x > \circ < k1^x >= \sum_{\alpha,\beta,\gamma} < k1^x/\alpha\beta \cdot k1^x/\alpha\gamma \cdot \beta \circ \gamma > \tag{16}$$

We need to determine the α, β, γ that give the term $< 1^2 >$ on the rhs of (16). At first it appears that there are four choices of α, β, γ

$$\alpha = k, 1^{x-1} \quad \beta = \gamma = 0 \qquad (i)$$

$$\alpha = k-1, 1^x \quad \beta = \gamma = 0 \qquad (ii)$$

$$\alpha = k-1, 1^{x-1} \quad \beta = 2 \quad \gamma = 1^2 \qquad (iii)$$

$$\alpha = k-1, 1^{x-1} \quad \beta = 1^2 \quad \gamma = 2 \qquad (iv)$$

remembering that $\{k1^x/k-1, 1^{x-1}\} = \{2\} + \{1^2\}$ and that $\{2\} \circ \{1^2\} = \{1^2\} \circ \{2\} \equiv \{1^2\}$. However, the modification rules (15) require that we also consider the terms $< n-1, 2 >$ and $< n-1, 1 >$. These terms arise when $x = n-2k$ they occur as

$< n-1, 2 > +4 < n-2, 1^2 >$ and modify via (15) to $-3\{n-2, 1^2\}$ and hence if $x = n - 2k$ we must count 1 rather than 4 for all other values of x.

All the above leads to the final algorithms for n even and n odd as

n Even

1. Count 1 for each $< 1^x >$ with $n - 2 \geq x \geq 1$ giving $n - 2$.
2. Count 1 for each $< x >$ with $\frac{n}{2} > x \geq 2$ giving $\frac{n-4}{2}$.
3. Count 4 for each $n - 2k - 1 \geq x \geq 1$ $(k > 1)$ giving $n^2 - 8n + 16$.
4. Count 1 for each $n - 2k = x$ $(k > 1)$ giving $\frac{n-4}{2}$.
5. Adding all the terms together gives $n^2 - 6n + 10$.

n Odd

1. Count 1 for each $< 1^x >$ with $n - 2 \geq x \geq 1$ giving $n - 2$.
2. Count 1 for each $< x >$ with $\frac{n+1}{2} > x \geq 2$ giving $\frac{n-3}{2}$.
3. Count 4 for each $n - 2k - 1 \geq x \geq 1$ $(k > 1)$ giving $n^2 - 8n + 15$.
4. Count 1 for each $n - 2k - 1 = x$ $(k > 1)$ giving $\frac{n-3}{2}$.
5. Adding all the terms together gives $n^2 - 6n + 10$.

Thus for general n the coefficient of $\{n - 2, 1^2\}$ is

$$n^2 - 6n + 10$$

Conjecture 1 If we expand

$$\chi_n^{ST}(M_{2,1}) = \sum_{\lambda \in H(2,1;n)} \chi_\lambda \otimes \chi_\lambda = \sum_\mu c_\mu \chi_\mu$$

for up to $n = 17, 18$, it is noticeable and it may stated as a conjecture, the stabilization of coefficients as the column length increases. Thus the coefficient $c_{(1^n)}$ stabilizes at $n = 3$, $c_{(2,1^{n-2})}$ at $n = 6$ and generally $c_{(k,1^{n-k})}$ stabilizes at $n = 3k$ and it is equal to $\frac{k}{3}(2k^2 - 3k + 4)$. One then notices that $c_{(k,2,1^{n-2-k})}$ stabilizes at $n = 3k + 2$, likewise $c_{(k,3,1^{n-3-k})}$ stabilizes for $k = 3$ at $n = 13$, $k = 4$ at $n = 16$. Steps of 3 seem to be relevant.

Remark 1 The computational aspects of this paper were made using SCHUR, an interactive program for calculating the properties of Lie groups and symmetric functions by Brian G. Wybourne [19].

References

1. Berele, A.: Approximate multiplicities in the trace cocharacters sequence of two three by three matrices. Comm. Algebra **25**, 1975–1983 (1997)
2. Berele, A.: Computing supermatrix invariants. Adv. Appl. Math. **48**, 273–289 (2012)
3. Berele, A.: Using hook schur functions to compute matrix cocharacters. Comm. Algebra **41**, 1123–1133 (2013)

4. Carini, L.: Computing cocharacters of sign trace identities in reduced notation. Linear Multilinear Algebra **54**, 147–156 (2006)
5. Carini, L., Carbonara, J.O., Remmel, J.B.: Trace cocharacters and the Kronecker products of Schur functions. J. Algebra **260**, 631–656 (2003)
6. Drensky, V., Genov, G.K.: Multiplicities in the trace cocharacter sequence of two 4x4 matrices. Mediterr. J. Math. **2**, 231–241 (2005)
7. Drensky, V., Genov, G.K., Valenti, A.: Multiplicities in the mixed trace cocharacter sequence of two 3x3 matrices. Internat. J. Algebra Comput. **2**, 275–285 (2006)
8. Littlewood, D.E.: The Theory of Group Characters and Matrix Representations of Groups, 2nd edn. Clarendon Press, Oxford (1950)
9. Littlewood, D.E.: The Kronecker product of symmetric group representations. J. London Math. Soc. **31**, 89–93 (1956)
10. Littlewood, D.E.: The inner plethysm of S-functions. Can. J. Math. **10**, 1–16 (1958)
11. Littlewood, D.E.: Products and plethysms of characters with orthogonal, symplectic and symmetric groups. Can. J. Math. **10**, 17–32 (1958)
12. Macdonald, I.J.: Symmetric Functions and Hall Polynomials. Clarendon Press, Oxford (1995)
13. Murnaghan, F.D.: The analysis of the Kronecker product of irreducible representations of symmetric groups. Amer. J. Math. **60**, 761–784 (1938)
14. Procesi, C.: The invariant theory of nxn matrices. Advances Math. **19**, 306–381 (1976)
15. Razmyslov, Y.P.: Trace identities of full matrix algebras over a field of characteristic zero. Izv. Akad. Nauk. SSSR Ser. Math. **38**, 337–360 (1974)
16. Razmyslov, Y.P.: Trace identities and central polynomials in matrix superalgebras $M_{n,k}$. Mat. Sb. (N.S.)128 (170) **2**, 194–215 (1985)
17. Regev, A.: Sign trace identities. Linear Multilinear Algebra **21**, 1–28 (1987)
18. Remmel, J.B.: Computing cocharacters of sign trace identities. Linear Multilinear Algebra **23**, 1–14 (1988)
19. Scharf, T., Thibon, J.-Y., Wybourne, B.: Reduced notation, inner plethysms and the symmetric group. J. Phys. A Math. Gen. **26**, 7461–7478 (1993)
20. Weyl, H.: The Classical Groups. Princeton University Press, Princeton (1946)

b-Generalized Skew Derivations on Multilinear Polynomials in Prime Rings

Vincenzo De Filippis, Giovanni Scudo, and Feng Wei

Abstract Let R be a prime ring of characteristic different from 2, Q_r be its right Martindale quotient ring and C be its extended centroid. In this paper we define b-generalized skew derivations of prime rings. Then we describe all possible forms of two b-generalized skew derivations F and G satisfying the condition $F(x)x - xG(x) = 0$, for all $x \in S$, where S is the set of the evaluations of a multilinear polynomial $f(x_1, \dots, x_n)$ over C with n non-commuting variables. Several potential research topics related to our current work are also presented.

Keywords Prime rings · Generalized skew derivations · Multilinear polynomials

1 Introduction

In this paper, unless otherwise mentioned, R always denotes a prime ring with center $Z(R)$. We denote the *right Martindale quotient ring* of R by Q_r. The center of Q_r is denoted by C, which is called *extended centroid* of R. We refer the reader to the book [4] for more details.

An additive mapping $d\colon R \longrightarrow R$ is said to be a *derivation* of R if

$$d(xy) = d(x)y + xd(y)$$

for all $x, y \in R$. An additive mapping $F\colon R \longrightarrow R$ is called a *generalized derivation* of R if there exists a derivation d of R such that

$$F(xy) = F(x)y + xd(y)$$

V. De Filippis (✉) · G. Scudo
Department of Engineering, University of Messina, Messina, Italy
e-mail: enzo.defilippis@unime.it

F. Wei
School of Mathematics and Statistics, Beijing Institute of Technology, Beijing, P.R. China
e-mail: gscudo@unime.it

O. M. Di Vincenzo, A. Giambruno (eds.), *Polynomial Identities in Algebras*, Springer INdAM Series 44, https://doi.org/10.1007/978-3-030-63111-6_7

for all $x, y \in R$. The derivation d is uniquely determined by F, which is called an *associated derivation* of F.

The definition of generalized skew derivation is a unified notion of skew derivation and generalized derivation, which are considered as classical additive mappings of non-commutative algebras, have been investigated by many people from various views, see [1, 9, 11–14, 16, 24, 25, 28, 29, 39, 42, 45]. Let R be an associative ring and α be an automorphism of R. An additive mapping $d\colon R \longrightarrow R$ is said to be a *skew derivation* of R if

$$d(xy) = d(x)y + \alpha(x)d(y)$$

for all $x, y \in R$. The automorphisms α is called an *associated automorphism* of d. An additive mapping $F\colon R \longrightarrow R$ is called a *generalized skew derivation* of R if there exists a skew derivation d of R with associated automorphism α such that

$$F(xy) = F(x)y + \alpha(x)d(y)$$

for all $x, y \in R$. In this case, d is called an *associated skew derivation* of F and α is called an *associated automorphism* of F. It was Chang who first introduced this notion and initiated the study of generalized skew derivations of (semi-)prime rings in [10]. Therein, he described the identity of the form $h(x) = af(x) + g(x)b$, where f, g and h are the so-called generalized (α, β)-derivations of a prime ring R, a and b are some fixed noncentral elements of R.

It is worth pointing out that many research papers are devoting to studying the additive mappings in the interfaces between algebra and operator algebra. In [7], Brešar and Villena investigate the automatic continuity of skew derivations on Banach algebras and gave the skew derivation version of noncommutative Singer-Wermer conjecture on Banach algebras. Various technical generalizations of derivations on (semi-)prime rings are used to discuss the range inclusion problems of generalized derivations on noncommutative Banach algebras, see [5, 8, 27, 46, 47]. More recently, Eremita et al determine the structure of generalized skew derivations implemented by elementary operators [30]. Liu and his students characterize a (generalized-)skew derivation F of Banach algebras so that the values of F on a left ideal are nilpotent [41, 43]. Qi and Hou in [45] study generalized skew derivations on nest algebras determined by acting on zero products.

Brešar in [6] gives a description of additive mappings which are commuting on a prime ring R. More precisely, he proves that if F is an additive mapping of R into itself which is centralizing on R and if either R has a characteristic different from 2 or F is commuting on R, then F is of the form $F(x) = \lambda x + \zeta(x)$, where λ is an element of the extended centroid C of R and ζ is an additive mapping of R into C. Moreover, the general situation when two additive mappings F and G of the ring R satisfy $F(x)x - xG(x) \in Z(R)$ for all x in a subset S of R is considered. In particular, it is showed that if $0 \neq F$ and G are both derivations of R and S is a nonzero left ideal of R, then R is commutative. Many researchers successfully extended this result concerning derivations, by replacing S with other subsets of R or replacing F and G with other types of additive mappings. In [49], Wong characterizes derivations F and G of R such that $F(x)x - xG(x) \in Z(R)$, for all $x \in S$, where S is the set of all the evaluations (in a non-zero ideal of R) of a non-central multilinear polynomial over C. Later, Lee and Shiue in [36] extend Wong's result to derivations acting on arbitrary polynomials. Then, in [40], Liu generalizes the theorem of Wong to one-sided ideals. More recently, Chen in [15] extends Lee and Shiue's result to generalized derivations.

In a recent paper [34], Koşan and Lee propose the following new definition. Let $d\colon R \longrightarrow Q_r$ be an additive mapping and $b \in Q_r$. An additive map $F\colon R \longrightarrow Q_r$ is called a *left b-generalized derivation*, with associated mapping d, if $F(xy) = F(x)y + bxd(y)$, for all $x, y \in R$. In the present paper this mapping F will be called *b-generalized derivation* with associated pair (b, d). Clearly, any generalized derivation with associated derivation d is a b-generalized derivation with associated pair $(1, d)$.

In view of this idea, we now give the following:

Definition 1 Let $b \in Q_r$, $d : R \longrightarrow Q_r$ an additive mapping and α be an automorphism of R. An additive mapping $F : R \longrightarrow Q_r$ is called a *b-generalized skew derivation* of R, with associated term (b, α, d) if

$$F(xy) = F(x)y + b\alpha(x)d(y)$$

for all $x, y \in R$.

According to the above definition, we can conclude that general results about b-generalized skew derivations may give useful and powerful corollaries about derivations, generalized derivations, skew derivations and generalized skew derivations.

The main goal of the present paper is to prove the following theorem. It characterizes b-generalized skew derivations which are commuting on multilinear polynomials in prime rings:

Theorem 1 *Let R be a prime ring of characteristic different from 2, Q_r be its right Martindale quotient ring and C be its extended centroid, $\alpha \in Aut(R)$, d and δ skew derivations of R with associated automorphism α, such that both d and δ are commuting with α. Suppose that F, G are b-generalized skew derivations of R, with associated terms (b, α, d) and (p, α, δ), respectively. Let $f(x_1, \ldots, x_n)$ be a non-central multilinear polynomial over C with n non-commuting variables. If*

$$F(f(r_1, \ldots, r_n))f(r_1, \ldots, r_n) - f(r_1, \ldots, r_n)G(f(r_1, \ldots, r_n)) = 0 \tag{1}$$

for all $r_1, \ldots, r_n \in R$, then one of the following statements holds:

1. *there exists $a' \in Q_r$ such that, $F(x) = xa'$ and $G(x) = a'x$ for all $x \in R$;*
2. *$f(x_1, \ldots, x_n)^2$ is central-valued on R and there exist $a', b' \in Q_r$ such that $F(x) = a'x + xb'$, $G(x) = b'x + xa'$, for all $x \in R$.*

Let us recall some results which will be useful in the sequel.

Note 1 Let R be a prime ring, then the following statements hold:

1. Every generalized derivation of R can be uniquely extended to Q_r [35, Theorem 3].
2. Any automorphism of R can be uniquely extended to Q_r [19, Fact 2].
3. Every generalized skew derivation of R can be uniquely extended to Q_r [10, Lemma 2].

Lemma 1 *Let R be a prime ring, $\alpha \in \mathrm{Aut}(R)$, $0 \neq b \in Q_r$, $d\colon R \longrightarrow R$ be an additive mapping of R and F be the b-generalized skew derivation of R with associated term (b, α, d). Then d is a skew derivation of R with associated automorphism α.*

Proof See [26, Lemma 3.2].

Lemma 2 *Let R be a prime ring, $\alpha \in Aut(R)$, $b \in Q_r$, $d\colon R \longrightarrow R$ be an additive mapping of R and F be the b-generalized skew derivation of R with associated term (b, α, d). Then F can be uniquely extended to Q_r and assumes the form $F(x) = ax + bd(x)$, where $a \in Q_r$.*

Proof See [26, Lemma 3.3].

2 Some Results on Differential Identities with Automorphisms

In order to proceed with our proofs, we need to recall some well-known results on skew derivations and automorphisms involved in generalized polynomial identities for prime rings.

Let us denote by $\mathrm{SDer}(Q_r)$ the set of all skew-derivations of Q_r. By a *skew-derivation word* we mean an additive mapping Δ of the form $\Delta = d_1 d_1 \ldots d_m$, where $d_i \in \mathrm{SDer}(Q_r)$. A *skew-differential polynomial* is a generalized polynomial with coefficients in Q_r of the form $\Phi(\Delta_j(x_i))$ involving noncommutative indeterminates x_i on which the skew derivation words Δ_j act as unary operations. The skew-differential polynomial $\Phi(\Delta_j(x_i))$ is said to be a *skew-differential identity* on a subset T of Q_r if it vanishes on any assignment of values from T to its indeterminates x_i.

Let R be a prime ring, $\mathrm{SD}_{\mathrm{int}}$ be the C-subspace of $\mathrm{SDer}(Q_r)$ consisting of all inner skew-derivations of Q_r, and let d and δ be two non-zero skew-derivations of Q_r. The following results follow as special cases from results in [18–21, 33].

Note 2 Let d and δ be skew derivations on R, associated with the same automorphism α of R. Assume that d and δ are $C-$linearly independent modulo $\mathrm{SD}_{\mathrm{int}}$.
If d and δ are commuting with the automorphism α and $\Phi(\Delta_j(x_i))$ is a skew-differential identity on R, where Δ_j are skew-derivations words from the set $\{d, \delta\}$,

then $\Phi(y_{ji})$ is a generalized polynomial identity of R, where y_{ji} are distinct indeterminates (see [33, Theorem 6.5.9]).

In particular, we have

Note 3 In [22] Chuang and Lee investigate polynomial identities with a single skew derivation. They prove that if $\Phi(x_i, D(x_i))$ is a generalized polynomial identity for R, where R is a prime ring and D is an outer skew derivation of R, then R also satisfies the generalized polynomial identity $\Phi(x_i, y_i)$, where x_i and y_i are distinct indeterminates. Furthermore, they observe [22, Theorem 1] that in the case $\Phi(x_i, D(x_i), \alpha(x_i))$ is a generalized polynomial identity for a prime ring R, D is an outer skew derivation of R and α is an outer automorphism of R, then R also satisfies the generalized polynomial identity $\Phi(x_i, y_i, z_i)$, where x_i, y_i, and z_i are distinct indeterminates.

Note 4 If d and δ are $C-$linearly dependent modulo $\mathrm{SD_{int}}$, then there exist $\lambda, \mu \in C$, $a \in Q_r$ and $\alpha \in \mathrm{Aut}(Q_r)$ such that $\lambda d(x) + \mu\delta(x) = ax - \alpha(x)a$ for all $x \in R$.

Note 5 By Chuang and Lee [22] we can state the following result. If d is a non-zero skew-derivation of R and

$$\Phi\left(x_1, \ldots, x_n, d(x_1), \ldots, d(x_n)\right)$$

is a skew-differential polynomial identity of R, then one of the following statements holds:

1. either $d \in \mathrm{SD_{int}}$;
2. or R satisfies the generalized polynomial identity

$$\Phi(x_1, \ldots, x_n, y_1, \ldots, y_n).$$

Note 6 Let R be a prime ring and I be a two-sided ideal of R. Then I, R, and Q_r satisfy the same generalized polynomial identities with coefficients in Q_r (see [18]). Furthermore, I, R, and Q_r satisfy the same generalized polynomial identities with automorphisms (see [20, Theorem 1]).

Note 7 Let R be a prime ring, $\mathrm{Inn}(Q_r)$ be the C-subspace of $\mathrm{Aut}(Q_r)$ consisting of all inner automorphisms of Q_r and let α and β be two non-trivial automorphisms of Q_r.
α and β are called *mutually outer* if $\alpha\beta^{-1}$ is not an inner automorphism of Q_r.
If α and β are mutually outer automorphisms of Q_r and $\Phi(x_i, \alpha(x_i), \beta(x_i))$ is an automorphic identity for R, then by Kharchenko [32, Theorem 4] we know that $\Phi(x_i, y_i, z_i)$ is a generalized polynomial identity for R, where x_i, y_i, z_i are distinct indeterminates.

Note 8 Let R be a prime ring, $\alpha, \beta \in \mathrm{Aut}(Q_r)$ and $d : R \to R$ be a skew derivation, associated with the automorphism α. If there exist $0 \neq \theta \in C$, $0 \neq \eta \in C$ and $u, b \in Q_r$ such that

$$d(x) = \theta\Big(ux - \alpha(x)u\Big) + \eta\Big(bx - \beta(x)b\Big), \quad \forall x \in R \tag{2}$$

then d is an inner skew derivation of R. More precisely, either $b = 0$ or $\alpha = \beta$.

Proof Starting from relation (2) we have

$$d(xy) = \theta\Big(uxy - \alpha(x)\alpha(y)u\Big) + \eta\Big(bxy - \beta(x)\beta(y)b\Big), \quad \forall x, y \in R. \tag{3}$$

On the other hand,

$$\begin{aligned} &d(xy) = d(x)y + \alpha(x)d(y) = \\ &\theta\Big(ux - \alpha(x)u\Big)y + \eta\Big(bx - \beta(x)b\Big)y + \\ &\alpha(x)\theta\Big(uy - \alpha(y)u\Big) + \alpha(x)\eta\Big(by - \beta(y)b\Big). \end{aligned} \tag{4}$$

Comparision of (3) with (4) leads to

$$\eta\Big(\beta(x)\beta(y)b - \beta(x)by + \alpha(x)by - \alpha(x)\beta(y)b\Big) = 0, \quad \forall x, y \in R. \tag{5}$$

Suppose first that α and β are mutually outer, in the sense of Note 7. Therefore, by (5) and since $\eta \neq 0$, it follows that

$$y_1y_2b - y_1by + x_1by - x_1y_2b = 0, \quad \forall x, y, x_1, y_1, y_2 \in R. \tag{6}$$

In particular, for $y_2 = x_1 = 0$ we get $y_1by = 0$, for any $y, y_1 \in R$ and, by the primeness of R, it follows $b = 0$, as required.
Now we assume that α and β are not mutually outer, that is there exists an invertible element $q \in Q_r$ such that $\alpha\beta^{-1}(x) = qxq^{-1}$, for any $x \in R$. Replacing x by $\beta(x)$, it follows easily that $\alpha(x) = q\beta(x)q^{-1}$. Hence by (5)

$$\beta(x)\beta(y)b - \beta(x)by + q\beta(x)q^{-1}by - q\beta(x)q^{-1}\beta(y)b = 0, \quad \forall x, y \in R$$

that is

$$\Big(q\beta(x)q^{-1} - \beta(x)\Big)\Big(\beta(y)b - by\Big) = 0, \quad \forall x, y \in R. \tag{7}$$

Now replace y by yz in (7), then

$$\left(q\beta(x)q^{-1} - \beta(x)\right)\left(\beta(y)\beta(z)b - byz\right) = 0, \quad \forall x, y, z \in R \tag{8}$$

and using (7) in (8) it follows

$$\left(q\beta(x)q^{-1} - \beta(x)\right)\beta(y)\left(\beta(z)b - bz\right) = 0, \quad \forall x, y, z \in R. \tag{9}$$

By the primeness of R, one has that either $\beta(z)b - bz = 0$, for any $z \in R$, or $q\beta(x)q^{-1} - \beta(x) = 0$, for any $x \in R$. In the first case $d(x) = \theta\left(ux - \alpha(x)u\right)$ and we are done. In the latter case, for any $x \in R$ we get $\beta(x) = q\beta(x)q^{-1} = \alpha(x)$ and we are done again.

Note 9 Assuming that $f(x_1, \ldots, x_n)$ is a multilinear polynomial over C and d is a skew derivation of R, associated with the automorphism α, we denote

$$f(x_1, \ldots, x_n) = \sum_{\sigma \in S_n} \gamma_\sigma x_{\sigma(1)} \cdot x_{\sigma(2)} \cdots x_{\sigma(n)}, \quad \gamma_\sigma \in C.$$

Let $f^d(x_1, \ldots, x_n)$ be the polynomial originated from $f(x_1, \ldots, x_n)$ by replacing each coefficient γ_σ with $d(\gamma_\sigma)$. Thus

$$d\left(\gamma_\sigma \cdot x_{\sigma(1)} \cdot x_{\sigma(2)} \cdots x_{\sigma(n)}\right) = d(\gamma_\sigma)x_{\sigma(1)} \cdot x_{\sigma(2)} \cdots x_{\sigma(n)} +$$

$$+\alpha(\gamma_\sigma)\sum_{j=0}^{n-1} \alpha(x_{\sigma(1)} \cdot x_{\sigma(2)} \cdots x_{\sigma(j)})d(x_{\sigma(j+1)})x_{\sigma(j+2)} \cdots x_{\sigma(n)}$$

and

$$d(f(x_1, \ldots, x_n)) = f^d(x_1, \ldots, x_n) +$$

$$+\sum_{\sigma \in S_n} \alpha(\gamma_\sigma)\sum_{j=0}^{n-1} \alpha(x_{\sigma(1)} \cdot x_{\sigma(2)} \cdots x_{\sigma(j)})d(x_{\sigma(j+1)})x_{\sigma(j+2)} \cdots x_{\sigma(n)}.$$

3 Commuting Generalized Derivations and Commuting Generalized Skew Derivations

Here we would like also to collect some results in literature concerning commuting generalized derivations and commuting generalized skew derivations. This section will be useful in the sequel in order to conclude the proof of our main results.

Proposition 1 ([2, Lemma 3]) *Let R be a prime ring, Q_r be its right Martindale quotient ring and C be its extended centroid, $f(x_1,\dots,x_n)$ be a multilinear polynomial over C, which is not central-valued on R. Suppose there exist $a,b,c,q\in Q_r$ such that*

$$\begin{aligned}&\Big(af(r_1,\dots,r_n)+f(r_1,\dots,r_n)b\Big)f(r_1,\dots,r_n)\\&-f(r_1,\dots,r_n)\Big(cf(r_1,\dots,r_n)+f(r_1,\dots,r_n)q\Big)=0\end{aligned}\tag{10}$$

for all $r_1,\dots,r_n\in R$. Then one of the following statements holds:

1. *$a,q\in C$, $q-a=b-c=\alpha\in C$;*
2. *$f(x_1,\dots,x_n)^2$ is central-valued on R and there exists $\alpha\in C$ such that $q-a=b-c=\alpha$;*
3. *$char(R)=2$ and R satisfies S_4.*

Corollary 1 *Let R be a prime ring and $f(x_1,\dots,x_n)$ be a multilinear polynomial over C with n non-commuting variables. Let $a,b\in R$ be such that*

$$af(r_1,\dots,r_n)^2+f(r_1,\dots,r_n)bf(r_1,\dots,r_n)=0$$

for all $r_1,\dots,r_n\in R$. If $f(x_1,\dots,x_n)$ is not central valued on R, then either $a=-b\in C$, or $char(R)=2$ and R satisfies S_4.

Lemma 3 ([2, Lemma 1]) *Let R be a prime ring and $f(x_1,\dots,x_n)$ be a polynomial over C with n non-commuting variables. Let $a,b\in R$ be such that $af(r_1,\dots,r_n)+f(r_1,\dots,r_n)b=0$ for all $r_1,\dots,r_n\in R$. If $f(x_1,\dots,x_n)$ is not a polynomial identity for R, then either $a=-b\in C$, or $f(x_1,\dots,x_n)$ is central-valued on R and $a+b=0$, unless $char(R)=2$ and $R\subseteq M_2(C)$, the 2×2 matrix ring over C.*

Corollary 2 *Let R be a prime ring of characteristic different from 2 and $f(x_1,\dots,x_n)$ be a polynomial over C with n non-commuting variables. Let $a\in R$ be such that $f(r_1,\dots,r_n)a=0$ (or $af(r_1,\dots,r_n)=0$) for all $r_1,\dots,r_n\in R$. If $f(x_1,\dots,x_n)$ is not a polynomial identity for R, then $a=0$.*

Theorem 2 ([2, Theorem 1]) *Let R be a prime ring, Q_r be its right Martindale quotient ring and C be its extended centroid, I a non-zero two-sided ideal of R, F and G non-zero generalized derivations of R. Suppose that $f(x_1,\dots,x_n)$ is a* *non-central multilinear polynomial over C such that*

$$F(f(r_1,\ldots,r_n))f(r_1,\ldots,r_n)-f(r_1,\ldots,r_n)G(f(r_1,\ldots,r_n))=0$$

for all $r_1,\ldots,r_n\in I$, *then one of the following statements holds:*

1. *there exists* $a\in Q_r$ *such that,* $F(x)=xa$ *and* $G(x)=ax$ *for all* $x\in R$;
2. $f(x_1,\ldots,x_n)^2$ *is central-valued on* R *and there exist* $a,b\in Q_r$ *such that* $F(x)=ax+xb$, $G(x)=bx+xa$, *for all* $x\in R$;
3. $char(R)=2$ *and* R *satisfies* S_4, *the standard identity of degree 4.*

4 Some Remarks on Matrix Algebras

Let us state some well-known facts concerning the case when $R=M_m(K)$ is the algebra of $m\times m$ matrices over a field K. Note that the set $f(R)=\{f(r_1,\ldots,r_n)|r_1,\ldots,r_n\in R\}$ is invariant under the action of all inner automorphisms of R. Let us write $r=(r_1,\ldots,r_n)\in R\times R\times\ldots\times R=R^n$. Then for any inner automorphism φ of $M_m(K)$, we get that $\underline{r}=(\varphi(r_1),\ldots,\varphi(r_n))\in R^n$ and $\varphi(f(r))=f(\underline{r})\in f(R)$. As usual, we denote the matrix unit having 1 in (i,j)-entry and zero elsewhere by e_{ij}.

Let us recall some results from [37]. Let T be a ring with 1 and let $e_{ij}\in M_m(T)$ be the matrix unit having 1 in (i,j)-entry and zero elsewhere. For a sequence $u=(A_1,\ldots,A_n)$ in $M_m(T)$, the value of u is defined to be the product $|u|=A_1A_2\cdots A_n$ and u is nonvanishing if $|u|\neq 0$. For a permutation σ of $\{1,2,\cdots,n\}$, we write $u^\sigma=(A_{\sigma(1)},\ldots,A_{\sigma(n)})$. We call u *simple* if it is of the form $u=(a_1e_{i_1j_1},\cdots,a_ne_{i_nj_n})$, where $a_i\in T$. A simple sequence u is called *even* if for some σ, $|u^\sigma|=be_{ii}\neq 0$, and *odd* if for some σ, $|u^\sigma|=be_{ij}\neq 0$, where $i\neq j$. In [37] it is proved that:

Note 10 Let T be a K-algebra with 1 and let $R=M_m(T)$, $m\geq 2$. Suppose that $g(x_1,\ldots,x_n)$ is a multilinear polynomial over K such that $g(u)=0$ for all odd simple sequences u. Then $g(x_1,\ldots,x_n)$ is central-valued on R.

Note 11 Let T be a K-algebra with 1 and let $R=M_m(T)$, $m\geq 2$. Suppose that $g(x_1,\ldots,x_n)$ is a multilinear polynomial over K. Let $u=(A_1,\ldots,A_n)$ be a simple sequence from R.

1. If u is even, then $g(u)$ is a diagonal matrix.
2. If u is odd, then $g(u)=ae_{pq}$ for some $a\in T$ and $p\neq q$.

We also notice that:

Note 12 Since $f(x_1,\ldots,x_n)$ is not central-valued on R, then by Note 10 there exists an odd simple sequence $r=(r_1,\ldots,r_n)$ from R such that $f(r)=f(r_1,\ldots,r_n)\neq 0$. By Note 11, $f(r)=\beta e_{pq}$, where $0\neq\beta\in C$ and $p\neq q$. Since $f(x_1,\ldots,x_n)$ is a multilinear polynomial and C is a field, we may assume that $\beta = 1$. Now, for distinct i, j, let $\sigma \in S_n$ be such that $\sigma(p) = i$ and $\sigma(q) = j$, and let ψ be the automorphism of R defined by $\psi(\sum_{s,t} \xi_{st} e_{st}) = \sum_{s,t} \xi_{st} e_{\sigma(s)\sigma(t)}$. Then $f(\psi(r)) = f(\psi(r_1), \ldots, \psi(r_n)) = \psi(f(r)) = \beta e_{ij} = e_{ij}$.

Note 13 By Note 11 and [37, Lemma 9], since $f(x_1, \ldots, x_n)$ is not central-valued on R, then there exists a sequence of matrices $r_1, \ldots, r_n \in R$ such that $f(r_1, \ldots, r_n) = \sum_i \alpha_i e_{ii} = D$ is a non-central diagonal matrix, for $\alpha_i \in C$. Suppose $r \neq s$ such that $\alpha_r \neq \alpha_s$. For all $l \neq m$, let $\psi \in \mathrm{Aut}_C(R)$ defined by $\psi(x) = \psi(\sum_{ij} \alpha_{ij} e_{ij}) = \sum_{ij} \alpha_{ij} e_{\sigma(i)\sigma(j)}$, where σ is a permutation in the symmetric group of n elements, such that $\sigma(r) = l$ and $\sigma(s) = m$. Thus $\psi(D)$ is an element of $f(R)$ and it is a diagonal matrix with (l, l) and (m, m) entries distinct.

Note 14 ([23, Lemma 1.5]) Let H be an infinite field and $n \geq 2$. If $A_1, \ldots, A_k$ are not scalar matrices in $M_m(\mathcal{H})$ then there exists some invertible matrix $P \in M_m(H)$ such that each matrix $PA_1P^{-1}, \ldots, PA_kP^{-1}$ has all non-zero entries.

5 Commuting Inner *b*-Generalized Skew Derivations

The present section is devoted to the proof of a reduced version of Theorem 1. More precisely, we prove the Theorem in the case α, β are automorphisms of R and F, G are inner b-generalized skew derivations of R respectively defined as follows:

$$F(x) = ax + b\alpha(x)c, \qquad G(x) = ux + p\beta(x)w$$

for all $x \in R$ and suitable fixed $a, b, c, u, p, w \in Q_r$.
We would like to remark that in this section F and G have not necessarily the same associated automorphism.

We start with the following case:

Lemma 4 *Let $R = M_m(C)$, $m \geq 2$ and let C be infinite. Suppose that F, G are inner b-generalized skew derivations of R respectively defined as follows:*

$$F(x) = ax + bqxq^{-1}c, \qquad G(x) = ux + pvxv^{-1}w$$

for all $x \in R$ and suitable fixed $a, b, c, u, p, w, q, v \in Q_r$, with invertible elements q, v of Q_r. Let $f(x_1, \ldots, x_n)$ be a non-central multilinear polynomial over C with n non-commuting variables. If

$$F(f(r_1, \ldots, r_n))f(r_1, \ldots, r_n) - f(r_1, \ldots, r_n)G(f(r_1, \ldots, r_n)) = 0 \qquad (11)$$

for all $r_1, \ldots, r_n \in R$, then the following statements hold simultaneously:

1. *either $bq \in Z(R)$ or $q^{-1}c \in Z(R)$.*
2. *either $pv \in Z(R)$ or $v^{-1}w \in Z(R)$.*

Proof We assume that $bq \notin Z(R)$ and $q^{-1}c \notin Z(R)$, that is both $q^{-1}c$ and bq are not scalar matrices, and prove that a contradiction follows. By Note 14, there exists some invertible matrix $P \in M_m(C)$ such that each matrix $PbqP^{-1}$, $P(q^{-1}c)P^{-1}$ has all non-zero entries. Denote by $\varphi(x) = PxP^{-1}$ the inner automorphism induced by P. Say $\varphi(bq) = \sum_{hl} q_{hl}e_{hl}$ and $\varphi(q^{-1}c) = \sum_{hl} c_{hl}e_{hl}$ for $0 \neq q_{hl}, 0 \neq c_{hl} \in C$. Without loss of generality, we may replace bq and $q^{-1}c$ with $\varphi(bq)$ and $\varphi(q^{-1}c)$, respectively. Hence, for $f(r_1, \ldots, r_n) = \lambda e_{ij} \neq 0$ in (11), we get that the (j, j)-entry in (11) is

$$q_{ji}c_{ji} = 0,$$

which is a contradiction.

Assume now that $pv \notin Z(R)$ and $v^{-1}w \notin Z(R)$, that is both $v^{-1}w$ and pv are not scalar matrices, and prove that a contradiction follows. As above, there exists $\chi(x) = QxQ^{-1}$ the inner automorphism induced by $Q \in R$, such that $\chi(pv) = \sum_{hl} p_{hl}e_{hl}$ and $\chi(v^{-1}w) = \sum_{hl} w_{hl}e_{hl}$ for $0 \neq p_{hl}, 0 \neq w_{hl} \in C$. Moreover we replace pv and $v^{-1}w$ with $\chi(pv)$ and $\chi(v^{-1}w)$, respectively. Hence, again for $f(r_1, \ldots, r_n) = \lambda e_{ij} \neq 0$ in (20), we observe that the (i, i)-entry in (11) is

$$p_{ji}w_{ji} = 0,$$

which is also a contradiction.

Lemma 5 *Let $R = M_m(C)$, $m \geq 2$ and let $char(C) \neq 2$. Suppose that F, G are inner b-generalized skew derivations of R respectively defined as follows:*

$$F(x) = ax + bqxq^{-1}c, \qquad G(x) = ux + pvxv^{-1}w$$

for all $x \in R$ and suitable fixed $a, b, c, u, p, w, q, v \in Q_r$, with invertible elements q, v of Q_r. Let $f(x_1, \ldots, x_n)$ be a non-central multilinear polynomial over C with n non-commuting variables. If

$$F(f(r_1, \ldots, r_n))f(r_1, \ldots, r_n) - f(r_1, \ldots, r_n)G(f(r_1, \ldots, r_n)) = 0$$

for all $r_1, \ldots, r_n \in R$, then one of the following assertions holds:

1. *$bq \in Z(R)$ and $pv \in Z(R)$;*
2. *$bq \in Z(R)$ and $v^{-1}w \in Z(R)$;*
3. *$q^{-1}c \in Z(R)$ and $pv \in Z(R)$;*
4. *$q^{-1}c \in Z(R)$ and $v^{-1}w \in Z(R)$.*

Proof If one assumes that C is infinite, the conclusion follows from Lemma 4.

Now let E be an infinite field which is an extension of the field C and let $\overline{R} = M_t(E) \cong R \otimes_C E$. Notice that the multilinear polynomial $f(x_1, \ldots, x_n)$ is central-valued on R if and only if it is central-valued on $\overline{R}$. Consider the generalized polynomial

$$\begin{aligned}&\Psi(x_1,\ldots,x_n)=\\&\left(af(x_1,\ldots,x_n)+bqf(x_1,\ldots,x_n)q^{-1}c\right)f(x_1,\ldots,x_n)\\&-f(x_1,\ldots,x_n)\left(uf(x_1,\ldots,x_n)+pvf(x_1,\ldots,x_n)v^{-1}w\right),\end{aligned}\tag{12}$$

which is a generalized polynomial identity for R. Moreover, it is multi-homogeneous of multi-degree $(2,\ldots,2)$ in the indeterminates $x_1,\ldots,x_n$. Hence the complete linearization of $\Psi(x_1,\ldots,x_n)$ is a multilinear generalized polynomial $\Theta(x_1,\ldots,x_n,y_1,\ldots,y_n)$. Moreover,

$$\Theta(x_1,\ldots,x_n,x_1,\ldots,x_n)=2^n\Psi(x_1,\ldots,x_n).$$

Clearly, the multilinear polynomial $\Theta(x_1,\ldots,x_n,y_1,\ldots,y_n)$ is a generalized polynomial identity for R and $\overline{R}$ too. Since $\mathrm{char}(C)\neq 2$, we obtain $\Psi(r_1,\ldots,r_n)=0$ for all $r_1,\ldots,r_n\in\overline{R}$, and the conclusion follows from Lemma 4.

Lemma 6 *Assume that*

$$\begin{aligned}&\Psi(x_1,\ldots,x_n)=\\&\left(af(x_1,\ldots,x_n)+bqf(x_1,\ldots,x_n)q^{-1}c\right)f(x_1,\ldots,x_n)\\&-f(x_1,\ldots,x_n)\left(uf(x_1,\ldots,x_n)+pvf(x_1,\ldots,x_n)v^{-1}w\right)\end{aligned}\tag{13}$$

is a generalized polynomial identity for R. If R does not satisfy any non-trivial generalized polynomial identity, then one of the following holds:

1. *$bq\in C$ and $p=0$;*
2. *$bq\in C$ and $v^{-1}w\in C$;*
3. *$q^{-1}c\in C$ and $p=0$;*
4. *$q^{-1}c\in C$ and $v^{-1}w\in C$;*
5. *$a=u\in C$, $q^{-1}c\in C$, $pv\in C$, $bc=0$ and $pw=0$.*

Proof We firstly assume that $a\notin C$.
If $\{a,bq,1\}$ is linearly C-independent and since $\Psi(x_1,\ldots,x_n)$ is a trivial generalized polynomial identity for R, then the component $af(x_1,\ldots,x_n)^2$ is also a trivial generalized identity for R, implying the contradiction $a=0$. Hence we assume there exist $\alpha,\gamma\in C$, such that $bq=\alpha a+\gamma$. In this case (13) reduces to

$$\begin{aligned}&af(x_1,\ldots,x_n)^2+(\alpha a+\gamma)f(x_1,\ldots,x_n)q^{-1}cf(x_1,\ldots,x_n)\\&-f(x_1,\ldots,x_n)uf(x_1,\ldots,x_n)-f(x_1,\ldots,x_n)pvf(x_1,\ldots,x_n)v^{-1}w.\end{aligned}\tag{14}$$

Since $\{1, a\}$ is linearly C-independent and (14) is a trivial generalized polynomial identity for R, then the components

$$af(x_1, \ldots, x_n)(1 + \alpha q^{-1}c) \tag{15}$$

and

$$\begin{aligned}&\gamma f(x_1, \ldots, x_n)q^{-1}cf(x_1, \ldots, x_n) - f(x_1, \ldots, x_n)uf(x_1, \ldots, x_n)\\&-f(x_1, \ldots, x_n)pvf(x_1, \ldots, x_n)v^{-1}w\end{aligned} \tag{16}$$

are also trivial generalized polynomial identities for R. By (15), we get $q^{-1}c \in C$. Thus, in the case $v^{-1}w \in C$ we are done. Here we assume that $v^{-1}w \notin C$, that is $\{1, v^{-1}w\}$ is linearly C-independent. Therefore, by (16) it follows that R satisfies $f(x_1, \ldots, x_n)pvf(x_1, \ldots, x_n)v^{-1}w$, which implies $pv = 0$, that is $p = 0$ (since v is invertible).

Assume now both $a \in C$ and $bq \in C$. Hence (13) reduces to

$$\begin{aligned}&f(x_1, \ldots, x_n)(a + bc)f(x_1, \ldots, x_n)\\&-f(x_1, \ldots, x_n)uf(x_1, \ldots, x_n) - f(x_1, \ldots, x_n)pvf(x_1, \ldots, x_n)v^{-1}w.\end{aligned} \tag{17}$$

Also in this case, if $v^{-1}w \in C$ we are done.

Assume that $\{1, v^{-1}w\}$ is linearly C-independent. Starting from (17) one has that the component $f(x_1, \ldots, x_n)pvf(x_1, \ldots, x_n)v^{-1}w$ must be a trivial generalized polynomial identity for R. This gives that $pv = 0$, that is $p = 0$.
Finally, we consider the case $a \in C$ and $bq \notin C$. Thus, by (13) we have that

$$\begin{aligned}&bqf(x_1, \ldots, x_n)q^{-1}cf(x_1, \ldots, x_n)\\&+f(x_1, \ldots, x_n)(a - u)f(x_1, \ldots, x_n) - f(x_1, \ldots, x_n)pvf(x_1, \ldots, x_n)v^{-1}w\end{aligned} \tag{18}$$

is a trivial generalized polynomial identity for R. Since $bq \notin C$ and by (18), it follows that $bqf(x_1, \ldots, x_n)q^{-1}cf(x_1, \ldots, x_n)$ is also a trivial generalized polynomial identity for R, implying $q^{-1}c \in C$ and $bc = 0$. As above, if $v^{-1}w \in C$ we are done. On the other hand, if $v^{-1}w \notin C$ and again by (18), one has that $f(x_1, \ldots, x_n)pvf(x_1, \ldots, x_n)v^{-1}w$ is a trivial generalized polynomial identity for R. This means that $pv \in C$ and $pw = 0$. In light of what has just been said and by (18), R satisfies

$$f(x_1, \ldots, x_n)(a - u)f(x_1, \ldots, x_n) \tag{19}$$

that is $a = u$.

Remark 1 We would like to remark that any conclusion of the previous Lemma implies that F and G are generalized derivations of R. Hence, in view of Theorem 2, the statement of Lemma 6 can be written as follows: there exists $a' \in Q_r$ such that $F(x) = xa'$ and $G(x) = a'x$ for all $x \in R$.

Proposition 2 *Let R be a prime ring of characteristic different from 2, Q_r be its right Martindale quotient ring and C be its extended centroid. Suppose that F, G are inner b-generalized skew derivations of R respectively defined as follows:*

$$F(x) = ax + bqxq^{-1}c, \qquad G(x) = ux + pvxv^{-1}w$$

for all $x \in R$ and suitable fixed $a, b, c, u, p, w, q, v \in Q_r$, with invertible elements q, v of Q_r. Let $f(x_1, \ldots, x_n)$ be a non-central multilinear polynomial over C with n non-commuting variables. If

$$F(f(r_1, \ldots, r_n))f(r_1, \ldots, r_n) - f(r_1, \ldots, r_n)G(f(r_1, \ldots, r_n)) = 0$$

for all $r_1, \ldots, r_n \in R$, then one of the following statements holds:

1. *$bq \in Z(R)$ and $pv \in C$;*
2. *$bq \in Z(R)$ and $v^{-1}w \in C$;*
3. *$q^{-1}c \in Z(R)$ and $pv \in C$;*
4. *$q^{-1}c \in Z(R)$ and $v^{-1}w \in C$.*

In other words, F and G are generalized derivations of R and one of the following statements holds:

1. *there exists $a' \in Q_r$ such that, $F(x) = xa'$ and $G(x) = a'x$ for all $x \in R$;*
2. *$f(x_1, \ldots, x_n)^2$ is central-valued on R and there exist $a', b' \in Q_r$ such that $F(x) = a'x + xb'$, $G(x) = b'x + xa'$, for all $x \in R$.*

Proof If R does not satisfy any non-trivial generalized polynomial identity, then the conclusion follows from Lemma 6. Therefore we may assume that

$$\begin{aligned}&\Psi(x_1, \ldots, x_n) = \\ &\left(af(x_1, \ldots, x_n) + bqf(x_1, \ldots, x_n)q^{-1}c\right)f(x_1, \ldots, x_n) \\ &-f(x_1, \ldots, x_n)\left(uf(x_1, \ldots, x_n) + pvf(x_1, \ldots, x_n)v^{-1}w\right)\end{aligned} \qquad (20)$$

is a non-trivial generalized polynomial identity for R.

By Chuang [18] it follows that $\Psi(x_1, \ldots, x_n)$ is a non-trivial generalized polynomial identity for Q_r. By the well-known Martindale's theorem of [44], Q_r is a primitive ring having nonzero socle with the field C as its associated division ring. By Jacobson [31, Page 75] Q_r is isomorphic to a dense subring of the ring of linear transformations of a vector space V over C, containing nonzero linear transformations of finite rank. Assume first that $\dim_C V = k \geq 2$ is a finite positive integer, then $Q \cong M_k(C)$ and the conclusion follows from Lemma 5.

Let us now consider the case of $\dim_C V = \infty$. As in [48, Lemma 2], the set $f(R) = \{f(r_1, \dots, r_n) | r_i \in R\}$ is dense on R. By the fact that $\Psi(r_1, \dots, r_n) = 0$ is a generalized polynomial identity of R, we know that R satisfies

$$\left(ax + bqxq^{-1}c\right)x - x\left(ux + pvxv^{-1}w\right). \tag{21}$$

Recall that if an element $r \in R$ centralizes the non-zero ideal $H = \text{soc}(RC)$, then $r \in C$.

Hence we may assume there exist $r_1, r_2, r_3, r_4 \in H = \text{soc}(RC)$ such that:

1. either $[bq, r_1] \neq 0$ or $[pv, r_1] \neq 0$;
2. either $[bq, r_2] \neq 0$ or $[v^{-1}w, r_2] \neq 0$
3. either $[q^{-1}c, r_3] \neq 0$ or $[pv, r_3] \neq 0$
4. either $[q^{-1}c, r_4] \neq 0$ or $[v^{-1}w, r_4] \neq 0$

and prove that a number of contradictions follows.

By Litoff's Theorem [31, Page 90] there exists $e^2 = e \in H$ such that

- $r_1, r_2, r_3, r_4 \in eRe$;
- $ar_1, r_1a, ar_2, r_2a, ar_3, r_3a, ar_4, r_4a \in eRe$;
- $br_1, r_1b, br_2, r_2b, br_3, r_3b, br_4, r_4b \in eRe$;
- $cr_1, r_1c, cr_2, r_2c, cr_3, r_3c, cr_4, r_4c \in eRe$;
- $qr_1, r_1q, qr_2, r_2q, qr_3, r_3q, qr_4, r_4q \in eRe$;
- $ur_1, r_1u, ur_2, r_2u, ur_3, r_3u, ur_4, r_4u \in eRe$;
- $pr_1, r_1p, pr_2, r_2p, pr_3, r_3p, pr_4, r_4p \in eRe$;
- $vr_1, r_1v, vr_2, r_2v, vr_3, r_3v, vr_4, r_4v \in eRe$;
- $wr_1, r_1w, wr_2, r_2w, wr_3, r_3w, wr_4, r_4w \in eRe$;
- $pvr_1, r_1pv, pvr_2, r_2pv, pvr_3, r_3pv, pvr_4, r_4pv \in eRe$;
- $bqr_1, r_1bq, bqr_2, r_2bq, bqr_3, r_3bq, bqr_4, r_4bq \in eRe$;
- $q^{-1}cr_1, r_1q^{-1}c, q^{-1}cr_2, r_2q^{-1}c, q^{-1}cr_3, r_3q^{-1}c, q^{-1}cr_4, r_4q^{-1}c \in eRe$;
- $v^{-1}wr_1, r_1v^{-1}w, v^{-1}wr_2, r_2v^{-1}w, v^{-1}wr_3, r_3v^{-1}w, v^{-1}wr_4, r_4v^{-1}w \in eRe$,

where $eRe \cong M_m(C)$, the matrix ring over the extended centroid C. Note that eRe satisfies (21). By the above Lemma 5, we have that one of the following assertions holds:

1. $ebqe \in C$ and $epve \in C$, which contradicts with the choice of $r_1 \in H$;
2. $ebqe \in C$ and $ev^{-1}we \in C$, which contradicts with the choice of $r_2 \in H$;
3. $eq^{-1}ce \in C$ and $epve \in C$, which contradicts with the choice of $r_3 \in H$;
4. $eq^{-1}ce \in C$ and $ev^{-1}we \in C$, which contradicts with the choice of $r_4 \in H$.

As an easy consequence of Proposition 2 we also obtain a reduced version of Theorem 1 for the case both F and G are inner b-generalized derivations of R:

Proposition 3 *Let R be a prime ring of characteristic different from 2, Q_r be its right Martindale quotient ring and C be its extended centroid. Suppose that F, G are inner b-generalized derivations of R respectively defined as follows:*

$$F(x) = ax + bxc, \qquad G(x) = px + qxv$$

for all $x \in R$ and suitable fixed $a, b, c, p, q, v \in Q_r$. Let $f(x_1, \dots, x_n)$ be a non-central multilinear polynomial over C with n non-commuting variables. If

$$F(f(r_1, \dots, r_n))f(r_1, \dots, r_n) - f(r_1, \dots, r_n)G(f(r_1, \dots, r_n)) = 0$$

for all $r_1, \dots, r_n \in R$, then one of the following holds:

1. *there exists $a' \in Q_r$ such that, $F(x) = xa'$ and $G(x) = a'x$ for all $x \in R$;*
2. *$f(x_1, \dots, x_n)^2$ is central-valued on R and there exist $a', b' \in Q_r$ such that $F(x) = a'x + xb'$, $G(x) = b'x + xa'$, for all $x \in R$.*

We are now ready to prove the more general result of this section.
We permit the following facts:

Note 15 Let R be a non-commutative prime ring, $a, b \in R$ such that $axb \in Z(R)$, for all $x \in R$. Then either $a = 0$ or $b = 0$.

Proof We assume that $a \neq 0$ and $b \neq 0$. For any $x \in R$ and by our assumption, both $a(xb) \in Z(R)$ and $a(xb)b \in Z(R)$. Thus we have that either $b \in Z(R)$ or $axb = 0$ for all $x \in R$. In the first case it follows that $aR \subseteq Z(R)$, which contradicts with the non-commutativity of R. In the latter case, by the primeness of R, we have the required conclusion.

Note 16 Let R be a non-commutative prime ring, $a, b \in R$, $f(x_1, \dots, x_n)$ a polynomial over C, which is not central valued on R. If $af(r_1, \dots, r_n)b \in Z(R)$, for all $r_1, \dots, r_n \in R$, then either $a = 0$ or $b = 0$.

Proof Let S be the additive subgroup of R generated by $\{f(y_1, \dots, y_n) : y_i \in R\}$. Since $f(y_1, \dots, y_n)$ is not central and $char(R) \neq 2$, it is well known that S contains a non-central Lie ideal L of R (see [17]). Moreover, since L is not central then there exists a non-central ideal I of R such that $[I, R] \subseteq L$. Therefore $a[i, r]b \in Z(R)$, for any $i \in I$, $r \in R$. Since I and Q_r satisfy the same generalized identities it follows that $a[x, y]b \in C$ for any $x, y \in Q_r$. In this situation we may apply the main result in [3] and one of the following holds: either $a = 0$ or $b = 0$ or Q_r is a central simple algebra of dimension at most 4 over C. Moreover, since Q_r is not commutative, then Q_r contains some non-trivial idempotent elements $e = e^2$. In this last case, by the main hypothesis, one has $a[e, x(1-e)]b \in C$, that is $aex(1-e)b \in C$, for all $x \in Q_r$. By Note 15, either $ae = 0$ or $(1 - e)b = 0$.
If $ae = 0$ and by $a[y, ex]b \in C$, we get $ayexb \in C$, for any $x, y \in Q_r$. Thus, using Note 15 and since $e \neq 0$, it follows that either $a = 0$ or $b = 0$, as required.

On the other hand, if $(1-e)b = 0$ and by $a[x, y(1-e)]b \in C$, we have that $ay(1-e)xb \in C$, for any $x, y \in Q_r$. Once again by Note 15 and since $e \neq 1$, we get their required conclusion.

Theorem 3 *Let R be a prime ring of characteristic different from 2, Q_r be its right Martindale quotient ring and C be its extended centroid. Suppose that F, G are inner b-generalized skew derivations of R respectively defined as follows:*

$$F(x) = ax + b\alpha(x)c, \qquad G(x) = ux + p\beta(x)w$$

for all $x \in R$ and suitable fixed $a, b, c, u, p, w \in Q_r$, and $\alpha, \beta \in \mathrm{Aut}(Q_r)$. Let $f(x_1, \ldots, x_n)$ be a non-central multilinear polynomial over C with n non-commuting variables. If

$$F(f(r_1, \ldots, r_n))f(r_1, \ldots, r_n) - f(r_1, \ldots, r_n)G(f(r_1, \ldots, r_n)) = 0 \tag{22}$$

for all $r_1, \ldots, r_n \in R$, then one of the following statements holds:

1. *$\alpha = \beta = id$, where id denotes the identical mapping on Q_r;*
2. *$\alpha = id$ and there exists an invertible element $v \in Q_r$ such that $\beta(x) = vxv^{-1}$, for all $x \in R$;*
3. *$\beta = id$ and there exists an invertible element $q \in Q_r$ such that $\alpha(x) = qxq^{-1}$, for all $x \in R$;*
4. *$\beta = id$ and $b = 0$;*
5. *$\beta = id$ and $c = 0$;*
6. *$\alpha = id$ and $p = 0$;*
7. *$\alpha = id$ and $w = 0$;*
8. *there exists an invertible element $q \in Q_r$ such that $\alpha(x) = qxq^{-1}$, for all $x \in R$, and either $p = 0$ or $w = 0$;*
9. *there exists an invertible element $v \in Q_r$ such that $\beta(x) = vxv^{-1}$, for all $x \in R$, and either $b = 0$ or $c = 0$;*
10. *$b = p = 0$;*
11. *$b = w = 0$;*
12. *$c = p = 0$;*
13. *$c = w = 0$;*
14. *there exist invertible elements $q, v \in Q_r$ such that $\alpha(x) = qxq^{-1}$ and $\beta(x) = vxv^{-1}$, for all $x \in R$.*

In other words one of the following occurs:

- *F and G are ordinary generalized derivations of R.*
- *F and G are inner b-generalized derivations;*
- *F and G are inner b-generalized skew derivations of R, associated with inner automorphisms;*

In any case, respectively in light of Propositions 1, 3 and 2, we have that one of the following statements holds:

1. *there exists* $a' \in Q_r$ *such that,* $F(x) = xa'$ *and* $G(x) = a'x$ *for all* $x \in R$*;*
2. $f(x_1, \ldots, x_n)^2$ *is central-valued on R and there exist* $a', b' \in Q_r$ *such that* $F(x) = a'x + xb'$, $G(x) = b'x + xa'$, *for all* $x \in R$.

Proof On the contrary, we assume that the following hold simultaneously:

- either $\alpha \neq id$ or $\beta \neq id$;
- either $\alpha \neq id$ or β is not an inner automorphism on Q_r;
- either $\beta \neq id$ or α is not an inner automorphism on Q_r;
- either $\alpha \neq id$ or $b \neq 0$;
- either $\alpha \neq id$ or $c \neq 0$;
- either $\beta \neq id$ or $p \neq 0$;
- either $\beta \neq id$ or $w \neq 0$;
- either α is not inner, or both $p \neq 0$ and $w \neq 0$;
- either β is not inner, or both $b \neq 0$ and $c \neq 0$;
- either $b \neq 0$ or $p \neq 0$;
- either $b \neq 0$ or $w \neq 0$;
- either $c \neq 0$ or $p \neq 0$;
- either $c \neq 0$ or $w \neq 0$;
- at least one among α and β is not an inner automorphism of R.

By our assumption R satisfies the following generalized polynomial

$$\Big(af(x_1, \ldots, x_n) + b\alpha\big(f(x_1, \ldots, x_n)\big)c\Big)f(x_1, \ldots, x_n) \\ -f(x_1, \ldots, x_n)\Big(uf(x_1, \ldots, x_n) + p\beta\big(f(x_1, \ldots, x_n)\big)w\Big). \tag{23}$$

In view of the Note 6, Q_r satisfies (23).

In case $\alpha = id$, then β is not inner. Thus, by (23), Q_r satisfies the generalized polynomial

$$\Big(af(x_1, \ldots, x_n) + b(f(x_1, \ldots, x_n)c\Big)f(x_1, \ldots, x_n) \\ -f(x_1, \ldots, x_n)\Big(uf(x_1, \ldots, x_n) + pf^{\beta}(y_1, \ldots, y_n)w\Big). \tag{24}$$

In particular, $pf^{\beta}(y_1, \ldots, y_n)w$ is a generalized polynomial identity for Q_r. It is easy to see that $pXw = 0$, for any $X \in S$, the additive subgroup of Q_r generated by $\{f^{\beta}(y_1, \ldots, y_n) : y_i \in Q_r\}$. Since $f^{\beta}(y_1, \ldots, y_n)$ is not central and $char(Q_r) \neq 2$, it is well known that S must contain a non-central Lie ideal L. This implies $pLw = (0)$ and, by the primeness of Q_r we get the contradiction that either $p = 0$ or $w = 0$.

Similarly, if we assume that $\beta = id$, then we obtain the contradiction that either $b = 0$ or $c = 0$.

Thus we may suppose both $\alpha \neq id$ and $\beta \neq id$. In what follows we denote $f^{\alpha}(x_1, \ldots, x_n) = \alpha\Big(f(x_1, \ldots, x_n)\Big)$.

If α and β are mutually outer, then by (23), Q_r satisfies

$$\begin{aligned}&\Big(af(x_1, \ldots, x_n) + bf^{\alpha}(y_1, \ldots, y_n)c\Big)f(x_1, \ldots, x_n)\\ &-f(x_1, \ldots, x_n)\Big(uf(x_1, \ldots, x_n) + pf^{\beta}(z_1, \ldots, z_n)w\Big).\end{aligned} \tag{25}$$

In particular, Q_r satisfies both

$$bf^{\alpha}(y_1, \ldots, y_n)cf(x_1, \ldots, x_n)$$

and

$$f(x_1, \ldots, x_n)pf^{\beta}(z_1, \ldots, z_n)w.$$

Applying twice Corollary 2 to both last relations yields that either $b = 0$ or $c = 0$ and simultaneously either $p = 0$ or $w = 0$, which is a contradiction.

Assume finally that α and β are not mutually outer, then exists an invertible element $q \in Q_r$ such that $\alpha\beta^{-1}(x) = qxq^{-1}$, for any $x \in R$. Therefore $\alpha(x) = q\beta(x)q^{-1}$ and by (23) it follows that Q_r satisfies

$$\begin{aligned}&\Big(af(x_1, \ldots, x_n) + bq\beta\big(f(x_1, \ldots, x_n)\big)q^{-1}c\Big)f(x_1, \ldots, x_n)\\ &-f(x_1, \ldots, x_n)\Big(uf(x_1, \ldots, x_n) + p\beta\big(f(x_1, \ldots, x_n)\big)w\Big).\end{aligned} \tag{26}$$

If β is an inner automorphism of Q_r, then the required conclusion follows from Proposition 2. On the other hand, if β is outer, then, by (26) we have that Q_r satisfies

$$\begin{aligned}&\Big(af(x_1, \ldots, x_n) + bqf^{\beta}(y_1, \ldots, y_n)q^{-1}c\Big)f(x_1, \ldots, x_n)\\ &-f(x_1, \ldots, x_n)\Big(uf(x_1, \ldots, x_n) + pf^{\beta}(y_1, \ldots, y_n)w\Big)\end{aligned} \tag{27}$$

and in particular

$$bqf^{\beta}(y_1, \ldots, y_n)q^{-1}cf(x_1, \ldots, x_n) - f(x_1, \ldots, x_n)pf^{\beta}(y_1, \ldots, y_n)w \tag{28}$$

is a generalized polynomial identity for Q_r. Since $f(x_1, \ldots, x_n)$ is not central valued and in light of Lemma 3, one has that $bqf^{\beta}(y_1, \ldots, y_n)q^{-1}c =$

$pf^{\beta}(y_1, \ldots, y_n)w \in C$ for any $y_1, \ldots, y_n \in Q_r$. Hence Note 16 implies that the following hold simultaneously:

- either $b = 0$ or $c = 0$;
- either $p = 0$ or $w = 0$

and in any case we get a contradiction.

6 Commuting *b*-Generalized Derivations on Multilinear Polynomials

In this section we provide a proof of Theorem 1 in the case both F and G are arbitrary b-generalized derivations (not necessarily inner) and prove the following:

Theorem 4 *Let R be a prime ring of characteristic different from 2, Q_r be its right Martindale quotient ring and C be its extended centroid, F and G non-zero b-generalized derivations of R. Suppose that $f(x_1, \ldots, x_n)$ is a non-central multilinear polynomial over C such that $F(f(X))f(X) - f(X)G(f(X)) = 0$, for all $X = (x_1, \ldots, x_n) \in R^n$, then one of the following statements holds:*

1. *there exists $u \in Q_r$ such that, $F(x) = xu$ and $G(x) = ux$ for all $x \in R$;*
2. *$f(x_1, \ldots, x_n)^2$ is central-valued on R and there exist $a, b \in Q_r$ such that $F(x) = ax + xb$, $G(x) = bx + xa$, for all $x \in R$.*

Hence F and G are generalized derivations of R.

Proof As mentioned in the Introduction, we can write $F(x) = ax + bd(x)$, $G(x) = px + q\delta(x)$ for all $x \in R$, where $a, b, p, q \in Q_r$ and d, δ are derivations of R. In light of Proposition 3, we may assume that:

- At least one among d and δ is not an inner derivation of R;
- At least one among b and q is not zero;
- If d is an inner derivation of R, then $\delta \neq 0$ and $q \neq 0$;
- If δ is an inner derivation of R then $d \neq 0$ and $b \neq 0$.

We will prove that, under these assumptions, a number of contradiction follows.

Assume first that d and δ are both non-zero derivations and linearly C-independent modulo Q_r-inner derivations. Since Q_r satisfies

$$\Big(af(x_1, \ldots, x_n) + bf^d(x_1, \ldots, x_n) + b\sum_{i=1}^n f(x_1, \ldots, d(x_i), \ldots, x_n)\Big)f(x_1, \ldots, x_n)$$
$$-f(x_1, \ldots, x_n)\Big(pf(x_1, \ldots, x_n) + qf^{\delta}(x_1, \ldots, x_n) + q\sum_{i=1}^n f(x_1, \ldots, \delta(x_i), \ldots, x_n)\Big) \tag{29}$$

and by Kharchenko [32], we arrive at that Q_r satisfies

$$\Big(af(x_1,\ldots,x_n)+bf^d(x_1,\ldots,x_n)+b\sum_{i=1}^n f(x_1,\ldots,y_i,\ldots,x_n)\Big)f(x_1,\ldots,x_n)$$
$$-f(x_1,\ldots,x_n)\Big(pf(x_1,\ldots,x_n)+qf^\delta(x_1,\ldots,x_n)+q\sum_{i=1}^n f(x_1,\ldots,z_i,\ldots,x_n)\Big). \tag{30}$$

In particular, Q_r satisfies the blended components

$$bf(y_1,x_2,\ldots,x_n)\cdot f(x_1,\ldots,x_n)$$

and

$$f(x_1,\ldots,x_n)\cdot q\cdot f(y_1,x_2,\ldots,x_n),$$

which imply the contradiction $b=0$ (by Corollary 2) and $q=0$ (by Corollary 1).

Assume now that d and δ are both non-zero derivations and C-dependent modulo Q_r-inner derivations. Without loss of generality, we assume that $\delta=\lambda d+ad_w$, that is $\delta(x)=\lambda d(x)+[w,x]$, for suitable $0\neq\lambda\in C$ and $w\in Q_r$. Moreover, in light of the previous remarks, d is not an inner derivation of R. By the hypothesis we have that

$$\Big(af(x_1,\ldots,x_n)+bf^d(x_1,\ldots,x_n)+b\sum_{i=1}^n f(x_1,\ldots,d(x_i),\ldots,x_n)\Big)f(x_1,\ldots,x_n)$$
$$-f(x_1,\ldots,x_n)\Big(pf(x_1,\ldots,x_n)+\lambda qf^d(x_1,\ldots,x_n)+$$
$$+\lambda q\sum_{i=1}^n f(x_1,\ldots,d(x_i),\ldots,x_n)+q[w,f(x_1,\ldots,x_n)]\Big) \tag{31}$$

is a differential polynomial identity for Q_r, and again by Kharchenko [32] it follows that Q_r satisfies

$$\Big(af(x_1,\ldots,x_n)+bf^d(x_1,\ldots,x_n)+b\sum_{i=1}^n f(x_1,\ldots,y_i,\ldots,x_n)\Big)f(x_1,\ldots,x_n)$$
$$-f(x_1,\ldots,x_n)\Big(pf(x_1,\ldots,x_n)+\lambda qf^d(x_1,\ldots,x_n)+$$
$$+\lambda q\sum_{i=1}^n f(x_1,\ldots,y_i,\ldots,x_n)+q[w,f(x_1,\ldots,x_n)]\Big). \tag{32}$$

In particular, Q_r satisfies the blended component

$$b\sum_i f(x_1,\ldots,y_i,\ldots,x_n)f(x_1,\ldots,x_n)-\lambda f(x_1,\ldots,x_n)q\sum_i f(x_1,\ldots,y_i,\ldots,x_n). \tag{33}$$

Let us choose $y_2 = y_3 = \ldots = y_n = 0$ and $y_1 = x_1$ in (33). This yields that Q_r satisfies

$$bf(x_1,\ldots,x_n)^2 - \lambda f(x_1,\ldots,x_n)qf(x_1,\ldots,x_n). \tag{34}$$

Moreover, for $z \notin C$ and $y_i = [z, x_i]$ for any $i = 1,\ldots,n$ in (33), we also have that

$$b[z, f(x_1,\ldots,x_n)]f(x_1,\ldots,x_n) - \lambda f(x_1,\ldots,x_n)q[z, f(x_1,\ldots,x_n)] \tag{35}$$

is a generalized polynomial identity for Q_r. Application of Proposition 1 to (34) implies that $b = \lambda q \in C$. Therefore, by (35) it follows that Q_r satisfies

$$b[z, f(x_1,\ldots,x_n)]_2.$$

Since $z \notin C$ and since neither $char(R) = 2$ nor $f(x_1,\ldots,x_n)$ is central-valued on R, by Liu [38] we get $b = 0$, and so also $q = 0$, which is a contradiction.

We finally consider the case either $d = 0$ or $\delta = 0$. Without loss of generality, we may assume $\delta = 0$ (the case $d = 0$ is similar and we omit it for brevity). By our assumption it follows that Q_r satisfies

$$\left(af(x_1,\ldots,x_n) + bf^d(x_1,\ldots,x_n) + b\sum_{i=1}^n f(x_1,\ldots,d(x_i),\ldots,x_n)\right)f(x_1,\ldots,x_n)$$
$$-f(x_1,\ldots,x_n)pf(x_1,\ldots,x_n). \tag{36}$$

Moreover, as above remarked, in this case d is not an inner derivation of R. In view of Kharchenko's theorem in [32], Q_r satisfies

$$\left(af(x_1,\ldots,x_n) + bf^d(x_1,\ldots,x_n) + b\sum_{i=1}^n f(x_1,\ldots,y_i,\ldots,x_n)\right)f(x_1,\ldots,x_n)$$
$$-f(x_1,\ldots,x_n)pf(x_1,\ldots,x_n). \tag{37}$$

Therefore

$$bf(y_1,x_2,\ldots,x_n)f(x_1,\ldots,x_n) \tag{38}$$

is a generalized polynomial identity for Q_r, implying again the contradiction $b = 0$.

7 The Main Result

The last part of our paper is dedicated to the proof of Theorem 1 in its most general form. For sake of clearness and completeness, we recall our hypothesis.

We assume that R is a prime ring of characteristic different from 2, Q_r its right Martindale quotient ring and C its extended centroid, $\alpha \in \mathrm{Aut}(R)$, d and δ skew derivations of R associated with α, such that both d and δ are commuting with α. We suppose that F, G are b-generalized skew derivations of R, respectively associated with terms (b, α, d) and (p, β, δ). We may write $F(x) = ax + bd(x)$ and $G(x) = ux + p\delta(x)$, for all $x \in R$ and suitable $a, u \in Q_r$. We assume that $f(x_1, \ldots, x_n)$ is a non-central multilinear polynomial over C with n non-commuting variables, such that

$$F(f(r_1, \ldots, r_n))f(r_1, \ldots, r_n) - f(r_1, \ldots, r_n)G(f(r_1, \ldots, r_n)) = 0 \tag{39}$$

for all $r_1, \ldots, r_n \in R$, that is R satisfies

$$\begin{aligned}&\Big(af(x_1, \ldots, x_n) + bd(f(x_1, \ldots, x_n))\Big)f(x_1, \ldots, x_n)\\ &-f(x_1, \ldots, x_n)\Big(uf(x_1, \ldots, x_n) + p\delta(f(x_1, \ldots, x_n))\Big).\end{aligned} \tag{40}$$

Under these assumptions, we'll prove that one of the following statements holds:

1. $d = \delta = 0$;
2. $\alpha = id$;
3. there exist $b', c' \in Q_r$ such that $d(x) = b'x - \alpha(x)b'$ and $\delta(x) = c'x - \alpha(x)c'$, for all $x \in R$;
4. $b = p = 0$;
5. $b = 0$ and $\delta = 0$;
6. $p = 0$ and $d = 0$.

In other words, either F and G are generalized derivations of R, or F and G are b-generalized derivations of R, or F and G are inner b-generalized skew derivations of R. Therefore, respectively in light of Theorems 2, 4 and 3, we have that one of the following holds:

1. there exists $a' \in Q_r$ such that, $F(x) = xa'$ and $G(x) = a'x$ for all $x \in R$;
2. $f(x_1, \ldots, x_n)^2$ is central valued on R and there exist $a', b' \in Q_r$ such that $F(x) = a'x + xb'$, $G(x) = b'x + xa'$, for all $x \in R$.

Proof of Theorem 1 By (40) and Note 9,

$$\begin{aligned}&\Big(af(x_1, \ldots, x_n) + bf^d(x_1, \ldots, x_n) +\\ &b\textstyle\sum_{\sigma \in S_n} \alpha(\gamma_\sigma) \sum_{j=0}^{n-1} \alpha(x_{\sigma(1)} \cdot x_{\sigma(2)} \cdots x_{\sigma(j)})d(x_{\sigma(j+1)})x_{\sigma(j+2)} \cdots x_{\sigma(n)}\Big)f(x_1, \ldots, x_n)\\ &-f(x_1, \ldots, x_n)\Big(uf(x_1, \ldots, x_n) + pf^\delta(x_1, \ldots, x_n)\\ &+p\textstyle\sum_{\sigma \in S_n} \beta(\gamma_\sigma) \sum_{j=0}^{n-1} \alpha(x_{\sigma(1)} \cdot x_{\sigma(2)} \cdots x_{\sigma(j)})\delta(x_{\sigma(j+1)})x_{\sigma(j+2)} \cdots x_{\sigma(n)}\Big)\end{aligned} \tag{41}$$

is a generalized identity for R.

On the contrary we assume that the following hold simultaneously:

- either $d \neq 0$ or $\delta \neq 0$;
- $\alpha \neq id$;
- at least one among d and δ is not an inner skew derivation of R;
- at least one among b and p is not zero;
- at least one among b or δ is not zero;
- at least one among p or d is not zero.

7.1 Let d and δ be C−Linearly Independent Modulo SD_{int}

In this case, in view of (41) we know that R satisfies the generalized polynomial

$$\begin{aligned}&\Big(af(x_1,\ldots,x_n)+bf^d(x_1,\ldots,x_n)+\\&b\textstyle\sum_{\sigma\in S_n}\alpha(\gamma_\sigma)\sum_{j=0}^{n-1}\alpha(x_{\sigma(1)}\cdot x_{\sigma(2)}\cdots x_{\sigma(j)})y_{\sigma(j+1)}x_{\sigma(j+2)}\cdots x_{\sigma(n)}\Big)f(x_1,\ldots,x_n)\\&-f(x_1,\ldots,x_n)\Big(uf(x_1,\ldots,x_n)+pf^\delta(x_1,\ldots,x_n)\\&+p\textstyle\sum_{\sigma\in S_n}\beta(\gamma_\sigma)\sum_{j=0}^{n-1}\alpha(x_{\sigma(1)}\cdot x_{\sigma(2)}\cdots x_{\sigma(j)})z_{\sigma(j+1)}x_{\sigma(j+2)}\cdots x_{\sigma(n)}\Big).\end{aligned} \tag{42}$$

In particular, R satisfies any blended component

$$b\Big(\sum_{\sigma\in S_n}\alpha(\gamma_\sigma)\sum_{i=1}^{n}\alpha(x_{\sigma(1)}\cdot x_{\sigma(2)}\cdots x_{\sigma(i-1)})y_{\sigma(i)}x_{\sigma(i+1)}\cdots x_{\sigma(n)}\Big)f(x_1,\ldots,x_n). \tag{43}$$

In light of the Note 6, Q_r satisfies (43).

Suppose there exists an invertible element $q\in Q_r$ such that $\alpha(x)=qxq^{-1}$ for all $x\in Q_r$. Since $\alpha\neq id\in \mathrm{Aut}(R)$, we may assume $q\notin C$. Moreover, it is clear that $\alpha(\gamma_\sigma)=\gamma_\sigma$ for all coefficients involved in $f(x_1,\ldots,x_n)$. If we replace each $y_{\sigma(i)}$ with $qx_{\sigma(i)}$ in (43), then Q_r satisfies the generalized polynomial

$$b\Big(q\sum_{\sigma\in S_n}\gamma_\sigma x_{\sigma(1)}\cdot x_{\sigma(2)}\cdots x_{\sigma(i-1)}x_{\sigma(i)}x_{\sigma(i+1)}\cdots x_{\sigma(n)}\Big)f(x_1,\ldots,x_n).$$

That is $bqf(x_1,\ldots,x_n)^2=0$, which implies $bq=0$. Since q is invertible, we obtain that $b=0$.

Finally, assume that α is outer. By (43) it follows that Q_r satisfies the generalized polynomial

$$b\Big(\sum_{\sigma\in S_n}\alpha(\gamma_\sigma)\sum_{i=1}^{n}z_{\sigma(1)}\cdot z_{\sigma(2)}\cdots z_{\sigma(i-1)}y_{\sigma(i)}x_{\sigma(i+1)}\cdots x_{\sigma(n)}\Big)f(x_1,\ldots,x_n). \tag{44}$$

For any $i = 1, \ldots, n$, Q_r also satisfies the generalized polynomial

$$b\Big(\sum_{\sigma \in S_{n-1}} \alpha(\gamma_\sigma) z_{\sigma(1)} \cdot z_{\sigma(2)} \cdots z_{\sigma(i-1)} \cdot z_{\sigma(i+1)} \cdots z_{\sigma(n)} \cdot y_i\Big) f(x_1, \ldots, x_n). \quad (45)$$

Let us write

$$\sum_{\sigma \in S_{n-1}} \alpha(\gamma_\sigma) x_{\sigma(1)} \cdots x_{\sigma(j-i)} x_{\sigma(j+1)} \cdots x_{\sigma(n)} = t_j(x_1, \ldots, x_{j-1}, x_{j+1}, \ldots, x_n),$$

where any t_j is a multilinear polynomial of degree $n-1$ and x_j never appears in any monomial of t_j. It follows from (45) that Q_r satisfies the generalized polynomial

$$bt_j(z_1, \ldots, z_{j-1}, z_{j+1}, \ldots, z_n) f(x_1, \ldots, x_n).$$

As a consequence of Lemma 3 and Corollary 2, either $b = 0$ or $t_j(z_1, \ldots, z_{j-1}, z_{j+1}, \ldots, z_n)$ is a generalized polynomial identity for Q_r for all $j = 1, \ldots, n$. Moreover, we also denote $f^\alpha(x_1, \ldots, x_n)$ the polynomial obtained from $f(x_1, \ldots, x_n)$ by replacing each coefficient γ_σ with $\alpha(\gamma_\sigma)$ and notice that $f^\alpha(r_1, \ldots, r_n) \neq 0$. Hence, in the case $t_j(z_1, \ldots, z_{j-1}, z_{j+1}, \ldots, z_n)$ is a generalized polynomial identity for Q_r for all $j = 1, \ldots, n$, and since

$$f^\alpha(x_1, \ldots, x_n) = \sum_j x_j t_j(x_1, \ldots, x_{j-1}, x_{j+1}, \ldots, x_n),$$

$f^\alpha(x_1, \ldots, x_n)$ is a generalized polynomial identity for Q_r, which is also a contradiction. Thus we conclude again that $b = 0$.
The previous argument shows that $b = 0$ in any case.
Moreover, by (42) it follows that Q_r satisfies

$$f(x_1, \ldots, x_n) p\Big(\sum_{\sigma \in S_n} \beta(\gamma_\sigma) \sum_{i=1}^{n} \alpha(x_{\sigma(1)} \cdot x_{\sigma(2)} \cdots x_{\sigma(i-1)}) z_{\sigma(i)} x_{\sigma(i+1)} \cdots x_{\sigma(n)}\Big).$$

By using the same above argument, one can show that $p = 0$, which is a contradiction. We omit the proof for brevity.

7.2 Let d and δ be C−Linearly Dependent Modulo SD_{int}

We firstly assume that there exist $0 \neq \lambda \in C$, $0 \neq \mu \in C$, $c \in Q_r$ and $\gamma \in \mathrm{Aut}(R)$ such that $\lambda d(x) + \mu\delta(x) = cx - \gamma(x)c$ for all $x \in R$. Denote $\eta = -\mu^{-1}\lambda$ and $q = \mu^{-1}c$. Thus $\delta(x) = \eta d(x) + qx - \gamma(x)q$ for all $x \in R$. Therefore by (40), Q_r

satisfies the generalized polynomial

$$\begin{aligned} &af(x_1,\ldots,x_n)^2+bd\big(f(x_1,\ldots,x_n)\big)f(x_1,\ldots,x_n)\\ &-f(x_1,\ldots,x_n)(u+pq)f(x_1,\ldots,x_n)-\eta f(x_1,\ldots,x_n)pd\big(f(x_1,\ldots,x_n)\big)\\ &+f(x_1,\ldots,x_n)p\gamma\big(f(x_1,\ldots,x_n)\big)q. \end{aligned} \tag{46}$$

That is, Q_r satisfies the generalized polynomial

$$\begin{aligned} &af(x_1,\ldots,x_n)^2+bf^d(x_1,\ldots,x_n)f(x_1,\ldots,x_n)\\ &+b\Big(\sum_{\sigma\in S_n}\alpha(\gamma_\sigma)\sum_{j=0}^{n-1}\alpha(x_{\sigma(1)}\cdot x_{\sigma(2)}\cdots x_{\sigma(j)})d(x_{\sigma(j+1)})x_{\sigma(j+2)}\cdots x_{\sigma(n)}\Big)f(x_1,\ldots,x_n)\\ &-f(x_1,\ldots,x_n)(u+pq)f(x_1,\ldots,x_n)-\eta f(x_1,\ldots,x_n)pf^d(x_1,\ldots,x_n)\\ &-\eta f(x_1,\ldots,x_n)p\Big(\sum_{\sigma\in S_n}\alpha(\gamma_\sigma)\sum_{j=0}^{n-1}\alpha(x_{\sigma(1)}\cdot x_{\sigma(2)}\cdots x_{\sigma(j)})d(x_{\sigma(j+1)})x_{\sigma(j+2)}\cdots x_{\sigma(n)}\Big)\\ &+f(x_1,\ldots,x_n)p\gamma\big(f(x_1,\ldots,x_n)\big)q. \end{aligned} \tag{47}$$

In case d is outer, by (47) Q_r satisfies the generalized polynomial

$$\begin{aligned} &af(x_1,\ldots,x_n)^2+bf^d(x_1,\ldots,x_n)f(x_1,\ldots,x_n)\\ &+b\Big(\sum_{\sigma\in S_n}\alpha(\gamma_\sigma)\sum_{j=0}^{n-1}\alpha(x_{\sigma(1)}\cdot x_{\sigma(2)}\cdots x_{\sigma(j)})y_{\sigma(j+1)}x_{\sigma(j+2)}\cdots x_{\sigma(n)}\Big)f(x_1,\ldots,x_n)\\ &-f(x_1,\ldots,x_n)(u+pq)f(x_1,\ldots,x_n)-\eta f(x_1,\ldots,x_n)pf^d(x_1,\ldots,x_n)\\ &-\eta f(x_1,\ldots,x_n)p\Big(\sum_{\sigma\in S_n}\alpha(\gamma_\sigma)\sum_{j=0}^{n-1}\alpha(x_{\sigma(1)}\cdot x_{\sigma(2)}\cdots x_{\sigma(j)})y_{\sigma(j+1)}x_{\sigma(j+2)}\cdots x_{\sigma(n)}\Big)\\ &+f(x_1,\ldots,x_n)p\gamma\big(f(x_1,\ldots,x_n)\big)q. \end{aligned} \tag{48}$$

In particular,

$$\begin{aligned} &b\Big(\sum_{\sigma\in S_n}\alpha(\gamma_\sigma)\sum_{j=0}^{n-1}\alpha(x_{\sigma(1)}\cdots x_{\sigma(j)})y_{\sigma(j+1)}x_{\sigma(j+2)}\cdots x_{\sigma(n)}\Big)f(x_1,\ldots,x_n)\\ &-\eta f(x_1,\ldots,x_n)p\Big(\sum_{\sigma\in S_n}\alpha(\gamma_\sigma)\sum_{j=0}^{n-1}\alpha(x_{\sigma(1)}\cdots x_{\sigma(j)})y_{\sigma(j+1)}x_{\sigma(j+2)}\cdots x_{\sigma(n)}\Big) \end{aligned} \tag{49}$$

is satisfied by R as well as Q_r (see Note 6 again).

Suppose there exists an invertible element $w\in Q_r$ such that $\alpha(x)=wxw^{-1}$ for all $x\in Q_r$. Since $\alpha\neq 1\in \mathrm{Aut}(R)$, we may assume $w\notin C$. As above, we remark that $\alpha(\gamma_\sigma)=\gamma_\sigma$ for all coefficients involved in $f(x_1,\ldots,x_n)$. Therefore, if we replace each $y_{\sigma(i)}$ with $wx_{\sigma(i)}$ in (49), we obtain that Q_r satisfies the generalized polynomial

$$\Big(bwf(x_1,\ldots,x_n)-f(x_1,\ldots,x_n)(\eta pw)\Big)f(x_1,\ldots,x_n).$$

Applying again Corollary 1 yields $bw=\eta pw\in C$. In particular $b=\eta p$. Let us now replace each $y_{\sigma(i)}$ with $w[z,x_{\sigma(i)}]$ in (49), for some element $z\notin C$. Thus we obtain that Q_r satisfies the generalized polynomial

$$bw\Big[z, f(x_1, \ldots, x_n)\Big]_2.$$

Since $f(x_1, \ldots, x_n)$ is not central-valued on Q_r and $z \notin C$, we get the contradiction $b = p = 0$.

Finally, assume that α is outer. By (49) we know that Q_r satisfies the generalized polynomial

$$\begin{aligned} &b\Big(\textstyle\sum_{\sigma\in S_n}\alpha(\gamma_\sigma)\sum_{i=1}^n z_{\sigma(1)}\cdot z_{\sigma(2)}\cdots z_{\sigma(i-1)}y_{\sigma(i)}x_{\sigma(i+1)}\cdots x_{\sigma(n)}\Big)f(x_1,\ldots,x_n)\\ &-\eta f(x_1,\ldots,x_n)p\Big(\textstyle\sum_{\sigma\in S_n}\alpha(\gamma_\sigma)\sum_{i=1}^n z_{\sigma(1)}\cdot z_{\sigma(2)}\cdots z_{\sigma(i-1)}y_{\sigma(i)}x_{\sigma(i+1)}\cdots x_{\sigma(n)}\Big)f(x_1,\ldots,x_n) \end{aligned} \tag{50}$$

and, for any $i = 1, \ldots, n$, Q_r also satisfies the generalized polynomial

$$\begin{aligned} &b\Big(\textstyle\sum_{\sigma\in S_{n-1}}\alpha(\gamma_\sigma)z_{\sigma(1)}\cdot z_{\sigma(2)}\cdots z_{\sigma(i-1)}\cdot z_{\sigma(i+1)}\cdots z_{\sigma(n)}\cdot y_i\Big)f(x_1,\ldots,x_n)\\ &-\eta f(x_1,\ldots,x_n)p\Big(\textstyle\sum_{\sigma\in S_{n-1}}\alpha(\gamma_\sigma)z_{\sigma(1)}\cdot z_{\sigma(2)}\cdots z_{\sigma(i-1)}\cdot z_{\sigma(i+1)}\cdots z_{\sigma(n)}\cdot y_i\Big). \end{aligned} \tag{51}$$

As above, let us write

$$\sum_{\sigma\in S_{n-1}}\alpha(\gamma_\sigma)x_{\sigma(1)}\cdots x_{\sigma(j-i)}x_{\sigma(j+1)}\cdots x_{\sigma(n)} = t_j(x_1,\ldots,x_{j-1},x_{j+1},\ldots,x_n),$$

where any t_j is a multilinear polynomial of degree $n-1$ and x_j never appears in any monomial of t_j. In view of (51), we get

$$\begin{aligned} &b\Big(t_j(z_1,\ldots,z_{j-1},z_{j+1},\ldots,z_n)y\Big)f(x_1,\ldots,x_n)\\ &-\eta f(x_1,\ldots,x_n)p\Big(t_j(z_1,\ldots,z_{j-1},z_{j+1},\ldots,z_n)y\Big). \end{aligned} \tag{52}$$

From Lemma 3 it follows that

$$bt_j(z_1,\ldots,z_{j-1},z_{j+1},\ldots,z_n)y = \eta pt_j(z_1,\ldots,z_{j-1},z_{j+1},\ldots,z_n)y \in C. \tag{53}$$

Suppose that $t_j(z_1,\ldots,z_{j-1},z_{j+1},\ldots,z_n)$ is central-valued on Q_r for all $j = 1,\ldots,n$. Since

$$f^\alpha(x_1,\ldots,x_n) = \sum_j x_jt_j(x_1,\ldots,x_{j-1},x_{j+1},\ldots,x_n),$$

it follows that $f^\alpha(x_1, \ldots, x_n)$ is a central-valued on Q_r, a contradiction. Therefore (53) forces $b = 0$ and $\eta p = 0$, which is again a contradiction.

Let us next start from (46) and consider the case when $d(x) = vx - \alpha(x)v$ for all $x \in R$ and for some fixed $v \in Q_r$. Hence, $\delta(x) = (\eta v + q)x - \alpha(x)\eta v - \gamma(x)q$. Therefore, by Note 8, F and G are simultaneously inner b-generalized skew derivations of R and, by Theorem 3 a number of contradictions follows.

We analyze now the last case. Let us start again from relation (40) and assume again that d and δ are $C-$linearly dependent modulo $\mathrm{SD_{int}}$. That is $\lambda d(x)+\mu\delta(x) = cx - \gamma(x)c$ for all $x \in R$. Moreover, in view of the previous argument, we have to assume now $\lambda = 0$. Thus $\delta(x) = qx-\gamma(x)q$ for all $x \in R$ and $q = \mu^{-1}c$. Therefore by (40), Q_r satisfies the generalized polynomial

$$\begin{aligned} &af(x_1, \ldots, x_n)^2 + bd\big(f(x_1, \ldots, x_n)\big)f(x_1, \ldots, x_n) \\ &-f(x_1, \ldots, x_n)(u + pq)f(x_1, \ldots, x_n) \\ &+f(x_1, \ldots, x_n)p\gamma\big(f(x_1, \ldots, x_n)\big)q. \end{aligned} \tag{54}$$

We finally observe that (54) is equivalent to (46) in case $\eta = 0$. Therefore the same above argument completes our proof.

8 Some Open Problems

In the light of the motivation and contents of this article, we will propose several topics for future research in this field. More precisely, some informations about the structure of a prime ring R and the description of all possible forms of a b-generalized skew derivation F of R can be obtained if one of the following conditions is satisfied:

1. $F(x)^n = 0$ for all $x \in L$, where n is a fixed positive integer and L is a noncommutative Lie ideal of R.
2. $F(x)^n \in Z(R)$ for all $x \in L$, where n is a fixed positive integer and L is a noncommutative Lie ideal of R.
3. $F(x)^n \in Z(R)$ for all $x \in I$, where n is a fixed positive integer and I is a non-zero one sided ideal of R.
4. $aF(x)^n = 0$ for all $x \in I$, where n is a fixed positive integer, I is an ideal of R and a is a non-zero element of R.

Acknowledgments This research was done when the first author visited the School of Mathematics at Beijing Institute of Technology in the winter of 2019. He takes this opportunity to express their sincere thanks to the School of Mathematics and the Office of International Affairs at Beijing Institute of Technology for the hospitality extended to him during his visit. All authors are deeply grateful to a special Training Program of International Exchange and Cooperation of the Beijing Institute of Technology.

References

1. Ali, A., Ali, S., De Filippis, V.: Generalized skew derivations with nilpotent values in prime rings. Commun. Algebra **42**, 1606–1618 (2014)
2. Argaç, N., De Filippis, V.: Actions of generalized derivations on multilinear polynomials in prime rings. Algebra Colloq. **18**, 955–964 (2011)
3. Argaç, N., De Filippis, V.: A note on certain central differential identities with generalized derivations. Hacettepe J. Math. **41**(6), 847–852 (2012)
4. Beidar, K.I., Martindale, W.S. III, Mikhalev, A.V.: Rings with Generalized Identities. Monographs and Textbooks in Pure and Applied Mathematics, vol. 196. Marcel Dekker, New York (1996)
5. Boudi, N., Ouchrif, S.: On generalized derivations in Banach algebras. Studia Math. **194**, 81–89 (2009)
6. Brešar, M.: Centralizing mappings and derivations in prime rings. J. Algebra **156**, 385–394 (1993)
7. Brešar, M., Villena, A.R.: The noncommutative Singer-Wermer conjecture and φ-derivations. J. Lond. Math. Soc. **66**(2), 710–720 (2002)
8. Carini, L., De Filippis, V.: Identities with generalized derivations on prime rings and Banach algebras. Algebra Colloq. **19**, 971–986 (2012)
9. Carini, L., De Filippis, V., Wei, F.: Generalized skew derivations cocentralizing multilinear polynomials. Mediterr. J. Math. **13**, 2397–2424 (2016)
10. Chang, J.-C.: On the identitity $h(x) = af(x) + g(x)b$. Taiwanese J. Math. **7**, 103–113 (2003)
11. Chang, J.-C.: Generalized skew derivations with annihilating Engel conditions. Taiwanese J. Math. **12**, 1641–1650 (2008)
12. Chang, J.-C.: Generalized skew derivations with nilpotent values on Lie ideals. Monatsh. Math. **161**, 155–160 (2010)
13. Chang, J.-C.: Generalized skew derivations with power central values on Lie ideals. Commun. Algebra **39**, 2241–2248 (2011)
14. Chang, J.-C.: Generalized skew derivations with Engel conditions on Lie ideals. Bull. Inst. Math. Acad. Sin. **6**, 305–320 (2011)
15. Chen, H.-Y.: Generalized derivations cocentralizing polynomials. Commun. Algebra **41**, 2873–2798 (2013)
16. Cheng, H.-W., Wei, F.: Generalized skew derivations of rings. Adv. Math. **35**, 237–243 (2006)
17. Chuang, C.-L.: The additive subgroup generated by a polynomial. Israel J. Math. **59**, 98–106 (1987)
18. Chuang, C.-L.: GPIs having coefficients in Utumi quotient rings. Proc. Am. Math. Soc. **103**, 723–728 (1988)
19. Chuang, C.-L.: Differential identities with automorphisms and antiautomorphisms I. J. Algebra **149**, 371–404 (1992)
20. Chuang, C.-L.: Differential identities with automorphisms and antiautomorphisms II. J. Algebra **160**, 130–171 (1993)
21. Chuang, C.-L.: Identities with skew derivations. J. Algebra **224**, 292–335 (2000)
22. Chuang, C.-L., Lee, T.-K.: Identities with a single skew derivation. J. Algebra **288**, 59–77 (2005)
23. De Filippis, V.: A product of two generalized derivations on polynomials in prime rings. Collect. Math. **61**, 303–322 (2010)
24. De Filippis, V., Di Vincenzo, O.M.: Hypercentralizing generalized skew derivations on left ideals in prime rings. Monatsh. Math. **173**, 315–341 (2014)
25. De Filippis, V., Wei, F.: Posner's second theorem and annihilator conditions with generalized skew derivations. Colloq. Math. **129**, 61–74 (2012)
26. De Filippis, V., Wei, F.: An Engel condition with X-Generalized Skew Derivations on Lie ideals. Commun. Algebra **46**(12), 5433–5446 (2018)

27. De Filippis, V., Scudo, G., Mohammad S. Tammam El-Sayiad, An identity with generalized derivations on Lie ideals, right ideals and Banach algebras. Czechoslovak Math. J. **62**(137), 453–468 (2012)
28. De Filippis, V., Fosner, A., Wei, F.: Identities with generalized skew derivations on Lie ideals. Algebra Represent. Theory **16**, 1017–1038 (2013)
29. Demir, Ç, Albaş, E., Argaç, N., De Filippis, V.: Generalized skew derivations with invertible values on multilinear polynomials. Commun. Algebra **40**, 4042–4059 (2012)
30. Eremita, D., Gogić, I., Ilišević, D.: Generalized skew derivations implemented by elementary operators. Algebr. Represent. Theory **17**, 983–996 (2014). http://dx.doi.org/10.1007/s10468-013-9429-8
31. Jacobson, N.: Structure of Rings. American Mathematical Society, Providence (1964)
32. Kharchenko, V.K.: Differential identities of prime rings. Algebra i Logika **17**, 220–238 (1978); Engl. Transl.: Algebra Logic **17**, 155–168 (1978)
33. Kharchenko, V.K.: Automorphisms and Derivations of Associative Rings. Kluwer Academic Publishers, Dordrecht (1991)
34. Koşan, M.T., Lee, T.-K.: b-generalized derivations of semiprime rings having nilpotent values. J. Austral. Math. Soc. **96**, 326–337 (2014). http://dx.doi.org/10.1017/S1446788713000670
35. Lee, T.-K.: Generalized derivations of left faithful rings. Commun. Algebra **27**(8), 4057–4073 (1999)
36. Lee, T.-K., Shiue, W.-K.: Derivations cocentralizing polynomials. Taiwanese J. Math. **2**, 457–467 (1998)
37. Leron, U.: Nil and power central polynomials in rings. Trans. Am. Math. Soc. **202**, 97–103 (1975)
38. Liu, C.-K.: Derivations with Engel and annihilator conditions on multilinear polynomials. Commun. Algebra **33**, 719–725 (2005)
39. Liu, K.-S.: Differential identities and constants of algebraic automorphisms in prime rings. Ph.D. Thesis, National Taiwan University (2006)
40. Liu, C.-K.: Derivations cocentralizing multilinear polynomials on left ideals. Monatsh. Math. **162**, 297–311 (2011)
41. Liu, C.-K.: Skew derivations with nilpotent values in rings and Banach algebras. Commun. Algebra **40**, 4336–4345 (2012)
42. Liu, C.-K.: On skew derivations in semiprime rings. Algebra Represent. Theory **16**, 1561–1576 (2013)
43. Liu, C.-K., Chen, H.-Y., Liau, P.-K.: Generalized skew derivations with nilpotent values on left ideals of rings and Banach algebras. Linear Multilinear Algebra **62**, 453–465 (2014). http://dx.doi.org/10.1080/03081087.2013.780604
44. Martindale, W.S., III: Prime rings satisfying a generalized polynomial identity. J. Algebra **12**, 576–584 (1969)
45. Qi, X.-F., Hou, J.-C.: Generalized skew derivations on nest algebras characterized by acting on zero products. Publ. Math. Debrecen **78**, 457–468 (2011)
46. Wei, F., Xiao, Z.-K.: Generalized derivations on (semi-)prime rings and noncommutative Banach algebras. Rend. Semin. Mat. Univ. Padova **122**, 171–190 (2009)
47. Wei, F., Xiao, Z.-K.: Generalized Jordan derivations on semiprime rings and its applications in range inclusion problems. Mediterr. J. Math. **8**, 271–291 (2011)
48. Wong, T.-L.: Derivations with power central values on multilinear polynomials. Algebra Colloq. **3**, 369–378 (1996)
49. Wong, T.-L.: Derivations cocentralizing multilinear polynomials. Taiwanese J. Math. **1**, 31–37 (1997)

Relatively Free Algebras of Finite Rank

Thiago Castilho de Mello and Felipe Yukihide Yasumura

Abstract Let $\mathbb{K}$ be a field of characteristic zero and $B = B_0 + B_1$ a finite dimensional associative superalgebra. In this paper we investigate the polynomial identities of the relatively free algebras of finite rank of the variety $\mathfrak{V}$ defined by the Grassmann envelope of B. We also consider the k-th Grassmann Envelope of B, $G^{(k)}(B)$, constructed with the k-generated Grassmann algebra, instead of the infinite dimensional Grassmann algebra. We specialize our studies for the algebras $UT_2(G)$ and $UT_2(G^{(k)})$, which can be seen as the Grassmann envelope and k-th Grassmann envelope, respectively, of the superalgebra $UT_2(\mathbb{K}[u])$, where $u^2 = 1$.

Keywords Polynomial Identities · Relatively free algebras · Grassmann envelope

1 Introduction

In this paper $\mathbb{K}$ will denote a field of characteristic 0. If $X = \{x_1, x_2, \dots\}$ is an infinite countable set, we denote by $\mathbb{K}\langle X\rangle$ the free associative unitary algebra freely generated by X. If A is an associative algebra, we say that it is an *algebra with polynomial identity* (PI-algebra, for short) if there exists a nonzero polynomial $f = f(x_1, \dots, x_n) \in \mathbb{K}\langle X\rangle$ such that $f(a_1, \dots, a_n) = 0$, for arbitrary $a_1, \dots, a_n \in A$. In this case, we say that f is a (polynomial) identity of A, or simply that A satisfies f.

If A is a PI-algebra the set $T(A) = \{f \in \mathbb{K}\langle X\rangle \mid f \text{ is an identity of } A\}$ is an ideal of $\mathbb{K}\langle X\rangle$ invariant under any endomorphism of the algebra $\mathbb{K}\langle X\rangle$. An ideal with this

T. C. de Mello (✉)
Universidade Federal de São Paulo, Instituto de Ciência e Tecnologia, São José dos Campos, Brazil
e-mail: tcmello@unifesp.br

F. Y. Yasumura
Universidade de São Paulo, Instituto de Matemática e Estatística, São Paulo, Brazil
e-mail: fyyasumura@ime.usp.br

O. M. Di Vincenzo, A. Giambruno (eds.), *Polynomial Identities in Algebras*, Springer INdAM Series 44, https://doi.org/10.1007/978-3-030-63111-6_8

property is called a *T-ideal* or *verbal ideal* of $\mathbb{K}\langle X\rangle$. We refer the reader to [5, 10] for the basic theory of PI-algebras.

The T-ideals play a central role in the theory of PI-algebras, and they are often studied through the equivalent notion of varieties of algebras. If $\mathcal{F}$ is a subset of $\mathbb{K}\langle X\rangle$, the class of all algebras satisfying the identities from $\mathcal{F}$ is called *the variety of (associative) algebras defined by* $\mathcal{F}$ and denoted by var($\mathcal{F}$). Given $\mathfrak{V}$ and $\mathfrak{W}$ varieties of algebras, we say that $\mathfrak{W}$ is a subvariety of $\mathfrak{V}$ if $\mathfrak{W} \subseteq \mathfrak{V}$. If $\mathfrak{V}$ is a variety of algebras, we denote by $T(\mathfrak{V})$ the set of identities satisfied by all algebras in $\mathfrak{V}$. One can easily see that $T(\mathfrak{V})$ is a T-ideal of $\mathbb{K}\langle X\rangle$. It is called *the T-ideal of* $\mathfrak{V}$. If $\mathfrak{V} = \text{var}(\mathcal{F})$, we say that the elements of $T(\mathfrak{V})$ are *consequences of* (or *follow from*) the elements of $\mathcal{F}$.

Let $\mathfrak{V}$ be a variety of associative algebras. In the theory of algebras with polynomial identities, an important role is played by the so called *relatively free algebras of* $\mathfrak{V}$. The relatively free algebra of $\mathfrak{V}$ freely generated by a set X is an algebra $F_X(\mathfrak{V}) \in \mathfrak{V}$, with an inclusion map $\iota : X \hookrightarrow F_X(\mathfrak{V})$, satisfying the following universal property:

Given any algebra $A \in \mathfrak{V}$, and a map $\varphi_0 : X \longrightarrow A$, there exists a unique algebra homomorphism $\varphi : F_X(\mathfrak{V}) \longrightarrow A$ such that $\varphi \circ \iota = \varphi_0$.

It is a simple exercise to show that

$$F_X(\mathfrak{V}) \cong \frac{\mathbb{K}\langle X\rangle}{T(\mathfrak{V})}.$$

Moreover, it is well known that for two given sets X and Y, the algebras $F_X(\mathfrak{V})$ and $F_Y(\mathfrak{V})$ are isomorphic if and only if X and Y have the same cardinality. Therefore if $|X| = k \in \mathbb{N}$, we denote $F_X(\mathfrak{V})$ simply by $F_k(\mathfrak{V})$ and if X is a countably infinite set we denote it simply by $F(\mathfrak{V})$.

The first studies about relatively free algebras are due to Procesi (see [19, 20]), when dealing with the so called *algebra of generic matrices*, which is isomorphic to the relatively free algebra in the variety generated by $M_n(\mathbb{K})$. This algebra is a fundamental object in invariant theory and has noteworthy properties. For instance, it has no zero divisors and one can work with its quotient ring, the so called *generic division algebra* (see [6]). Another interesting property is that given a polynomial $f(x_1, \ldots, x_k)$, we have that f is a central polynomial for $M_n(\mathbb{K})$ if and only if $f(x_1, \ldots, x_k) + T(M_n(\mathbb{K}))$ is a central element of $F_k(M_n(\mathbb{K}))$. Such properties do not hold in any variety. A simple example can be seen in the variety generated by the algebra $M_{1,1} = \begin{pmatrix} G_0 & G_1 \\ G_1 & G_0 \end{pmatrix}$, where $f(x, y) = [x, y]^2$ is a central element in $F_2(M_{1,1})$, but it is not a central polynomial for $M_{1,1}$ (see [14, 18]).

We say that a variety of algebras $\mathfrak{V}$ has a finite basic rank if $\mathfrak{V} = \text{var}(A)$, where A is a finitely generated algebra. The minimal number of generators of such an algebra is the *basic rank of the variety* $\mathfrak{V}$.

Of course that the variety generated by the algebra $F(\mathfrak{V})$ is $\mathfrak{V}$ itself, and that for any k, $F_k(\mathfrak{V}) \in \mathfrak{V}$, but it is not true in general that there exists k such that $F_k(\mathfrak{V})$ generates $\mathfrak{V}$.

The basic rank of a variety $\mathfrak{V}$ can be characterized in terms of its relatively free algebras, as we can see in the following easy-to-prove proposition.

Proposition 1 *The basic rank of a variety $\mathfrak{V}$ is the least integer k such that $\mathfrak{V} = var(F_k(\mathfrak{V}))$*

As examples, we mention that the variety generated by the Grassmann algebra of an infinite dimensional vector space has infinite basic rank, while the algebra of $n \times n$ matrices over the field ($n > 1$) generates a basic rank 2 variety (since $M_n(\mathbb{K})$ is a 2-generated algebra).

A natural problem in the theory of PI-algebras is to classify in some sense the subvarieties in a given variety of algebras $\mathfrak{V}$. A very important role in this direction is played by the exponent of a variety $\mathfrak{V}$. Proving a conjecture of Amitsur, Giambruno and Zaicev showed that for any variety of associative algebras over a field of characteristic zero the exponent exists and is an integer [7, 8] (see also [10]). Therefore, it was natural to classify varieties of algebras in terms of its exponents. A successful approach was the classification in terms of forbidden algebras. For example, Kemer showed that the varieties of exponent 1 are exactly those varieties not containing the infinite dimensional Grassmann algebra and the 2×2 algebra of upper triangular matrices. Similar results were given for varieties of exponent 2, with a list of 5 forbidden algebras (see [9]).

The classification of subvarieties of important varieties of algebras were also studied. For instance, the classification of the subvarieties of the variety $\mathfrak{G}$, generated by the infinite dimensional Grassmann algebra, was given by La Mattina [16], and subvarieties of the variety generated by $M_2(\mathbb{K})$ were studied by Drensky [4]. The classification of subvarieties of a given variety is a difficult task. A less accurate, but interesting approach, is the classification up to asymptotic equivalence of varieties. This notion was introduced by Kemer in [12], where he classified subvarieties of a variety satisfying the identity St_4 (the standard identity of degree 4), up to asymptotic equivalence. We say that two T-ideals are *asymptotically equivalent* if they satisfy the same proper polynomials from a certain degree on. We recall that proper polynomials are those which are linear combinations of products of commutators. In a similar way, two varieties are asymptotically equivalent if their T-ideals are so. This notion was also used to classify, up to asymptotic equivalence, the subvarieties of the variety $\mathfrak{M}_5$, generated by the identity $[x_1, x_2][x_3, x_4, x_5]$ (see [3]). Such variety can be realized as the variety generated by the algebra $A = \begin{pmatrix} G_0 & G \\ 0 & G \end{pmatrix}$. This is one of the forbidden algebras in the classification of varieties of exponent 2.

If $\mathfrak{V}$ is a variety of associative algebras of infinite basic rank, we have a lattice of T-ideals

$$T(F_1(\mathfrak{V})) \supseteq T(F_2(\mathfrak{V})) \supseteq \cdots T(F_k(\mathfrak{V})) \supseteq T(F_{k+1}(\mathfrak{V})) \supseteq \cdots \supseteq T(F(\mathfrak{V})) = T(\mathfrak{V}).$$

As a consequence of Lemma 3 below, one can easily see that there is an infinite number of proper inclusions above.

A natural but difficult problem in general is to describe for all k, the T-ideals $T(F_k(\mathfrak{V}))$. This task was realized only for a small list of varieties of infinite basic rank, namely: the variety $\mathfrak{G}$, the variety $\mathfrak{M}_5$ [11], and the variety generated by the algebra $M_{1,1}(G)$. The last only for $k = 2$ (see [15]).

In the mentioned examples, the knowledge of the identities of the relatively free algebras of finite rank, were useful to give an alternative description to the subvarieties of the given variety. For instance, if $A \in \mathfrak{G}$ is a unitary algebra, it is PI-equivalent to $\mathbb{K}$, G, or $F_{2k}(\mathfrak{G})$, for some k, and if $A \in \mathfrak{M}_5$, then, it is PI-equivalent to $\mathbb{K}$, $UT_2(\mathbb{K})$, E, $F_{2k}(\mathfrak{M}_5)$ or $F_{2k}(\mathfrak{M}_5) \oplus G$, for some k.

We believe that the knowledge of the identities of the relatively free algebras of finite rank of a given variety of infinite basic rank may play an important role in the description of its subvarieties. This is a motivation to the study of such identities.

From Kemer's theory [13] we know that every finitely generated algebra satisfies the same identities of a finite dimensional algebra. In light of this, given a variety $\mathfrak{V}$, it is interesting to find finite dimensional algebras $A_k \in \mathfrak{V}$ such that $T(A_k) = T(F_k(\mathfrak{V}))$, for all k. This was done to the above-mentioned examples. In those cases, it was verified that for all k, the algebra A_k was obtained with the construction we describe below.

It is well known from Kemer's theory that the variety $\mathfrak{V}$ is generated by the *Grassmann envelope* of a suitable finite dimensional superalgebra $B = B_0 + B_1$. Recall that the Grassmann envelope of B is given by $G(B) = G_0 \otimes B_0 + G_1 \otimes B_1$, i.e., the even part of the superalgebra $G \otimes B$. Similarly, one can define the k-th Grassmann envelope of B as $G^{(k)}(B) = G_0^{(k)} \otimes B_0 + G_1^{(k)} \otimes B_1$, where $G^{(k)}$ is the Grassmann algebra of a k-dimensional vector space over K.

If $B = B_0 + B_1$ is the superalgebra (which exists by Kemer theory) satisfying $T(G(B)) = T(\mathfrak{V})$, the above-mentioned examples satisfy the following interesting property:

$$T(F_k(\mathfrak{V})) = T(G^{(k)}(B)), \tag{1}$$

for any k, in the case $\mathfrak{V} = \mathfrak{G}$ or $\mathfrak{V} = \mathfrak{M}_5$. In the case $\mathfrak{V} = var(M_{1,1})$ we only know it for $k = 2$ (see [11]).

In light of the these results, it is an interesting problem to compare the T-ideals of $T(G^{(k)}(B))$ and $T(F_k(G(B)))$ for a given finite dimensional superalgebra B.

In the present paper, we obtain partial results on this problem for the variety $\mathfrak{G}_2$, generated by $UT_2(G)$. We show that the equality (1) does not hold for this variety.

We divide this paper as follows. We construct different models for the relatively free algebras in Sect. 2, which will give different approaches for the problem. In Sect. 3 we prove general facts that hold for the relatively free algebras of finite rank. In Sect. 4, we investigate the polynomial identities of $UT_2(G^{(k)})$, and exhibit a basis of identities when $2 \leq k \leq 5$. Finally, in Sect. 5, we investigate the polynomial identities of $F_k(UT_2(G))$.

2 Models for Relatively Free Algebras

The relatively free algebras are quotients of the polynomial algebra $\mathbb{K}\langle X\rangle$. In particular, its elements are cosets of noncommutative polynomials. In order to have a more concrete object to work with, we will present some models of these relatively free algebras, which can simplify the problem of working with quotient classes.

The most simple example of a model for a relatively free algebra is the *algebra of generic matrices* (for the variety generated by an $n \times n$ matrix algebra over an infinite field $\mathbb{K}$). By a *model*, we mean an algebra isomorphic to the given relatively free algebra.

Let n be a positive integer, $X = \{x_{ij}^{(k)} \mid i, j \in \{1, \dots, n\}, k \in \mathbb{N}\}$ and $\mathbb{K}[X]$ be the algebra of commutative polynomials on the variables of X. The algebra of generic $n \times n$ matrices is the subalgebra of $M_n(\mathbb{K}[X])$ generated by the matrices

$$\xi_k = \begin{pmatrix} x_{11}^{(k)} & x_{12}^{(k)} & \cdots & x_{1n}^{(k)} \\ x_{21}^{(k)} & x_{22}^{(k)} & \cdots & x_{2n}^{(k)} \\ \vdots & \vdots & \ddots & \vdots \\ x_{n1}^{(k)} & x_{n2}^{(k)} & \cdots & x_{nn}^{(k)} \end{pmatrix}, \quad \text{for } k \in \mathbb{N}.$$

In a similar way, one can construct a model for a relatively free algebra of a variety generated by a finite dimensional algebra A. For, one only needs to fix a basis $B = \{v_1, \dots, v_n\}$ of A, and consider a subalgebra of $\mathbb{K}[X] \otimes A$, (where $X = \{x_i^{(k)} \mid i \in \{1, \dots, n\}, k \in \mathbb{N}\}$) generated by the elements

$$\xi_k = \sum_{i=1}^{n} x_i^{(k)} \otimes v_i, \quad \text{for } k \in \mathbb{N}.$$

On the other hand, when dealing with a variety of infinite basic rank, the above construction is not possible.

Examples of models for relatively free algebras of infinite basic rank varieties were given by Berele in [1]. More specifically, Berele constructed models for the relatively free algebras of varieties generated by $M_n(G)$ and $M_{a,b}(G)$ (the so called T-prime varieties), as algebras of matrices over the free supercommutative algebra $\mathbb{K}[X; Y]$.

We say that a superalgebra $A = A_0 + A_1$ is supercommutative if its Grassmann envelope is commutative, i.e., if for any $a, b \in A_0 \cup A_1$, one has $ab = (-1)^{\deg a \deg b} ba$.

Considering $\mathbb{K}$ as an infinite field, we proceed with a construction of a free supercommutative superalgebra.

Let X and Y be countably infinite sets. We build the algebra $\mathbb{K}\langle X \cup Y\rangle$ and induce on it a $\mathbb{Z}_2$-grading by defining $\deg x = 0, x \in X$, and $\deg y = 1, y \in Y$. The algebra $\mathbb{K}\langle X \cup Y\rangle$ with such grading is called the *free* $\mathbb{Z}_2$*-graded algebra*. If I is the ideal generated by the elements $ab - (-1)^{\deg a \deg b} ba$, $a, b \in X \cup Y$, we define the *free supercommutative algebra*, denoted by $\mathbb{K}[X; Y]$, as the quotient algebra

$$\mathbb{K}[X; Y] = \frac{\mathbb{K}\langle X \cup Y\rangle}{I}.$$

One can easily verify that given any supercommutative superalgebra

$$A = A_0 + A_1,$$

and a map $\varphi_0 : X \cup Y \longrightarrow A$ such that $\varphi_0(x) \in A_0$ if $x \in X$ and $\varphi_0(y) \in A_1$ if $y \in Y$, there exists a unique homomorphism $\varphi : \mathbb{K}[X; Y] \longrightarrow A$ which extends φ_0.

Now given a finite dimensional superalgebra $B = B_0 + B_1$, we proceed with a construction of a model for the relatively free algebra of the variety generated by $G(B)$. We remark that such is a completely general construction, since any variety of associative algebras is generated by $G(B)$, for some B, although given an arbitrary variety it is not a simple task to determine one such B.

Let us fix $\{u_1, \ldots, u_r\}$ a basis of B_0 and $\{v_1, \ldots, v_s\}$ a basis of B_1 and let us consider the sets $X = \{x_j^{(i)} \mid i \in \mathbb{N}, j \in \{1, \ldots, r\}\}$ and $Y = \{y_j^{(i)} \mid i \in \mathbb{N}, j \in \{1, \ldots, s\}\}$.

We consider the free supercommutative algebra $\mathbb{K}[X; Y]$ and for each $i \in \mathbb{N}$, we define $\xi_i \in B \otimes \mathbb{K}[X; Y]$ as

$$\xi_i = \sum_{j=1}^{r} u_j \otimes x_j^{(i)} + \sum_{j=1}^{s} v_j \otimes y_j^{(i)}$$

Then we have:

Proposition 2 *Let $n \in \mathbb{N}$, and define $\mathbb{K}[\xi_1, \xi_2, \ldots]$ and $\mathbb{K}[\xi_1 \ldots, \xi_n]$ as the subalgebras of $B \otimes \mathbb{K}[X; Y]$ generated by the elements $\xi_1, \xi_2, \ldots$ and by the elements $\xi_1, \ldots, \xi_n$, respectively. Then, the following isomorphisms hold:*

$$\mathbb{K}[\xi_1, \xi_2, \ldots] \cong F(G(B))$$

$$\mathbb{K}[\xi_1, \ldots, \xi_n] \cong F_n(G(B))$$

Proof Define the algebra homomorphism $\eta : \mathbb{K}\langle t_1, t_2 \ldots\rangle \longrightarrow \mathbb{K}[\xi_1, \xi_2, \ldots]$ by $\eta(t_i) = \xi_i$. In particular, if $f(t_1, \ldots, t_k) \in \mathbb{K}\langle t_1, t_2 \ldots\rangle$, then $\eta(f) = f(\xi_1, \ldots, \xi_k)$.

Of course η is surjective. Once we show that $\ker \eta = T(G(B))$, the result is proved.

Suppose $f \in \ker \eta$. This means that $f(\xi_1, \ldots, \xi_k) = 0$.

For each $i \in \{1, \ldots, k\}$ we consider arbitrary elements a_i of $G(B)$. These can be written as

$$a_i = \sum_{j=1}^{r} u_j \otimes g_j^{(i)} + \sum_{j=1}^{s} v_j \otimes h_j^{(i)}$$

where $g_j^{(i)}$ and $h_j^{(i)}$ are arbitrary even and odd elements of the Grassmann algebra respectively. Since $\mathbb{K}[X; Y]$ is the free supercommutative algebra, there exists a homomorphism $\varphi : \mathbb{K}[X; Y] \longrightarrow G$ extending the map $\varphi_0 : X \cup Y \longrightarrow G$, given by $\varphi_0(x_j^{(i)}) = g_j^{(i)}$ and $\varphi_0(y_j^{(i)}) = h_j^{(i)}$.

From this, we define the homomorphism of algebras $\phi : B\otimes\mathbb{K}[X; Y] \longrightarrow B\otimes G$, given by φ in $\mathbb{K}[X; Y]$ and fixing B. Then, for each i,

$$a_i = \phi\left(\sum_{j=1}^{r} u_j \otimes x_j^{(i)} + \sum_{j=1}^{s} v_j \otimes y_j^{(i)}\right) = \phi(\xi_i)$$

As a consequence,

$$f(a_1, \ldots, a_k) = \phi(\eta(f)) = 0,$$

which means $f \in T(G(B))$.

Conversely, suppose $f \in T(G(B))$. We will show that $f(\xi_1, \ldots, \xi_k) = 0$.

Write

$$f(\xi_1, \ldots, \xi_k) = \sum_{j=1}^{r} u_j \otimes m_j + \sum_{j=1}^{s} v_j \otimes n_j$$

where m_j and n_j are $\mathbb{Z}_2$-graded polynomials of even and odd degree respectively in the commutative and anticommutative variables $x_p^{(q)}$ and $y_p^{(q)}$ of $\mathbb{K}[X; Y]$.

As we have already shown, if

$$a_i = \sum_{j=1}^{r} u_j \otimes g_j^{(i)} + \sum_{j=1}^{s} v_j \otimes h_j^{(i)}$$

are arbitrary elements of $G(B)$, we have $f(a_1, \ldots, a_k) = \phi(f(\xi_1, \ldots, \xi_k)) = \sum_j u_j \otimes m_j(g; h) + \sum_j v_j \otimes n_j(g; h)$. Since the a_is are arbitrary, so are the homogeneous elements g and h of even and odd homogeneous degree in G. As a consequence, since $f \in T(G(B))$, m_j and n_j are $\mathbb{Z}_2$-graded identities of G and since $\mathbb{K}[X; Y]$ is the free supercomutative algebra, it follows that $m_j = n_j = 0$ in $\mathbb{K}[X; Y]$. What means that $f(\xi_1, \ldots, \xi_k) = 0$, finishing the proof.

The case of $F_n(G(B))$ is analogous. □

It should be remarked that such model has appeared in [10, Section 3.8].

Another possible model for relatively free algebras of some special kind of varieties is presented now.

Consider R a PI-algebra and let A be a subalgebra of $M_n(R)$ generated by matrix units e_{ij}. We give a model for the relatively free algebra of the variety generated by A as a subalgebra of matrices over the relatively free algebra of R. The general construction can be find in the paper [2, Lemma 6]. Here we present it, as an example, for the particular case of $A = UT_2(G)$, which we will use below in the paper.

Example Let $\mathcal{U}$ and $\mathcal{U}_k$ be the subalgebras of $UT_2(F(\mathfrak{G}))$ generated by the generic matrices $\xi_1, \xi_2, \ldots$ and $\xi_1, \ldots, \xi_k$, respectively, where

$$\xi_i = \begin{pmatrix} x_{11}^{(1)} + T(G) & x_{12}^{(i)} + T(G) \\ 0 & x_{22}^{(2)} + T(G) \end{pmatrix}$$

Then

$$\mathcal{U} \cong F(UT_2(G)))$$

$$\mathcal{U}_k \cong F_k(UT_2(G))), \text{ for } k \in \mathbb{N}$$

It is interesting to observe that when dealing with this model, one transfers the problem of dealing with cosets of matrices, to dealing with cosets of elements in the entries of such matrices. When the relatively free algebra of R is well known, this construction is very useful. For instance, if R is the field, its relatively free algebra is the polynomial algebra in commuting variables, and we are in the classical case of generic matrices. In this paper we will deal with this model when R is the Grassmann algebra, but since its relatively free algebra is easy to handle with, this will help us to obtain our results.

3 General Remarks

Let $\mathcal{A}$ be a finite-dimensional associative superalgebra.

As mentioned in [11], if $k_1 \leq k_2$, then

$$T(\mathcal{A}) \subseteq T(F_{k_2}(\mathcal{A})) \subseteq T(F_{k_1}(\mathcal{A})). \tag{2}$$

Moreover, in [11, Lemma 8] the authors prove:

Lemma 3 $T(F_n(\mathcal{A})) \cap \mathbb{K}\langle x_1, \ldots, x_n \rangle = T(\mathcal{A}) \cap \mathbb{K}\langle x_1, \ldots, x_n \rangle$. □

As a consequence, we have the following:

Proposition 4 $T(\mathcal{A}) = \bigcap_{n\geq 1} T(F_n(\mathcal{A}))$. *In particular,* $F(\mathcal{A})$ *is a subdirect product of the* $\{F(F_n(\mathcal{A}))\}_{n\in\mathbb{N}}$.

Proof Clearly $T(\mathcal{A}) \subseteq \bigcap_{n\geq 1} T(F_n(\mathcal{A}))$. Conversely, given

$$f = f(x_1, \ldots, x_k) \in \bigcap_{n\geq 1} T(F_n(\mathcal{A})),$$

by Lemma 3, we have $f \in T(F_k(\mathcal{A})) \cap \mathbb{K}\langle x_1, \ldots, x_k \rangle \subseteq T(\mathcal{A})$. □

Finally, we have the following alternative description of $F(\mathcal{A})$. We start with a lemma:

Lemma 5 *If $i \leq j$, then there exists an algebra monomorphism $u_{ij} : F_i(\mathcal{A}) \to F_j(\mathcal{A})$. Moreover, if $i \leq j \leq k$, then $u_{ik} = u_{jk}u_{ij}$.*

Proof Since $F_i(\mathcal{A})$ is free in var$(\mathcal{A})$, a homomorphism from $F_i(\mathcal{A})$ to an algebra in this variety is defined by a choice of images of the free generators of $F_i(\mathcal{A})$. So we can let u_{ij} send the free generators $\xi_1, \ldots, \xi_i$ of $F_i(\mathcal{A})$ to the first i free generators of $F_j(\mathcal{A})$. If the image of some element is zero in $F_j(\mathcal{A})$, then it is a polynomial identity of $F_i(\mathcal{A})$, so it will be zero in $F_i(\mathcal{A})$. Thus, this map is injective. The last assertion is immediate from the construction of the u_{ij}. □

The last lemma says that the pair $\big((\mathcal{A}_i)_{i\in\mathbb{N}}, (u_{ij})_{i\leq j}\big)$ is a direct system.

Proposition 6 $F(\mathcal{A}) = \varinjlim F_i(\mathcal{A})$.

Proof For each i, let $u_i : F_i(\mathcal{A}) \to F(\mathcal{A})$ be the map sending the free generators of $F_i(\mathcal{A})$ to the first i free generators of $F(\mathcal{A})$. Clearly $u_j u_{ij} = u_i$, for all $i \leq j$.

Now, consider a target $(\mathcal{B}, (\phi_i)_{i\in\mathbb{N}})$, that is, an algebra $\mathcal{B}$ together with homomorphisms $\phi_i : F_i(\mathcal{A}) \to \mathcal{B}$ such that $\phi_j u_{ij} = \phi_i$. We define $u : F(\mathcal{A}) \to \mathcal{B}$ in the free generators ξ_i via $u(\xi_i) := \phi_i(\xi_i)$ (the same image of ϕ_i applied in the last generator of $F_i(\mathcal{A})$). So clearly $uu_i = \phi_i$, for each $i \in \mathbb{N}$. This proves that $\varinjlim F_i(\mathcal{A}) = F(\mathcal{A})$. □

Corollary 7 *For any $\mathcal{B} \in var(\mathcal{A})$, one has*

$$Hom(F(\mathcal{A}), \mathcal{B}) = \varprojlim Hom(F_i(\mathcal{A}), \mathcal{B}).$$

The former corollary has an intuitive (and somewhat obvious) interpretation. Let $f \in F(\mathcal{A})$. It is known that the following three assertions are equivalent:

(i) f is a polynomial identity of $\mathcal{B}$,
(ii) $f \in \operatorname{Ker}\Psi$, for all $\Psi \in \operatorname{Hom}(F(\mathcal{A}), \mathcal{B})$.
(iii) $f \in \operatorname{Ker}\Psi$, for all $\Psi \in \operatorname{Hom}(F_j(\mathcal{A}), \mathcal{B})$, for a sufficiently large j (indeed, it is enough to take a j greater or equal to the number of variables of f).

The last corollary states the equivalence between (ii) and (iii).

4 Polynomial Identities for $UT_2(G^{(k)})$

In this section, we investigate the polynomial identities of $UT_2(G^{(k)})$. We find a explicit set of polynomials that, together with some class of polynomials, generate the T-ideal of polynomial identities of $UT_2(G^{(k)})$.

One may notice that if A and B are algebras such that $R = \begin{pmatrix} A & M \\ 0 & B \end{pmatrix}$ is an algebra, then $T(A)T(B) \subseteq T(R)$. Verifying if the above inclusion is an equality is a more difficult task. In some cases, the approach of Lewin's Theorem applies (see [17] or [10, Corollary 1.8.2] for a more suitable version). Some results in this direction are given in the paper [2], where the authors describe conditions under which the T-ideal of a block-triangular matrix algebra over an algebra A factors as the product of the ideals of the blocks. But one can see that the algebras $G^{(k)}$ do not satisfy the necessary hypothesis to that, namely the existence of a partially multiplicative basis for its relatively free algebras, so in this paper we try a different approach.

We let $\mathbb{K}$ be a field of characteristic zero.

Lemma 8 *For any $t \in \mathbb{N}$, the following are consequences of the identity $[x_1, x_2, x_3][x_4, x_5, x_6] = 0$:*

1. $[y_1, y_2, y_3]p[z_1, z_2, z_3] = 0,$
2. $([y_1, y_2][y_3, y_4] + [y_1, y_3][y_2, y_4])p[z_1, z_2, z_3] = 0,$
3. $[y_1, y_2, y_3]p([z_1, z_2][z_3, z_4] + [z_1, z_3][z_2, z_4]) = 0,$
4. $([y_1, y_2][y_3, y_4] + [y_1, y_3][y_2, y_4])p([z_1, z_2][z_3, z_4] + [z_1, z_3][z_2, z_4]) = 0.$

where $p = p(x_1, \ldots, x_t)$ is any multilinear polynomial.

Proof The first one follows from

$$\begin{aligned}[y_1, y_2, y_3]p(x_1, \ldots, x_t)[z_1, z_2, z_3] =& [[y_1, y_2, y_3], p(x_1, \ldots, x_t)][z_1, z_2, z_3] \\ &+ p(x_1, \ldots, x_t)[y_1, y_2, y_3][z_1, z_2, z_3].\end{aligned}$$

Working modulo the identity $[x_1, x_2, x_3][x_4, x_5, x_6]$, for the second one we have

$$\begin{aligned}0 &= [y_1, y_2^2, y_3]p(x_1, \ldots, x_t)[z_1, z_2, z_3] \\ &= ([[y_1, y_2]y_2, y_3] + [y_2[y_1, y_2], y_3])p(x_1, \ldots, x_t)[z_1, z_2, z_3] \\ &= ([y_1, y_2][y_2, y_3] + [y_1, y_2, y_3]y_2 + [y_2, y_3][y_1, y_2] \\ &\quad + y_2[y_1, y_2, y_3])p(x_1, \ldots, x_t)[z_1, z_2, z_3] \\ &= ([y_1, y_2][y_2, y_3] + [y_1, y_2][y_2, y_3] + [[y_2, y_3], [y_1, y_2]])p(x_1, \ldots, x_t)[z_1, z_2, z_3] \\ &= 2[y_1, y_2][y_2, y_3]p(x_1, \ldots, x_t)[z_1, z_2, z_3].\end{aligned}$$

Linearizing the above identity, we obtain (2). Analogously we obtain (3) and (4). □

We fix $m \in \mathbb{N}$.

Lemma 9 *The polynomials*

1. $[x_1, x_2, x_3][x_4, x_5, x_6] = 0,$

2. $[x_1, x_2] \ldots [x_{2m+3}, x_{2m+4}] = 0$,

are polynomial identities for $UT_2(G^{(2m)})$ *and* $UT_2(G^{(2m+1)})$.

Proof We know that $T(G^{(2m)}) = T(G^{(2m+1)})$. So, by [2, Lemma 10], we have $T(UT_2(G^{(2m)})) = T(UT_2(G^{(2m+1)}))$. Thus, we only need to check the statement for $UT_2(G^{(2m)})$. It is well-known that $[x_1, x_2, x_3][x_4, x_5, x_6]$ is a polynomial identity for $UT_2(G)$, hence, so is for $UT_2(G^{(2m)})$ as well.

Now, consider the polynomial q of (2). Since q is multilinear, we only need to check evaluations of q on matrix units multiplied by elements of $G^{(2m)}$. An evaluation will be automatically zero if two or more variables are substituted by a multiple of e_{12}. If all variables assume diagonal values, then we obtain zero again, since the diagonal of $UT_2(G^{(2m)})$ is $G^{(2m)} \oplus G^{(2m)}$.

So, assume that $x_i = ge_{12}$, for some g, and let x_j be the variable appearing together with x_i. So $[x_i, x_j] = g'e_{12}$. Next, the variables that appear before $[x_i, x_j]$ must be evaluated on some multiple of e_{11}, and the variables after $[x_i, x_j]$ must be evaluated on a multiple of e_{22}; otherwise we certainly obtain zero. So we have that $q = w_1 g' w_2 e_{12}$, where $w_1 g' w_2$ is a product of elements of $G^{(2m)}$, containing at least $m + 1$ commutators of elements of $G^{(2m)}$. So $w_1 g' w_2 = 0$, and q is a polynomial identity of $UT_2(G^{(2m)})$. □

Before we proceed, we recall the following classical result:

Theorem 10 (Theorem 5.2.1(ii) of [5]) *Let* $\mathbb{K}$ *be any infinite field, and* $n \in \mathbb{N}$. *The relatively free algebra of the variety generated by the identity*

$$[x_1, x_2] \cdots [x_{2n-1}, x_{2n}] = 0$$

has a basis consisting of all products

$$x_1^{a_1} \ldots x_m^{a_m} [x_{i_{11}}, x_{i_{21}}, \ldots, x_{i_{p_1 1}}] \ldots [x_{i_{1r}}, x_{i_{2r}}, \ldots, x_{i_{p_r r}}],$$

where the number r *of participating commutators is* $\leq n - 1$ *and the indices in each commutator* $[x_{i_{1s}}, x_{i_{2s}}, \ldots, x_{i_{p_s s}}]$ *satisfy* $i_{1s} > i_{2s} \leq \cdots \leq i_{p_s s}$. □

Remark Now, let $\mathcal{T}_m$ be the T-ideal generated by the identities of Lemma 9. We notice the following fact. Assume we have a multilinear polynomial of the following kind:

$$[x_{i_1}, x_{i_2}] \ldots [x_{2r-1}, x_{2r}][y_1, \ldots, y_s][x_{j_1}, x_{j_2}] \ldots [x_{j_{2t-1}}, x_{j_{2t}}],$$

where $s \geq 3$. Then, using the identities of Lemma 8, modulo the polynomial $[x_1, x_2, x_3][x_4, x_5, x_6]$, we can order $i_1 < \cdots < i_{2r}$, and $j_1 < \cdots < j_{2t}$.

Consider the following family of polynomials:

$$[x_{i_1}, x_{i_2}] \dots [x_{i_{2r-1}}, x_{i_{2r}}][x_{j_1}, \dots, x_{j_s}][x_{k_1}, x_{k_2}] \dots [x_{k_{2t-1}}, x_{k_{2t}}],$$
$$r \geq 0, t \geq 0, \quad r+t \leq m, \quad i_1 < i_2 < \dots < i_{2r}, \quad k_1 < k_2 < \dots < k_{2t}, \qquad (3)$$
$$s > 2, \quad j_1 > j_2 < j_3 < \dots < j_s.$$

Also, for each $t \in \mathbb{N}$, let $\mathfrak{B}_m^{(t)}$ be a basis of

$$\mathrm{Span}\{[x_{\sigma(1)}, x_{\sigma(2)}] \cdots [x_{\sigma(2t-1)}, x_{\sigma(2t)}] \mid \sigma \in S_{2t}\} + T(UT_2(G^{(2m)}))/T(UT_2(G^{(2m)})).$$

Denote $\mathfrak{B}_m = \bigcup_{t \in \mathbb{N}} \mathfrak{B}_m^{(t)}$. Let $\mathfrak{I}_m$ be the set of polynomial identities of $UT_2(E^{(2m)})$ given by a linear combination of product of commutators of length 2 (note that $\mathfrak{I}_m \neq 0$, since it contains identity (d) of Lemma 8).

Lemma 11 *The polynomials* (3) *and* $\mathfrak{B}_m$ *generate the proper multilinear polynomials in* $\mathbb{K}\langle X \rangle$ *modulo* $\mathcal{T}_m + \mathfrak{I}_m$.

Proof Since $[x_1, x_2] \dots [x_{2m+3}, x_{2m+4}] \in \mathcal{T}_m$, it is enough to write the elements of the relatively free algebra $F(UT_{m+2}(\mathbb{K}))$ as a linear combination of polynomials of kind (3), and elements of $\mathfrak{B}_m$. From Theorem 10, and since $[x_1, x_2, x_3][x_4, x_5, x_6] = 0$, it is enough to consider a polynomial q of kind

$$q = [x_{i_1}, x_{i_2}] \dots [x_{i_{2r-1}}, x_{i_{2r}}][x_{j_1}, \dots, x_{j_s}][x_{k_1}, x_{k_2}] \dots [x_{k_{2t-1}}, x_{k_{2t}}].$$

If $s > 2$, then from the remark above, we can order $i_1 < \dots < i_{2r}$, and $k_1 < \dots < k_{2t}$, and we are done. If $s = 2$, then q is a product of commutators of length 2. So, q is a linear combination of elements $\mathfrak{B}_m$ modulo $\mathfrak{I}_m$, by definition. □

Lemma 12 *The family of polynomials given by* (3) *and* $\mathfrak{B}_m$ *are linearly independent modulo* $T(UT_2(G^{(2m)}))$.

Proof Consider a multilinear polynomial identity $f \in T(UT_2(G^{(2m)}))$, and write $f = f_1 + f_2$, where f_1 is a linear combination of the polynomials (3), and f_2 is a linear combination of polynomials in $\mathfrak{B}_m$. For some $s > 2$, consider the following evaluation ψ:

$$x_{i_1} = g_1 e_{11}, \dots, x_{i_{2r}} = g_{2r} e_{11},$$
$$x_{j_2} = e_{12},$$
$$x_{j_1} = x_{j_3} = \dots = x_{j_s} = e_{11},$$
$$x_{k_1} = g_{2r+1} e_{22}, \dots, x_{k_{2t}} = g_{2r+2t} e_{22}.$$

Then, any polynomial which is the product of more than $r + t + 1$ commutators annihilate. Note that, since $s > 2$, this evaluation gives $\psi(f_2) = 0$. Among the polynomials of type (3), there is a single polynomial having a nonzero evaluation, namely

$$[x_{i_1}, x_{i_2}]\dots[x_{i_{2r-1}}, x_{i_{2r}}][x_{j_1}, \dots, x_{j_s}][x_{k_1}, x_{k_2}]\dots[x_{k_{2t-1}}, x_{k_{2t}}].$$

This proves that $f_1 = 0$. So $f = f_2$. By the choice of $\mathfrak{B}_m$, we obtain $f_2 = 0$ and we are done. □

As a consequence, we have the following.

Theorem 13 *For $m \in \mathbb{N}$, set*

$$\mathcal{T}_m = \langle [x_1, x_2, x_3][x_4, x_5, x_6], [x_1, x_2]\dots[x_{2m+3}, x_{2m+4}] \rangle.$$

Then,

$$T(UT_2(G^{(2m)})) = \mathcal{T}_m + \mathfrak{I}_m,$$

where $\mathfrak{I}_m$ is the set of polynomial identities given by linear combination of product of commutators of length 2. □

4.1 The Case $UT_2(G^{(2m)})$, for $1 \leq m \leq 2$

For small k, we can prove that we do not need the set $\mathfrak{I}_m$ of the previous result.

Lemma 14 *Consider the following family of polynomials:*

$$\begin{gathered}[x_{i_1}, x_{i_2}]\cdots[x_{i_{2t-1}}, x_{i_{2t}}],\\ i_1 > i_2, \quad i_3 > i_4, \dots, \quad i_{2t-1} > i_{2t}.\end{gathered} \tag{4}$$

Fix any $m \in \mathbb{N}$, and let $t \leq \min\{3, m\}$. Then, the polynomials (4) of degree $2t$ generate the subspace spanned by all (multilinear) product of commutators of length 2 of degree $2t$ of $\mathbb{K}\langle X\rangle$ modulo $\mathcal{T}_m$, and they are linearly independent modulo $T(UT_2(G^{(2m)}))$.

Proof The assertion that these polynomials generate all multilinear product of commutators of length 2 modulo $\mathcal{T}_m$ is direct from Theorem 10. So we only need to prove the linearly independence part. If $t = 1$, then there is nothing to do.

So, let f be a linear combination of polynomials (4), $\deg f = 4$. The evaluation

$$\begin{aligned} x_{i_1} &= g_1 e_{11}, \\ x_{i_2} &= g_2 e_{11}, \\ x_{i_3} &= e_{12}, \\ x_{i_4} &= e_{22}, \end{aligned}$$

will make all product of commutators zero, but $[x_{i_1}, x_{i_2}][x_{i_3}, x_{i_4}]$. Thus, the elements of degree 4 are linearly independent.

Similarly, if $\deg f = 6$, then the evaluation

$$
\begin{aligned}
x_{i_1} &= g_1 e_{11}, \\
x_{i_2} &= g_2 e_{11}, \\
x_{i_3} &= e_{12}, \\
x_{i_4} &= e_{22}, \\
x_{i_5} &= g_3 e_{22}, \\
x_{i_6} &= g_4 e_{22},
\end{aligned}
$$

will make all product of commutators zero, but $[x_{i_1}, x_{i_2}][x_{i_3}, x_{i_4}][x_{i_5}, x_{i_6}]$. This concludes the proof. □

As a consequence, if $1 \leq m \leq 2$, then $\mathfrak{I}_m \subseteq \langle [x_1, x_2] \ldots [x_{2m-1}, x_{2m}] \rangle$. Thus, using the lemmas from the previous section, we see that

$$\mathcal{T}_m \subseteq T(UT_2(G^{(2m)})) \subseteq \mathcal{T}_m.$$

We proved:

Theorem 15 *For* $1 \leq m \leq 2$, *set*

$$\mathcal{T}_m = \langle [x_1, x_2, x_3][x_4, x_5, x_6], [x_1, x_2] \ldots [x_{2m+3}, x_{2m+4}] \rangle.$$

Then, $T(UT_2(G^{(2m)})) = T(UT_2(G^{(2m+1)})) = \mathcal{T}_m$.

5 Polynomial Identities for $F_k(UT_2(G))$

Let us denote by T_1 the T-ideal generated by $[x_1, x_2, x_3]$ and by T_2 the T-ideal generated by $[x_1, x_2, x_3][x_4, x_5, x_6]$.

Lemma 16 *Let* $n \geq 1$. *If* f *is defined as*

$$f = u_0[v_1, v_2, v_3]u_1[w_{\sigma(1)}, w_{\sigma(2)}]u_2 \cdots u_n[w_{\sigma(2n-1)}, w_{\sigma(2n)}]u_{n+1},$$

with $u_i, v_i, w_i \in K\langle X \rangle$ *for all* i *and* $\sigma \in S_{2n}$, *then*

$$f = (-1)^{\sigma} u_0[v_1, v_2, v_3][w_1, w_2] \cdots [w_{2n-1}, w_{2n}]u_1 \cdots u_{n+1} \mod T_2.$$

Proof After using the identity

$$c[a, b] = [a, b]c - [a, b, c],$$

n times, we obtain

$$f = u_0[v_1, v_2, v_3][w_{\sigma(1)}, w_{\sigma(2)}] \cdots [w_{\sigma(2n-1)}, w_{\sigma(2n)}]u_1 \cdots u_{n+1} \mod T_2. \tag{5}$$

Since

$$[x_1, x_2][x_3, x_4] = -[x_1, x_3][x_2, x_4] \mod T_1,$$

the identity

$$[w_{\sigma(1)}, w_{\sigma(2)}] \cdots [w_{\sigma(2n-1)}, w_{\sigma(2n)}] = (-1)^{\sigma}[w_1, w_2] \cdots [w_{2n-1}, w_{2n}] \mod T_1$$

holds. The above identity and (5) imply

$$f = (-1)^{\sigma} u_0[v_1, v_2, v_3][w_1, w_2] \cdots [w_{2n-1}, w_{2n}]u_1 \cdots u_{n+1} \mod T_2.$$

□

Remark An analogous version of the above lemma, is also true, if one considers the factor $[v_1, v_2, v_3]$ at the end of the monomial. The proof is completely analogous.

Lemma 17 *If $m \geq 1$, the following polynomials are identities for $F_k(UT_2(G))$, for $k \leq 2m + 1$:*

1. $[x_1, x_2, x_3][x_4, x_5, x_6]$;
2. $[x_1, x_2, x_3][x_4, x_5] \cdots [x_{2m+4}, x_{2m+5}]$;
3. $[x_1, x_2] \cdots [x_{2m+1}, x_{2m+2}][x_{2m+3}, x_{2m+4}, x_{2m+5}]$;
4. $[x_1, x_2] \cdots [x_{4m+3}, x_{4m+4}]$.

Proof First we observe that it is enough to prove the result for $k = 2m + 1$. We use the model for the relatively free algebra of rank $2m + 1$ of $UT_2(G)$ given in section 2, i.e., the subalgebra $\mathcal{U}_{2m+1}$ of $UT_2(F(G))$ generated by the generic matrices $\xi_1, \ldots, \xi_{2m+1}$, where

$$\xi_i = \begin{pmatrix} x_{11}^{(1)} + T(G) & x_{12}^{(i)} + T(G) \\ 0 & x_{22}^{(2)} + T(G) \end{pmatrix}$$

We observe that the set $A_{1,1} = \{p \mid p \text{ is the entry (1,1) of some element of } \mathcal{U}_{2m+1}\}$ is an algebra, isomorphic to $F_{2m+1}(UT_2(G))$ of G, in the variables $x_{11}^{(1)}, \ldots, x_{11}^{(2m+1)}$. Analogously, the set $A_{2,2} = \{p \mid p \text{ is the entry (2,2) of some element of } \mathcal{U}_{2m+1}\}$ is an algebra, isomorphic to the relatively free algebra of rank $2m + 1$ of G, in the variables $x_{22}^{(1)}, \ldots, x_{22}^{(2m+1)}$. In particular, they satisfy the polynomial identities

$$[x_1, x_2, x_3] \text{ and } [x_1, x_2] \cdots [x_{2m+1}, x_{2m+2}].$$

It is clear that (1) is a polynomial identity, since it is an identity for $UT_2(G)$.

To show that (2) and (3) are identities, it is enough to verify they vanish under substitution of variables by monomials in the variables ξ_i. By Lemma 16 and using the identity $[ab, c] = a[b, c] + [a, c]b$, it is enough to show that they vanish under substitution of variables by the generic elements ξ_i, $i \in \{1, \dots, 2m+1\}$.

Now one verifies that the substitution of such elements in the polynomials $[x_1, x_2, x_3]$ and $[x_4, x_5] \cdots [x_{2m+4}, x_{2m+5}]$ yields matrices which are multiples of the unit matrix e_{12} by an element of $F(G)$, since these polynomials are identities for $A_{1,1}$ and for $A_{2,2}$. As a consequence, the product of the evaluations of such polynomials in both orders vanishes, showing that (2) and (3) are identities for $F_{2m+1}(UT_2(G))$.

Again, to prove that (4) is an identity, it is enough to verify it vanishes under substitution of variables by monomials in the variables ξ_i. After using several times the identity $[ab, c] = a[b, c] + [a, c]b$, one obtains a linear combination of elements of the form

$$u_0[y_1, y_2]u_1[y_3, y_4]u_2 \cdots u_{2m+1}[y_{4m+3}, y_{4m+4}]u_{2m+2},$$

where the u_i are elements of $\mathcal{U}_{2m+1}$ and the y_i are generic matrices ξ_j.

If $0 < i \leq 2m+1$, then, by using the identity $c[a, b] = [a, b]c - [a, b, c]$ in the factor $u_i[y_{2i+1}, y_{2i+2}]$, it turns into $[y_{2i+1}, y_{2i+2}]u_i - [y_{2i+1}, y_{2i+2}, u_i]$. Now, since $i \leq 2m+1$ one observes that using Lemma 16 and the fact that (2) is an identity, the component of the sum corresponding to the triple commutator vanishes. Applying such procedure several times, we obtain that the elements $u_1, \dots, u_{2m+1}$ can be moved to the middle of the monomial (just after the $(m+1)$-th commutator). In a analogous way, using the remark after Lemma 16 and the fact that (3) is an identity, we obtain that if $m+1 < i \leq 2m+1$, the elements u_i can also be moved to the middle of the monomial, i.e.,

$$u_0[y_1, y_2]u_1[y_3, y_4]u_2 \cdots u_{2m+1}[y_{4m+3}, y_{4m+4}]u_{2m+2} =$$

$$= u_0[y_1, y_2] \cdots [y_{2m+1}, y_{2m+2}]u_1 \cdots u_{2m+1}[y_{2m+3}, y_{2m+4}] \cdots [y_{4m+3}, y_{4m+4}]u_{m+2}$$

Since the product of $2m+1$ commutators is a multiple of e_{12}, and the above is a product of two multiples of e_{12}, we obtain that the above element is zero in $\mathcal{U}_{2m+1}$.

□

6 Conclusion

Describing the ideal of identities of the relatively free algebras of finite rank of a given variety may be a very difficult problem. Even the simple case of $UT_2(G)$ is still open even though it seems to be possible to prove it with the canonical techniques.

The role played by the relatively free algebras of finite rank in the description of the subvarieties of a given variety (at least up to asymptotic equivalence) must be studied.

An interesting problem is to consider $\mathfrak{V}$ a variety of algebras generated by $G(B)$, where $B = B_0 + B_1$ is a finite dimensional superalgebra and to investigate if, given an n, there exists an m such that $T(F_n(G(B))) = T(G^{(m)}(B))$ (since we have verified that in the variety generated by $UT_2(G)$, $n = m$ does not hold, as in the previously known cases).

In order to know if such questions are true in some generality, it is necessary first to study it for some simpler examples.

For varieties that we know the structure of the S_n-module $P_n(\mathcal{V})$ (or $\Gamma_n(\mathcal{V})$), of multilinear (or proper multilinear) polynomials modulo the identities of $\mathfrak{V}$, this may be approached by verifying which of the generators of such modules are identities of $F_m(\mathfrak{V})$. We will investigate this problem in future projects.

Acknowledgments T. C. de Mello was supported by grants #2018/15627-2 and #2018/23690-6, São Paulo Research Foundation (FAPESP), Brazil.

F. Y. Yasumura was supported in part by the Coordenação de Aperfeiçoamento de Pessoal de Nível Superior—Brasil (CAPES)—Finance Code 001.

References

1. Berele, A.: Generic verbally prime PI-algebras and their GK-dimensions. Commun. Algebra **21**(5), 1487–1504 (1993)
2. Centrone, L., de Mello, T.C.: On the factorization of T_G-ideals of graded matrix algebras. Beitr. Algebra Geom. **59**(3), 597–615 (2018)
3. Di Vincenzo, O.M., Drensky, V., Nardozza, V.: Algebras satisfying the polynomial identity $[x_1, x_2][x_3, x_4, x_5] = 0$. J. Algebra Appl. **3**(2), 121–142 (2004)
4. Drenski, V.S.: Polynomial identities of finite-dimensional algebras. Serdica **12**(3), 209–216 (1986)
5. Drensky, V.: Free Algebras and PI-Algebras. Springer, Singapore (2000). Graduate course in algebra
6. Formanek, E.: The polynomial identities and invariants of $n \times n$ matrices. In: CBMS Regional Conference Series in Mathematics. Published for the Conference Board of the Mathematical Sciences, Washington, DC, vol. 78. American Mathematical Society, Providence (1991)
7. Giambruno, A., Zaicev, M.: On codimension growth of finitely generated associative algebras. Adv. Math. **140**(2), 145–155 (1998)
8. Giambruno, A., Zaicev, M.: Exponential codimension growth of PI algebras: an exact estimate. Adv. Math. **142**(2), 221–243 (1999)
9. Giambruno, A., Zaicev, M.: A characterization of varieties of associative algebras of exponent two. Serdica Math. J. **26**(3), 245–252 (2000)
10. Giambruno, A., Zaicev, M.: Polynomial Identities and Asymptotic Methods. Mathematical Surveys and Monographs, vol. 122. American Mathematical Society, Providence (2005)
11. Gonçalves, D.J., de Mello, T. C.: Minimal varieties and identities of relatively free algebras. Commun. Algebra **43**(12), 5217–5235 (2015)
12. Kemer, A.R.: Asymptotic basis of identities of algebras with unities from the variety Var($M_2(F)$). Izv. Vyssh. Uchebn. Zaved. Mat. **6**, 71–76 (1989)

13. Kemer, A.R.: Ideals of Identities of Associative Algebras. Translations of Mathematical Monographs, vol. 87. American Mathematical Society, Providence (1991). Translated from the Russian by C. W. Kohls
14. Koshlukov, P., de Mello, T.C.: The centre of generic algebras of small PI algebras. J. Algebra **375**, 109–120 (2013)
15. Koshlukov, P., de Mello, T.C.: On the polynomial identities of the algebra $M_{11}(E)$. Linear Algebra Appl. **438**(11), 4469–4482 (2013)
16. La Mattina, D.: Varieties of almost polynomial growth: classifying their subvarieties. Manuscripta Math. **123**(2), 185–203 (2007)
17. Lewin, J.: A matrix representation for associative algebras. I, II. Trans. Am. Math. Soc. **188**, 293–308; ibid. 188 (1974), 309–317 (1974)
18. Popov, A.P.: Identities of the tensor square of a Grassmann algebra. Algebra i Logika **21**(4), 442–471 (1982)
19. Procesi, C.: Rings with Polynomial Identities. Pure and Applied Mathematics, vol. 17. Marcel Dekker, New York (1973)
20. Procesi, C.: The invariant theory of $n \times n$ matrices. Adv. Math. **19**(3), 306–381 (1976)

Graded Algebras, Algebraic Functions, Planar Trees, and Elliptic Integrals

Vesselin Drensky

Dedicated to the 70-th anniversary of Antonio Giambruno, a mathematician, person and friend.

Abstract This article surveys results on graded algebras and their Hilbert series. We give simple constructions of finitely generated graded associative algebras R with Hilbert series $H(R,t)$ very close to an arbitrary power series $a(t)$ with exponentially bounded nonnegative integer coefficients. Then we summarize some related facts on algebras with polynomial identity. Further we discuss the problem how to find power series $a(t)$ which are rational/algebraic/transcendental over $\mathbb{Q}(t)$. Applying a classical result of Fatou we conclude that if a finitely generated graded algebra has a finite Gelfand-Kirillov dimension, then its Hilbert series is either rational or transcendental. In particular the same dichotomy holds for the Hilbert series of a finitely generated algebra with polynomial identity. We show how to use planar rooted trees to produce algebraic power series. Finally we survey some results on noncommutative invariant theory which show that we can obtain as Hilbert series various algebraic functions and even elliptic integrals.

Keywords Graded algebras · Monomial algebras · Algebras with polynomial identity · Hilbert series · Algebraic series · Transcendental series · Free magma · Free nonassociative algebras · Planar trees · Elliptic integrals

1 Introduction

We consider algebras R over a field K. Except Sects. 6 and 7 all algebras are finitely generated and associative. The field K is arbitrary of any characteristic except in Sect. 7 when it is of characteristic 0.

V. Drensky (✉)
Institute of Mathematics and Informatics, Bulgarian Academy of Sciences, Sofia, Bulgaria
e-mail: drensky@math.bas.bg

O. M. Di Vincenzo, A. Giambruno (eds.), *Polynomial Identities in Algebras*, Springer INdAM Series 44, https://doi.org/10.1007/978-3-030-63111-6_9

The purpose of this article is to survey some results, both old and recent, on graded algebras and their Hilbert series. In Sect. 2 we discuss the growth of algebras, and graded algebras and their Hilbert series. Then in Sect. 3 we give constructions of graded algebras with prescribed Hilbert series. Section 4 is devoted to algebras with polynomial identities, or PI-algebras. We survey some results concerning basic properties and the growth of such algebras. Section 5 deals with power series with nonnegative integer coefficients. We consider methods to produce series which are transcendental over $\mathbb{Q}(t)$ and graded algebras with transcendental Hilbert series. Combining a classical result of Fatou from 1906 with a theorem of Shirshov from 1957 we obtain immediately that the Hilbert series of a finitely generated graded PI-algebra is either rational or transcendental. We also survey some constructions of algebraic power series based on automata theory and theory of formal languages. In the next Sect. 6 we consider a method for construction of algebraic power series with nonnegative integer coefficients. The main idea is to combine results on planar rooted trees with number of leaves divisible by a given integer with the fact that submagmas of free Ω-magmas are also free. Finally, in Sect. 7 we use methods of noncommutative invariant theory to construct free graded algebras (also nonassociative and not finitely generated) with Hilbert series which are either algebraic or transcendental. In particular, we give simple examples of free nonassociative algebras with Hilbert series which are elliptic integrals.

If not explicitly stated otherwise, all power series in our exposition will have nonnegative integer coefficients. Usually, when we state theorems about power series we do not present them in the most general form and restrict the considerations to the case of nonnegative integer coefficients.

2 Growth of Algebras and Hilbert Series

If R is a finite dimensional algebra we can measure how big it is using its dimension $\dim(R)$ as a vector space. But how to measure infinite dimensional algebras? If R is an algebra (not necessarily associative) generated by a finite dimensional vector space V, then the *growth function* of R is defined by

$$g_V(n) = \dim(R^n), \quad R^n = V^0 + V^1 + V^2 + \cdots + V^n, \quad n = 0, 1, 2, \ldots.$$

This definition has the disadvantage that depends on the generating vector space V. For example, the algebra of polynomials in d variables $K[X_d] = K[x_1, \ldots, x_d]$ is generated by the vector space V with basis $X_d = \{x_1, \ldots, x_d\}$ and the growth function $g_V(n)$ is

$$g_V(n) = \binom{n+d}{d} = \frac{(n+d)(n+d-1)\cdots(n+1)}{d!} = \frac{n^d}{d!} + O(n^{d-1}).$$

The algebra $K[X_d]$ is generated also by the monomials of first and second degree, i.e. by the vector space $W = V + V^2$. Then

$$g_W(n) = \binom{2n+d}{d} = \frac{2^d n^d}{d!} + O(n^{d-1}).$$

What is common between both generating functions? There is a standard method to compare eventually monotone increasing and positive valued functions $f : \mathbb{N}_0 = \mathbb{N} \cup \{0\} \to \mathbb{R}$. This class of functions consists of all functions f such that there exists an $n_0 \in \mathbb{N}$ such that $f(n_0) \geq 0$ and $f(n_2) \geq f(n_1) \geq f(n_0)$ for all $n_2 \geq n_1 \geq n_0$. We define a preorder $\preceq$ and equivalence $\sim$ on the set of such functions. We write that $f_1 \preceq f_2$ for two functions f_1 and f_2 (and say that f_2 *grows faster than* f_1) if and only if there exist positive integers a and p such that for all sufficiently large n the inequality $f_1(n) \leq af_2(pn)$ holds and $f_1 \sim f_2$ if and only if $f_1 \preceq f_2$ and $f_2 \preceq f_1$. This allows to obtain some invariant of the growth because $g_V(n) \sim g_W(n)$ for any generating vector spaces V and W of the algebra R. The equivalence is expressed in the following notion. The limit superior

$$\text{GKdim}(R) = \limsup_{n\to\infty} \log_n(g_V(n))$$

is called the *Gelfand-Kirillov dimension* of R. It is known that GKdim(R) does not depend on the system of generators of the algebra R.

Below we give a brief information for the values of the Gelfand-Kirillov dimension of finitely generated associative algebras. For details we refer to the book by Krause and Lenagan [60].

Theorem 1

(i) If R is commutative then GKdim(R) is an integer equal to the transcendence degree of the algebra R.

(ii) If R is associative then GKdim$(R) \in \{0, 1\} \cup [2, \infty]$ and every of these reals is realized as a Gelfand-Kirillov dimension.

Part (i) of Theorem 1 is a classical result. The restriction GKdim$(R) \notin (1, 2)$ in part (ii) is the *Bergman Gap Theorem* [14]. Algebras R with GKdim$(R) \in [2, \infty)$ are realized by Borho and Kraft [16], see also the modification of their construction in the book of the author [32, Theorem 9.4.11]. We shall mimic these constructions in the next section.

In the sequel we shall work with graded algebras. The algebra R is *graded* if it is a direct sum of vector subspaces $R_0, R_1, R_2, \ldots$ called *homogeneous components* of R and

$$R_m R_n \subset R_{m+n}, \quad m, n = 0, 1, 2, \ldots.$$

It is convenient to assume that $R_0 = 0$ or $R_0 = K$. In most of our considerations the generators of R are of first degree. The formal power series

$$H(R,t) = \sum_{n\geq 0} \dim(R_n)t^n,$$

is called the *Hilbert series* (or *Poincaré series*) of R.

We often shall work with power series with nonnegative integer coefficients

$$a(t) = \sum_{n\geq 0} a_n t^n, \quad a_n \in \mathbb{N}_0.$$

The advantage of studying such power series instead of the sequence a_n, $n = 0, 1, 2, \ldots$, of the coefficients of $a(t)$ is that we may apply the theory of analytic functions or to encode some recurrence relations. In particular, we may find a closed formula for a_n or may estimate its asymptotic behavior.

The formal power series $a(t)$ is a *rational function* if it converges in a neighborhood of 0 to a fraction of two polynomials with rational coefficients, i.e. to an element of the field $\mathbb{Q}(t)$. Similarly, algebraic and transcendental power series are also over $\mathbb{Q}(t)$. *Algebraic power series* $a(t)$ have the property that $p(t, a(t)) = 0$ for some polynomial $p(t, z) \in \mathbb{Q}[t, z]$ and *transcendental power series* do not satisfy any polynomial equation with rational coefficients.

Algebraic power series have a nice characterization given by the Abel-Tannery-Cockle-Harley-Comtet theorem [1, p. 287], [20–22, 49, 94] (see [4] for comments).

Theorem 2 *An algebraic power series*

$$f(t) = \sum_{n\geq 0} a_n t^n$$

is D-finite, i.e. it satisfies a linear differential equation with coefficients which are polynomials in t. Equivalently, its coefficients a_n satisfy a linear recurrence with coefficients which are polynomials in n.

We shall recall the usual definition of different kinds of growth of a sequence a_n, $n = 0, 1, 2, \ldots$, of complex numbers. If there exist positive b and c such that $|a_n| \leq bn^c$ for all n, we say that the sequence is of *polynomial growth*. (We use this terminology although it is more precise to say that the sequence a_n, $n = 0, 1, 2, \ldots$, is polynomially bounded.) If there exist $b_1, b_2 > 0$ and $c_1, c_2 > 1$ such that $|a_n| \leq b_2 c_2^n$ for all n and $b_1 c_1^{n_k} \leq |a_{n_k}|$ for a subsequence a_{n_k}, $k = 0, 1, 2, \ldots$, then the sequence is of *exponential growth*. Finally, if for any $b, c > 0$ there exists a subsequence a_{n_k}, $k = 0, 1, 2, \ldots$, such that $|a_{n_k}| > bn_k^c$ and for any $b_1 > 0$, $c_1 > 1$ the inequality $|a_n| < b_1 c_1^n$ holds for all sufficiently large n, then the sequence is of *intermediate growth*.

The following statement is well known, see e.g. [41, Theorem VII.8, p. 501] for more precise asymptotics of the coefficients.

Proposition 1 *The coefficients of an algebraic power series*

$$a(t) = \sum_{n\geq 0} a_n t^n$$

are either of polynomial or of exponential growth.

Every algebra R generated by a finite set $\{r_1, \dots, r_d\}$ is a homomorphic image of the free associative algebra $K\langle X_d\rangle = K\langle x_1, \dots, x_d\rangle$. The map $\pi_0 : x_i \to r_i$, $i = 1, \dots, d$, is extended to a homomorphism $\pi : K\langle X_d\rangle \to R$ and $R \cong K\langle X_d\rangle/I$, $I = \ker(\pi)$. If the ideal I of $K\langle X_d\rangle$ is finitely generated, then the algebra R is *finitely presented.* An important special case of graded algebras is the class of *monomial algebras.* Monomial algebras are factor algebras of $K\langle X_d\rangle$ modulo an ideal generated by monomials, i.e. by elements of the free unitary semigroup $\langle X_d\rangle$.

Below we give some properties of Hilbert series. We start with commutative graded algebras.

Theorem 3 *Let R be a finitely generated graded commutative algebra. Then:*

(i) (Theorem of Hilbert-Serre) The Hilbert series $H(R,t)$ is a rational function with denominator which is a product of binomials $1 - t^m$.

(ii) If

$$H(R,t) = p(t)\prod \frac{1}{(1-t^{m_i})^{a_i}}, \quad a_i \geq 1, \quad p(t) \in \mathbb{Z}[t],$$

then the Gelfand-Kirillov dimension GKdim(R) is equal to the multiplicity of 1 as a pole of $H(R,t)$: If $p(1) \neq 0$, then GKdim$(R) = \sum a_i$.

The coefficients of the Hilbert series of a finitely generated commutative algebras are a subject of many additional restrictions, see Macaulay [73]. The picture for noncommutative graded algebras is more complicated than in the commutative case. Govorov [43] proved that if the set of monomials U is finite, then the Hilbert series of the monomial algebra $R = K\langle X\rangle/(U)$ is a rational function. He conjectured [43, 44] that the same holds for the Hilbert series of finitely presented graded algebras. By a theorem of Backelin [3] this holds when the ideal (U) is generated by a single homogeneous polynomial. On the other hand Shearer [88] presented an example of a finitely presented graded algebra with algebraic nonrational Hilbert series. As he mentioned his construction gives also an example with a transcendental Hilbert series. Another simple example of a finitely presented algebra with algebraic Hilbert series was given by Kobayashi [56]. It is interesting to mention that the rationality of the Hilbert series may depend on the base field K. The following theorem is from the recent paper by Piontkovski [84].

Theorem 4 *Let K be a field of positive characteristic p and let the coefficients of the Hilbert series $H(R,t)$ of the finitely generated graded algebra R be bounded by a constant. If $H(R,t)$ is transcendental, then the base field K contains an element* *which is not algebraic over the prime subfield $\mathbb{F}_p$ of K. For every such field K there exist graded algebras R with transcendental Hilbert series $H(R, t)$ with coefficients bounded by a constant.*

In the next sections we shall discuss the problem how to construct more algebras with algebraic and nonrational Hilbert series.

By Proposition 1 if the Hilbert series $H(R, t)$ is algebraic, then its coefficients grow either exponentially or polynomially. Hence a power series with intermediate growth of the coefficients is transcendental. In [43] Govorov constructed a two-generated monomial algebra with Hilbert series with intermediate growth of the coefficients.

A very natural class of finitely generated graded algebras with Hilbert series with coefficients of intermediate growth are universal enveloping algebras of infinite dimensional Lie algebras of subexponential (i.e. slower than exponential) growth. The first example of this kind was given by Smith [93]:

Theorem 5

(i) *If L is an infinite dimensional graded Lie algebra with subexponential growth of the coefficients of its Hilbert series, then the Hilbert series of its universal enveloping algebra $U(L)$ is with intermediate growth of the coefficients.*
(ii) *There exists a two-generated infinite dimensional graded Lie algebra L with Hilbert series*

$$H(L, t) = t + \frac{1}{1-t}.$$

Then the Hilbert series of $U(L)$ is with intermediate growth of the coefficients:

$$H(U(L), t) = \frac{1}{1-t}\prod_{n\geq 1}\frac{1}{1-t^n}.$$

The Lie algebra L in Theorem 5 (ii) has a basis $\{x, y_1, y_2, \ldots\}$, $\deg(x) = 1$, $\deg(y_i) = i$, $i = 1, 2, \ldots$, and the defining relations of L are

$$[x, y_i] = y_{i+1}, \quad [y_i, y_j] = 0, \quad i, j = 1, 2, \ldots.$$

Lichtman [70] generalized the result of Smith for different classes of Lie algebras. Later Petrogradsky [82, 83] developed the theory of functions with intermediate growth of the coefficients which are realized as Hilbert series in the known examples of algebras with intermediate growth. In this way he introduced a detailed scale to measure the growth of algebras which reflected also on the growth of the coefficients of the Hilbert series of graded associative and Lie algebras.

The algebras in the examples of Smith [93], Lichtman [70], and Petrogradsky [82, 83] are not finitely presented. Borho and Kraft [16] conjectured that finitely presented associative algebras cannot be of intermediate growth. For a counterexample it is sufficient to show that there exists a finitely presented and infinite dimensional Lie algebra with polynomial growth. Leites and Poletaeva [68] showed that over a field of characteristic 0 the classical Lie algebras W_d, H_d, S_d, K_d of polynomial vector fields are finitely presented. Recall that the algebra $W_d = \mathrm{Der}(K[X_d])$ consists of the derivations of the polynomial algebra $K[X_d]$. The special algebra $S_d \subset W_{d+1}$ and the Hamiltonian algebra $H_d \subset W_{2d}$ annihilate suitable exterior differential forms, and the contact algebra $K_d \subset W_{2d-1}$ multiplies a certain form. The easiest example is the Witt algebra W_1 of the derivations of $K[x]$.

The first example of a finitely presented graded algebra with Hilbert series with intermediate growth of the coefficients was given by Ufnarovskij [95]. In his example the algebra is two-generated by elements of degree 1 and 2. The Lie algebra W_1 of the derivations of the polynomial algebra in one variable over a field K of characteristic 0 has a graded basis

$$\left\{\delta_{i-1} = x^i\frac{d}{dx} \mid i \geq 0\right\},\quad \deg\left(x^i\frac{d}{dx}\right) = i-1,$$

and multiplication

$$[\delta_{i-1},\delta_{j-1}] = \left[x^i\frac{d}{dx}, x^j\frac{d}{dx}\right] = (j-i)x^{i+j-1}\frac{d}{dx} = (j-i)\delta_{i+j-2}.$$

Hence for $i \geq 2$ the derivations δ_{i+1} may be defined inductively by

$$\delta_{i+1} = \frac{1}{i-1}[\delta_1,\delta_i].$$

Theorem 6 *Let L be the Lie subalgebra of W_1 generated by δ_1 and δ_2. It has a basis $\{\delta_i \mid i = 1,2,\ldots\}$ and defining relations*

$$[\delta_2,\delta_3] = \delta_5 \textit{ and } [\delta_2,\delta_5] = 3\delta_7.$$

The universal enveloping algebra $U(L)$ of L is generated by $f_1 = x$ and $f_2 = y$, where

$$f_{i+1} = \frac{1}{i-1}(f_1f_i - f_if_1),\quad i = 2,3,\ldots.$$

It is a factor algebra of the free algebra $K\langle x,y\rangle$ modulo the ideal generated by

$$(f_2f_3 - f_3f_2) - f_5 \textit{ and } (f_2f_5 - f_5f_2) - 3f_7.$$

If $\deg(f_i) = i$, $i = 1, 2, \ldots$, *then*

$$H(U(L), t) = \prod_{n\geq 1} \frac{1}{1-t^n}.$$

In a note added in the proofs Shearer [88] gave two more examples of finitely presented graded algebras with Hilbert series which also have an intermediate growth of the coefficients. His algebras are generated by three elements and have three defining relations but, as in the example of Ufnarovskij [95] one of the generators is of second degree.

Theorem 7 *Let* $R = K\langle x_1, x_2, y\rangle/(U)$, *where*

$$\deg(x_1) = \deg(x_2) = 1, \deg(y) = 2,$$

$$U = \{x_1y - yx_1, x_1x_2x_1 - x_2y, x_2^2y\}.$$

Then the Hilbert series of R is

$$H(R, t) = \frac{1}{(1-t)(1-t^2)} \prod_{n\geq 1} \frac{1}{1-t^n}.$$

If in U we replace x_2^2y *with* x_2^2, *then*

$$H(R, t) = \frac{1}{(1-t)(1-t^2)} \prod_{n\geq 1} (1+t^n).$$

Koçak [57] modified the construction of Shearer [88] such that the three generators are of first degree:

Theorem 8 *Let*

$$U = \{x_2^2x_1 - x_1x_2^2, x_2^2x_3 - x_1x_3x_1, x_1x_3^2, x_1x_2x_1, x_1x_2x_3, x_3x_2x_1, x_3x_2x_3\}.$$

Then the coefficients of the Hilbert series of the algebra $R = K\langle x_1, x_2, x_3\rangle/(U)$ *are of intermediate growth.*

Koçak [57] also constructed a graded algebra with quadratic defining relations and intermediate growth of the coefficients of its Hilbert series.

Theorem 9 *Let the Lie algebra L be generated by two elements* x_1 *and* x_2 *of first degree with defining relations*

$$[[[x_1, x_2], x_2], x_2] = [[[x_2, x_1], x_1], x_1] = 0,$$

and let $U(L) = K \oplus U(L)_1 \oplus U(L)_2 \oplus \cdots$ be its universal enveloping algebra. Then the coefficients of the Hilbert series of the algebra R are of intermediate growth, where R is generated by the homogeneous component $U(L)_4$ of degree 4, and R is a quadratic algebra with 14 generators and 96 quadratic relations. Its growth function $g(n)$ satisfies $g(n) \sim \exp(\sqrt{n})$.

The Lie algebra L in Theorem 9 is isomorphic to the Lie algebra of 2×2 matrices with coefficients from $K[z]$ generated by

$$x_1 = \begin{pmatrix} 0 & 1 \\ 0 & 0 \end{pmatrix} \text{ and } x_2 = \begin{pmatrix} 0 & 0 \\ z & 0 \end{pmatrix}.$$

The series

$$\prod_{n\geq 1} \frac{1}{1-t^n} = \sum_{n\geq 0} p_n t^n$$

and

$$\prod_{n\geq 1} (1+t^n) = \sum_{n\geq 0} \rho_n t^n$$

play very special rôles in combinatorics: p_n is equal to the number of partitions of n and ρ_n is the number of partitions of n in different parts. Recall that $\lambda = (\lambda_1, \ldots, \lambda_k)$ is a partition of n, if the parts λ_i are integers such that $\lambda_1 + \cdots + \lambda_k = n$ and $\lambda_1 \geq \cdots \geq \lambda_k \geq 0$; for ρ_n we assume that $\lambda_1 > \cdots > \lambda_k \geq 0$. The asymptotics of p_n and ρ_n was found by Hardy and Ramanujan [47] in 1918 and independently by Uspensky [101] in 1920:

$$p_n \approx \frac{1}{4n\sqrt{3}} \exp\left(\pi\sqrt{\frac{2}{3}n}\right), \quad \rho_n \approx \frac{1}{4\sqrt[4]{3n^3}} \exp\left(\pi\sqrt{\frac{1}{3}n}\right).$$

See also the recent paper by Koçak [58] for more examples and a survey on finitely presented algebras of intermediate growth.

For further reading, including theory of Gröbner bases and other combinatorial properties of algebras we refer e.g. Herzog and Hibi [50] for commutative algebras and Ufnarovskij [99] and Belov, Borisenko, Latyshev [9] for noncommutative algebras.

3 Algebras with Prescribed Hilbert Series

In this section we shall discuss the following problem.

Problem 1 Given a power series

$$a(t) = \sum_{n\geq 0} a_n t^n, \quad a_n \in \mathbb{N}_0,$$

does there exist a finitely generated graded algebra R with Hilbert series equal to $a(t)$ or at least very close to $a(t)$?

We shall recall the construction of Borho and Kraft [16] of a finitely generated graded algebra with Gelfand-Kirillov dimension equal to $\beta \in [2, \infty)$. If R is a finitely generated graded algebra with $\mathrm{GKdim}(R) = \alpha \in [2, 3)$ and $m \in \mathbb{N}$, then the tensor product $K[y_1, \dots, y_m] \otimes_K R$ is of Gelfand-Kirillov dimension $\alpha + m$. Hence for the construction of an algebra R with $\mathrm{GKdim}(R) \in [2, \infty)$ it is sufficient to handle the case $\mathrm{GKdim}(R) = \alpha \in [2, 3)$. Let $S \subset \mathbb{N}_0$ be a set of nonnegative integers and let

$$a(t) = \sum_{s\in S} t^s.$$

We shall construct a two-generated monomial algebra R with Hilbert series

$$H(R,t) = \frac{1}{1-t} + \frac{t}{(1-t)^2} + \frac{a(t)t^2}{(1-t)^2}.$$

We fix the set $U \subset \langle x, y\rangle$

$$U = \{yx^iyx^jy,\, yx^ky \mid i, j \geq 0, k \in \mathbb{N}_0 \setminus S\}.$$

Then the factor algebra $R = K\langle x, y\rangle/(U)$ of the free algebra $K\langle x, y\rangle$ modulo the ideal generated by U has a basis

$$\{x^i, x^iyx^j, x^iyx^syx^j \mid i, j \geq 0, s \in S\}$$

and hence R has the desired Hilbert series.

Pay attention that in the above example the cube $(y)^3$ of the ideal (y) generated by y is equal to zero in R. A similar construction of a two-generated monomial algebra R is given in [32, Theorem 9.4.11]. Assuming that $(y)^{k+1} = 0$ in R, we construct a two-generated monomial algebra R with Hilbert series

$$H(R,t) = \sum_{i=0}^{k-1} \frac{t^i}{(1-t)^{i+1}} + \frac{a(t)t^k}{(1-t)^k},$$

A similar approach was used in the recent paper [33]:

Theorem 10 *Let*

$$a(t) = \sum_{n\geq 0} a_n t^n$$

be a power series with nonnegative integer coefficients.

(i) If d is a positive integer such that $a_n \leq d^n$, $n = 0, 1, 2, \ldots$, then for any integer $p = 0, 1, 2$, there exists a $(d+1)$-generated monomial algebra R such that its Hilbert series is

$$H(R,t) = \frac{1}{1-dt} + \frac{t}{(1-dt)^2} + \frac{t^2 a(t)}{(1-dt)^p}.$$

(ii) If $a_n \leq \binom{d+n-1}{n-1}$, $n = 0, 1, 2, \ldots$, for some positive integer d, then for any integer $p = 0, 1, 2$, there exists a $(d+1)$-generated graded algebra R such that its Hilbert series is

$$H(R,t) = \frac{1}{(1-t)^d} + \frac{t}{(1-t)^{2d}} + \frac{t^2 a(t)}{(1-t)^{dp}}.$$

Under the assumptions of Theorem 10 (i) a modification of the proof gives that for any nonnegative integers $p, q, p+q \leq 2$, there exists a $(d+1)$-generated graded algebra R such that its Hilbert series is

$$H(R,t) = \frac{1}{1-dt} + \frac{t}{(1-dt)^2} + \frac{t^2 a(t)}{(1-dt)^p(1-t)^{dq}}.$$

In the same way we can construct a monomial algebra R with Hilbert series

$$H(R,t) = \frac{1+2t}{1-dt} - t + t^2 a(t).$$

In all these constructions it is clear that if the power series $a(t)$ is rational, algebraic or transcendental, the same property has the Hilbert series of the algebra R.

4 PI-Algebras

Let R be an algebra and let $f(x_1, \ldots, x_n) \in K\langle X\rangle = K\langle x_1, x_2, \ldots\rangle$. We say that $f(x_1, \ldots, x_n)$ is a *polynomial identity* for the algebra R if $f(r_1, \ldots, r_n) = 0$ for all $r_1, \ldots, r_n \in R$. If R satisfies a nontrivial polynomial identity it is called a *PI-algebra*.

The study of PI-algebras is an important part of ring theory with a rich structural and combinatorial theory. PI-algebras form a reasonably big class containing the finite dimensional and the commutative algebras and enjoying many of their properties. In this section we shall discuss only the growth and the Hilbert series of finitely generated PI-algebras. For more details we refer to the survey article [31].

One of the main combinatorial theorems for finitely generated PI-algebras is the *Shirshov Height Theorem* [91].

Theorem 11 *Let R be a PI-algebra generated by d elements $r_1, \dots, r_d$ and satisfying a polynomial identity of degree k. Then there exists a positive integer $h = h(d, k)$ such that as a vector space R is spanned on the products $u_1^{n_1} \cdots u_h^{n_h}$, $n_i \geq 0$, $i = 1, \dots, h$, and every u_i is of the form $u_i = r_{j_1} \cdots r_{j_p}$ with $p \leq k - 1$.*

The integer h is called the *height* of R.

Corollary 1 *Let R be a d-generated PI-algebra satisfying a polynomial identity of degree k. Then the growth function of R is bounded by a polynomial of degree h where $h = h(d, k)$ is the height in the theorem of Shirshov.*

Proof Let the algebra R be generated by $r_1, \dots, r_d$. Then the number of all words $u = r_{j_1} \cdots r_{j_p}$ of length p is equal to d^p. Hence all words of length $\leq k - 1$ are $1 + d + d^2 + \cdots + d^{k-1}$. If we extend the generating set of R to the set of all words of length $\leq k - 1$, Theorem 11 implies that as a vector space R behaves as a finite sum of polynomial algebras $K[u_{i_1}, \dots, u_{i_h}]$. Hence the growth function of R is bounded by a polynomial of degree h. □

As an immediate consequence we obtain the following theorem of Berele [12].

Theorem 12 *Every finitely generated PI-algebra R is of finite Gelfand-Kirillov dimension. If R is d-generated and satisfies a polynomial identity of degree k, then $\mathrm{GKdim}(R) \leq h$, where $h = h(d, k)$ is the height in the Shirshov Height Theorem.*

The original estimate for the height h in terms of the number of generators d of R and the degree k of the satisfied polynomial identity can be derived from a lemma of Shirshov on combinatorics of words. There are many attempts to improve the estimates for h and to decrease the length $p \leq k - 1$ of the words $u_i = r_{j_1} \cdots r_{j_p}$ in the Shirshov Height Theorem 11. Shestakov conjectured (see the abstract of the talk of Lvov [72]) that the bound $k - 1$ for the length can be reduced to $\lceil k/2 \rceil$, where $\lceil \alpha \rceil$, $\alpha \in \mathbb{R}$, is the integer part of α. Lvov added some additional arguments which replace $\lceil k/2 \rceil$ with the PI-degree $\mathrm{PIdeg}(R)$ of R in the conjecture of Shestakov. Recall that a PI-algebra R is of *PI-degree* c (or of *complexity* c), if c is the largest integer such that all multilinear polynomial identities of R follow from the multilinear identities of the $c \times c$ matrix algebra $M_c(K)$. The conjecture of Shestakov was confirmed by Ufnarovskij [97], Belov [7] and Chekanu [18]. Other proofs are given in the survey article by Belov, Borisenko and Latyshev [9] and in the book by the author and Formanek [34]. Concerning the height h the original proof of Shirshov [91] gives primitive recursive estimates. Later it was shown that h is exponentially bounded in terms of the number of generators d of the algebra R and the degree k of the polynomial identity, see the references in the paper by Belov and Kharitonov [10]. In the same paper Belov and Kharitonov found a subexponential bound for h: For a fixed d and k sufficiently large

$$h < k^{12(1+o(1))\log_3 k}.$$

Theorems 11 and 12 confirm that from many points of view finitely generated PI-algebras are similar to commutative algebras. There are also essential differences. The Gelfand-Kirillov dimension of a finitely generated commutative algebra is an integer. The discussed in Sect. 3 examples of two-generated PI-algebras R of Gelfand-Kirillov dimension $\alpha \in [2, 3)$ and the tensor products $K[y_1, \dots, y_m] \otimes_K R$ from [16] satisfy the polynomial identity

$$(x_1x_2 - x_2x_1)(x_3x_4 - x_4x_3)(x_5x_6 - x_6x_5) = 0.$$

The examples in [32, Theorem 9.4.11] are two-generated and satisfy the polynomial identity

$$(x_1x_2 - x_2x_1) \cdots (x_{2m-1}x_{2m} - x_{2m}x_{2m-1}) = 0$$

for a suitable m. Another difference is that the Hilbert series of a finitely generated commutative graded algebra R is rational and for PI-algebras R it may be also transcendental. In the next section we shall see that for graded PI-algebras $H(R, t)$ cannot be algebraic and nonrational.

On the other hand, there is an important class of PI-algebras which play the same rôle as the polynomial algebras in commutative algebra and the free associative algebras in the theory of associative algebras.

Definition 1 Let $I(R) \subset K\langle X\rangle$ be the ideal of all polynomial identities of the algebra R (such ideals are called *T-ideals*). The factor algebra

$$F_d(\text{var}R) = K\langle X_d\rangle/(K\langle X_d\rangle \cap I(R))$$

is called the *relatively free algebra of rank d in the variety of algebras* var$R$ *generated by* R.

Kemer developed the structure theory of T-ideals in the free algebra $K\langle X\rangle$ over a field K of characteristic 0 in the spirit of classical ideal theory in commutative algebras, which allowed him to solve several outstanding open problems in the theory of PI-algebras, see [53] for an account and [51] for further references. It is well known that over an infinite field K all relatively free algebras are graded and it is a natural question to study their Hilbert series. Using the results of Kemer, Belov [8] established the following theorem which shows that relatively free algebras share many nice properties typical for commutative algebra.

Theorem 13 *Let K be a field of characteristic* 0 *and let R be a PI-algebra. Then the Hilbert series $H(F_d(\mathrm{var}R), t)$ is a rational function with denominator similar to the denominators of the Hilbert series of finitely generated graded commutative algebras.*

5 Algebraic and Transcendental Power Series

The following partial case of a classical theorem of Fatou [39] from 1906 shows that the condition that a power series with nonnegative integer coefficients is algebraic is very restrictive.

Theorem 14 *If the coefficients of a power series are nonnegative integers and are bounded polynomially, then the series is either rational or transcendental.*

The coefficients of the Hilbert series of graded algebras of finite Gelfand-Kirillov dimension grow polynomially. Hence we obtain immediately the following consequence of Theorem 14.

Theorem 15 *The Hilbert series of a finitely generated graded algebra of finite Gelfand-Kirillov dimension is either rational or transcendental.*

Corollary 1 and Theorem 12 imply that the same dichotomy holds also for finitely generated graded PI-algebras.

Theorem 16 *The Hilbert series of a finitely generated graded PI-algebra is either rational or transcendental.*

In order to construct the graded algebras with algebraic or transcendental Hilbert series in Sect. 3 we need algebraic and transcendental power series with nonnegative integer coefficients. We shall survey several methods for construction of transcendental power series. We already discussed in Sect. 2 that the power series with intermediate growth of the coefficients are transcendental.

Recall that the power series $a(t)$ is *lacunary*, if

$$a(t) = \sum_{k\geq 1} a_{n_k} t^{n_k}, \quad a_{n_k} \neq 0, \quad \lim_{k\to\infty} (n_{k+1} - n_k) = \infty.$$

Maybe the best known example of such series is

$$a(t) = \sum_{n\geq 1} t^{n!}$$

which produces the first explicitly given transcendental number

$$a\left(\frac{1}{10}\right) = \sum_{n\geq 1} \frac{1}{10^{n!}},$$

the constant of Liouville [71]. The following theorem is due to Mahler [75, p. 42].

Theorem 17 *Lacunary series with nonnegative integer coefficients are transcendental.*

Example 1 The following power series satisfies the conditions in Theorem 17. A direct proof of their transcendency is given in the book of Nishioka [81, Theorem 1.1.2]:

$$a(t) = \sum_{n\geq 0} t^{d^n}, \quad d \geq 2.$$

In the definition of lacunary series we do not restrict the growth of the coefficients although in the above given examples the nonzero coefficients are equal to 1. Another way to construct transcendental series with polynomial or exponential growth of the coefficients uses completely (or strongly) multiplicative functions, i.e. functions $\alpha : \mathbb{N} \to \mathbb{N}_0$ satisfying $\alpha(n_1)\alpha(n_2) = \alpha(n_1 n_2)$, $n_1, n_2 \in \mathbb{N}$. Sárközy [86] described the functions α such that the generating function $a(t) = \sum_{n\geq 1} a_n t^n$ of the sequence $a_n = \alpha(n)$, $n = 1, 2, \ldots$, is rational. Later Bézivin [15] extended this result for algebraic generating functions. Recently, another, more number-theoretic proof of the theorem of Bézivin was given by Bell, Bruin, and Coons [6].

Theorem 18 *Let $\alpha : \mathbb{N} \to \mathbb{N}_0$ be a multiplicative function such that its generating function*

$$a(t) = \sum_{n\geq 1} \alpha(n) t^n$$

is algebraic. Then either $\alpha(n) = 0$ for all sufficiently large n, i.e. $a(t)$ is a polynomial, or there exists a nonnegative integer k and a multiplicative periodic function $\chi : \mathbb{N} \to \mathbb{Q}$ such that $\alpha(n) = n^k \chi(n)$.

The multiplicative periodic functions which appear in Theorem 18 of Bézivin were described by Leitmann and Wolke [69].

The proof of the following corollary can be found in [6]. Here we give simplified arguments.

Corollary 2 *If $\alpha : \mathbb{N} \to \mathbb{N}_0$ is a multiplicative function, then the generating function*

$$a(t) = \sum_{n\geq 1} \alpha(n) t^n$$

is either rational or transcendental.

Proof Let the generating function $a(t)$ of the multiplicative function $\alpha : \mathbb{N} \to \mathbb{N}_0$ be algebraic. By Theorem 18 $a(t)$ is either a polynomial (hence a rational function) or α is of the form $\alpha(n) = n^k\chi(n)$, $n = 1, 2, \ldots$, where $k \in \mathbb{N}_0$ and χ is a multiplicative periodic function. The periodicity of χ implies that it is bounded. Hence $\alpha(n) \leq n^k c$ for some constant $c > 0$ and the coefficients of the power series $a(t)$ grow polynomially. By Theorem 14 of Fatou the power series $a(t)$ cannot be algebraic and nonrational. □

Now it is easy to construct multiplicative functions with transcendental generating function. The following simple example is from [33].

Example 2 If $\alpha : \mathbb{N} \to \mathbb{N}_0$ is a multiplicative function, it is completely determined by its values on the prime numbers p. Let $\alpha(p) = q$, where the q's are pairwise different primes and $\alpha(p) \neq p$ for all prime p. If the generating function $a(t) = \sum_{n\geq 1} \alpha(n)t^n$ is rational, then there exists a positive integer k and a periodic multiplicative function $\chi : \mathbb{N} \to \mathbb{Q}$ such that

$$\alpha(p) = p^k\chi(p) = q, \quad \chi(p) = \frac{q}{p^k}.$$

Therefore the multiplicative function χ is not periodic and this implies that $a(t)$ cannot be rational.

By the theorem of Govorov [43] the Hilbert series $H(R, t)$ of a finitely presented monomial algebra R is a rational function. Ufnarovskij [96] gave a construction which associates to R a finite oriented graph $\Gamma(R)$.

Definition 2 Let

$$R = K\langle X_d\rangle/(U), \quad U \subset \langle X_d\rangle, |U| < \infty,$$

be a finitely presented monomial algebra and let $k+1$ be the maximum of the degrees of the monomials in the set U. The following graph $\Gamma(R)$ is called the *Ufnarovskij graph*. The set of the vertices of $\Gamma(R)$ consists of all monomials of degree k which are not divisible by a monomial in U. Two vertices v_1 and v_2 are connected by an oriented edge from v_1 to v_2 if and only if there are two elements $x_i, x_j \in X_d$ such that $v_1x_i = x_jv_2 \notin U$. Then the edge is labeled by x_i. (Multiple edges and loops are allowed.) The generating function

$$g(\Gamma(R), t) = \sum_{n\geq 1} g_nt^n,$$

of the graph $\Gamma(R)$ has coefficients g_n equal to the number of paths of length n.

The algebra R in the above definition has a basis consisting of all monomials in $\langle X_d\rangle$ which are not divisible by a monomial in U. The edges of $\Gamma(R)$ are in a bijective correspondence with the basis elements of degree $k + 1$ of R and the paths of length n are in bijection with the monomials of degree $n + k$ in the basis.

Ufnarovskij [96] gave simple arguments (based on the Cayley-Hamilton theorem only) for the proof of the following result.

Theorem 19 *Let $R = K\langle X_d\rangle/(U)$ be a finitely presented monomial algebra and let the maximum of the degrees of the monomials in U is equal to $k+1$. Then the generating function $g(\Gamma(R),t)$ of the graph $\Gamma(R)$ is a rational function. The Hilbert series $H(R,t)$ of R and the generating function $g(\Gamma(R),t)$ are related by*

$$H(R,t) = \sum_{n\geq 0} a_n t^n = \sum_{n=1}^{k} a_n t^n + t^k g(\Gamma(R),t).$$

Hence $H(R,t)$ is a rational function.

Now the theorem of Govorov [43] is an obvious consequence of Theorem 19. Additionally, the growth of the finitely presented monomial algebra R can be immediately determined from purely combinatorial properties of its graph $G(R)$—the existence of cycles and their disposition.

The construction of Ufnarovskij can be translated in terms of automata theory and theory of formal languages.

A *language L on the alphabet X_d* is a subset of $\langle X_d\rangle$. The language L is *regular* if it is obtained from a finite subset of $\langle X_d\rangle$ applying a finite number of operations of union, multiplication, and the operation $*$ defined by $T^* = \bigcup_{n\geq 1} T^n$, $T \subset \langle X_d\rangle$.

In the theory of computation a *deterministic finite automaton* is a five-tuple $A = (S, X_d, \delta, s_0, F)$, where S is a finite set of *states*, X_d is a finite alphabet, $\delta : S \times X_d \to S$ is a *transition function*, s_0 is the *initial* or the *start state*, and $F \subseteq S$ is the (possible empty) set of the *final states*. The automaton A can be identified with a *finite directed graph* $\Gamma(A)$. The set of states S is identified with the set of vertices of $\Gamma(A)$. Each vertex $v \in S$ is an *origin* of d edges labeled by the elements of X_d and v_2 is the *destination* of the edge from v_1 to v_2 labeled by x_i if $\delta(v_1, x_i) = v_2$. The language $L(A)$ *recognized by the automaton* A consists of all words $x_{i_1}\cdots x_{i_n}$ such that starting from the initial state s_0 and following the edges labeled by $x_{i_1},\ldots,x_{i_n}$ we reach a vertex f from the set of final states F. The theorem of Kleene connects deterministic finite automata and regular languages.

Theorem 20 *A language L is regular if and only if it is recognized by a deterministic finite automaton.*

For a background on the topic we refer e.g. to the book by Lallement [63]. Ufnarovskij [98] introduced the notion of an automaton monomial algebra.

Definition 3 Let

$$R = K\langle X_d\rangle/(U), \quad U \subset \langle X_d\rangle,$$

be a monomial algebra. It is called *automaton* if the set of monomials in $\langle X_d\rangle$ not divisible by a monomial from U (which form a basis of R) is a regular language. Equivalently, if U is a minimal set of relations, then U is also a regular language.

It is known that when $L \subset \langle X_d\rangle$ is a regular language, then the generating function $g(L,t)$ of the sequence of the numbers of its words of length n is a rational function. Since finite sets $U \subset \langle X_d\rangle$ are regular languages, this gives one more proof of the theorem of Govorov [43]. Involving methods of graph theory Ufnarovskij [98] showed how to construct a basis of the automaton algebra R and to compute efficiently its growth and Hilbert series.

For further results, see e.g. the paper by Månsson and Nordbeck [78] where the authors introduce the generalized Ufnarovski graph and as an application show how this construction can be used to test Noetherianity of automaton algebras. Another application is given by Cedó and Okniński [17] who proved that every finitely generated algebra which is a finitely generated module of a finitely generated commutative subalgebra is automaton. See also Ufnarovski [100] and Månsson [77] for applying computers for explicit calculations.

The above discussions show that it is relatively easy to construct algebras with rational Hilbert series. It is more difficult to construct algebras with algebraic and nonrational Hilbert series. Now we shall survey some constructions of algebraic power series using automata theory and theory of formal languages. Recently there are new applications of the theory of regular languages and the theory of finite-state automata which give new results and new proofs of old results providing algebras with rational and algebraic nonrational Hilbert series, see La Scala [65], La Scala, Piontkovski and Tiwari [67] and La Scala and Piontkovski [66] and the references there.

6 Planar Rooted Trees and Algebraic Series

In this section we shall present another method for construction of algebraic power series with nonnegative integer coefficients. The leading idea is to start with a sequence of finite sets of objects A_n, $n = 0, 1, 2, \ldots$, for which we know (or can prove), that the generating function

$$a(z) = \sum_{n\geq 0} |A_n| z^n$$

of the sequence $|A_n|$, $n = 0, 1, 2, \ldots$, is algebraic and nonrational.

A motivating example are the Catalan numbers. The n-th Catalan number c_n is equal to the number of planar binary rooted trees with n leaves. We may introduce the operation *gluing of trees* in the set of planar binary rooted trees which gives it the structure of a nonassociative groupoid (or a nonassociative *magma*). Clearly, this magma is isomorphic to the one-generated free magma $\{x\}$. For example, the tree

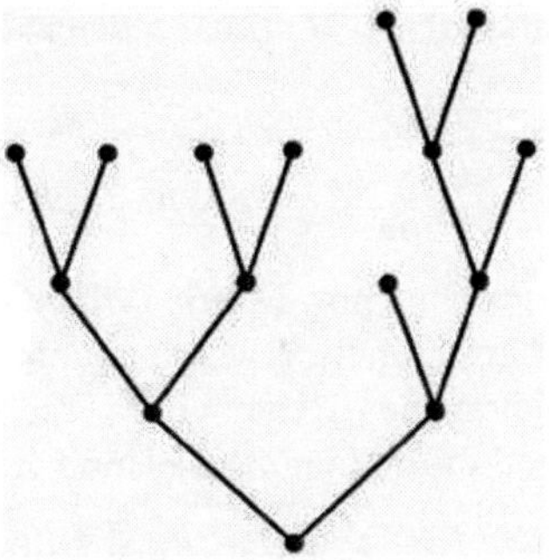

Fig. 1 The tree corresponding to the monomial $((xx)(xx))(x((xx)x))$

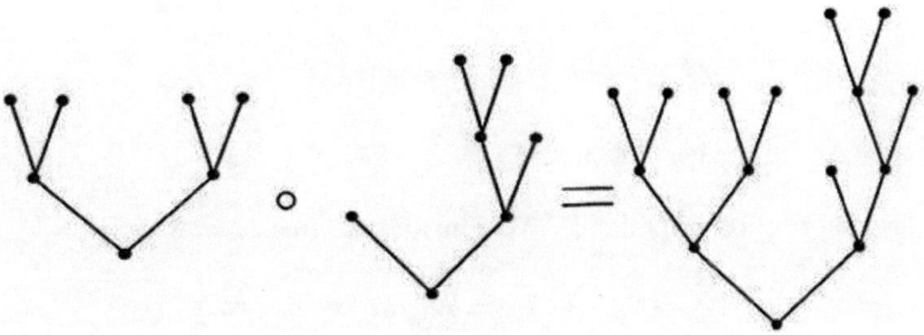

Fig. 2 Concatenation of the monomials $(xx)(xx)$ and $x((xx)x)$

in Fig. 1 corresponds to the nonassociative monomial $((xx)(xx))(x((xx)x))$ and the gluing of the trees in Fig. 2 can be interpreted as the concatenation of the monomials $(xx)(xx)$ and $x((xx)x)$ preserving the parentheses:

$$(xx)(xx) \circ x((xx)x) = ((xx)(xx))(x((xx)x)).$$

Hence, as it is well known, the Catalan numbers satisfy the recurrence relation

$$c_n = \sum_{k=1}^{n-1} c_k c_{n-k}, \quad n = 2, 3, \dots,$$

which implies that their generating function

$$c(t) = \sum_{n\geq 1} c_n t^n$$

satisfies the quadratic equation $c^2(t) = c(t) - t$. This also gives the formulas

$$c(t) = \frac{1-\sqrt{1-4t}}{2}, \quad c_n = \frac{1}{n}\binom{2n-2}{n-1}, \quad n = 1, 2, \ldots.$$

In this way we obtain a nonrational power series which is algebraic. More generally we may consider the generating function which counts the planar rooted trees with fixed number of outcoming branches in each vertex, see, e.g. Drensky and Holtkamp [36]. This can be formalized in the language of universal algebra in the following way.

We start with a set

$$\Omega = \Omega_2 \cup \Omega_3 \cup \cdots$$

which is a union of finite sets of n-ary operations

$$\Omega_n = \{\nu_{ni} \mid i = 1, \ldots, p_n\}, \quad n \geq 2,$$

and an arbitrary set of variables Y. We consider the free Ω-magma

$$\{Y\}_\Omega = \mathcal{M}_\Omega(Y).$$

The elements of $\{Y\}_\Omega$ are the Ω-monomials which are built inductively. We assume that $Y \subset \{Y\}_\Omega$ and if $u_1, \ldots, u_n \in \{Y\}_\Omega$, then $\nu_{ni}(u_1, \ldots, u_n)$ also belongs to $\{Y\}_\Omega$. In the same way as one constructs the free associative algebra $K\langle Y\rangle$ as the vector space with basis the elements of the free semigroup $\langle Y\rangle$ and the free nonassociative algebra $\{Y\}$ starting with the free magma $\{Y\}$, one can construct the free Ω-algebra $K\{Y\}_\Omega$. This allows to use methods and ideas of ring theory for the study of free Ω-magmas. The elements of $\{Y\}_\Omega$ can be described in terms of labeled reduced planar rooted trees in a way similar to the way we identify the free magma $\{x\}$ with the set of planar binary rooted trees.

If T is a planar rooted tree we orient the edges in direction from the root to the leaves. We assume that the tree is reduced, i.e. from each vertex which is not a leaf there are at least two outcomming edges. Then we label the leaves with the elements of Y and if a vertex is with n outcomming edges we label it with an n-ary operation ν_{ni}. We call such trees Ω-trees with labeled leaves. There is a one-to-one correspondence between the Ω-monomials and the Ω-trees with labeled leaves. For example, if $Y = X = \{x_1, x_2, \ldots\}$, then the monomial

$$\nu_{31}(\nu_{23}(x_1, x_1), x_3, \nu_{32}(x_2, x_1, x_4))$$

corresponds to the tree in Fig. 3:

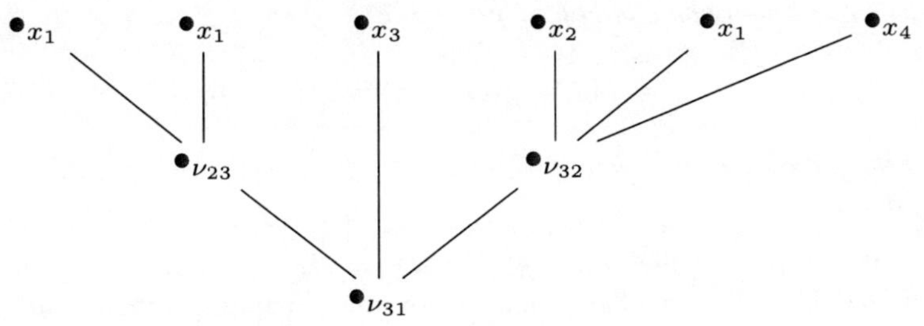

Fig. 3 The tree corresponding to $\upsilon_{31}(\upsilon_{23}(x_1, x_1), x_3, \upsilon_{32}(x_2, x_1, x_4))$

The set of Ω-trees with labeled leaves inherits the natural grading of the free Ω-magma $\{Y\}_\Omega$:

$$\deg(\nu_{ni}(u_1, \dots, u_n)) = \sum_{k=1}^{n} \deg(u_k).$$

The following proposition describes the generating function of the free Ω-magma $\{Y\}_\Omega$ and the Hilbert series of the free Ω-algebra $K\{Y\}_\Omega$.

Proposition 2 *Let*

$$p(t) = \sum_{n\geq 2} p_n y^n = \sum_{n\geq 2} |\Omega_n| t^n$$

be the generating function of the set of operations $\Omega = \Omega_2 \cup \Omega_3 \cup \cdots$.

(i) When $Y = \{x\}$ *consists of one element, then the generating function of the free* Ω*-magma* $\{x\}_\Omega$ *(and the Hilbert series of the free* Ω*-algebra* $K\{x\}_\Omega$*)*

$$g(\{x\}_\Omega, t) = H(K\{x\}_\Omega, t) = \sum_{n\geq 1} |\{x\}_\Omega^{(n)}| t^n$$

is the only solution $z = f(t)$ *of the equation* $p(z) - z + t = 0$ *which satisfies the condition* $f(0) = 0$.

(ii) In the general case, if

$$Y = Y^{(1)} \cup Y^{(2)} \cup \cdots, \text{ where } Y^{(k)} = \{y \in Y \mid \deg(y) = k\},$$

and

$$a(t) = \sum_{k\geq 1} |Y^{(k)}| t^k$$

is the generating function of the graded set Y, then

$$z = f(t) = g(\{Y\}_\Omega, t) = H(K\{Y\}_\Omega, t)$$

is the solution of the equation $p(z) - z + a(t) = 0$ *satisfying the condition* $f(0) = 0$.

The problem when the series $g(\{x\}_\Omega, t) = H(K\{x\}_\Omega, t)$ is algebraic and nonrational depending on the properties of the generating function $p(t)$ from Proposition 2 is studied in the forthcoming paper by Drensky and Lalov [37]. As an immediate consequence of Proposition 2 we obtain:

Corollary 3 *If* $p(t)$ *is a polynomial (with nonnegative integer coefficients), then* $g(\{x\}_\Omega, t)$ *is an algebraic nonrational function.*

Under some mild conditions the same conclusion holds when $p(t)$ is a rational function. The following remark is based on arguments from [37].

Remark 1 Let the function $p(t)$ from Proposition 2 be algebraic and let $b(t, p(t)) = 0$ for some polynomial $b(t, z) \in \mathbb{Q}[t, z]$. Hence $g(\{x\}_\Omega, t)$ is equal to the solution $z = f(t)$ of the equation $b(z, p(z)) = b(f(t), p(f(t))) = 0$. Since $p(f(t)) = f(t) - t$, we obtain that $b(f(t), f(t) - t) = 0$. Hence when the function $p(t)$ is algebraic then this gives an algorithm which has as an input the polynomial equation $b(t, z) = 0$ with coefficients in $\mathbb{Q}[t]$ satisfied by $p(t)$ and as an output the polynomial equation $b(z, z - t) = 0$, again with coefficients in $\mathbb{Q}[t]$, satisfied by $g(\{x\}_\Omega, t)$.

Remark 2 Up till now in this section we start with an algebraic series with nonnegative integer coefficients and obtain an algebraic equation satisfied by $g(\{x\}_\Omega, t)$. Then we want to obtain conditions which guarantee that the series $g(\{x\}_\Omega, t)$ is not rational. We can apply a similar strategy working with the free Ω-magma $\{Y\}_\Omega$ with larger graded generating sets Y. Depending on the properties of the generating function $a(t)$ of the set Y from Proposition 2 (ii) we can handle the following three cases:

(1) Both $p(t)$ and $a(t)$ are polynomials in $\mathbb{Q}[t]$. Then $g(\{Y\}_\Omega, t)$ is equal to the solution $z = f(t)$ of the equation $p(z) - z + a(t) = 0$ with $f(0) = 0$.
(2) Let $p(t) \in \mathbb{Q}[t]$ be a polynomial and let $a(t)$ be algebraic satisfying the polynomial equation $q(t, a(t)) = 0$, $q(t, z) \in \mathbb{Q}[t, z]$. Then $g(\{Y\}_\Omega, t)$ is the solution $z = f(a(t))$ of the equation $p(z) - z + a(t) = 0$. Replacing $a(t) = f(a(t)) - p(f(a(t)))$ in $q(t, a(t)) = 0$ we obtain that $z = f(a(t))$ is a solution of the polynomial equation $q(t, z - p(z)) = 0$, and $q(t, z - p(z)) \in \mathbb{Q}[t, z]$.
(3) Both $p(t)$ and $a(t)$ are algebraic functions and $b(t, p(t)) = q(t, a(t)) = 0$ for some polynomials $b(t, z), q(t, z) \in \mathbb{Q}[t, z]$. Applying the arguments in Remark 1 we obtain that $g(\{x\}_\Omega, t) = f(t)$ is a solution u of the polynomial equation $b(u, u - t) = 0$. Hence $g(\{Y\}_\Omega, t) = f(a(t))$ is a solution u of the equation $b(u, u - a(t)) = 0$. Since $q(t, a(t)) = 0$, the polynomial equations $b(u, u - z) = 0$ and $q(t, z) = 0$ have a common solution $z = a(t)$.

Hence the resultant $r(t,u) = \mathrm{Res}_z(q(t,z), b(u,u-z))$ of the polynomials $q(t,z), b(u,u-z) \in (\mathbb{Q}[t,u])[z]$ is equal to 0 which gives a polynomial equation $r(t,u) = 0$ with a solution $u = f(a(t))$.

A variety of algebraic systems satisfies the *Schreier property* if the subsystems of the free systems are also free. This holds for example for free groups (the *Nielsen-Schreier theorem* [80, 87], two different proofs can be found in [52, 76]), for free Lie algebras (the *Shirshov theorem* [90]), for free nonassociative and free Ω-algebras (*theorems of Kurosh* [61, 62]). It is folklore that *any Ω-submagma of the free Ω-magma $\{Y\}_\Omega$ is also free*. A proof can be found e.g. in Feigelstock [40]. (This can be derived also from the theorems of Kurosh [61, 62].)

We shall give an example considered in Drensky and Holtkamp [36]. The subset S of the magma $\{x\}$ consisting of all nonassociative monomials of even degree is closed under multiplication and hence forms a free submagma of $\{x\}$. It is easy to see that the set of free generators of S consists of all monomials of the form $u = u_1u_2$, where both u_1 and u_2 are of odd degree. Let

$$a(t) = \sum_{n\geq 1} a_{2n}t^{2n}$$

be the generating function of the free generating set of S. The generating function $g(S,t)$ of S is expressed in terms of the generating function of the Catalan numbers

$$g(S,t) = \sum_{n\geq 2} c_{2n}t^{2n} = \frac{1}{2}(c(t) + c(-t)).$$

From the equation

$$g^2(S,t) - g(S,t) + a(t) = c^2(a(t)) - c(a(t)) + a(t) = 0$$

we obtain that $a(t)$ satisfies the quadratic equation

$$4a^2(t) - a(t) + t^2 = 0 \text{ and } a(t) = \frac{1}{4}c(4t^2), \quad a_{2n} = 4^{n-1}c_n.$$

Applying the Stirling formula for $n!$ after some calculations we obtain

$$\frac{a_{2n}}{c_{2n}} \approx \frac{1}{2}\sqrt{\frac{2n-1}{n-1}}, \quad \lim_{n\to\infty} \frac{a_{2n}}{c_{2n}} = \frac{\sqrt{2}}{2} \approx 0.707105.$$

Every monomial u of even degree in $\{x\}$ is a product of two submonomials u_1 and u_2 where both u_1 and u_2 are either of even or of odd degree. The above calculations show that the monomials $u = u_1u_2$ with u_1 and u_2 of odd degree are much more than those of even degree. This can be translated in the language of planar binary rooted trees with even number of leaves. Every such tree has two branches which

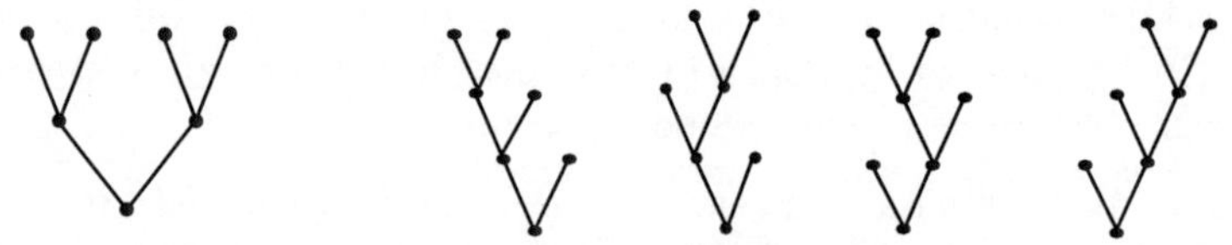

Fig. 4 Trees with 4 leaves with even and odd branches

both are of the same parity of the number of leaves. The trees in Fig. 4 correspond, respectively, to the monomials

$$(xx)(xx) \text{ (even branches)}, ((xx)x)x,\ (x(xx))x,\ x((xx)x),\ x(x(xx)) \text{ (odd branches)}.$$

It turns out that the trees with branches with odd number of leaves are more than 70% of all trees with even number of leaves which, at least for the authors of [36], was quite surprising.

The above observation was the starting point of the project of Drensky and Lalov [37]. One of the first results there is the following.

Theorem 21 *Let Ω be a set of operations with algebraic generating function $p(t)$ and let $\{x\}_\Omega$ be the one-generated free Ω-magma. For a fixed positive integer s consider the Ω-submagma S_Ω consisting of all monomials of degree divisible by s. Then the generating function $a(t)$ of the free generating set of S_Ω is algebraic.*

The following lemma answers the problem when the set S is nonempty.

Lemma 1 *Let the number of the n-ary operations in Ω is equal to p_n and let d be the greatest common divisor of all numbers $n-1$, for which p_n is different from 0. Then S_Ω is nonempty if and only if d and s are relatively prime. Moreover, the set S_Ω is either empty or is infinite.*

One of the main problems in this direction is the following.

Problem 2 If in the notation of Theorem 21 we know the polynomial equation $b(t,z) \in \mathbb{Q}[t,z]$ satisfied by $p(t)$, how to find the equation satisfied by the generating function $a(t)$ of the free generating set of S_Ω?

In [37] we have found an algorithm which solves this problem. In particular, we have the following statement which gives more examples of algebraic power series with nonnegative integer coefficients.

Theorem 22 *If the generating function $p(t)$ of the operations in Ω is a polynomial and the set S_Ω is nonempty, then the generating function $a(t)$ of the free set of generators of S_Ω is algebraic and nonrational.*

Example 3 Let Ω consist of one binary operation only and let $s = 3$. This corresponds to the set S of binary planar rooted trees with number of leaves divisible by 3. Applying the algorithm in [37] we obtain that the generating function $a(t)$ of the free generating set of S satisfies

$$729a^4(t) - 486a^3(t) + 108a^2(t)^2 - (64t^3 + 8)a(t) + 16t^3 = 0.$$

Solving this equation we obtain four possibilities for $a(t)$. We expand each of them in series and since only one solution has nonnegative coefficients of the first powers, we obtain the value of the desired generating function:

$$a(t) = \frac{1}{6} - \frac{1}{18}\sqrt{1 + 4t + 16t^2} - \frac{\sqrt{1 - 2t - 8t^2 + \frac{1-64t^3}{\sqrt{1+4t+16t^2}}}}{9\sqrt{2}}$$

$$= 2t^3 + 38t^6 + 1262t^9 + 51302t^{12} + 2319176t^{15} + 111964106t^{18} + 5652760340t^{21} + \cdots .$$

Example 4 Let $\Omega = \Omega_2 \cup \Omega_3 \cup \cdots$ and let $|\Omega| = 1$ for all $n = 2, 3, \ldots$. Its generating function is

$$p(t) = \frac{t^2}{1 - t}.$$

Then the one-generated free Ω-magma can be identified with the set of all planar rooted reduced trees and the generating function of $\{x\}_\Omega$ is equal to the generating function of the super-Catalan numbers (see [92, sequence A001003]). Let S be the set of all monomials of even degree. The calculations in [37] give that the generating function $a(t)$ of the set of free generators of S satisfies the equation

$$36a^4(t) - 12(t^2 + 1)a^3(t) + (19t^2 + 1)a^2(t) + 3t^2(t^2 - 1)a(t) + 2t^4 = 0$$

Only two of the solutions have nonnegative coefficients of the first few powers. However, the correct solution is chosen taking into account the coefficient of the 4th power and it is

$$a(t)) = \frac{1}{12}(1 + t^2) - \frac{1}{12}\sqrt{1 - 34t^2 + t^4} - \frac{\sqrt{t^2 + t^4 - \frac{t^2-34t^4+t^6}{\sqrt{1-34t^2+t^4}}}}{6\sqrt{2}}$$

$$= t^2 + 10t^4 + 174t^6 + 3730t^8 + 89158t^{10} + 2278938t^{12} + 60962718t^{14} + 1685358882t^{16} + \cdots .$$

7 Noncommutative Invariant Theory

In this section we shall follow the traditions of classical invariant theory and shall work over the complex field $\mathbb{C}$ although most of our results are true for any field K is of characteristic 0. In classical invariant theory one considers the canonical action of the general linear group $\mathrm{GL}_d(\mathbb{C})$ on the d-dimensional vector space V_d with basis $\{v_1, \ldots, v_d\}$. The algebra $\mathbb{C}[X_d]$ consists of the polynomial functions $f(X_d) = f(x_1, \ldots, x_d)$, where

$$x_i(v) = \xi_i \text{ for } v = \xi_1 v_1 + \cdots + \xi_d v_d \in V_d, \xi_1, \ldots, \xi_d \in \mathbb{C}.$$

The group $\mathrm{GL}_d(\mathbb{C})$ acts on $\mathbb{C}[X_d]$ by the rule

$$g(f)(v) = f(g^{-1}(v)), \quad g \in \mathrm{GL}_d(\mathbb{C}), f \in \mathbb{C}[X_d], v \in V_d.$$

If G is a subgroup of $\mathrm{GL}_d(\mathbb{C})$, then the algebra $\mathbb{C}[X_d]^G$ of G-invariants consists of all $f(X_d) \in \mathbb{C}[X_d]$ such that

$$g(f) = f \text{ for all } g \in G.$$

For a background on classical invariant theory see, e.g. some of the books by Derksen and Kemper [23], Dolgachev [25] or Procesi [85].

One possible noncommutative generalization is to replace the polynomial algebra with the free associative algebra $\mathbb{C}\langle X_d\rangle$ under the natural restriction $d \geq 2$. It is more convenient to assume that $\mathrm{GL}_d(\mathbb{C})$ acts canonically on the vector space $\mathbb{C}X_d$ with basis X_d and to extend diagonally its action on $\mathbb{C}\langle X_d\rangle$ by the rule

$$g(f(x_1, \ldots, x_d)) = f(g(x_1), \ldots, g(x_d)), \quad g \in \mathrm{GL}_d(\mathbb{C}), f \in \mathbb{C}\langle X_d\rangle.$$

Then, for a subgroup G of $\mathrm{GL}_d(\mathbb{C})$ the algebra of G-invariants is

$$\mathbb{C}\langle X_d\rangle^G = \{f(X_d) \in \mathbb{C}\langle X_d\rangle \mid g(f) = f \text{ for all } g \in G\}.$$

The algebras of invariants in the commutative case have a lot of nice properties. For example, the algebra $\mathbb{C}[X_d]^G$ is finitely generated for a large class of groups including all reductive groups, when G is a maximal unipotent subgroup of a reductive group (see Hadžiev [46] or Grosshans [45, Theorem 9.4]), and consequently when G is a Borel subgroup of a reductive group. Since the algebra $\mathbb{C}[X_d]^G$ is graded, the Hilbert-Serre theorem (Theorem 3 (i)) gives that for such groups G the Hilbert series $H(\mathbb{C}[X_d]^G, t)$ is a rational function. In this case the algebra $\mathbb{C}[X_d]^G$ is a homomorphic image of a polynomial algebra $\mathbb{C}[Y_p]$ modulo some ideal I. But it is quite rare when the algebra $\mathbb{C}[X_d]^G$ is isomorphic to the polynomial algebra $\mathbb{C}[Y_p]$. By the theorem of Shephard and Todd [89] and Chevalley [19] *if G is finite then $\mathbb{C}[X_d]^G \cong \mathbb{C}[X_d]$ if and only if G is generated by pseudoreflections.*

The picture of invariant theory for the free algebra $\mathbb{C}\langle X_d\rangle$ is quite different. The algebra $\mathbb{C}\langle X_d\rangle^G$ is very rarely finitely generated.

Theorem 23

(i) *(Dicks and Formanek [24] and Kharchenko [55]) If G is a finite group then $\mathbb{C}\langle X_d\rangle^G$ is finitely generated if and only if G is cyclic and acts on the vector space $\mathbb{C}X_d$ by scalar multiplication.*

(ii) *(Koryukin [59]) Let G be an arbitrary subgroup of $GL_d(\mathbb{C})$ and let $\mathbb{C}\langle X_d\rangle^G$ be finitely generated. Assume that the vector space $\mathbb{C}X_d$ does not have a proper subspace $\mathbb{C}Y_e$, $Y_e = \{y_1, \dots, y_e\}$, $e < d$, such that $\mathbb{C}\langle X_d\rangle^G \subseteq \mathbb{C}\langle Y_e\rangle$. Then G is a finite and acts on $\mathbb{C}X_d$ by scalar multiplication.*

On the other hand the following theorem of Koryukin [59] implies something positive.

Theorem 24 *Let us equip the homogeneous component of degree n of the free algebra $\mathbb{C}\langle X_d\rangle$ with the action of the symmetric group S_n by permuting the positions of the variables:*

$$\left(\sum \alpha_i x_{i_1}\cdots x_{i_n}\right)\sigma^{-1} = \sum \alpha_i x_{i_{\sigma(1)}}\cdots x_{i_{\sigma(n)}}, \quad \sigma \in S_n.$$

Then under this additional action the algebra $\mathbb{C}\langle X_d\rangle^G$ is finitely generated for any reductive group G.

The analogue of the Shephard-Todd-Chevalley theorem sounds also very different for $K\langle X_d\rangle$. It turns out that the algebra $\mathbb{C}\langle X_d\rangle^G$ is always free. Additionally, when G is finite, then there is a Galois correspondence between the subgroups of G and the free subalgebras of $\mathbb{C}\langle X_d\rangle$ which contain $\mathbb{C}\langle X_d\rangle^G$.

Theorem 25 (Lane [64] and Kharchenko [54]) *For every subgroup G of $GL_d(\mathbb{C})$ the algebra of invariants $\mathbb{C}\langle X_d\rangle^G$ is free.*

Theorem 26 (Kharchenko [54]) *Let G be a finite subgroup of $GL_d(\mathbb{C})$. The map $H \longrightarrow \mathbb{C}\langle X_d\rangle^H$ gives a one-to-one correspondence between the subgroups of G and the free subalgebras of $\mathbb{C}\langle X_d\rangle$ containing $\mathbb{C}\langle X_d\rangle^G$.*

Comparing with the commutative case, the behavior of the Hilbert series of $\mathbb{C}\langle X_d\rangle^G$ depends surprisingly very much on the properties of the group G. For example, the classical Molien formula [79] for the Hilbert series of the algebra of invariants $\mathbb{C}[X_d]^G$ for a finite group G states that

$$H(\mathbb{C}[X_d]^G, t) = \frac{1}{|G|}\sum_{g\in G}\frac{1}{\det(1-tg)},$$

where $\det(1-tg)$ is the determinant of the matrix $I_d - tg \in \mathrm{GL}_d(\mathbb{C})$ (and I_d is the identity $d \times d$ matrix). The analogue of the Molien formula for $H(\mathbb{C}\langle X_d\rangle^G, t)$ is due to Dicks and Formanek [24]:

Theorem 27 *For a finite subgroup G of $GL_d(\mathbb{C})$*

$$H(\mathbb{C}\langle X_d\rangle^G, t) = \frac{1}{|G|}\sum_{g\in G}\frac{1}{1-tr(tg)},$$

where $tr(tg)$ is the trace of the matrix tg, $g \in GL_d(\mathbb{C})$.

Corollary 4 *If G is a finite subgroup of $GL_d(\mathbb{C})$, then the Hilbert series of the free algebra $\mathbb{C}\langle X_d\rangle^G$ and the generating function $a(t)$ of its set of homogeneous free generators are rational functions.*

We shall illustrate Corollary 4 with two examples.

Example 5 Let $d = 2$ and $G = S_2$ be the symmetric group of degree 2. It consists of the matrices

$$I_2 = \begin{pmatrix} 1 & 0 \\ 0 & 1 \end{pmatrix}, \quad \sigma = \begin{pmatrix} 0 & 1 \\ 1 & 0 \end{pmatrix}.$$

Hence

$$\det(I_2 - tI_2) = \begin{vmatrix} 1-t & 0 \\ 0 & 1-t \end{vmatrix} = (1-t)^2, \det(I_2 - t\sigma) = \begin{vmatrix} 1 & -t \\ -t & 1 \end{vmatrix} = 1 - t^2.$$

The Molien formula gives

$$H(\mathbb{C}[x_1, x_2]^{S_2}, t) = \frac{1}{2}\left(\frac{1}{\det(I_2 - tI_2)} + \frac{1}{\det(I_2 - t\sigma)}\right) = \frac{1}{(1-t)(1-t^2)},$$

which expresses the fact that the algebra $\mathbb{C}[x_1, x_2]^{S_2}$ is isomorphic to the polynomial algebra generated by the elementary symmetric functions

$$e_1 = x_1 + x_2 \text{ and } e_2 = x_1x_2.$$

Since $\text{tr}(I_2) = 2$, $\text{tr}(\sigma) = 0$, by the Dicks-Formanek formula we obtain

$$H(\mathbb{C}\langle x_1, x_2\rangle^{S_2}, t) = \frac{1}{2}\left(\frac{1}{1-\ \text{tr}(tI_2)} + \frac{1}{1-\text{tr}(t\sigma)}\right)$$

$$= \frac{1}{2}\left(\frac{1}{1-2t} + 1\right) = \frac{1-t}{1-2t} = 1 + t + 2t^2 + 4t^3 + \cdots.$$

As in the case of free nonassociative algebras, there is a formula for the Hilbert series of the algebra $\mathbb{C}\langle Y\rangle$ for an arbitrary graded set Y of free generators. If the generating function of Y is $a(t)$, then

$$H(\mathbb{C}\langle Y\rangle, t) = \frac{1}{1 - a(t)}.$$

Easy computations give for the free generators of $\mathbb{C}\langle x_1, x_2\rangle^{S_2}$

$$a(t) = \frac{t}{1-t}.$$

This shows that the free homogeneous set of generators of the algebra $\mathbb{C}\langle x_1, x_2\rangle^{S_2}$ consists of one polynomial for each degree $n \geq 1$. This example is a partial case of a result of Wolf [104] where she studied the symmetric polynomials in the free associative algebra $\mathbb{C}\langle X_d\rangle$, $d \geq 2$.

Example 6 Let $G = \langle \sigma \rangle \subset GL_3(\mathbb{C})$ be the cyclic group of order 3 which permutes the variables x_1, x_2, x_3. It is generated by the matrix

$$\sigma = \begin{pmatrix} 0 & 0 & 1 \\ 1 & 0 & 0 \\ 0 & 1 & 0 \end{pmatrix},$$

$$\det(I_3 - tI_3) = (1-t)^3, \det(I_3 - t\sigma) = \det(I_3 - t\sigma^2) = 1 - t^3,$$

$$H(\mathbb{C}[X_3]^G, t) = \frac{1}{3}\left(\frac{1}{(1-t)^3} + \frac{2}{1-t^3}\right) = \frac{1+t^3}{(1-t)(1-t^2)(1-t^3)}$$

and this is a confirmation of the well known fact that $\mathbb{C}[X_3]^G$ is a free $\mathbb{C}[e_1, e_2, e_3]$-module generated by 1 and $x_1^2x_2 + x_2^2x_3 + x_3^2x_1$. Here, as usual,

$$e_1 = x_1 + x_2 + x_3, \quad e_2 = x_1x_2 + x_1x_3 + x_2x_3, \quad e_3 = x_1x_2x_3.$$

Since $\operatorname{tr}(\sigma) = 0$, for $H(\mathbb{C}\langle X_3\rangle^G, t)$ we obtain

$$H(\mathbb{C}\langle X_3\rangle^G, t) = \frac{1}{3}\left(\frac{1}{1-3t} + 2\right) = \frac{1-2t}{1-3t} = 1 + t + 3t^2 + 9t^3 + \cdots .$$

For the generating function $a(t)$ of the free generators of $\mathbb{C}\langle X_3\rangle^G$ we have

$$\frac{1}{1-a(t)} = \frac{1-2t}{1-3t}, \quad a(t) = \frac{t}{1-2t}.$$

The situation changes drastically when we consider arbitrary reductive groups G. In the commutative case the Hilbert series of the algebra of invariants $\mathbb{C}[X_d]^G$ is always rational. Surprisingly even in the simplest noncommutative case we obtain an algebraic Hilbert series which is not rational.

Example 7

(i) Let the special linear group $SL_2 = SL_2(\mathbb{C})$ act canonically on the two-dimensional vector space with basis X_2. Almkvist, Dicks and Formanek [2] showed that

$$H(\mathbb{C}\langle X_2\rangle^{SL_2}, t) = \frac{1-\sqrt{1-4t^2}}{2t^2}.$$

This means that homogeneous invariants exist for even degree only and their dimension of degree $2(n-1)$ is equal to the nth Catalan number c_n. As in Example 5 we use the formula relating the Hilbert series $H(\mathbb{C}\langle X_2\rangle^{SL_2}, t)$ and the generating function $a_{SL_2}(t)$ of the free homogeneous generating set $\mathbb{C}\langle X_2\rangle^{SL_2}$:

$$H(\mathbb{C}\langle X_2\rangle^{SL_2}, t) = \frac{1}{1-a_{SL_2}(t)}.$$

This implies that

$$a_{SL_2}(t) = 1 - \frac{2t^2}{1-\sqrt{1-4t^2}}.$$

(ii) Drensky and Gupta [35] computed the Hilbert series of the algebra of invariants $\mathbb{C}\langle X_2\rangle^{UT_2}$ of the unitriangular group $UT_2 = UT_2(\mathbb{C})$:

$$H(\mathbb{C}\langle X_2\rangle^{UT_2}, t) = \frac{1-\sqrt{1-4t^2}}{t(2t-1+\sqrt{1-4t^2})}.$$

As in the case of $SL_2(\mathbb{C})$ we can obtain for the free generating set of the algebra of $UT_2(\mathbb{C})$-invariants

$$a_{UT_2}(t) = 1 - \frac{t(2t-1+\sqrt{1-4t^2})}{1-\sqrt{1-4t^2}} = t + a_{SL_2}(t).$$

Since $\mathbb{C}\langle X_2\rangle^{SL_2} \subset \mathbb{C}\langle X_2\rangle^{UT_2}$, this equality suggests that the set of free generators of $\mathbb{C}\langle X_2\rangle^{UT_2}$ consists the free generators of $\mathbb{C}\langle X_2\rangle^{SL_2}$ and one more generator of first degree which was confirmed in [35]. The paper [35] contains also a procedure which constructs inductively a free generating set of $\mathbb{C}\langle X_2\rangle^{SL_2}$.

Invariant theory of $SL_2(\mathbb{C})$ and $UT_2(\mathbb{C})$ considered, respectively, as subgroups of $SL_d(\mathbb{C})$ and $UT_d(\mathbb{C})$, acting on the polynomial algebra $\mathbb{C}[X_d]$ and the free associative algebra $\mathbb{C}\langle X_d\rangle$ can be translated in the language of derivations. We shall restrict our considerations for the case $d = 2$ only. Recall that the linear operator δ

acting on an algebra R is called a *derivation* if

$$\delta(r_1r_2) = \delta(r_1)r_2 + r_1\delta(r_2) \text{ for all } r_1, r_2 \in R.$$

The derivation is *locally nilpotent* if for any $r \in R$ there exists an n such that $\delta^n(r) = 0$. The kernel R^δ of δ is called its *algebra of constants*. It is well known, see e.g. Bedratyuk [5] for comments, references and applications, that there is a one-to-one correspondence between the $\mathbb{G}_a$-actions (the actions of the additive group $(\mathbb{C}, +)$) on $\mathbb{C}X_d$ and the linear locally nilpotent derivations on $\mathbb{C}[X_d]$.

If $\text{UT}_2(\mathbb{C})$ acts on $\mathbb{C}[X_2]$ and on $\mathbb{C}\langle X_2\rangle$ by the rule

$$g(x_1) = x_1, \quad g(x_2) = x_2 + \alpha x_1, \quad g \in \text{UT}_2(\mathbb{C}), \alpha \in \mathbb{C},$$

then $\mathbb{C}[X_2]^{\text{UT}_2}$ and $\mathbb{C}\langle X_2\rangle^{\text{UT}_2}$ coincide, respectively, with the algebras of constants $\mathbb{C}[X_2]^{\delta_1}$ and $\mathbb{C}\langle X_2\rangle^{\delta_1}$ of the derivation δ_1 defined by

$$\delta_1(x_1) = 0, \quad \delta_1(x_2) = x_1.$$

Equivalently,

$$\mathbb{C}[X_2]^{\text{UT}_2} = \{f(x_1, x_2) \in \mathbb{C}[X_2] \mid f(x_1, x_2 + x_1) = f(x_1, x_2)\},$$

$$\mathbb{C}\langle X_2\rangle^{\text{UT}_2} = \{f(x_1, x_2) \in \mathbb{C}\langle X_2\rangle \mid f(x_1, x_2 + x_1) = f(x_1, x_2)\}.$$

Similarly, $\mathbb{C}[X_2]^{\text{SL}_2}$ and $\mathbb{C}\langle X_2\rangle^{\text{SL}_2}$ coincide, respectively, with the subalgebras of $\mathbb{C}[X_2]^{\text{UT}_2}$ and $\mathbb{C}\langle X_2\rangle^{\text{UT}_2}$ consisting of all $f(x_1, x_2)$ in $\mathbb{C}[X_2]^{\text{UT}_2}$ and $\mathbb{C}\langle X_2\rangle^{\text{UT}_2}$ such that

$$f(x_1 + x_2, x_2) = f(x_1, x_2).$$

Up till now we discussed Hilbert series of algebras of invariants which are subalgebras of polynomial algebras and free associative algebras. Instead we may consider free algebras in other classes. One of the most important algebras from this point of view are relatively free algebras of varieties of associative or nonassociative algebras. We shall restrict our considerations to varieties of associative algebras over $\mathbb{C}$.

Let R be an associative PI-algebra and let $F_d(\text{var}R)$ be the relatively free algebra of rank d in the variety $\text{var}R$ generated by R. Again, we assume that the general linear group $\text{GL}_d(\mathbb{C})$ acts canonically on the vector space $\mathbb{C}X_d$ and extend this action diagonally on the whole algebra $F_d(\text{var}R)$. (Equipped with this action, in the case when $\text{var}R$ is the variety $\mathfrak{A}$ of all commutative associative algebras, we do not consider polynomials as functions. The algebra $F_d(\mathfrak{A})$ is isomorphic to the symmetric algebra $S(\mathbb{C}X_d)$ of the vector space $\mathbb{C}X_d$.) For a subgroup G of $\text{GL}_d(\mathbb{C})$ the algebra of G-invariants $F_d^G(\text{var}R)$ is defined in an obvious way as in the case of $\mathbb{C}[X_d]^G$ and $\mathbb{C}\langle X_d\rangle^G$. For a background on invariant theory of relatively free algebras we refer to the survey articles by Formanek [42] and the author [30], see also the references in Domokos and Drensky [28, 29].

Although PI-algebras are considered to have many similar properties with commutative algebras, from the point of view of invariant theory they behave quite differently. For example, the finite generation of $F_d^G(\mathrm{var}R)$ for all finite groups forces very strong restrictions on the polynomial identities of R, and the restrictions are much stronger when we assume that $F_d^G(\mathrm{var}R)$ is finitely generated for all reductive groups, see the surveys [30, 42] and Kharlampovich and Sapir [48] where the finite generation is related also with algorithmic problems. As an illustration we shall mention only a result in Domokos and Drensky [27]. *The algebra* $F_d^G(\mathrm{var}R)$ *is finitely generated for all reductive groups* G *if and only* R *satisfies the identity of Lie nilpotency* $[x_1, \dots, x_c] = 0$ *for some* $c \geq 2$. Also, the analogue of the theorem of Shephard and Todd [89] and Chevalley [19] holds for a very limited class of varieties. By a theorem of Domokos [26] *if* G *is finite then the algebra* $F_d^G(\mathrm{var}R)$ *is relatively free in* $\mathrm{var}R$ *if and only if* G *is generated by pseudoreflections and* R *satisfies the polynomial identity* $[x_1, x_2, x_3] = 0$.

If we consider Hilbert series of relatively free algebras, they are of the same kind as in the commutative case. Hence we cannot obtain nonrational algebraic or trascendental power series in this way. The following theorem was established in Domokos and Drensky [28]. A key ingredient of its proof is the result of Belov [8] for the rationality of the Hilbert series $F_d(\mathrm{var}R)$ and its extension by Berele [13].

Theorem 28 *Let* G *be a subgroup of* $GL_d(\mathbb{C})$ *such that for any finitely generated* $\mathbb{N}_0$*-graded commutative algebra* A *with* $A_0 = \mathbb{C}$ *on which* $GL_d(\mathbb{C})$ *acts rationally via graded algebra automorphisms, the subalgebra* A^G *of* G*-invariants is finitely generated. Then for every PI-algebra* R *the Hilbert series of the relatively free algebra* $F_d^G(varR)$ *is a rational function with denominator similar to the denominators of the Hilbert series of the algebras of* G*-invariants in the commutative case.*

More applications for computing Hilbert series of invariants of classical groups and important numerical invariants of PI-algebras can be found in the paper by Benanti, Boumova, Drensky, Genov and Koev [11]. Here the usage of derivations is combined with the classical method for solving in nonnegative integers systems of linear Diophantine equations and inequalities discovered by Elliott [38] from 1903 and its further development by MacMahon [74] in his "Ω-Calculus" or Partition Analysis.

If we go to free nonassociative algebras, the Hilbert series of the algebras of invariants may be even more far from rational than in the case of free associative algebras. We shall complete our article with the following result in Drensky and Holtkamp [36].

Theorem 29 *Let $\mathbb{C}\{X_2\}$ be the free two-generated nonassociative algebra. Then the Hilbert series of the algebras of invariants $\mathbb{C}\{X_2\}^{SL_2}$ and $\mathbb{C}\{X_2\}^{UT_2}$ are elliptic integrals:*

$$H(\mathbb{C}\{X_2\}^{SL_2}, t) = \int_0^1 \sin^2(2\pi u)\left(1 - \sqrt{1 - 8t\sin(2\pi u)}\right) du,$$

$$H(\mathbb{C}\{X_2\}^{UT_2}, t) = \int_0^1 \cos^2(\pi u)\left(1 - \sqrt{1 - 8t\cos(2\pi u)}\right) du.$$

The proof uses a noncommutative analogue of the Molien-Weyl integral formula for the Hilbert series in classical invariant theory (which is an integral version of the Molien formula for finite groups [102, 103]).

It would be interesting to obtain the Hilbert series for algebras of invariants for the groups $SL_2(K)$ and $UT_2(K)$ acting on other free Ω-algebras, as well for the invariants of other important groups.

Acknowledgments The author is deeply obliged to the anonymous referee for the careful reading and the numerous valuable suggestions for improving the exposition of the paper.

Partially supported by Grant KP-06 N 32/1 of 07.12.2019 “Groups and Rings – Theory and Applications” of the Bulgarian National Science Fund.

References

1. Abel, N.H.: Œuvres complètes, Tome II. Éditions J. Gabay, Sceaux (1992), Reprint of the second edition of 1881
2. Almkvist, G., Dicks, W., Formanek, E.: Hilbert series of fixed free algebras and noncommutative classical invariant theory. J. Algebra **93**, 189–214 (1985). https://doi.org/10.1016/0021-8693(85)90183-8
3. Backelin, J.: La série de Poincaré-Betti d'une algèbre graduee de type fini à une rélation est rationnelle. C. R. Acad. Sci. Paris Sér. A **287**, 843–846 (1978)
4. Banderier, C., Drmota, M.: Formulae and asymptotics for coefficients of algebraic functions. Comb. Probab. Comput. **24**(1), 1–53 (2015). https://doi.org/10.1017/S0963548314000728
5. Bedratyuk, L.: Weitzenböck derivations and classical invariant theory: I. Poincaré series. Serdica Math. J. **36**(2), 99–120 (2010). http://www.math.bas.bg/serdica/2010/2010-099-120.pdf
6. Bell, J.P., Bruin, N., Coons, M.: Transcendence of generating functions whose coefficients are multiplicative. Trans. Am. Math. Soc. **364**(2), 933–959 (2012). https://doi.org/10.1090/S0002-9947-2011-05479-6
7. Belov, A.Ya.: On a Shirshov basis of relatively free algebras of complexity n (Russian). Mat. Sb. **135**, 373–384 (1988). Translation: Math. USSR Sb. **63**, 363–374 (1988). https://doi.org/10.1070/SM1989v063n02ABEH003279
8. Belov, A.Ya.: Rationality of Hilbert series of relatively free algebras (Russian). Uspekhi Mat. Nauk **52**(2), 153–154 (1997). Translation: Russian Math. Surv. **52**, 394–395 (1997) https://doi.org/10.1070/RM1997v052n02ABEH001786
9. Belov, A.Ya., Borisenko, V.V., Latyshev, V.N.: Monomial algebras, Algebra, 4. J. Math. Sci. (New York) **87**(3), 3463–3575 (1997). Russian version: Algebra – 4. Itogi Nauki Tekh.,

Ser. Sovrem. Probl. Mat. i ee pril. Temat. obz. **26**, 35–214 (2002). https://doi.org/10.1007/BF02355446
10. Belov, A.Ya., Kharitonov, M.I.: Subexponential estimates in Shirshov's theorem on height (Russian). Mat. Sb. **203**(4), 81–102 (2012). Translation: Sb. Math. **203**(4), 534–553 (2012). https://doi.org/10.1070/SM2012v203n04ABEH004233
11. Benanti, F., Boumova, S., Drensky, V., Genov, G.K., Koev, K.: Computing with rational symmetric functions and applications to invariant theory and PI-algebras. Serdica Math. J. **38**(1–3), 137–188 (2012). http://www.math.bas.bg/serdica/2012/2012-137-188.pdf
12. Berele, A.: Homogeneous polynomial identities. Isr. J. Math. **42**, 258–272 (1982). https://doi.org/10.1007/BF02802727
13. Berele, A.: Applications of Belov's theorem to the cocharacter sequence of p.i. algebras. J. Algebra **298**(1), 208–214 (2006). https://doi.org/10.1016/j.jalgebra.2005.09.011
14. Bergman, G.M.: A note on growth functions of algebras and semigroups. Research Note, Univ. of California, Berkeley, unpublished mimeographed notes (1978). https://math.berkeley.edu/~gbergman/papers/unpub/growth.pdf
15. Bézivin, J.-P.: Fonctions multiplicatives et équations différentielles. Bull. Soc. Math. Fr. **123**(3), 329–349 (1995). https://doi.org/10.24033/bsmf.2262
16. Borho, W., Kraft, H.-P.: Über die Gelfand-Kirillov-Dimension. Math. Ann. **220**, 1–24 (1976). https://doi.org/10.1007/BF01354525
17. Cedó, F., Okniński, J.: On a class of automaton algebras. Proc. Edinb. Math. Soc. II. Ser. **60**(1), 31–38 (2017). https://doi.org/10.1017/S0013091516000134
18. Chekanu, G.P.: On local finiteness of algebras (Russian). Mat. Issled. **105**, 153–171 (1988). https://eudml.org/doc/190041
19. Chevalley, C.: Invariants of finite groups generated by reflections. Am. J. Math. **77**, 778–782 (1955). https://doi.org/10.2307/2372597
20. Cockle, J.: On transcendental and algebraic solution. Philosophical Magazine **XXI**, 379–383 (1861). https://doi.org/10.1080/14786446108643074
21. Comtet, L.: Calcul pratique des coefficients de Taylor d'une fonction algébrique. Enseignement Math. (2) **10**, 267–270 (1964). https://www.e-periodica.ch/digbib/view?pid=ens-001:1964:10#353
22. Comtet, L.: Advanced Combinatorics. D. Reidel Publishing Co., Dordrecht (1974). Enlarged edition of the 2-volumes Analyse combinatoire, published in French in 1970, by Presses Universitaires de France
23. Derksen, H., Kemper, G.: Computational Invariant Theory. With two appendices by Vladimir L. Popov and an addendum by Nobert A. Campo and Vladimir L. Popov. 2nd enlarged edition. Encyclopedia of Mathematical Sciences, vol. 130. Invariant Theory and Algebraic Transformation Groups 8. Springer, Berlin (2015). https://doi.org/10.1007/978-3-662-48422-7
24. Dicks, W., Formanek, E.: Poincaré series and a problem of S. Montgomery. Lin. Multilin. Algebra **12**, 21–30 (1982). https://doi.org/10.1080/03081088208817467
25. Dolgachev, I.: Lectures on Invariant Theory. London Math. Soc. Lecture Note Series, vol. 296. Cambridge University Press, Cambridge (2003)
26. Domokos, M.: Relatively free invariant algebras of finite reflection groups. Trans. Am. Math. Soc. **348**(6), 2217–2234 (1996). https://doi.org/10.1090/S0002-9947-96-01687-X
27. Domokos, M., Drensky, V.: A Hilbert-Nagata theorem in noncommutative invariant theory. Trans. Am. Math. Soc. **350**, 2797–2811 (1998). https://doi.org/10.1090/S0002-9947-98-02208-9
28. Domokos, M., Drensky, V.: Rationality of Hilbert series in noncommutative invariant theory. Int. J. Algebra Comput. **27**(7), 831–848 (2017). https://doi.org/10.1142/S0218196717500394
29. Domokos, M., Drensky, V.: Constructive noncommutative invariant theory. Transformation Groups (to appear). arXiv:1811.06342 [math.RT]
30. Drensky, V.: Commutative and noncommutative invariant theory. In: Math. and Education in Math., Proc. of the 24-th Spring Conf. of the Union of Bulgar. Mathematicians, Svishtov, April 4–7, 1995, Sofia, pp. 14–50 (1995)

31. Drensky, V.: Gelfand-Kirillov dimension of PI-algebras. In: Methods in Ring Theory, Proc. of the Trento Conf., Lect. Notes in Pure and Appl. Math. vol. 198, Dekker, pp. 97–113 (1998)
32. Drensky, V.: Free Algebras and PI-Algebras. Graduate Course in Algebra. Springer, Singapore (2000)
33. Drensky, V.: Graded algebras with prescribed Hilbert series. arXiv:2001.01064v1 [math.RA]
34. Drensky, V., Formanek, E.: Polynomial Identity Rings. Advanced Courses in Mathematics, CRM Barcelona. Birkhäuser, Basel-Boston (2004)
35. Drensky, V., Gupta, C.K.: Constants of Weitzenböck derivations and invariants of unipotent transformations acting on relatively free algebras. J. Algebra **292**, 393–428 (2005). https://doi.org/10.1016/j.jalgebra.2005.07.004
36. Drensky, V., Holtkamp, R.: Planar trees, free nonassociative algebras, invariants, and elliptic integrals. Algebra Discrete Math. **7**(2), 1–41 (2008) http://admjournal.luguniv.edu.ua/index.php/adm/article/view/806
37. Drensky, V., Lalov, Ch.: Free magmas, planar trees, and their generating functions. preprint
38. Elliott, E.B.: On linear homogeneous diophantine equations. Q. J. Pure Appl. Math. **34**, 348–377 (1903)
39. Fatou, P.: Séries trigonométriques et séries de Taylor. Acta Math. **30**, 335–400 (1906). https://doi.org/10.1007/BF02418579
40. Feigelstock, S.: A universal subalgebra theorem. Am. Math. Monthly **72**, 884–888 (1965). https://doi.org/10.2307/2315039
41. Flajolet, P., Sedgewick, R.: Analytic Combinatorics. Cambridge University Press (2009)
42. Formanek, E.: Noncommutative invariant theory. Contemp. Math. **43**, 87–119 (1985)
43. Govorov, V.E.: Graded algebras (Russian). Mat. Zametki **12**(2), 197–204 (1972). Translation: Math. Notes **12**, 552–556 (1972). https://doi.org/10.1007/BF01095017
44. Govorov, V.E.: On the dimension of graded algebras (Russian). Mat. Zametki **14**(2), 209–216 (1973). Translation: Math. Notes **14**, 678–682 (1973). https://doi.org/10.1007/BF01147113
45. Grosshans, F.D.: Algebraic Homogeneous Spaces and Invariant Theory. Lecture Notes in Mathematics, vol. 1673. Springer, Berlin (1997) https://doi.org/10.1007/BFb0093525
46. Hadžiev, Dž.: Some questions in the theory of vector invariants (Russian). Mat. Sb. (NS) **72**(114), 420–435 (1967). Translation: Math. USSR, Sb. **1**(3), 383–396 (1967). https://doi.org/10.1070/SM1967v001n03ABEH001988
47. Hardy, G.H., Ramanujan, S.: Asymptotic formulae in combinatory analysis. Proc. Lond. Math. Soc. (2) **17**, 75–115 (1918). https://doi.org/10.1112/plms/s2-17.1.75
48. Kharlampovich, O.G., Sapir, M.V.: Algorithmic problems in varieties. Int. J. Algebra Comput. **5**(4–5), 379–602 (1995). https://doi.org/10.1142/S0218196795000227
49. Harley, R.: On the theory of the transcendental solution of algebraic equations. Q. J. Pure Appl. Math. **5**, 337–361 (1862)
50. Herzog, J., Hibi, T.: Monomial Ideals. Graduate Texts in Mathematics, vol. 260. Springer, London (2011). https://doi.org/10.1007/978-0-85729-106-6
51. Kanel-Belov, A., Rowen, L.H.: Computational Aspects of Polynomial Identities. Research Notes in Mathematics, vol. 9. A K Peters, Wellesley, MA (2005)
52. Kargapolov, M.I., Merzlyakov, Yu.I.: Fundamentals of the Theory of Groups (Russian). "Nauka", Moscow, Third edition, 1982, Fourth Edition, 1996. Translation from the Second Edition: Grad. Texts in Math., vol 62. Springer, New York-Berlin (1979)
53. Kemer, A.R.: Ideals of Identities of Associative Algebras. Translations of Math. Monographs, vol. 87. American Mathematical Society (AMS), Providence, RI (1991)
54. Kharchenko, V.K.: Algebra of invariants of free algebras (Russian). Algebra i Logika **17**, 478–487 (1978). Translation: Algebra and Logic **17**, 316–321 (1978). https://doi.org/10.1007/BF01674783
55. Kharchenko, V.K.: Noncommutative invariants of finite groups and Noetherian varieties. J. Pure Appl. Algebra **31**, 83–90 (1984). https://doi.org/10.1016/0022-4049(84)90079-3
56. Kobayashi, Y.: Another graded algebra with a nonrational Hilbert series. Proc. Am. Math. Soc. **81**, 19–22 (1981). https://doi.org/10.1090/S0002-9939-1981-0589129-3

57. Koçak, D.: Finitely presented quadratic algebras of intermediate growth. Algebra Discrete Math. **20**(1), 69–88 (2015). http://admjournal.luguniv.edu.ua/index.php/adm/article/view/100
58. Koçak, D.: Intermediate growth in finitely presented algebras. Int. J. Algebra Comput. **27**(4), 391–401 (2017). https://doi.org/10.1142/S0218196717500205
59. Koryukin, A.N.: Noncommutative invariants of reductive groups (Russian). Algebra i Logika **23**(4), 419–429 (1984). Translation: Algebra Logic **23**, 290–296 (1984). https://doi.org/10.1007/BF02071789
60. Krause, G.R., Lenagan, T.H.: Growth of Algebras and Gelfand-Kirillov Dimension. Revised ed. Graduate Studies in Mathematics, vol. 22. American Mathematical Society (AMS), Providence, RI (2000)
61. Kurosh, A.G.: Non-associative free algebras and free products of algebras (Russian). Mat. Sbornik **20**, 239–260 (English summary, 260–262) (1947). http://mi.mathnet.ru/msb6215, https://eudml.org/doc/65356
62. Kurosh, A.G.: Free sums of multiple operator algebras (Russian). Sib. Mat. Zh. **1**, 62–70 (1960)
63. Lallement, G.: Semigroups and Combinatorial Applications. Pure and Applied Mathematics. A Wiley-Interscience Publication, Wiley, New York-Chichester-Brisbane (1979)
64. Lane, D.R.: Free Algebras of Rank Two and Their Automorphisms. Ph.D. Thesis, Bedford College, London (1976)
65. La Scala, R.: Monomial right ideals and the Hilbert series of noncommutative modules. J. Symb. Comput. **80**(Part 2), 403–415 (2017). https://doi.org/10.1016/j.jsc.2016.07.001
66. La Scala, R., Piontkovski, D.: Context-free languages and associative algebras with algebraic Hilbert series. arXiv:2001.09112 [math.RA]
67. La Scala, R., Piontkovski, D., Tiwari, S.K.: Noncommutative algebras, context-free grammars and algebraic Hilbert series. J. Symb. Comput. **101**, 28–50 (2020). https://doi.org/10.1016/j.jsc.2019.06.005
68. Leites, D., Poletaeva, E.: Defining relations for classical Lie algebras of polynomial vector fields. Math. Scand. **81**(1), 5–19 (1997). https://doi.org/10.7146/math.scand.a-12863
69. Leitmann, D., Wolke, D.: Periodische und multiplikative zahlentheoretische Funktionen. Monatsh. Math. **81**, 279–289 (1976). https://doi.org/10.1007/BF01387755
70. Lichtman, A.I.: Growth in enveloping algebras. Isr. J. Math. **47**, 296–304 (1984). https://doi.org/10.1007/BF02760603
71. Liouville, J.: Sur des classes très étendues de quantités dont valeur n'est ni algébrique, ni même réducible à des irrationelles algébriques. C.R. Acad. Sci., Paris, Sér. A **18**, 883–885 (1844). J. Math. Pures Appl. **16**, 133–142 (1851). https://eudml.org/doc/235459
72. Lvov, I.V.: On the Shirshov height theorem (Russian). In: Fifth All-Union Symposium on the Theory of Rings, Algebras and Modules (Held at Novosibirsk, September, 21st–23rd, 1982). Abstracts of communications, Institute of Mathematics SO AN SSSR, Novosibirsk, pp. 89–90 (1982)
73. Macaulay, F.S.: Some properties of enumeration in the theory of modular systems. Proc. Lond. Math. Soc. (2) **26**, 531–555 (1927). https://doi.org/10.1112/plms/s2-26.1.531
74. MacMahon, P.A.: Combinatory Analysis, vols. 1 and 2, Cambridge Univ. Press (1915, 1916). Reprinted in one volume: Chelsea, New York (1960)
75. Mahler, K.: Lectures on Transcendental Numbers. Edited and completed by B. Diviš and W.J. Le Veque. Lecture Notes in Mathematics, vol. 546. Springer, Berlin-Heidelberg-New York (1976). https://doi.org/10.1007/BFb0081107
76. Magnus, W., Karrass, A., Solitar, D.: Combinatorial Group Theory. Interscience. Wiley, New York-London-Sydney (1966)
77. Månsson, J.: On the computation of Hilbert series and Poincaré series for algebras with infinite Gröbner bases. Comput. Sci. J. Mold. **8**(1), 42–63 (2000). http://www.math.md/publications/csjm/issues/v8-n1/8204/
78. Månsson, J., Nordbeck, P.: A generalized Ufnarovski graph. Appl. Algebra Eng. Commun. Comput. **16**(5), 293–306 (2005). https://doi.org/10.1007/s00200-005-0178-8

79. Molien, T.: Über die Invarianten der linearen Substitutionsgruppen. Sitz. König Preuss. Akad. Wiss. **52**, 1152–1156 (1897)
80. Nielsen, J.: Die Isomorphismengruppe der freien Gruppen. Math. Ann. **91**, 169–209 (1924). https://doi.org/10.1007/BF01556078
81. Nishioka, K.: Mahler Functions and Transcendence. Lecture Notes in Mathematics, vol. 1631. Springer, Berlin (1996). https://doi.org/10.1007/BFb0093672
82. Petrogradsky, V.M.: Intermediate growth in Lie algebras and their enveloping algebras. J.Algebra **179**, 459–482 (1996). https://doi.org/10.1006/jabr.1996.0020
83. Petrogradsky, V.M.: Growth of finitely generated polynilpotent Lie algebras and groups, generalized partitions, and functions analytic in the unit circle. Int. J. Algebra Comput. **9**(2), 179–212 (1999). https://doi.org/10.1142/S0218196799000138
84. Piontkovski, D.: Algebras of linear growth and the dynamical Mordell-Lang conjecture. Adv. Math. **343**, 141–156 (2019). https://doi.org/10.1016/j.aim.2018.11.016
85. Procesi, C.: Lie Groups (An Approach through Invariants and Representations). Springer, New York (2007)
86. Sárközy, A.: On multiplicative arithmetic functions satisfying a linear recursion. Stud. Sci. Math. Hung. **13**, 79–104 (1978)
87. Schreier, O.: Die Untergruppen der freien Gruppen. Abh. Math. Semin. Univ. Hambg. **5**, 161–183 (1927). https://doi.org/10.1007/BF02952517
88. Shearer, J.B.: A graded algebra with a non-rational Hilbert series. J. Algebra **62**, 228–231 (1980). https://doi.org/10.1016/0021-8693(80)90213-6
89. Shephard, G.C., Todd, J.A.: Finite unitary reflection groups. Can. J. Math. **6**, 274–304 (1954). https://doi.org/10.4153/CJM-1954-028-3
90. Shirshov, A.I.: Subalgebras of free Lie algebras (Russian). Mat. Sb. **33**, 441–452 (1953). http://mi.mathnet.ru/msb5381
91. Shirshov, A.I.: On rings with identity relations (Russian). Mat. Sb. **41**, 277–283 (1957). http://mi.mathnet.ru/msb5082
92. Sloane, N. (ed.): The On-Line Encyclopedia of Integer Sequences. https://oeis.org/
93. Smith, M.K.: Universal enveloping algebras with subexponential but not polynomially bounded growth. Proc. Am. Math. Soc. **60**, 22–24 (1976). https://doi.org/10.1090/S0002-9939-1976-0419534-5
94. Tannery, J.: Propriétés des intégrales des équations differentielles linéaires à coefficients variables. Thèse de doctorat és sciences mathématiques, Faculté des Sciences de Paris, 1874. (Available at http://gallica.bnf.fr.) http://jubilotheque.upmc.fr/ead.html?id=TH_000285_001#
95. Ufnarovskij, V.A.: Poincaré series of graded algebras (Russian). Mat. Zametki **27**, 21–32 (1980). Translation: Math. Notes **27**, 12–18 (1980). https://doi.org/10.1007/BF01149808
96. Ufnarovskij, V.A.: A growth criterion for graphs and algebras defined by words (Russian). Mat. Zametki **31**, 465–472 (1982). Translation: Math. Notes **31**, 238–241 (1982). https://doi.org/10.1007/BF01145476
97. Ufnarovskij, V.A.: An independence theorem and its consequences (Russian). Mat. Sb. **128**, 124–132 (1985). Translation: Math. USSR Sb. **56**, 121–129 (1985). https://doi.org/10.1070/SM1987v056n01ABEH003027
98. Ufnarovskij, V.A.: On the use of graphs for calculating basis, growth, and Hilbert series of associative algebras (Russian). Mat. Sb. **180**(11), 1548–1560 (1989). Translation: Math. USSR, Sb. **68**(2), 417–428 (1989). https://doi.org/10.1070/SM1991v068n02ABEH001373
99. Ufnarovskij, V.A.: Combinatorial and asymptotic methods in algebra (Russian). Itogi Nauki Tekh., Ser. Sovrem. Probl. Mat., Fundam. Napravleniya **57**, 5–177 (1990). Translation: In: Kostrikin, A.I., Shafarevich, I.R. (eds.), Algebra VI, Encyclopaedia of Mathematical Sciences, vol. 57, pp. 1–196. Springer, Berlin (1995). http://mi.mathnet.ru/intf160
100. Ufnarovski, V.A.: Calculations of growth and Hilbert series by computer. In: Fischer, Klaus G., et al., (ed.) Computational Algebra. Papers from the Mid-Atlantic Algebra Conference, Held at George Mason University, Fairfax, VA, USA, May 20–23, 1993. Lect. Notes Pure Appl. Math., vol. 151, pp. 247–256. Dekker, New York (1993)

101. Uspensky, J.V.: Asymptotic formulae for numerical functions which occur in the theory of partitions (Russian). Bull. Acad. Sci. URSS **14**, 199–218 (1920)
102. Weyl, H.: Zur Darstellungstheorie und Invariantenabzählung der projektiven, der Komplex- und der Drehungsgruppe. Acta Math. **48**, 255–278 (1926); reprinted in Gesammelte Abhandlungen, Band III, pp. 1–25. Springer, Berlin-Heidelberg-New York (1968). https://doi.org/10.1007/BF02565334
103. Weyl, H.: The Classical Groups, Their Invariants and Representations. Princeton Univ. Press, Princeton, N.J. (1946). New Edition (1997)
104. Wolf, M.C.: Symmetric functions of non-commutative elements. Duke Math. J. **2**(4), 626–637 (1936). https://doi.org/10.1215/S0012-7094-36-00253-3

Central Polynomials of Algebras and Their Growth

Antonio Giambruno and Mikhail Zaicev

Abstract A polynomial in noncommutative variables taking central values in an algebra A is called a central polynomial of A. For instance the algebra of $k \times k$ matrices has central polynomials. For general algebras the existence of central polynomials is not granted. Nevertheless if an algebra has such polynomials, how can one measure how many are there?

The growth of central polynomials for any algebra satisfying a polynomial identity over a field of characteristic zero was started in recent years and here we shall survey the results so far obtained.

It turns out that one can prove the existence of two limits called the central exponent and the proper central exponent of A. They give a measure of the exponential growth of the central polynomials and the proper central polynomials of any algebra A satisfying a polynomial identity. They are closely related to $exp(A)$, the PI-exponent of the algebra.

Keywords Central polynomial · Polynomial identity · Codimension · Exponential growth

1 Introduction

In the 30s Wagner noticed that the polynomial in noncommutative variables $[[x_1, x_2]^2, x_3]$ vanishes when evaluated in $M_2(F)$, the algebra of 2×2 matrices over a field F, or in other words, the polynomial $[x_1, x_2]^2$ takes values in the scalars, the center of $M_2(F)$. Recall that a polynomial in non commuting variables taking

A. Giambruno (✉)
Dipartimento di Matematica e Informatica, Università di Palermo, Palermo, Italy
e-mail: antonio.giambruno@unipa.it,

M.Zaicev
Department of Algebra, Faculty of Mathematics and Mechanics, Moscow State University, Moscow, Russia
e-mail: zaicevmv@mail.ru

O. M. Di Vincenzo, A. Giambruno (eds.), *Polynomial Identities in Algebras*,
Springer INdAM Series 44, https://doi.org/10.1007/978-3-030-63111-6_10

central values when evaluated in an algebra A is called a central polynomial. Also if it takes at least one non-zero value we say that it is a proper central polynomial of A.

It is easily checked that the above polynomial $[x_1, x_2]^2$ is a proper central polynomial of $M_2(F)$, but it turns out that it is peculiar of 2×2 matrices, meaning that it does not have an obvious generalization to matrices of higher size. Nevertheless in the 50s Kaplansky conjectured the existence of proper central polynomials for the algebra $M_k(F)$ of $k \times k$ matrices over F, for any $k \geq 3$ (see [18]). This conjecture was verified in the early 70s independently by Formanek and Razmyslov [6, 22]. They constructed proper central polynomials for $M_k(F)$ using completely different methods. It is worth noticing that generally, even if an algebra A has a non-zero center, the existence of proper central polynomials is not granted.

Recall that a central polynomial for an algebra A which is not proper is called a polynomial identity of A. An extensive study of the polynomial identities satisfied by an algebra has been carried out in the past, and one may ask whether some of the results so far obtained can be extended to central polynomials. To this end, a first natural question might be: how many proper central polynomials exist compared to the polynomial identities of an algebra?

Such question can be reformulated in terms of codimension growth by comparing the growth of the spaces of central polynomials, proper central polynomials and polynomial identities of an algebra in the following sense.

Let A be an associative algebra over a field F and let $F\langle X\rangle$ be the free associative algebra on $X = \{x_1, x_2, \ldots\}$ over F. For every $n \geq 1$ let $P_n(A)$ be the space of multilinear polynomials in the variables $x_1, \ldots, x_n$ modulo the polynomial identities of A. Also, let $P_n^z(A)$ be the space of multilinear polynomials in the variables $x_1, \ldots, x_n$ modulo the central polynomials of A. Then one defines two numerical sequences $c_n(A) = \dim P_n(A)$ and $c_n^z(A) = \dim P_n^z(A)$ $n = 1, 2, \ldots$, called the sequence of codimensions and the sequence of central codimensions of A, respectively. They are related to the polynomial identities and to the central polynomials of A, respectively. Clearly the sequence $\delta_n(A) = c_n(A) - c_n^z(A)$, $n = 1, 2, \ldots$, corresponds to the proper central polynomials of A and is called the sequence of proper central codimensions of A.

On a first attempt one may try to compute the values of these three sequences but this can be achieved in a very few cases. Then one can try to compare their asymptotics. This can be done in some cases.

The sequence of codimensions was first defined by Regev in [23], and in [24] he was able to compute its precise asymptotics for the algebra $M_k(F)$, when F is a field of characteristic zero. The sequences of central codimensions and proper central codimensions were introduced in [25] and the asymptotics of the two sequences for the algebra $M_k(F)$, char $F = 0$, were computed in [4].

It is well known that if A is an algebra satisfying a non-trivial polynomial identity (PI-algebra), then the sequence of codimensions $c_n(A)$, $n = 1, 2, \ldots$, is exponentially bounded [23]. Moreover, if A is an algebra over a field of characteristic zero the limit

$$exp(A) = \lim_{n\to\infty} \sqrt[n]{c_n(A)}$$

always exists and is a non-negative integer called the PI-exponent of A [7, 8]. Since $c_n(A) = c_n^z(A) + \delta_n(A)$ it follows that if A is a PI-algebra, the sequences $c_n^z(A)$ and $\delta_n(A)$, $n = 1, 2, \ldots$, are also exponentially bounded and it is worth asking if the corresponding limits

$$exp^z(A) = \lim_{n\to\infty} \sqrt[n]{c_n^z(A)}, \quad exp^\delta(A) = \lim_{n\to\infty} \sqrt[n]{\delta_n(A)} \tag{1}$$

exist.

Here we shall survey on the results obtained in recent years on this subject. Since the main tool for computing asymptotics is the representation theory of the symmetric group that is well understood in characteristic zero, one restricts himself to algebras over a field of characteristic zero.

In this framework we shall see that for any associative PI-algebra A, the central exponent $exp^z(A)$ and the proper central exponent $exp^\delta(A)$ always exist and are non-negative integers. Moreover they can be characterized as the dimension of suitable semisimple algebras related to A. In particular if $exp(A) \geq 2$, the central exponent $exp^z(A)$ and the PI-exponent $exp(A)$ coincide. Concerning the proper central exponent $exp^\delta(A)$ examples can be exhibited showing that it can take any value smaller than $exp(A)$.

One can consider the same kind of questions for non associative algebras. It is well-known that in this setting the codimensions of a PI-algebra can be overexponential. Nevertheless they are exponentially bounded for finite dimensional algebras. In [16] it was shown that even if the codimensions of an algebra are exponentially bounded and the PI-exponent exists, it is not necessarily an integer. Here we shall see that the same phenomenon can appear with respect to the sequence of proper central codimensions.

2 A General Setting

Throughout F will be a field of characteristic zero and $F\langle X\rangle$ the free associative algebra over F on a countable set $X = \{x_1, x_2, \ldots\}$. Recall that a polynomial $f(x_1, \ldots, x_n) \in F\langle X\rangle$ is a central polynomial of an associative algebra A if $f(a_1, \ldots, a_n) \in Z(A)$, the center of A, for any $a_1, \ldots, a_n \in A$. If f takes non-zero values in $Z(A)$, we say that f is a proper central polynomial whereas if f takes only the zero value, f is a polynomial identity of A.

The set $Id(A)$ of polynomial identities of A and the set $Id^z(A)$ of central polynomials of A are an ideal and a subalgebra of $F\langle X\rangle$, respectively. Since they are invariant under the endomorphisms of $F\langle X\rangle$, we say that they are a T-ideal and a T-subalgebra, respectively.

Regev in [23] and [25] introduced the notions of codimension and central codimension of an algebra A as follows. Let P_n be the space of multilinear polynomials in the variables $x_1, \dots, x_n$, and define

$$P_n(A) = \frac{P_n}{P_n \cap Id(A)}, \quad P_n^z(A) = \frac{P_n}{P_n \cap Id^z(A)}.$$

Then

$$\Delta_n(A) = \frac{P_n \cap Id^z(A)}{P_n \cap Id(A)}$$

corresponds to the space of proper central polynomials in n fixed variables.

The sequences

$$c_n(A) = \dim P_n(A), \quad c_n^z(A) = \dim P_n^z(A), \quad \delta_n(A) = \dim \Delta_n(A), \quad n = 1, 2, \dots,$$

are called the sequence of codimensions, central codimensions and proper central codimensions of A, respectively. The relation among them is given by the equality

$$c_n(A) = \delta_n(A) + c_n^z(A). \tag{2}$$

This is a special case of a more general relation among polynomials. In fact, let the symmetric group S_n act on P_n via $\sigma f(x_1, \dots, x_n) = f(x_{\sigma(1)}, \dots, x_{\sigma(n)})$, for $f(x_1, \dots, x_n) \in P_n$ and $\sigma \in S_n$. Then the three spaces $P_n(A)$, $P_n^z(A)$, $\Delta_n(A)$ have an induced structure of S_n-modules and we denote by $\chi_n(A)$, $\chi_n^z(A)$, $\chi(\Delta_n(A))$ the corresponding characters (called cocharacters), respectively.

Since char $F = 0$, we can decompose such characters into irreducibles:

$$\chi_n(A) = \sum_{\lambda \vdash n} m_\lambda \chi_\lambda, \quad \chi_n^z(A) = \sum_{\lambda \vdash n} m'_\lambda \chi_\lambda, \quad \chi(\Delta_n(A)) = \sum_{\lambda \vdash n} m''_\lambda \chi_\lambda, \tag{3}$$

where χ_λ is the irreducible character of S_n corresponding to the partition λ of n and $m_\lambda, m'_\lambda, m''_\lambda$ are the multiplicities. Then

$$\chi_n(A) = \chi(\Delta_n(A)) + \chi_n^z(A),$$

and in (3) we have $m_\lambda = m'_\lambda + m''_\lambda$, for all $\lambda \vdash n$.

Recall that an algebra satisfying a non-trivial polynomial identity is called a PI-algebra. Also we say that two algebras A and B are PI-equivalent if they have the same identities, i.e., $Id(A) = Id(B)$. Another useful relation among polynomial identities and central polynomials is the following.

Remark 1 Let A and B be two PI-algebras. If they are PI-equivalent, then $Id^z(A) = Id^z(B)$ and $\Delta_n(A) = \Delta_n(B)$, for all $n \geq 1$.

Recall that if $A = A^{(0)} \oplus A^{(1)}$ is a superalgebra and $G = G^{(0)} \oplus G^{(1)}$ is the infinite dimensional Grassmann algebra with its canonical $\mathbb{Z}_2$-grading, then the algebra

$$G(A) = A^{(0)} \otimes G^{(0)} \oplus A^{(1)} \otimes G^{(1)} \subseteq A \otimes G,$$

is called the Grassmann envelope of A.

By a fundamental result of Kemer [19] any associative PI-algebra is PI-equivalent to the Grassmann envelope of a finite dimensional superalgebra.

In [7] and [8] it was proved that if A is any PI-algebra,

$$C_1 n^{t_1} d^n \leq c_n(A) \leq C_2 n^{t_2} d^n,$$

for some constants $C_1 > 0, C_2, t_1, t_2$, where $d = exp(A)$ is an integer called the PI-exponent of A that can be characterized as follows.

By the result of Kemer mentioned above we may assume that the algebra A is the Grassmann envelope $G(B)$ of a finite dimensional superalgebra B. Also, by extending the base field, since codimensions do not change, we may assume that F is algebraically closed. Then $exp(A) = exp(G(B))$ can be characterized as the dimension of a suitable semisimple subalgebra of B, a so-called admissible subalgebra of B of maximal dimension. We refer the reader to [11] for the corresponding definitions and an account of results on polynomial identities and their numerical invariants. We should mention that the asymptotics of the codimensions of a PI-algebra were later obtained in [3] and [2] for algebras with unity. For algebras without unity the asymptotics were obtained up to a constant (see also [13]).

In what follows we shall describe the results obtained in recent years about the central codimensions and proper central codimensions. Since the three codimensions do not change by extension of the base field, throughout we shall assume that our algebras are over an algebraically closed field F of characteristic zero.

3 Examples of Central Polynomials

In this section we present some examples of proper central polynomials of associative algebras.

Example 1 This example was already mentioned in the introduction. Let $A_1 = M_2(F)$ be the algebra of of 2×2 matrices over F. Its center is the one-dimensional space of scalar matrices, $Z(A_1) = F(e_{11} + e_{22})$ and a corresponding central polynomial is $[x_1, x_2]^2$ (or its linearization).

Example 2 Recall that $M_{1,1}(G)$ is the subalgebra of the algebra $M_2(G)$ of 2×2 matrices over the infinite dimensional Grassmann algebra consisting of the matrices of the type

$$\begin{pmatrix} a & b \\ c & d \end{pmatrix}, \text{ where } a, d \in G^{(0)},\ b, c \in G^{(1)}.$$

The center of $A_2 = M_{1,1}(G)$ consists of the "scalar" matrices,

$$Z(A_2) = \left\{ \begin{pmatrix} a & 0 \\ 0 & a \end{pmatrix} \mid a \in G^{(0)} \right\},$$

and $f(x_1, \ldots, x_4) = [[x_1, x_2], [x_3, x_4]]$ is a proper central polynomial for A_2.

Example 3 Denote by A_3 the subalgebra of $M_3(F)$ spanned by the matrix units e_{22}, e_{12}, e_{23} and e_{13}. Then $Z(A_3) = \mathrm{ann}_{A_3}(A_3) = \mathrm{span}\{e_{13}\}$ and the polynomials

$$[x_1, \ldots, x_k][x_{k+1}, \ldots, x_m],\ k \geq 2, m \geq k + 2, \tag{4}$$

are proper central polynomials for A_3. Here $[x_1, \ldots, x_k]$ denotes the left-normed commutator of $x_1, \ldots, x_k$. Similarly, one can check that the polynomials

$$[x_1, x_2]y_1 \cdots y_l[x_3, x_4] \tag{5}$$

with $l \geq 0$, are also proper central polynomials of A_3.

Since A_3 satisfies the identity

$$y_1[x_1, x_2]y_2 \equiv 0, \tag{6}$$

it is easy to see that any polynomial in (4) is a linear combination of polynomials of the type (5), modulo $Id(A_3)$.

Example 4 By slightly modifying the previous example we can consider the algebra

$$A_4 = \left\{ \begin{pmatrix} a & b & c \\ 0 & e & d \\ 0 & 0 & a \end{pmatrix} \mid a, b, c, d, e \in F \right\} = \mathrm{span}\{e_{11}+e_{33}, e_{22}, e_{12}, e_{23}, e_{13}\} \subset M_3(F).$$

The center of A_4 is $Z(A_4) = \mathrm{span}\{e_{13}\}$, a one-dimensional space.

Notice that the algebras A_3 and A_4 are not PI-equivalent. For example A_4 does not satisfy the identity (6). Nevertheless all the polynomials in (4) and (5) are proper central polynomials for A_4. Perhaps A_3 and A_4 have the same proper central polynomials.

Example 5 In order to get more complex examples of nontrivial central polynomials one can consider the algebra

$$A_5 = \left\{ \begin{pmatrix} 0 & B & C \\ 0 & A & D \\ 0 & 0 & 0 \end{pmatrix} | A, B, C, D \in M_t(F) \right\} \subset M_{3t}(F),$$

which is isomorphic to $A_3 \otimes M_t(F)$. This algebra has the following proper central polynomial

$$St_{2t}(x_1, \ldots, x_{2t}) St_{2t}(y_1, \ldots, y_{2t}),$$

where

$$St_m(x_1, \ldots, x_m) = \sum_{\sigma \in Sym_m} (\text{sgn } \sigma) x_{\sigma(1)} \cdots x_{\sigma(m)}$$

is the standard polynomial on the variables $x_1, \ldots, x_m$.

4 Algebras Without Proper Central Polynomials

Although the algebra $M_k(F)$ has proper central polynomials, there are several examples of algebras with non-zero center but with no proper central polynomials.

An algebra of interest in PI-theory is the algebra of upper block triangular matrices $UT(d_1, \ldots, d_k)$. Recall that $UT(d_1, \ldots, d_k)$ is a subalgebra of $M_{d_1 + \cdots + d_k}(F)$ defined as follows.

$$UT(d_1, \ldots, d_k) = \begin{pmatrix} A_1 & A_{12} & \cdots & A_{1k} \\ & A_2 & & \vdots \\ & & \ddots & A_{k-1k} \\ 0 & & & A_k \end{pmatrix},$$

where $A_i \cong M_{d_i}(F)$, $1 \leq i \leq k$, and $A_{ij} \cong M_{d_i \times d_j}(F)$, the space of $d_i \times d_j$ matrices over F, $1 \leq i < j \leq k$.

The interest in the algebras of upper block triangular matrices relies on several useful properties: they are explicit examples of the so called fundamental algebras introduced by Kemer (see [19]), their T-ideal of identities is a product of ideals of identities of matrix algebras (see [9, 20]), the asymptotics of their codimensions can be computed.

Unfortunately such algebras do not have proper central polynomials. In fact we have the following (see [14, Lemma 1]).

Lemma 1 *If $k > 1$, the algebra $UT(d_1, \ldots, d_k)$ has no proper central polynomials.*

Another useful example, which is actually a generalization of the above is the following.

First recall the definition of minimal superalgebra [10]. Let A be a finite dimensional superalgebra over F. Since F is algebraically closed and of characteristic zero, we can decompose $A = \bar{A} + J$, where $\bar{A} = A_1 \oplus \cdots \oplus A_m$ is a semisimple subalgebra and $J = J(A)$ is the Jacobson radical of A. It is well-known (see for instance [11]) that J is a homogeneous ideal and $\bar{A}$ can be chosen to be a superalgebra, i.e., homogeneous in the $\mathbb{Z}_2$-grading. It follows that we may assume that $A_1, \ldots, A_m$ are simple superalgebras. Then recall that each A_i is of the type either $M_k(F)$, or $M_{k,l}(F)$, $k \geq l \geq 1$ or $M_k(F \oplus cF)$, with $c^2 = 1$ (see for instance [11]). We call the diagonal matrix units $e_{11}, \ldots, e_{kk}$ of homogeneous degree zero of $M_k(F)$ (resp. $M_{k,l}(F)$ and $M_k(F \oplus cF)$), minimal graded idempotents.

We say that A is a minimal superalgebra if either A is simple or there exist homogeneous elements $w_{12}, \ldots, w_{m-1,m} \in J^{(0)} \cup J^{(1)}$ and minimal graded idempotents $e_1 \in A_1, \ldots, e_m \in A_m$ such that

$$e_i w_{i,i+1} = w_{i,i+1} e_{i+1} = w_{i,i+1}, \quad 1 \leq i \leq m-1,$$

$$w_{12} w_{23} \cdots w_{m-1,m} \neq 0,$$

and $w_{12}, \ldots, w_{m-1,m}$ generate J as a two-sided ideal of A.

The minimal superalgebras are strictly related to the so called minimal varieties of exponential growth. Recall that if $\mathcal{V}$ is a variety of algebras, its exponent is the PI-exponent of a generating algebra. Then a variety $\mathcal{V}$ is minimal of exponent $d \geq 2$ if $exp(\mathcal{V}) = d$ and $exp(\mathcal{W}) < d$, for any proper subvariety of $\mathcal{V}$. It was proved in [10] that any variety minimal of exponential growth is generated by the Grassmann envelope of a minimal superalgebra (see [11, Proposition 8.5.6]).

Simple superalgebras are examples of minimal superalgebras and it is not hard to see that they have proper central polynomials. Unfortunately this is a special case. In fact the following result can be proved [15, Lemma 4].

Lemma 2 *Let A be a minimal superalgebra. If A is not a simple superalgebra, then $G(A)$ has no proper central polynomials.*

5 The Proper Central Exponent

Let R be an associative algebra over F. As we mentioned above, since we are interested in computing the three codimensions, we may assume that R is the Grassmann envelope $G(A)$ of a finite dimensional superalgebra A. Also, since F is algebraically closed and of characteristic zero, we can decompose $A = \bar{A} + J$, where $\bar{A}$ is a semisimple subalgebra and $J = J(A)$ is the Jacobson radical of A.

Also we can decompose $\bar{A}$ into a sum $\bar{A} = A_1 \oplus \cdots \oplus A_m$, where $A_1, \ldots, A_m$ are simple superalgebras.

Now, if $G(A)$ has no proper central polynomials, then $\delta_n(G(A)) = 0$, for all n. In case $G(A)$ has proper central polynomials we make the following definition.

Definition 1 A semisimple subalgebra $B = A_{i_1} \oplus \cdots \oplus A_{i_k} \subseteq \bar{A}$, where $i_1, \ldots, i_k \in \{1, \ldots, m\}$ are distinct, is a centrally admissible subalgebra for $G(A)$ if there exists a multilinear proper central polynomial $f = f(x_1, \ldots, x_s)$ of $G(A)$ with $s \geq k$, such that $f(a_1, \ldots, a_k, b_1, \ldots, b_{s-k}) \neq 0$, for some $a_1 \in G(A_{i_1}), \ldots, a_k \in G(A_{i_k})$, $b_1, \ldots, b_{s-k} \in G(A)$.

As an illustration next we give examples of centrally admissible subalgebras for the algebras of Examples 1–5 of Sect. 3. Notice that since the algebras A_1, A_3, A_4, A_5 of those examples are finite dimensional we search for centrally admissible subalgebra contained in them. Now, for the algebra $A_1 = M_2(F)$ the whole of A_1 is a centrally admissible subalgebra. For the algebra $A_2 = M_{1,1}(G)$ a centrally admissible subalgebra is

$$M_{1,1}(F) = M_2(F).$$

In the Examples 3, 4 and 5 a centrally admissible subalgebra $S(A_i)$ coincides with a corresponding maximal semisimple subalgebra:

$$S(A_3) = \text{span}\{e_{22}\}, \quad S(A_4) = \text{span}\{e_{11} + e_{33}, e_{22}\}, \quad S(A_5) = \text{span}\{e_{22}\} \otimes M_t(F).$$

Centrally admissible subalgebras do not necessarily exist (see the example at the end of the section). Hence what can we say if $G(A)$ has proper central polynomials but no centrally admissible subalgebras?

In this case if A is nilpotent, then $\delta_N(G(A)) = 0$, for N large, and if A is not nilpotent, we let f be a proper central polynomial of $G(A)$. If $a_1, \ldots, a_n \in G(A)$ are such that $f(a_1, \ldots, a_n) \neq 0$, then $a_1, \ldots, a_n$ must lie in $G(J)$ and since $f(a_1, \ldots, a_n) \neq 0$ we have $J^n \neq 0$. It follows that $\delta_N(G(A)) = 0$ as soon as $J^N = 0$. In conclusion $\delta_N(G(A)) = 0$, for N large enough.

When $G(A)$ has centrally admissible subalgebras, then one can actually compute an upper and a lower bound of $\delta_n(G(A))$. In fact we have the following [15, Theorem 1, Proposition 3].

Theorem 1 *Let $G(A)$ be the Grassmann envelope of a finite dimensional superalgebra A over an algebraically closed field of characteristic zero.*

(1) If $G(A)$ has no proper central polynomials, then $\delta_n(G(A)) = 0$, for all n.
(2) If $G(A)$ has proper central polynomials but no centrally admissible subalgebras, then $\delta_n(G(A)) = 0$, for all n large enough.
(3) If $G(A)$ has centrally admissible subalgebras then, for all $n \geq 1$,

$$C_1 n^{t_1} d^n \leq \delta_n(G(A)) \leq C_2 n^{t_2} d^n, \tag{7}$$

for some constants $C_1 > 0, C_2, t_1, t_2$, where d is the maximal dimension of a centrally admissible subalgebra for $G(A)$.

It is clear that the inequalities in (7) still hold for any PI-algebra even if we also relax the hypothesis on F being algebraically closed. As a special case we can get the following.

Corollary 1 *If R is a PI-algebra over a field of characteristic zero, then the proper central exponent $exp^\delta(R) = \lim_{n\to\infty} \sqrt[n]{\delta_n(R)}$ exists and is a non-negative integer. Moreover $exp^\delta(R) \leq exp(R)$.*

Another easy consequence of the above theorem is the following.

Corollary 2 *Let R be a PI-algebra over a field of characteristic zero. Then the sequence $\delta_n(R)$, $n = 1, 2, \ldots$, is either polynomially bounded or grows as an exponential function a^n with $a \geq 2$.*

There is a close relation between centrally admissible subalgebra of $G(A)$ of maximal dimension and reduced algebras. Recall that a finite dimensional superalgebra $A = A_1 \oplus \cdots \oplus A_m + J$ is reduced if, after a reordering of the simple components, we have $A_1 J A_2 J \cdots J A_m \neq 0$ (see [11, Definition 9.4.2]). In fact we have.

Remark 2 If $B \subset \bar{A}$ is a centrally admissible subalgebra for $G(A)$ of maximal dimension, then the superalgebra $\hat{B} = B + J$ is reduced.

Proof Let $B = A_1 \oplus \cdots \oplus A_k$ with $A_1, \ldots, A_k$ simple superalgebras, and let $f = f(x_1, \ldots, x_s)$ be a multilinear proper central polynomial of $G(A)$. Let $a_i \otimes g_i \in G(A_i)$, $1 \leq i \leq k$, and $b_j \otimes h_j \in G(A)$, $1 \leq j \leq s-k$ be such that

$$f(a_1 \otimes g_1, \ldots, a_k \otimes g_k, b_1 \otimes h_1, \ldots, b_{s-k} \otimes h_{s-k}) \neq 0.$$

Since $G(A_i)G(A_j) = 0$, for any $i \neq j$ then

$$G(A_{i_1})G(J)G(A_{i_2})G(J)\cdots G(J)G(A_{i_k}) \neq 0,$$

for some permutation $(i_1, \ldots, i_k)$ of $(1, \ldots, k)$. This implies that $A_{i_1} J A_{i_2} J \cdots J A_{i_k} \neq 0$, i.e., the superalgebra $\hat{B} = B + J$ is reduced. □

Next we give an example of an algebra with proper central polynomials but no centrally admissible subalgebras.

Example 6 Let $A = B \oplus R$ where B is the subalgebra of $M_3(F)$ consisting of all matrices whose third row is zero and let $R = F\langle X_m\rangle / F\langle X_m\rangle^{m+1}$, where $X_m = \{x_1, \ldots, x_m\}$. We regard A as a superalgebra with trivial grading so that $G(A) = G^{(0)} \otimes A$ has the same identities as A. Clearly the center of A coincides with its annihilator and equals R^m. Hence any polynomial identity of B of degree m is a proper central polynomial of A. On the other hand, any central polynomial of A vanishes under any evaluation such that at least one variable is evaluated in a maximal semisimple subalgebra of A.

We know by (2) that for any PI-algebra A, $exp(A) \geq exp^{\delta}(A)$. But how far apart can be the two exponents? This question is answered in the following result [14, Corollary 4].

Theorem 2 *For any integer $N \geq 0$ there exists a finite dimensional algebra R such that $exp^{\delta}(R) \neq 0$ and $exp(R) - exp^{\delta}(R) > N$.*

Regarding the algebra $M_k(F)$ of $k \times k$ matrices over F, Regev in [24] computed the precise asymptotics. Recall that two functions $f(x)$ and $g(x)$ of a real variable are asymptotic equal, and we write $f(x) \simeq g(x)$, if $\lim_{x\to\infty} \frac{f(x)}{g(x)} = 1$. Then Regev's result is that $c_n(M_k(F)) \simeq Cn^{-\frac{k^2-1}{2}} k^{2n}$, where C is an explicitly computed constant. The asymptotics of the other two codimensions are strictly related to the above asymptotic, in fact in [4] it was proved that $c_n^z(M_k(F)) \simeq \frac{1}{k^2} c_n(M_k(F))$ and, so, $\delta_n(M_k(F)) \simeq \frac{k^2-1}{k^2} c_n(M_k(F))$.

6 The Central Exponent

In this section we focus on the computation of the central exponent $\exp^z(R)$ for any associative PI-algebra R.

Recall that the codimensions $c_n(R)$ are sandwiched between two exponential functions

$$C_1 n^{t_1} exp(R)^n \leq c_n(R) \leq C_2 n^{t_2} exp(R)^n, \tag{8}$$

for some constants $C_1 > 0, C_2, t_1, t_2$. Hence, since by (2) $c_n^z(R) \leq c_n(R)$, we get that

$$c_n^z(R) \leq C_2 n^{t_2} exp(R)^n,$$

Now, if $exp(R) = 0$ then R is a nilpotent algebra and, so, $exp^z(R) = 0$. In case $exp(R) = 1$, then either $exp^z(R) = 1$ or $exp^z(R) = 0$. In fact, $exp(R) = 1$ says that R is not nilpotent and the sequence of codimensions is polynomially bounded. Clearly the same holds for the sequence of central codimensions. Thus $exp^z(A) = 1$ provided $c_n^z(A) \neq 0$ for all n.

By making use of the basic properties of minimal varieties of exponential growth we can also compute a lower bound of the central codimensions. In fact we have the following [15, Theorem 2].

Theorem 3 *Let R be a PI-algebra over a field of characteristic zero. If* $exp(R) \geq 2$ *then*

$$C_1 n^{t_1} exp(R)^n \leq c_n^z(R) \leq C_2 n^{t_2} exp(R)^n,$$

for some constants $C_1 > 0, C_2, t_1, t_2$*. Hence* $exp^z(R)$ *exists and* $exp^z(R) = exp(R)$*. If* $exp(R) = 0$ *then* $exp^z(R) = 0$ *and if* $exp(R) = 1$ *then* $exp^z(R) = 0$ *or* 1.

Proof Assume as we may that F is algebraically closed and $R = G(A)$ with A is a finite dimensional superalgebra.

Suppose $exp(R) \geq 2$ and let $\mathcal{V} = var(G(A))$ be the variety of algebras generated by the algebra $G(A)$. By the solution of the Specht problem due to Kemer (see [19]), every T-ideal is finitely generated as a T-ideal. Hence the variety $\mathcal{V}$ contains a subvariety $\mathcal{W}$ which is minimal of exponent $exp(R) = \exp(\mathcal{W}) = exp(\mathcal{V})$. Hence there exists a minimal superalgebra B such that $\mathcal{W} = var(G(B))$.

Suppose that B is not a simple superalgebra. If $f \in Id^z(G(A))$ is a central polynomial of $G(A)$, by Lemma 2 f is an identity of $G(B)$. Hence $Id^z(G(A)) \subseteq Id^z(G(B)) = Id(G(B))$. This says that $c_n^z(G(A)) \geq c_n(G(B))$, and since $c_n(G(B))$ has a lower bound as in (8), we get the desired lower bound of $c_n^z(G(A))$.

When B is a simple superalgebra, we refer the reader to the proof in [15]. □

We remark that by [11, Theorem 7.2.4] when $exp(R) \leq 1$, the algebra R has the same identities as a finite dimensional algebra A and the case $exp^z(R) = exp^z(A) = 0$ can be characterized as follows [14, Proposition 1].

Proposition 1 *Let A be a finite dimensional algebra such that* $exp^z(A) = 0$*. Then* $A = A_1 \oplus A_2$ *where* A_1 *is a nilpotent algebra and* A_2 *is a commutative algebra.*

Proof If $c_n^z(A) = 0$ for some $n \geq 2$, then any monomial of degree n is a central polynomial of A. In particular, $x_1 \cdots x_n$ is a central polynomial.

Write $A = \bar{A} + J$ where $\bar{A} = A_1 \oplus \cdots \oplus A_m$ is a sum of simple algebras and J is the Jacobson radical. Since $x_1 \cdots x_n$ is central, we get that $A_1 \cong \cdots \cong A_m \cong F$ and they are central subalgebras.

Let e be the unity of A_1. Write $J = J_0 \oplus J_1$ where $xe = ex = 0$ if $x \in J_0$ and $ye = ey = y$ if $y \in J_1$. Also for $i = 2, \ldots, m$, since $eA_i = 0$ we get that $A_i J_1 = 0$. It follows that $A = (A_1 + J_1) \oplus (A_2 + \cdots + A_m + J_0)$. Notice that $A_1 + J_1$ is commutative since $y = e^{n-1}y$ lies in the center of A for any $y \in J_1$. Repeating this procedure we get $A = C_1 \oplus \cdots \oplus C_m \oplus I$ where all $C_1, \ldots, C_m$ are commutative and $I \subset J$ is nilpotent. □

7 Non Associative Algebras

One may wonder if the results about the growth of central polynomials still holds for non associative algebras. Starting with the free non associative algebra of countable rank $F\{X\}$ over a field F of characteristic zero, we consider the space of non associative multilinear polynomials in the first n variables. If A is an algebra over F, we defines similarly the spaces $P_n(A)$, $P_n^z(A)$, $\Delta_n(A)$ and the corresponding dimensions $c_n(A), c_n^z(A), \delta_n(A)$.

It is well-known that for a non associative PI-algebra A the codimensions are not necessarily exponentially bounded. In [16] it was shown that even if the codimensions of an algebra A are exponentially bounded and the PI-exponent $exp(A)$ exists, $exp(A)$ is not necessarily an integer. In fact, for any real number $\alpha > 1$ an algebra R_α was constructed such that $exp(R_\alpha)$ exists and equals α.

Now, for any finite dimensional algebra A the sequence $c_n(A)$, $n = 1, 2, \ldots$, is exponentially bounded (see [1, 12]). Hence from the equality in (2) for nonassociative polynomials, it follows that in this case also the sequences $c_n^z(A)$ and $\delta_n(A)$, $n = 1, 2, \ldots$, are exponentially bounded.

Here we want to point out that even for finite dimensional algebras the central exponent and the proper central exponent, even if they exist, can be non integer.

To this end we let A be the algebra over F with basis $\{e_{-1}, e_0, e_1, e_2, z\}$ and multiplication table given by

$$e_{-1}e_0 = e_{-1},\ e_{-1}e_1 = e_0,\ e_{-1}e_2 = e_0e_1 = e_1,\ e_0e_2 = e_2,\ e_2e_2 = z,$$

and all the other products equal to zero.

Clearly A is a five-dimensional non associative algebra and it can be checked that its center is $Z(A) = Fz = \mathrm{ann}_A(A)$.

Also A is a $\mathbb{Z}$-graded algebra: $A = \bigoplus_{i\in\mathbb{Z}} A_i$, where $A_{-1} = Fe_{-1}$, $A_0 = Fe_0$, $A_1 = Fe_1$, $A_2 = Fe_2$, $A_4 = Fz$, and $A_i = 0$, for all other $i \in \mathbb{Z}$.

Notice that if f is a multilinear polynomial and we evaluate f in the given basis of A, then the corresponding value will lie in a homogeneous component A_i.

In what follows we shall use the notation $a_1a_2\cdots a_n$ to indicate the left-normed product of $a_1, a_2, \ldots, a_n$, i.e., $a_1a_2a_3 = (a_1a_2)a_3$ and inductively $a_1a_2\cdots a_m = (a_1a_2\cdots a_{m-1})a_m$. In this notation we can write

$$St_4(x_1, \ldots, x_4) = \sum_{\sigma\in Sym_4} (\mathrm{sgn}\,\sigma) x_{\sigma(1)}\cdots x_{\sigma(4)}, \tag{9}$$

where all monomials on the right hand side of (9) are left normed.

Remark 3 ([17, Remark 1])

$$St_4(x_1, \ldots, x_4)St_4(y_1, \ldots, y_4) \tag{10}$$

is a multilinear proper central polynomial of A.

It can be shown that if χ_λ is an S_n-character appearing with non-zero multiplicity in one of the cocharacters of A (see (3)), then λ has at most four parts. Then for a partition $\lambda = (\lambda_1, \dots, \lambda_4) \vdash n$ we define the function

$$\Phi(\lambda) = \left(\frac{\lambda_1}{n}\right)^{-\frac{\lambda_1}{n}} \left(\frac{\lambda_2}{n}\right)^{-\frac{\lambda_2}{n}} \left(\frac{\lambda_3}{n}\right)^{-\frac{\lambda_3}{n}} \left(\frac{\lambda_4}{n}\right)^{-\frac{\lambda_4}{n}}.$$

Here some of the parts λ_k, $2 \le k \le 4$, could be zero and in this case we set $(\frac{\lambda_k}{n})^{-\frac{\lambda_k}{n}} = 1$.

We remark that the definition of $\Phi(\lambda)$ can be generalized to the real numbers $0 \le \alpha_1, \dots, \alpha_4 \le 1$ such that $\alpha_1 + \cdots + \alpha_4 = 1$ by setting

$$\Phi(\alpha_1, \dots, \alpha_4) = \alpha_1^{-\alpha_1} \cdots \alpha_4^{-\alpha_4}. \tag{11}$$

We are now ready to introduce the candidates for $exp(A)$, $exp^z(A)$ and $exp^\delta(A)$. Let T be the domain of $\mathbb{R}^4$ defined by the conditions

$$\begin{cases} \alpha_1 + \cdots + \alpha_4 = 1, \\ -\alpha_1 + \alpha_3 + 2\alpha_4 = 0, \\ \alpha_1 \ge \cdots \ge \alpha_4 \ge 0. \end{cases} \tag{12}$$

Since the function Φ is continuous, it takes a maximal value Φ_{max} on the compact set T.

We start by stating a technical result which is a consequence of [21] and [5] (see [17, Proposition 2]).

Proposition 2 *The function $\Phi(\alpha_1, \dots, \alpha_4)$ defined in (11) reaches the maximal value on the compact set T in the point $\beta = (\beta_1, \beta_2, \beta_3, \beta_4)$ where $\beta_1 \approx 0.421350946$, $\beta_2 \approx 0.276953179$, $\beta_3 \approx 0.182040800$, $\beta_4 \approx 0.119655073$, and $\Phi_{\max} \approx 3.610718614$.*

Remark 4 Notice that $\beta_1 + \cdots + \beta_4 \approx 1$ and $\beta_2 - \beta_3 < \beta_4$.

The final result of this section is the following [17, Theorem 2, Theorem 3].

Theorem 4 *The exponents $exp(A)$, $exp^z(A)$ and $exp^\delta(A)$ exist. Moreover,*

$$exp(A) = exp^z(A) = exp^\delta(A) = \Phi_{max}.$$

The above Theorem 4 and Proposition 2 say that the five-dimensional non-associative algebra A has the property that the three exponents , exp, exp^z, exp^δ exist and their value is $exp(A) = exp^z(A) = exp^\delta(A) \approx 3.610718614$, a non integer.

Acknowledgments The first author "Antonio Giambruno" was partially supported by the GNSAGA of INDAM. The second author "Mikhail Zaicev" was supported by the Russian Science Foundation, grant 16-11-10013.

References

1. Bahturin, Y., Drensky, V.: Graded polynomial identities of matrices. Linear Algebra Appl. **357**, 15–34 (2002)
2. Berele, A.: Properties of hook Schur functions with applications to P.I. algebras. Adv. Appl. Math. **41**(1), 52–75 (2008)
3. Berele, A., Regev, A.: Asymptotic behaviour of codimensions of P.I. algebras satisfying Capelli identities. Trans. Am. Math. Soc. **360**(10), 5155–5172 (2008)
4. Berele, A., Regev, A.: Growth of central polynomials of verbally prime algebras. Isr. J. Math. **228**, 201–210 (2018)
5. Bogdanchuk, O.A., Mishchenko, S.P., Verëvkin, A.B.: On Lie algebras with exponential growth of the codimensions. Serdica Math. J. **40**, 209–240 (2014)
6. Formanek, E.: Central polynomials for matrix rings. J. Algebra **23**, 129–132 (1972)
7. Giambruno, A., Zaicev, M.: On codimension growth of finitely generated associative algebras. Adv. Math. **140**, 145–155 (1998)
8. Giambruno, A., Zaicev, M.: Exponential codimension growth of P.I. algebras: An exact estimate. Adv. Math. **142**, 221–243 (1999)
9. Giambruno, A., Zaicev, M.: Minimal varieties of algebras of exponential growth. Adv. Math. **174**, 310–323 (2003)
10. Giambruno, A., Zaicev, M.: Codimension growth and minimal superalgebras. Trans. Am. Math. Soc. **355**, 5091–5117 (2003)
11. Giambruno, A., Zaicev, M.: Polynomial Identities and Asymptotic Methods, Mathematical Surveys and Monographs, vol. 122. American Mathematical Society, Providence, RI (2005)
12. Giambruno, A., Zaicev, M.: Codimension growth of special simple Jordan algebras. Trans. Am. Math. Soc., **362**, 3107–3123 (2010)
13. Giambruno, A., Zaicev, M.: Growth of polynomial identities: is the sequence of codimensions eventually non-decreasing? Bull. Lond. Math. Soc. **46**(4), 771–778 (2014)
14. Giambruno, A., Zaicev, M.: Central polynomials and growth functions. Isr. J. Math. **226**(1), 15–28 (2018)
15. Giambruno, A., Zaicev, M.: Central polynomials of associative algebras and their growth. Proc. Am. Math. Soc. **147**, 909–919 (2019)
16. Giambruno, A., Mishchenko, S., Zaicev, M.: Codimensions of algebras and growth functions. Adv. Math. **217**(3), 1027–1052 (2008)
17. Giambruno, A., Mishchenko, S., Zaicev, M.: Non integral exponential growth of central polynomials. Arch. Math. **112**, 149–160 (2019)
18. Kaplansky, I.: *Problems in the Theory of Rings*. Report of a conference on linear algebras, June, 1956, Publ. 502, pp. 1–3. National Academy of Sciences-National Research Council, Washington (1957)
19. Kemer, A.: Ideals of Identities of Associative Algebras. Translations of Mathematical Monographs, Vol. 87. American Mathematical Society, Providence, RI (1988)
20. Lewin, J.: A matrix representation for associative algebras I. Trans. Am. Math. Soc. **188**, 293–308 (1974)
21. Mishchenko, S.P., Verevkin, A.B., Zaitsev, M.A.: On sufficient condition for existence of the exponent of the linear algebras variety. Mosc. Univ. Math. Bull. **66**, 86–89 (2011)
22. Razmyslov, Ju.P.: A certain problem of Kaplansky. (Russian) Izv. Akad. Nauk SSSR Ser. Mat. **37**, 483–501 (1973)
23. Regev, A.: Existence of identities in $A \otimes B$. Isr. J. Math. **11**, 131–152 (1972)
24. Regev, A.: Codimensions and trace codimensions of matrices are asymptotically equal. Isr. J. Math. **47**, 246–250 (1984)
25. Regev, A.: Growth of the central polynomials. Comm. Algebra **44**, 4411–4421 (2016)

Trace Identities on Diagonal Matrix Algebras

Antonio Ioppolo, Plamen Koshlukov, and Daniela La Mattina

Dedicated to our dear colleague and friend Professor Antonio Giambruno on the occasion of his seventieth anniversary.

Abstract Let D_n be the algebra of $n \times n$ diagonal matrices. On such an algebra it is possible to define very many trace functions. The purpose of this paper is to present several results concerning trace identities satisfied by this kind of algebras.

Keywords Polynomial identities · Traces · Diagonal matrices · Codimensions

1 Introduction

The invariant theory of $n \times n$ matrices and the consequent theory of trace identities represent an interesting object of study and an important area of modern Mathematics. The methods and the main results in this field, obtained independently by Procesi [22] and Razmyslov [23], are one of the basic tools needed in order to develop the theory of PI-algebras.

Concerning the theory of polynomial identities, a key year is the 1972, when Regev introduced the famous codimension sequence of an associative algebra, a function measuring, in some sense, the growth of the identities of the algebra. He proved [24] that for a PI-algebra (an algebra satisfying a non-trivial identity) such a sequence is exponentially bounded. More than 20 years later, Giambruno and Zaicev [5, 6] proved that every variety of associative algebras over a field of characteristic zero has an integral exponential growth, answering positively to a famous conjecture

A. Ioppolo (✉) · P. Koshlukov
IMECC, UNICAMP, Campinas, SP, Brazil
e-mail: ioppolo@unicamp.br; plamen@unicamp.br

D. La Mattina
Department of Mathematics, University of Palermo, Palermo, Italy
e-mail: daniela.lamattina@unipa.it

O. M. Di Vincenzo, A. Giambruno (eds.), *Polynomial Identities in Algebras*, Springer INdAM Series 44, https://doi.org/10.1007/978-3-030-63111-6_11

posed by Amitsur in the early '80. We recall an important result obtained in 1979 by Kemer. He characterized the varieties of associative algebras having a polynomially bounded codimension sequence [21].

In recent years a lot of papers concerning associative algebras endowed with some additional structure, and their polynomial identities, have been published. In them it is possible to find several analogues of the celebrated results mentioned above (see for instance [1, 2, 4, 7–19, 25]).

In this paper we are interested in trace polynomial identities for matrix algebras. The algebra $M_n(F)$ of $n \times n$ matrices endowed with the usual trace represents one of the most famous examples of algebras with trace. Procesi and Razmyslov, in the papers mentioned above, described the trace identities of such an algebra over a field of characteristic 0. It turns out that they are consequences of the well-known Cayley–Hamilton polynomial written in terms of the traces of the matrix and its powers (and then linearised). We must note here that as it often happens, the simplicity of the statement of the theorem due to Razmyslov and Procesi is largely misleading, and that the proofs are very sophisticated and extensive. We also recall that the theorem of Razmyslov and Procesi is one of the most general results in PI theory. No analogues of it are known for the ordinary identities for the $n \times n$ matrices with $n > 2$, and it seems to us that with the methods at hand nowadays it would be hardly possible to obtain such analogues.

Among matrix algebras, in this paper we focus our attention on the algebra D_n of $n \times n$ diagonal matrices. On such an algebra it is possible to define very many traces: in fact, since D_n is commutative, a trace on it is just a linear function. This is in sharp contrast with the situation of full matrix algebras: in this case, every trace function is a scalar multiple of the usual trace.

The paper is organized in the following way. After two sections of preliminaries, we present results concerning the algebra D_n endowed with a particular trace. Then, in the last sections dedicated to D_2 and D_3, we study the polynomial identities satisfied by these algebras endowed with all possible traces.

2 Preliminaries

Throughout this paper F will denote a field of characteristic zero and A a unitary associative F-algebra. We say that A is an algebra with trace if it is endowed with a linear map $\mathrm{tr}\colon A \to F$ such that for all $a, b \in A$ one has

$$\mathrm{tr}(ab) = \mathrm{tr}(ba).$$

Accordingly, one can construct $F\langle X, \mathrm{Tr}\rangle$, the free algebra with trace on the countable set $X = \{x_1, x_2, \ldots\}$ where Tr is a formal trace. Let $\mathcal{M}$ denote the set of all monomials in the elements of X. Then $F\langle X, \mathrm{Tr}\rangle$ is the algebra generated by the free algebra $F\langle X\rangle$ together with the set of central (commuting) indeterminates $\mathrm{Tr}(M)$, $M \in \mathcal{M}$, subject to the conditions that $\mathrm{Tr}(MN) = \mathrm{Tr}(NM)$, and

$\mathrm{Tr}(\mathrm{Tr}(M)N) = \mathrm{Tr}(M)\mathrm{Tr}(N)$, for all $M, N \in \mathcal{M}$. In other words,

$$F\langle X, \mathrm{Tr}\rangle \cong F\langle X\rangle \otimes F[\mathrm{Tr}(M) \mid M \in \mathcal{M}].$$

The elements of the free algebra with trace are called trace polynomials.

A trace polynomial $f(x_1, \ldots, x_n, \mathrm{Tr}) \in F\langle X, \mathrm{Tr}\rangle$ is a trace identity for A if, after substituting the variables x_i with arbitrary elements $a_i \in A$ and Tr with the trace tr, we obtain $0 \in A$. We denote by $\mathrm{Id}^{tr}(A)$ the set of trace identities of A, which is a trace T-ideal of the free algebra with trace, i.e., an ideal invariant under all endomorphisms of $F\langle X, \mathrm{Tr}\rangle$.

As in the ordinary case, $\mathrm{Id}^{tr}(A)$ is completely determined by its multilinear polynomials.

Definition 1 The vector space of multilinear elements of the free algebra with trace in the first n variables is called the space of multilinear trace polynomials in $x_1, \ldots,$ x_n and it is denoted by MT_n (MT comes from *mixed trace*). Its elements are linear combinations of expressions of the type

$$\mathrm{Tr}(x_{i_1} \cdots x_{i_a}) \cdots \mathrm{Tr}(x_{j_1} \cdots x_{j_b}) x_{l_1} \cdots x_{l_c}$$

where $\{i_1, \ldots, i_a, \ldots, j_1, \ldots, j_b, l_1, \ldots, l_c\} = \{1, \ldots, n\}$.

Remark 1 It is well known that $\dim_F MT_n = (n+1)!$.

The non-negative integer

$$c_n^{tr}(A) = \dim_F \frac{MT_n}{MT_n \cap \mathrm{Id}^{tr}(A)}$$

is called the n-th trace codimension of A.

A prominent role among the elements of MT_n is played by the so-called pure trace polynomials, i.e., polynomials such that all the variables $x_1, \ldots, x_n$ appear inside a trace.

Definition 2 The vector space of multilinear pure trace polynomials in $x_1, \ldots, x_n$ is the space

$$PT_n = \mathrm{span}_F \left\{\mathrm{Tr}(x_{i_1} \cdots x_{i_a}) \cdots \mathrm{Tr}(x_{j_1} \ldots x_{j_b}) : \{i_1, \ldots, j_b\} = \{1, \ldots, n\}\right\}.$$

For a permutation $\sigma \in S_n$ we write

$$\sigma^{-1} = \left(i_1 \cdots i_{r_1}\right)\left(j_1 \cdots j_{r_2}\right) \cdots \left(l_1 \cdots l_{r_t}\right)$$

as a product of disjoint cycles, including one-cycles and let us assume that $r_1 \geq r_2 \geq \cdots \geq r_t$. In this case we say that σ is of cyclic type $\lambda = (r_1, \ldots, r_t)$. Assume further that each cycle has in its leftmost position the least integer that it moves. That is i_1 is the least element in the first cycle, j_1 in the second, and so on. We then define the pure trace monomial $ptr_\sigma \in PT_n$ as

$$ptr_\sigma(x_1,\ldots,x_n) = \mathrm{Tr}\left(x_{i_1}\cdots x_{i_{r_1}}\right)\mathrm{Tr}\left(x_{j_1}\cdots x_{j_{r_2}}\right)\cdots\mathrm{Tr}\left(x_{l_1}\cdots x_{l_{r_t}}\right).$$

Moreover, we define the so-called trace monomial $mtr_\sigma \in MT_{n-1}$ so that

$$ptr_\sigma(x_1,\ldots,x_n) = \mathrm{Tr}\left(mtr_\sigma(x_1,\ldots,x_{n-1})x_n\right).$$

3 Matrix Algebras with Trace

In this section we study matrix algebras with trace. Let $M_n := M_n(F)$ be the algebra of $n \times n$ matrices over F. One can endow such an algebra with the usual trace on matrices, denoted t_1, and defined as

$$t_1(a) = t_1\begin{pmatrix} a_{11} & \cdots & a_{1n} \\ \vdots & \ddots & \vdots \\ a_{n1} & \cdots & a_{nn}\end{pmatrix} = a_{11} + \cdots + a_{nn} \in F.$$

The following lemma is a well known result of elementary linear algebra.

Lemma 1 *Let $f\colon M_n \to F$ be a trace. Then there exists $\alpha \in F$ such that $f = \alpha t_1$.*

In what follows we shall use the notation t_α to indicate a trace on M_n such that $t_\alpha = \alpha t_1$. Moreover, $M_n^{t_\alpha}$ will denote the algebra of $n \times n$ matrices endowed with the trace t_α.

In sharp contrast with the above result, there are very many traces on the algebra $D_n = D_n(F)$ of diagonal matrices over F. The following lemma shows this situation.

Lemma 2 *If tr is a trace on D_n then there exist scalars $\alpha_1, \ldots, \alpha_n \in F$ such that for each diagonal matrix $a = diag(a_{11},\ldots,a_{nn}) \in D_n$ one has*

$$tr(a) = \alpha_1 a_{11} + \cdots + \alpha_n a_{nn}.$$

Proof The algebra $D_n \cong F^n$ is commutative, and $D_n \cong F^n$ with component-wise operations. Hence a linear function $\mathrm{tr}\colon D_n \to F$ must be of the form stated in the lemma. Clearly for each choice of the scalars α_i one obtains a trace on D_n.

We shall denote the trace tr such that, for all $a = \mathrm{diag}(a_{11},\ldots,a_{nn}) \in D_n$, $\mathrm{tr}(a) = \alpha_1 a_{11} + \cdots + \alpha_n a_{nn}$, for some fixed scalars $\alpha_1,\ldots,\alpha_n \in F$, with the symbol $t_{\alpha_1,\ldots,\alpha_n}$. At the same time, to indicate that the algebra D_n is endowed with such a trace, we shall write $D_n^{t_{\alpha_1,\ldots,\alpha_n}}$.

We have the following remark.

Remark 2 Let S_n be the symmetric group of order n on the set $\{1, 2, \dots, n\}$. For every $\sigma \in S_n$, the algebras $D_n^{t_{\alpha_1,\dots,\alpha_n}}$ and $D_n^{t_{\alpha_{\sigma(1)},\dots,\alpha_{\sigma(n)}}}$ are isomorphic, as algebras with trace.

Proof We need only to observe that the linear map $\varphi\colon D_n^{t_{\alpha_1,\dots,\alpha_n}} \to D_n^{t_{\alpha_{\sigma(1)},\dots,\alpha_{\sigma(n)}}}$, defined by $\varphi(e_{ii}) = e_{\sigma(i)\sigma(i)}$, for all $i = 1, \dots, n$, is an isomorphism of algebras with trace.

Recall that a trace function tr on an algebra A is said to be degenerate if there exists a non-zero element $a \in A$ such that

$$\mathrm{tr}(ab) = 0$$

for every $b \in A$. This means that the bilinear form $f(x, y) = \mathrm{tr}(xy)$ is degenerate on A. In the following lemma we describe the non-degenerate traces on D_n.

Lemma 3 *Let $D_n^{t_{\alpha_1,\dots,\alpha_n}}$ be the algebra of $n \times n$ matrices endowed with the trace $t_{\alpha_1,\dots,\alpha_n}$. Such a trace is non-degenerate if and only if all the scalars α_i are non-zero.*

Proof Let $t_{\alpha_1,\dots,\alpha_n}$ be non-degenerate and suppose that there exists i such that $\alpha_i = 0$. Consider the matrix unit e_{ii}. It is easy to see that we reach a contradiction since, for any element $\mathrm{diag}(a_{11}, \dots, a_{nn}) \in D_n$, we get

$$t_{\alpha_1,\dots,\alpha_n}(e_{ii}\mathrm{diag}(a_{11}, \dots, a_{nn})) = t_{\alpha_1,\dots,\alpha_n}(e_{ii}a_{ii}) = \alpha_i a_{ii} = 0.$$

In order to prove the opposite direction, let us assume that all the scalars α_i are non-zero. Suppose, by contradiction, that the trace $t_{\alpha_1,\dots,\alpha_n}$ is degenerate. Hence there exists a non-zero element $a = \mathrm{diag}(a_{11}, \dots, a_{nn}) \in D_n$ such that $t_{\alpha_1,\dots,\alpha_n}(ab) = 0$, for any $b \in D_n$. In particular, let $b = e_{ii}$, for $i = 1, \dots, n$. We have that

$$t_{\alpha_1,\dots,\alpha_n}(ae_{ii}) = t_{\alpha_1,\dots,\alpha_n}(\mathrm{diag}(a_{11}, \dots, a_{nn})e_{ii}) = t_{\alpha_1,\dots,\alpha_n}(a_{ii}e_{ii}) = \alpha_i a_{ii} = 0.$$

Since $\alpha_i \neq 0$, for all $i = 1, \dots, n$, we get that $a_{ii} = 0$ and so $a = 0$, a contradiction. The proof is complete.

4 Some Results on D_n

In this section we deal with the algebra $D_n^{t_{\alpha_1,\dots,\alpha_n}}$, i.e., the algebra D_n endowed with the trace $t_{\alpha_1,\dots,\alpha_n}$, defined as

$$t_{\alpha_1,\dots,\alpha_n}\begin{pmatrix} a_{11} & \dots & 0 \\ \vdots & \ddots & \vdots \\ 0 & \dots & a_{nn} \end{pmatrix} = \alpha_1 a_{11} + \dots + \alpha_n a_{nn}.$$

Here we want to highlight that if $\alpha_1 = \cdots = \alpha_n = 0$, then D_n is just an ordinary commutative algebra (with a zero trace). As D_n is unitary it is not nilpotent. Hence the T-ideal of identities is generated by just the commutator and its codimension sequence is equal to 1.

The first goal of this section is to find the generators of the T^{tr}-ideal of identities of the algebra $D_n^{t_{\alpha,\dots,\alpha}}$. Here $\alpha_1, \dots, \alpha_n$ are all equal to a non-zero scalar α.

To this end, given a trace polynomial $f(x_1, \dots, x_k, \mathrm{Tr}) \in MT_k$, we start by constructing a new trace polynomial $f^\alpha(x_1, \dots, x_k, \mathrm{Tr})$ in the following way. For every monomial $M_s(x_1, \dots, x_k)$ of $f(x_1, \dots, x_k, \mathrm{Tr})$ containing s traces, $f^\alpha(x_1, \dots, x_k, \mathrm{Tr})$ contains the monomial

$$\alpha^{-s} M_s(x_1, \dots, x_k).$$

Lemma 4 *Let the algebra D_n be endowed with the usual trace $t_{1,\dots,1}$ or with a trace $t_{\alpha,\dots,\alpha}$, for some $\alpha \in F \setminus \{0\}$. A trace polynomial $f(x_1, \dots, x_k, Tr)$ is an identity of $D_n^{t_{1,\dots,1}}$ if and only if $f^\alpha(x_1, \dots, x_k, Tr)$ is an identity of $D_n^{t_{\alpha,\dots,\alpha}}$.*

Proof We need only to observe that, for any $a_1, \dots, a_k \in D_n$, the following evaluations coincide:

$$f(a_1, \dots, a_k, t_{1,\dots,1}) = f^\alpha(a_1, \dots, a_k, t_{\alpha,\dots,\alpha}).$$

Let now

$$C_k(x_1, \dots, x_k) = \sum_{\sigma \in S_{k+1}} (-1)^\sigma mtr_\sigma(x_1, \dots, x_k)$$

be the k-th Cayley–Hamilton polynomial, $k \geq 2$.

Theorem 1 *Let $\alpha \in F \setminus \{0\}$. The trace T-ideal $Id^{tr}(D_n^{t_{\alpha,\dots,\alpha}})$ is generated, as a trace T-ideal, by the polynomials:*

1. $[x_1, x_2]$,
2. $C_n^\alpha(x_1, \dots, x_n)$.

Proof For $\alpha = 1$, the result was proved by Berele in [3, Theorem 2.1]. Using the same idea one deals with the case $\alpha \neq 1$.

We conclude this section with the following result in which the algebra D_n is endowed with a trace $t_{\alpha_1,\dots,\alpha_n}$.

Theorem 2 *Let $D_n^{t_{\alpha_1,\dots,\alpha_n}}$ be the algebra of $n \times n$ diagonal matrices endowed with the trace $t_{\alpha_1,\dots,\alpha_n}$. If there exist i, j such that $\alpha_i, \alpha_j \neq 0$ and $\alpha_i \neq \alpha_j$, then $D_n^{t_{\alpha_1,\dots,\alpha_n}}$ does not satisfy any multilinear trace identity of degree n which is not a consequence of the identity $[x_1, x_2] \equiv 0$.*

Proof Modulo the identity $[x_1, x_2] \equiv 0$, a generic multilinear trace identity f of degree n should be of the type

$$f(x_1, \ldots, x_n) = \sum a^{I_1,\ldots,I_s}_{\mu_1,\ldots,\mu_s} \operatorname{Tr}\left(x_{i_1^{(1)}} \cdots x_{i_{\mu_1}^{(1)}}\right) \cdots \operatorname{Tr}\left(x_{i_1^{(s-1)}} \cdots x_{i_{\mu_s}^{(s-1)}}\right) x_{i_1^{(s)}} \cdots x_{i_{\mu_s}^{(s)}}$$

where

- $s = 0, \ldots, n$ and $\mu_1 + \cdots + \mu_s = n$,
- $0 \le \mu_1 \le \cdots \le \mu_s \le n$,
- $I_j = \{i_1^{(j)}, \ldots, i_{\mu_j}^{(j)}\}, i_1^{(j)} < \cdots < i_{\mu_j}^{(j)}, j = 1, \ldots, s$.

Notice that two traces with the same number of elements are ordered in the following way: we put first the trace with the least index of the first variable.

In order to prove the result we will show that actually f is the zero polynomial. We assume that all the scalars $\alpha_1, \ldots, \alpha_n$ are non-zero (with few changes it is easy to deal with the general case).

First we introduce the following notation. We say that a monomial M of f is of the type

$$(\mu_1, \ldots, \mu_s), \quad \mu_1 \le \cdots \le \mu_s,$$

if the variables of M are divided in s groups of length $\mu_1, \ldots, \mu_s$, respectively, which are inside s traces (in this case, in M there are no variables outside the traces) or inside $s-1$ traces and a group of variables outside the traces. For example, the following monomials are both of type $(\mu_1, \ldots, \mu_s)$:

$$\operatorname{Tr}\left(x_{i_1^{(1)}} \cdots x_{i_{\mu_1}^{(1)}}\right) \cdots \operatorname{Tr}\left(x_{i_1^{(s)}} \cdots x_{i_{\mu_s}^{(s)}}\right),$$

$$\operatorname{Tr}\left(x_{i_1^{(1)}} \cdots x_{i_{\mu_1}^{(1)}}\right) \cdots \operatorname{Tr}\left(x_{i_1^{(s-1)}} \cdots x_{i_{\mu_{s-1}}^{(s-1)}}\right) x_{i_1^{(s)}} \cdots x_{i_{\mu_s}^{(s)}}.$$

Moreover, we shall introduce the following order on the types of a monomial. We say that the type $(\mu_1, \ldots, \mu_s)$ is greater than the type $(\eta_1, \ldots, \eta_r)$, and we write $(\mu_1, \ldots, \mu_s) \succ (\eta_1, \ldots, \eta_r)$, if

- $s > r$ or
- $s = r$ and $\mu_s > \eta_s$.

Now we can start by considering the greatest type, which is $(1, \dots, 1)$. The monomials of f of this type, with the corresponding scalars, are the following:

$$a_{1,\dots,1,0}^{\{x_1\},\dots,\{x_n\},\emptyset}\mathrm{Tr}(x_1)\cdots\mathrm{Tr}(x_n),$$
$$a_{1,\dots,1}^{\{x_1\},\dots,\{x_n\}}\mathrm{Tr}(x_1)\cdots\mathrm{Tr}(x_{n-1})x_n,$$
$$\vdots$$
$$a_{1,\dots,1}^{\{x_2\},\dots,\{x_n\},\{x_1\}}\mathrm{Tr}(x_2)\cdots\mathrm{Tr}(x_n)x_1.$$

Let us consider the evaluation $x_i = e_{ii}$, $i = 1, \dots, n$. It is clear that all the monomials of f which are not of the type $(1, \dots, 1)$ vanish under this evaluation. For any $i = 1, \dots, n$, it follows that

$$a_{1,\dots,1}^{\{x_1\},\dots,\{x_n\},\{x_i\}} = -\alpha_i\, a_{1,\dots,1,0}^{\{x_1\},\dots,\{x_n\},\emptyset}.$$

By hypothesis, there exist i, j such that $\alpha_i \neq \alpha_j$. We shall consider another evaluation: $x_i = e_{jj}$, $x_j = e_{ii}$ and $x_l = e_{ll}$, for all $l \notin \{i, j\}$. As before, we get that

$$a_{1,\dots,1}^{\{x_1\},\dots,\{x_n\},\{x_i\}} = -\alpha_j\, a_{1,\dots,1,0}^{\{x_1\},\dots,\{x_n\},\emptyset}.$$

Since $\alpha_i \neq \alpha_j$, we get that $a_{1,\dots,1,0}^{\{x_1\},\dots,\{x_n\},\emptyset} = 0$. As a consequence, we have also that

$$a_{1,\dots,1}^{\{x_1\},\dots,\{x_n\}} = \cdots = a_{1,\dots,1}^{\{x_2\},\dots,\{x_n\},\{x_1\}} = 0.$$

In conclusion, we have proved that all the scalars corresponding to the type $(1, \dots, 1)$ are actually zero.

The proof now continues in the same way. We consider the greatest remaining type $(\mu_1, \dots, \mu_s)$ (at the first step it will be $(1, \dots, 1, 2)$) and we shall prove that all the scalars corresponding to such monomials are actually zero.

Let M be the following monomial of f of type $(\mu_1, \dots, \mu_s)$ with no variables outside the traces:

$$M = a_{\mu_1,\dots,\mu_s,0}^{J_1,\dots,J_s,\emptyset}\,\mathrm{Tr}\left(x_{j_1^{(1)}}\cdots x_{j_{\mu_1}^{(1)}}\right)\cdots\mathrm{Tr}\left(x_{j_1^{(s)}}\cdots x_{j_{\mu_s}^{(s)}}\right),$$

with fixed $J_1 = \{j_1^{(1)}, \dots, j_{\mu_1}^{(1)}\}, \dots, J_s = \{j_1^{(s)}, \dots, j_{\mu_s}^{(s)}\}$.

We consider the evaluation:

$$x_{j_h^{(l)}} = e_{ll}, \quad h = 1, \dots, \mu_l, \quad l = 1, \dots, s. \tag{1}$$

As before, all the monomials of f of type less than $(\mu_1, \dots, \mu_s)$ vanish under this substitution. Moreover, the same will happen to any other monomial of f, distinct from M, of type $(\mu_1, \ldots, \mu_s)$ and with no variables outside the traces (in fact, some variables will be in other traces, compared to where they were in M).

Let now M' be a generic monomial of f of type $(\mu_1, \ldots, \mu_s)$ with variables outside the traces

$$a^{I_1,\ldots,I_s}_{\eta_1,\ldots,\eta_s} \operatorname{Tr}\left(x_{i_1^{(1)}} \cdots x_{i_{\eta_1}^{(1)}}\right) \cdots \operatorname{Tr}\left(x_{i_1^{(s-1)}} \cdots x_{i_{\eta_{s-1}}^{(s-1)}}\right) x_{i_1^{(s)}} \cdots x_{i_{\eta_s}^{(s)}}.$$

Here $\{\eta_1, \ldots, \eta_s\} = \{\mu_1, \ldots, \mu_s\}$ and $I_j = \{i_1^{(j)}, \ldots, i_{\eta_j}^{(j)}\}$, $i_1^{(j)} \leq \cdots \leq i_{\eta_j}^{(j)}$, $j = 1, \ldots, s$.

It is not difficult to see that, under the above evaluation, the monomial M' does not vanish if and only if

$$\{I_1, \ldots, I_s\} = \{J_1, \ldots, J_s\}.$$

In conclusion, under this evaluation, we obtain

$$\begin{aligned} f = {} & a^{J_1,\ldots,J_s,\emptyset}_{\mu_1,\ldots,\mu_s,0} \alpha_1 \cdots \alpha_s I_n + a^{J_1,\ldots,J_s}_{\mu_1,\ldots,\mu_s} \alpha_1 \cdots \alpha_{s-1} e_{ss} + \\ & a^{J_1,\ldots,J_s,J_{s-1}}_{\mu_1,\ldots,\mu_s,\mu_{s-1}} \alpha_1 \cdots \alpha_{s-2} \alpha_s e_{s-1s-1} + \cdots + \\ & a^{J_1,J_3\ldots,J_s,J_2}_{\mu_1,\mu_3\ldots,\mu_s,\mu_2} \alpha_1 \alpha_3 \cdots \alpha_s e_{22} + a^{J_2,\ldots,J_s,J_1}_{\mu_2,\ldots,\mu_s,\mu_1} \alpha_2 \cdots \alpha_s e_{11} = 0. \end{aligned}$$

Recalling that $I_n = e_{11} + \cdots + e_{nn}$ and that $n > s$ (in fact $s = n$ only in the first step of the proof), since $\alpha_1, \ldots, \alpha_s \neq 0$, we immediately get that

$$a^{J_1,\ldots,J_s,\emptyset}_{\mu_1,\ldots,\mu_s,0} = a^{J_1,\ldots,J_s}_{\mu_1,\ldots,\mu_s} = a^{J_1,\ldots,J_s,J_{s-1}}_{\mu_1,\ldots,\mu_s,\mu_{s-1}} = \cdots = a^{J_1,J_3\ldots,J_s,J_2}_{\mu_1,\mu_3\ldots,\mu_s,\mu_2} = a^{J_2,\ldots,J_s,J_1}_{\mu_2,\ldots,\mu_s,\mu_1} = 0.$$

We are left to deal with the remaining monomials of type $(\mu_1, \ldots, \mu_s)$. The procedure is the same. We consider one monomial with no variables outside the traces and make a suitable evaluation like that one in (1). As before, it follows that all the scalars corresponding to monomials of type $(\mu_1, \ldots, \mu_s)$ are actually zero.

Now it is sufficient to deal with the new greater type and the theorem is proved.

5 Trace Identities on D_2

In this section we deal with the algebra D_2 of 2×2 diagonal matrices over the field F. In accordance with the results of Sect. 3, we can define on D_2, up to isomorphism, the following kinds of trace functions:

1. $t_{\alpha,0}$, for any $\alpha \in F$,
2. $t_{\alpha,\alpha}$, for any non-zero $\alpha \in F$,
3. $t_{\alpha,\beta}$, for any distinct non-zero $\alpha, \beta \in F$.

Our goal is to find the generators of the T^{tr}-ideal of the identities of the algebra D_2 endowed with a trace of the above kinds.

Let us start with the case of $D_2^{t_{\alpha,0}}$. Recall that, if $\alpha = 0$, then D_2 is just an ordinary algebra with T-ideal of identities generated by the commutator and codimensions sequence equal to 1.

For $\alpha \neq 0$, we have the following result.

Theorem 3 *Let $\alpha \in F \setminus \{0\}$. The trace T-ideal $Id^{tr}(D_2^{t_{\alpha,0}})$ is generated, as a trace T-ideal, by the polynomials:*

- $f_1 = [x_1, x_2]$,
- $f_2 = Tr(x_1)Tr(x_2) - \alpha Tr(x_1x_2)$.

Moreover

$$c_n^{tr}(D_2^{t_{\alpha,0}}) = 2^n.$$

Proof It is clear that $T = \langle f_1, f_2 \rangle_{T^{tr}} \subseteq \mathrm{Id}^{tr}(D_2^{t_{\alpha,0}})$.

We need to prove the opposite inclusion. First we claim that the polynomials

$$\mathrm{Tr}(x_{i_1} \cdots x_{i_k})x_{j_1} \cdots x_{j_{n-k}}, \tag{2}$$

where $\{i_1, \ldots, i_k, j_1, \ldots, j_{n-k}\} = \{1, \ldots, n\}$, $i_1 < \cdots < i_k$, $j_1 < \cdots < j_{n-k}$, span MT_n modulo $MT_n \cap T$, for every $n \geq 1$. In fact, because of the identity $f_2 \equiv 0$, we can kill all products of two traces (and more than two traces). So we may consider only monomials with either no trace or with one trace. Clearly the identity f_1 implies that we can assume all of these monomials ordered, outside and also inside each trace. The claim is proved.

Our next goal is to show that the polynomials in (2) are linearly independent modulo $\mathrm{Id}^{tr}(D_2^{t_{\alpha,0}})$. To this end, let $g(x_1, \ldots, x_n, \mathrm{Tr})$ be a linear combination of the above polynomials which is a trace identity:

$$g(x_1, \ldots, x_n, \mathrm{Tr}) = \sum_{I,J} a_{I,J}\mathrm{Tr}(x_{i_1} \cdots x_{i_k})x_{j_1} \cdots x_{j_{n-k}},$$

where $I = \{x_{i_1}, \ldots, x_{i_k}\}$, $J = \{x_{j_1}, \ldots, x_{j_{n-k}}\}$ and $i_1 < \cdots < i_k$, $j_1 < \cdots < j_{n-k}$.

We claim that g is actually the zero polynomial. Suppose that, for some fixed $I = \{x_{i_1}, \ldots, x_{i_k}\}$ and $J = \{x_{j_1}, \ldots, x_{j_{n-k}}\}$, one has that $a_{I,J} \neq 0$. We consider the following evaluation:

$$x_{i_1} = \cdots = x_{i_k} = e_{11}, \qquad x_{j_1} = \cdots = x_{j_{n-k}} = e_{22}, \qquad \mathrm{Tr} = t_{\alpha,0}.$$

It follows that $g(e_{11}, \ldots, e_{11}, e_{22}, \ldots, e_{22}, t_{\alpha,0}) = a_{I,J}\alpha e_{22} = 0$. Hence $a_{I,J} = 0$. The claim is proved and so

$$\mathrm{Id}^{tr}(D_2^{t_{\alpha,0}}) = T.$$

Finally, in order to compute the codimension sequence of our algebra, we have only to count the number of elements in (2). Fixed k, there are exactly $\binom{n}{k}$ elements of the type $\mathrm{Tr}(x_{i_1}\cdots x_{i_k})x_{j_1}\cdots x_{j_{n-k}}$, $i_1<\cdots<i_k$, $j_1<\cdots<j_{n-k}$. Hence the number of elements in (2) is $\sum_{k=0}^{n}\binom{n}{k}=2^n$ and the proof is complete.

The T^{tr}-ideal of identities of the algebra $D_2^{t_{\alpha,\alpha}}$ is given in Theorem 1 for $n=2$. However, a complete proof of the next result can be found in [20].

Theorem 4 *Let $\alpha \in F\setminus\{0\}$. The trace T-ideal $Id^{tr}(D_2^{t_{\alpha,\alpha}})$ is generated, as a trace T-ideal, by the polynomials:*

- $f_1=[x_1,x_2]$,
- $f_3=C_2^{\alpha}(x_1,x_2)$.

Moreover

$$c_n^{tr}(D_2^{t_{\alpha,\alpha}})=\sum_{k=0}^{n}\binom{n}{k}=2^n.$$

We conclude this section by considering the case of $D_2^{t_{\alpha,\beta}}$. The following theorem is proved in [20].

Theorem 5 *Let $\alpha, \beta \in F\setminus\{0\}$, $\alpha\neq\beta$. The trace T-ideal $Id^{tr}(D_2^{t_{\alpha,\beta}})$ is generated, as a trace T-ideal, by the polynomials:*

- $f_1=[x_1,x_2]$,
- $f_4=-x_1Tr(x_2)Tr(x_3)+(\alpha+\beta)x_1Tr(x_2x_3)+x_3Tr(x_1)Tr(x_2)-(\alpha+\beta)x_3Tr(x_1x_2)-Tr(x_1)Tr(x_2x_3)+Tr(x_3)Tr(x_1x_2)$,
- $f_5=Tr(x_1)Tr(x_2)Tr(x_3)-(\alpha\beta^2+\alpha^2\beta)x_1x_2x_3+\alpha\beta x_1x_2Tr(x_3)+\alpha\beta x_1x_3Tr(x_2)+\alpha\beta x_2x_3Tr(x_1)-(\alpha+\beta)x_1Tr(x_2)Tr(x_3)+(\alpha^2+\alpha\beta+\beta^2)x_1Tr(x_2x_3)-\alpha\beta x_2Tr(x_1x_3)-\alpha\beta x_3Tr(x_1x_2)+\alpha\beta Tr(x_1x_2x_3)-(\alpha+\beta)Tr(x_1)Tr(x_2x_3)$.

Moreover

$$c_n^{tr}(D_2^{t_{\alpha,\beta}})=2^{n+1}-n-1.$$

6 Trace Identities on D_3

In this section, we focus our attention on the trace identities of the algebra D_3 of 3×3 diagonal matrices over the field F. By taking into account the results of Sect. 3, it is easy to see that on D_3, up to isomorphism, it is possible to define the following kinds of trace functions:

1. $t_{\alpha,0,0}$, for any $\alpha\in F$,
2. $t_{\alpha,\alpha,0}$, for any non-zero $\alpha\in F$,
3. $t_{\alpha,\beta,0}$, for any distinct non-zero $\alpha,\beta\in F$.

4. $t_{\alpha,\alpha,\alpha}$, for any non-zero $\alpha \in F$,
5. $t_{\alpha,\beta,\beta}$, for any distinct non-zero $\alpha, \beta \in F$,
6. $t_{\alpha,\beta,\gamma}$, for any distinct non-zero $\alpha, \beta, \gamma \in F$.

Let us start with the case of $D_3^{t_{\alpha,0,0}}$. Recall that, if $\alpha = 0$, then D_3 is just an ordinary algebra with T-ideal of identities generated by the commutator. Moreover its codimension sequence is constant equal to 1.

If $\alpha \neq 0$, the case of $D_3^{t_{\alpha,0,0}}$ is solved with the same approach with which we have found out the generators of $\mathrm{Id}^{tr}(D_2^{t_{\alpha,0}})$. Actually, exactly the same proof allows us to state the following general result.

Theorem 6 *Let $\alpha \in F \setminus \{0\}$. Then $Id^{tr}(D_n^{t_{\alpha,0,\dots,0}})$ is generated, as a trace T-ideal, by the polynomials:*

- $f_1 = [x_1, x_2]$,
- $f_2 = Tr(x_1)Tr(x_2) - \alpha Tr(x_1x_2)$.

Moreover

$$c_n^{tr} D_n^{t_{\alpha,0,\cdots,0}} = 2^n.$$

Concerning the algebra $D_3^{t_{\alpha,\alpha,0}}$ we have the following result.

Theorem 7 *Let $\alpha \in F \setminus \{0\}$. The trace T-ideal $Id^{tr}(D_3^{t_{\alpha,\alpha,0}})$ is generated, as a trace T-ideal, by the polynomials:*

- $f_1 = [x_1, x_2]$,
- $f_6 = Tr(x_1)Tr(x_2)Tr(x_3) + 2\alpha^2 Tr(x_1x_2x_3) - \alpha Tr(x_1)Tr(x_2x_3) - \alpha Tr(x_2)Tr(x_1x_3) - \alpha Tr(x_3)Tr(x_1x_2)$.

Proof Write $T = \langle f_1, f_6\rangle_{T^{tr}}$. It is easy to see that $T \subseteq \mathrm{Id}^{tr}(D_3^{t_{\alpha,\alpha,0}})$.

In order to prove the opposite inclusion let $f \in MT_n$ be a multilinear trace polynomial of degree n. Hence it is a linear combination of polynomials of the type

$$\mathrm{Tr}(x_{i_1} \cdots x_{i_a}) \cdots \mathrm{Tr}(x_{j_1} \cdots x_{j_b}) x_{l_1} \cdots x_{l_c},$$

where $\{i_1, \dots, i_a, \dots, j_1, \dots, j_b, l_1, \dots, l_c\} = \{1, \dots, n\}$. Because of the identity $f_6 \equiv 0$, we can kill all products of three traces (and more than three traces). Considering also the identity f_1, which implies that all the variables (inside or outside a trace) are ordered, we have proved that, for every $n \geq 1$, the following polynomials span MT_n modulo $MT_n \cap T$:

$$\mathrm{Tr}(x_{i_1} \cdots x_{i_k}) \mathrm{Tr}(x_{j_1} \cdots x_{j_h}) x_{l_1} \cdots x_{l_{n-k-h}}, \tag{3}$$

where $\{i_1, \dots, i_k, j_1, \dots, j_h, l_1, \dots, l_{n-k-h}\} = \{1, \dots, n\}$, and $i_1 < \cdots < i_k$, $j_1 < \cdots < j_h$, $l_1 < \cdots < j_{n-k-h}$.

By using the same idea employed in Theorem 2, we obtain that the above polynomials are linearly independent modulo $\mathrm{Id}^{tr}(D_3^{t_{\alpha,\alpha,0}})$.

Hence $\mathrm{Id}^{tr}(D_3^{t_{\alpha,\alpha,0}}) \subseteq T$ and so we have proved that

$$\mathrm{Id}^{tr}(D_3^{t_{\alpha,\alpha,0}}) = T.$$

Let us now focus our attention to the case of $D_3^{t_{\alpha,\beta,0}}$, $\alpha, \beta \neq 0, \alpha \neq \beta$.

By Theorem 2, we already know that such an algebra does not satisfy any multilinear trace identity of degree 3 which is not a consequence of the commutator. Moreover, thanks to the use of the Computer Algebra System Maxima, we have the following result.

Lemma 5 *Let $\alpha, \beta \in F$ be two distinct non-zero elements. For any $r_1, r_2, r_3 \in F$, the following polynomial is an identity of $D_3^{t_{\alpha,\beta,0}}$:*

$$\begin{aligned}
f_7(r_1, r_2, r_3) = &+ r_1\alpha\beta Tr(x_1)Tr(x_2x_3x_4) - (r_1(\alpha+\beta) + r_2 + r_3)Tr(x_1)Tr(x_4)Tr(x_2x_3)\\
&+ r_1\alpha\beta Tr(x_2)Tr(x_1x_3x_4) - (r_1(\alpha+\beta) + r_2 + r_3)Tr(x_2)Tr(x_3)Tr(x_1x_4)\\
&+ r_3 Tr(x_2)Tr(x_4)Tr(x_1x_3) - (\alpha(r_1\beta + r_3) + r_3\beta)Tr(x_1x_3)Tr(x_2x_4)\\
&+ r_2 Tr(x_3)Tr(x_4)Tr(x_1x_2) - (\alpha(r_1\beta + r_2) + r_2\beta)Tr(x_1x_2)Tr(x_3x_4)\\
&+ (r_1\beta^2 + \alpha(r_1\beta + r_2 + r_3) + \beta(r_2 + r_3) + r_1\alpha^2)Tr(x_1x_4)Tr(x_2x_3)\\
&+ r_3 Tr(x_1)Tr(x_3)Tr(x_2x_4) - r_1(\alpha\beta^2 + \alpha^2\beta)Tr(x_1x_2x_3x_4)\\
&+ r_1 Tr(x_1)Tr(x_2)Tr(x_3)Tr(x_4) + r_2 Tr(x_1)Tr(x_2)Tr(x_3x_4)\\
&+ r_1\alpha\beta Tr(x_3)Tr(x_1x_2x_4) + r_1\alpha\beta Tr(x_4)Tr(x_1x_2x_3).
\end{aligned}$$

Conjecture 1 Let $\alpha, \beta \in F \setminus \{0\}$, $\alpha \neq \beta$. Then $\mathrm{Id}^{tr}(D_3^{t_{\alpha,\beta,0}})$ is generated, as a trace T-ideal, by the polynomials:

$$f_1 = [x_1, x_2], \qquad f_7(1,0,0), \qquad f_7(0,1,0), \qquad f_7(0,0,1).$$

The T^{tr}-ideal of identities of the algebra $D_3^{t_{\alpha,\alpha,\alpha}}$ is given in Theorem 1 for $n = 3$.

Theorem 8 *Let $\alpha \in F \setminus \{0\}$. The trace T-ideal $Id^{tr}(D_3^{t_{\alpha,\alpha,\alpha}})$ is generated, as a trace T-ideal, by the polynomials:*

1. $[x_1, x_2]$,
2. $C_3^{\alpha}(x_1, x_2, x_3)$.

We conclude this paper by dealing with the case of $D_3^{t_{\alpha,\beta,\beta}}$ and $D_3^{t_{\alpha,\beta,\gamma}}$, where α, β, γ are distinct non-zero elements of the field F. The situation in these two cases is much more complicated and we have only partial results.

First, let us consider the algebra $D_3^{t_{\alpha,\beta,\beta}}$. By Theorem 2 we already know that such an algebra does not satisfy any multilinear trace identity of degree 3 which is not a consequence of the commutator. Concerning the trace identities of degree 4 we have the following results.

Lemma 6 *The modified Cayley–Hamilton polynomial $C_4^{\beta}(x_1, x_2, x_3, x_4)$ is a trace identity of $D_3^{t_{2\beta,\beta,\beta}}$.*

Conjecture 2 The algebra $D_3^{t_{\alpha,\beta,\beta}}$, with $\alpha \neq 2\beta$, does not satisfy any multilinear trace identity of degree 4 which is not a consequence of the commutativity.

Finally, for the algebra $D_3^{t_{\alpha,\beta,\gamma}}$, where α, β, γ are distinct non-zero elements of the field F, we have no identities of degree 3 which are not a consequence of the commutator (see Theorem 2). Moreover, according to several computations made on Maxima, we can conjecture that the same happens also at degree 4.

Thanks to the software Maxima we have the following result.

Lemma 7 *Let $\alpha, \beta, \gamma \in F$ be three distinct non-zero elements. Then the algebra $D_3^{t_{\alpha,\beta,\gamma}}$ satisfies a trace identity of degree* 5.

Acknowledgments The authors express their thanks to Antonio Giambruno for his very useful and helpful comments and suggestions.

Antonio Ioppolo was partially supported by FAPESP grant No.2018/17464-3. Plamen Koshlukov was partially supported by CNPq grant No.302238/2019-0, and by FAPESP grant No.2018/23690-6. Daniela La Mattina was partially supported by GNSAGA-INDAM.

References

1. Aljadeff, E., Giambruno, A.: Multialternating graded polynomials and growth of polynomial identities. Proc. Am. Math. Soc. **141**, 3055–3065 (2013)
2. Aljadeff, E., Giambruno, A., La Mattina, D.: Graded polynomial identities and exponential growth. J. Reine Angew. Math. **650**, 83–100 (2011)
3. Berele, A.: Trace identities for diagonal and upper triangular matrices. Internat. J. Algebra Comput. **6**, 645–654 (1996)
4. dos Santos, R. B.: $*$-Superalgebras and exponential growth. J. Algebra **473**, 283–306 (2017)
5. Giambruno, A., Zaicev, M.: On codimension growth of finitely generated associative algebras. Adv. Math. **140**, 145–155 (1998)
6. Giambruno, A., Zaicev, M.: Exponential codimension growth of PI algebras: an exact estimate. Adv. Math. **142**, 221–243 (1999)
7. Giambruno, A., Zaicev, M.: Involutions codimensions of finite dimensional algebras and exponential growth. J. Algebra **222**, 474–484 (1999)
8. Giambruno, A., Mishchenko, S.: On star-varieties with almost polynomial growth. Algebra Colloq. **8**, 33–42 (2001)
9. Giambruno, A., La Mattina, D.: Graded polynomial identities and codimensions: computing the exponential growth. Adv. Math. **225**, 859–881 (2010)
10. Giambruno, A., Mishchenko, S., Zaicev, M.: Polynomial identities on superalgebras and almost polynomial growth. Commun. Algebra **29**, 3787–3800 (2001)

11. Giambruno, A., dos Santos, R. B., Vieira, A. C.: Identities of $*$-superalgebras and almost polynomial growth. Linear Multilinear Algebra **64**, 484–501 (2016)
12. Giambruno, A., Ioppolo, A., La Mattina, D.: Varieties of algebras with superinvolution of almost polynomial growth. Algebr. Represent. Theory **19**, 599–611 (2016)
13. Giambruno, A., Polcino Milies, C., Valenti, A.: Star-polynomial identities: computing the exponential growth of the codimensions. J. Algebra **469**, 302–322 (2017)
14. Giambruno, A., Ioppolo, A., La Mattina, D.: Superalgebras with involution or superinvolution and almost polynomial growth of the codimensions, Algebr. Represent. Theory **22**, 961–976 (2019)
15. Ioppolo, A.: The exponent for superalgebras with superinvolution. Linear Algebra Appl. **555**, 1–20 (2018)
16. Ioppolo, A.: A Characterization of Superalgebras with Pseudoinvolution of Exponent 2. Algebr. Represent. Theory (2020), https://doi.org/10.1007/s10468-020-09996-4
17. Ioppolo, A., La Mattina, D.: Polynomial codimension growth of algebras with involutions and superinvolutions, J. Algebra **472**, 519–545 (2017)
18. Ioppolo, A., Martino, F.: Varieties of algebras with pseudoinvolution and polynomial growth. Linear Multilinear Algebra **66**, 2286–2304 (2018)
19. Ioppolo, A., Martino, F.: Classifying G-graded algebras of exponent two. Israel J. Math. **229**, 341–356 (2019)
20. Ioppolo, A., Koshlukov, P., La Mattina, D.: Trace identities and almost polynomial growth. J. Pure Appl. Algebra 225(2), 106501 (2021)
21. Kemer, A.R.: Varieties of finite rank. In: Proceedings of the 15th All the Union Algebraic Conference (1979)
22. Procesi, C.: The invariant theory of $n \times n$ matrices. Adv. Math. **19**, 306–381 (1976)
23. Razmyslov, Ju.P.: Identities with trace in full matrix algebras over a field of characteristic zero. Izv. Akad. Nauk SSSR Ser. Mat. **38**, 723–756 (1974)
24. Regev, A.: Existence of identities in $A \otimes B$. Israel J. Math. **11**, 131–152 (1972)
25. Valenti, A.: Group graded algebras and almost polynomial growth. J. Algebra **334**, 247–254 (2011)

Codimension Growth for Weak Polynomial Identities, and Non-integrality of the PI Exponent

Plamen Koshlukov and David Levi da Silva Macêdo

Dedicated to Professor Antonio Giambruno on the occasion of his seventieth anniversary.

Abstract Let K be a field of characteristic zero. We study the asymptotic behavior of the codimensions for polynomial identities of representations of Lie algebras, also called weak identities. These identities are related to pairs of the form (A, L) where A is an associative enveloping algebra for the Lie algebra L. First we obtain a characterization of ideals of weak identities with polynomial growth of the codimensions in terms of their cocharacter sequence. Moreover we obtain examples of pairs that generate varieties of pairs of almost polynomial growth. Second we show that any variety of pairs of associative type is generated by the Grassmann envelope of a finitely generated superpair. As a corollary we obtain that any special variety of pairs which does not contain pairs of type (R, sl_2), consists of pairs with a solvable Lie algebra. Here sl_2 denotes the Lie algebra of the 2×2 traceless matrices. Finally we give an example of a pair that contradicts a conjecture due to Amitsur.

Keywords Polynomial identities · Codimension growth · Identities of representations of Lie algebras · PI exponent

P. Koshlukov (✉)
Department of Mathematics, State University of Campinas, Campinas, SP, Brazil
e-mail: plamen@unicamp.br

D. L. S. Macedo
Department of Mathematics, Federal Rural University of Pernambuco, Recife, Brazil
e-mail: david.levi@ufrpe.br

O. M. Di Vincenzo, A. Giambruno (eds.), *Polynomial Identities in Algebras*, Springer INdAM Series 44, https://doi.org/10.1007/978-3-030-63111-6_12

1 Introduction

Let K be a field of characteristic zero. A pair (A, L) where A is an associative enveloping algebra for the Lie algebra L is said to be an associative–Lie pair. If $\rho\colon L \to \mathfrak{gl}(V)$ is a representation of the Lie algebra L one obtains an associative–Lie pair $(A, \rho(L))$. Here A is the associative subalgebra of $End_K(V)$ generated by $\rho(L)$. The polynomial identities of the representation ρ are the identities of the pair $(A, \rho(L))$. Sometimes these are also called weak identities. They were introduced by Razmyslov in [25] in his research which led to the description of the identities of the matrix algebra $M_2(K)$. Razmyslov obtained a finite basis for the identities of the Lie algebra $sl_2(K)$, as well as the weak identities of the pair $(M_2(K), sl_2(K))$, and in this way he managed to describe the identities of $M_2(K)$.

Denote by P_n the vector space of all multilinear polynomials of degree n in the variables $x_1, \ldots, x_n$ in the free associative algebra $K\langle X\rangle$ freely generated over K by $X = \{x_1, x_2, \ldots\}$. As in the case of ordinary identities for associative or Lie algebras, since $char K = 0$ it suffices to study only multilinear polynomial identities of (A, L). If (A, L) is a pair and $Id(A, L)$ is its weak ideal, that is the ideal of its weak polynomial identities in $K\langle X\rangle$, then $Id(A, L)$ is generated as a weak ideal by the elements in $P_n \cap Id(A, L)$ for $n \geq 1$. The vector space P_n is a left module over the symmetric group S_n in a natural way, and it is isomorphic to the left regular S_n-module KS_n. Moreover $P_n \cap Id(A, L)$ is its submodule. It is more convenient to consider the factor module $P_n(A, L) = P_n/(P_n \cap Id(A, L))$ instead of $P_n \cap Id(A, L)$. Following this line one applies the theory of representations of the symmetric group to the study of weak identities, and in an equivalent form, the representation of the general linear group. Hence it is important to known the decomposition of $P_n(A, L)$ into irreducible modules, its character, the generators of the irreducible modules and so on. One of the most important numerical invariants of a pair satisfying a non-trivial weak identity is its codimension sequence $c_n(A, L) = \dim P_n(A, L)$.

In the same manner one defines the S_n-module $P_n(B)$ and its codimension sequence $c_n(B) = \dim P_n(B)$ for an associative or Lie algebra B. The exact computation of the associative, Lie, or weak codimensions is extremely difficult; it has been done for very few algebras and pairs. That is why one is led to study the asymptotic behavior of the codimensions. In the associative case, Regev in [26] proved that the codimension sequence of a PI-algebra A is exponentially bounded. Moreover, in [8, 9] Giambruno and Zaicev proved that the sequence $(c_n(A))^{1/n}$ converges, and its limit is always an integer, called the PI exponent of A. Concerning Lie algebras, Zaicev in [28] established that the exponent exists and is an integer for finite dimensional Lie algebras. For pairs of the form $(A, \rho(L))$ where ρ is a finite dimensional representation of L, Gordienko in [12] also proved the existence and integrality of the exponent. The above three results concerning the integrality of the PI exponent give positive answers to a conjecture due to S.

Amitsur (stated for the associative case only but easily reformulated for any classes of algebras).

Given a variety of pairs $\mathcal{V}$, the growth of $\mathcal{V}$ is the growth of the codimension sequence $c_n(\mathcal{V})$ associated with the weak ideal $Id(\mathcal{V})$. Of special interest are the varieties of polynomial growth, that is the varieties of pairs such that the sequence $c_n(\mathcal{V})$ is polynomially bounded. The definitions for varieties of associative or Lie algebras are the same. In the associative and Lie cases, characterizations of the varieties with polynomial growth of codimensions in terms of their cocharacter sequence are well known [2, 15, 22].

It is important to study varieties satisfying extremal properties. One such property is the almost polynomial growth: that is the varieties whose growth is not polynomial but every proper subvariety has polynomial growth. In the associative case, by the results of Kemer in [16], the only associative algebras generating varieties of almost polynomial growth are E and UT_2. Recall here that E is the Grassmann (or exterior) algebra of an infinite dimensional vector space, and UT_2 stands for the algebra of the 2×2 upper triangular matrices. For Lie algebras, Drensky in [4] proved that sl_2 generates a variety of almost polynomial growth, and this is the only known non-soluble Lie algebra with this property.

It is well known that if A is an associative PI algebra then its cocharacter is contained in some hook $H(d, l)$. In other words all irreducible S_n-modules that appear in the decomposition of $P_n(A)$ correspond to Young diagrams that are contained in the hook $H(d, l)$ for appropriate integers d and l. Recall that a partition $\lambda = (\lambda_1, \dots, \lambda_r) \vdash n$ of n corresponds to a Young diagram in the hook $H(d, l)$ whenever $\lambda_{d+1} \leq l$. The pair (A, L) is special whenever A is an associative PI algebra. The pair (A, L) is of associative type if its cocharacter is contained in a hook $H(d, l)$ for some d and l. Clearly a special pair is of associative type.

In this paper we study at first the polynomial growth for weak codimensions. We characterize the varieties of pairs of polynomial growth in terms of their cocharacter sequence. By the results of Gordienko in [12], such a characterization is already valid for identities of finite dimensional representations. Afterwards we are interested in the pairs formed by the algebras UT_2, E and M_2. Here and in what follows $M_2 = M_2(K)$ stands for the matrix algebra of order 2 over the base field K, and sl_2 is the Lie algebra of the traceless 2×2 matrices over K. Hence we obtain the codimensions, cocharacters and exponents for the pairs $(UT_2, UT_2^{(-)})$, $(E, E^{(-)})$, and (M_2, sl_2). Moreover we prove that they generate varieties of pairs of almost polynomial growth. Notice that these are pairs associated to representations of Lie algebras, see Examples 1 and 2 below.

Furthermore we establish a partial analogue of Kemer's theory for pairs. To this end we introduce the notion of a graded pair and of a superpair. For the latter we define its Grassmann envelope and relate the identities satisfied by a pair and by its Grassmann envelope. This enables us to deduce that if $\mathcal{V}$ is a variety of pairs consisting of pairs of associative type then there exists a superpair whose Grassmann envelope generates $\mathcal{V}$. As a corollary to the theorem we prove that if

$\mathcal{V}$ is a special variety of pairs over an algebraically closed field, and $\mathcal{V}$ does not contain any pair corresponding to a representation of the Lie algebra sl_2 then $\mathcal{V}$ is soluble.

As another consequence of the method we are able to construct an example of a pair whose PI exponent (if it exists) lies in the open interval (6, 7), and thus cannot be an integer.

2 Preliminaries

2.1 *Generalities*

Throughout K stands for a field of characteristic 0. All algebras, vector spaces, and tensor products we consider will be over K.

If A is an associative algebra then considering the vector space of A together with the Lie bracket $[a, b] = ab - ba$, $a, b \in A$ one obtains a Lie algebra $A^{(-)}$. An associative algebra A is said to be an *enveloping* algebra for a Lie algebra L if L is a Lie subalgebra of $A^{(-)}$ and A (as an associative algebra) is generated by its vector subspace L. The pair (A, L) where L is a Lie algebra and A is an enveloping algebra for L is called an *associative–Lie pair* (or simply pair). The notions of subpair, homomorphism of pairs and direct product of pairs are defined in the natural way. If A is an associative algebra then $(A, A^{(-)})$ is an associative–Lie pair. Another important example is obtained from a representation $\rho\colon L \to \mathfrak{gl}(V)$ of a Lie algebra L. We consider the associative-Lie pair $(A, \rho(L))$ where A is the associative subalgebra of $End_K(V)$ generated by the image $\rho(L)$. In this case we say that the pair is *associated* to the representation ρ.

Let L be a Lie algebra, we denote by $\mathrm{ad} : L \to \mathfrak{gl}(L)$ the adjoint representation $x \mapsto \mathrm{ad}\, x$. For our purpose in this case we consider the action on the right, that is if $x \in L$ we write $y \mathrm{ad}\, x = [y, x]$, $y \in L$. Moreover we use L' to denote the commutator subalgebra (or derived algebra) of L.

We denote by $K\langle X\rangle$ the free associative algebra on the countable set $X = \{x_1, x_2, \dots\}$ and by $\mathcal{L}(X)$ the Lie subalgebra of $K\langle X\rangle^{(-)}$ generated by the set X. By the well known theorem of Witt, $\mathcal{L}(X)$ is the free Lie algebra. Combining these two free algebras we obtain the pair $(K\langle X\rangle, \mathcal{L}(X))$ called the *free pair* generated by X. The polynomials in $K\langle X\rangle$ and $\mathcal{L}(X)$ are called *associative and Lie polynomials*, respectively.

2.2 *Polynomial Identities*

Let (A, L) be an associative–Lie pair. A polynomial $f = f(x_1, x_2, \dots, x_n) \in K\langle X\rangle$ is called a *(weak) identity* of the pair (A, L) if $f(a_1, a_2, \dots, a_n) = 0$ in the algebra A for every $a_1, a_2, \ldots, a_n \in L$. In this case we say that (A, L) satisfies f or that $f \equiv 0$ on (A, L). The set

$$Id(A, L) = \{f \in K\langle X\rangle \mid f \equiv 0 \text{ in } (A, L)\}$$

is a weak ideal, that is an ideal of $K\langle X\rangle$ invariant under all endomorphisms of the free pair $(K\langle X\rangle, \mathcal{L}(X))$. In other words if $f(x_1, \ldots, x_n) \equiv 0$ is an identity of (A, L) then $af(g_1, \ldots, g_n)b \equiv 0$ is also an identity of (A, L) for any $g_1, \ldots, g_n \in \mathcal{L}(X)$ and $a, b \in K\langle X\rangle$. Let $S \subseteq K\langle X\rangle$ be a non-empty set. The *weak ideal generated by* S, denoted by $\langle S\rangle^W$, is the intersection of all weak ideals of $K\langle X\rangle$ that contain S. We denote by $\mathcal{V} = \mathcal{V}(S)$ the variety of pairs determined by the set $\langle S\rangle^W$, that is the class of all pairs (A, L) such that $f \equiv 0$ on (A, L) for all $f \in S$, and we write $I = \langle S\rangle^W = Id(\mathcal{V})$. The relatively free pair in $\mathcal{V}$ is the pair $(K\langle X\rangle/I, \mathcal{L}(X)/\mathcal{L}(X)\cap I)$, it is denoted by $\mathcal{F}_X(\mathcal{V})$. If $Id(\mathcal{V}) = Id(A, L)$ then we say that $\mathcal{V}$ is generated by the pair (A, L) and we write $\mathcal{V} = var(A, L)$.

The weak identities of a pair $(A, \rho(L))$ associated to a representation ρ are called *identities of the representation* ρ and we denote its ideal of identities by $Id(\rho)$.

We list several important examples of pairs that are associated to representations.

Example 1 The pairs $(UT_2, UT_2^{(-)})$ and $(E, E^{(-)})$ are associated to representations. Indeed, for the pair $(UT_2, UT_2^{(-)})$ it is enough to consider the identity $Id\colon UT_2^{(-)} \to \mathfrak{gl}(V)$ where $\dim V = 2$.

For the pair $(E, E^{(-)})$ let $\rho\colon E^{(-)} \to \mathfrak{gl}(E)$ be the map given by $\rho(x)y = xy$ for every $x, y \in E$. It is a well defined injective homomorphism of Lie algebras. Moreover ρ also defines an injective homomorphism of associative algebras from E to $End_K(E)$. Hence the associative subalgebra generated by the image of ρ is $\rho(E^{(-)}) \simeq E$; it follows that $(E, E^{(-)})$ is associated to the (infinite dimensional) representation ρ.

Example 2 Given $\rho\colon sl_2 \to \mathfrak{gl}(V)$ a nontrivial representation of the Lie algebra sl_2, we have that $\rho(sl_2) \simeq sl_2$ since sl_2 is simple. Hence, identifying sl_2 with its image, we associate to this representation the pair (A, sl_2) where A is the associative subalgebra generated by sl_2. When $\dim V = 2$, we have the pair (M_2, sl_2); it is well known that the associated representation is irreducible.

The polynomial $[x \circ y, z] \equiv 0$ where $x \circ y = xy + yx$, is a weak identity for the pair (M_2, sl_2). In other words, we can say that $[x \circ y, z] \equiv 0$ is a polynomial identity of a representation of sl_2 of dimension 2. Moreover, when $char K = 0$, this identity generates all weak identities of the pair (M_2, sl_2), that is $Id(M_2(K), sl_2(K)) = \langle [x \circ y, z]\rangle^W$. In fact this identity generates $Id(M_2(K), sl_2(K))$ in the more general case when K is an infinite field of characteristic different from 2, see [18]. It follows from the results in [6] that in characteristic 0, this identity together with the standard polynomial of degree 4, St_4, generate $Id(M_2(K), sl_2(K))$ in a weaker sense: that is when one is allowed to substitute the variables of a polynomial by linear combinations of variables only, and to multiply an identity on both sides.

Analogously to the associative–Lie case we define the polynomial identities for associative and Lie algebras. For these cases we shall use similar notation for the ideal of identities. An (associative or Lie) algebra B satisfying a non-trivial identity is called a PI-algebra and its ideal of identities $Id(B)$ is a T-ideal, that is, it is invariant by endomorphisms of $K\langle X\rangle$ and of $\mathcal{L}(X)$, respectively. The associative or Lie varieties are defined in the same way as in the case of pairs and we shall use analogous notation.

2.3 Multilinear Polynomials. Modules over the Symmetric Group

As mentioned in Sect. 1 above, in studying the polynomial identities of an algebra (or a pair) over a field of characteristic 0 one can consider the multilinear polynomials only. We denote by P_n the vector space of the polynomials in $K\langle X\rangle$ which are multilinear in $x_1, \ldots, x_n$. The symmetric group acts on the left-hand side by permuting the variables and so P_n is a left S_n-module; it is isomorphic to the regular module KS_n. Thus we can identify P_n and KS_n as S_n-modules. If (A, L) is a pair with $Id(A, L)$ its ideal of weak identities, $P_n \cap Id(A, L)$ is a submodule of P_n, and the factor $P_n(A, L) = P_n/(P_n \cap Id(A, L))$ is also an S_n-module. We recall briefly some of the notation and facts we shall need from the representation theory of S_n. The interested reader could consult the monographs [5, Ch. 12] and [9, Ch. 2] for a detailed treatment of these topics. Here we stick to the notation from [9], and the notions not explicitly defined here can be found in [9, Ch. 2]. The facts we recall below are quite well known in the case of identities of an algebra. We shall need them in the context of pairs, and that is why we decided to include the main statements we shall need, as well as a good part of the terminology we will use. The proofs for pairs are literally the same, and that is why we omit them but cite the corresponding results for algebras. Let $\lambda \vdash n$ be a partition of n, that is $\lambda = (\lambda_1, \ldots, \lambda_r)$, $\lambda_1 \geq \cdots \geq \lambda_r$, and $\lambda_1 + \cdots + \lambda_r = n$. We denote by D_λ its Young diagram. It is an array of r rows, having λ_i boxes in its ith row and the rows are aligned at the left. The diagram $D_{\lambda'}$ is the conjugate (or transpose) of D_λ, it corresponds to the conjugate (transpose) partition λ' of λ. The Young tableau T_λ is obtained by filling in the boxes of D_λ with the integers from 1 to n without repetition. The tableau T_λ is standard if these integers increase along the rows and also along the columns. Let R_{T_λ} and C_{T_λ} stand for the row and for the column stabilizer of T_λ. Put $e_{T_\lambda} = \sum (-1)^\pi \sigma\pi \in KS_n$ where $\sigma \in R_{T_\lambda}$, $\pi \in C_{T_\lambda}$ and $(-1)^\pi$ is the sign of the permutation π. Then e_{T_λ} is a scalar multiple of an idempotent (a semi-idempotent) in the group algebra KS_n and the submodule $M_\lambda = KS_n e_{T_\lambda}$ is irreducible. Also if $\mu \vdash n$ then $M_\lambda \cong M_\mu$ if and only if $\lambda = \mu$.

The degree of the representation M_λ, $\dim M_\lambda$, equals the number of standard tableaux corresponding to the partition λ. It can be computed by using the hook formula as follows. Fix a box at position (i, j) in D_λ, then its hook number h_{ij}

equals the quantity of boxes of the ith line on the right of the given box, plus those of the jth column below that box, plus the box (i, j). Then $\dim M_\lambda = n!/\prod h_{ij}$ where (i, j) runs over all boxes of D_λ. We denote by d_λ the degree of the module M_λ and by χ_λ its character, then $\chi_\lambda(1) = d_\lambda$. Let $e_1, \dots, e_{d_\lambda}$ be the semi-idempotents obtained from the standard λ-tableaux. If M is a submodule in KS_n isomorphic to M_λ with semi-idempotent e then e is a linear combination of the e_i's. Moreover if M is an S_n-module with character χ then $\chi = \sum m_\lambda \chi_\lambda$ where the sum is over all $\lambda \vdash n$, and the multiplicities m_λ are nonnegative integers. Given two characters $\chi = n_1\chi_1 + n_2\chi_2 + \cdots + n_{q_1}\chi_{q_1}$ and $\theta = m_1\chi_1 + m_2\chi_2 + \cdots + m_{q_2}\chi_{q_2}$ we denote $\chi \subseteq \theta$ if $q_1 \le q_2$ and $n_i \le m_i$, $i = 1, \dots, q_1$.

We shall need also the notion of a hook. Let d and l be nonnegative integers, then $H(d, l)$ is the set of all partitions λ such that $\lambda = (\lambda_1, \lambda_2, \dots)$, and $\lambda_{d+1} \le l$. So drawing an infinite hook area having an "arm" of height d and a "leg" of width l, the diagrams of these λ lie in the hook. With certain abuse of notation we write $\lambda \in H(d, l)$ or $D_\lambda \subseteq H(d, l)$, and also $\chi_\lambda \subseteq H(d, l)$. Analogously if M is an S_n-module with character $\chi(M) = \sum m_\lambda \chi_\lambda$ then M (and $\chi(M)$) lies in $H(d, l)$ whenever $\lambda \in H(d, l)$ for each λ with $m_\lambda > 0$. The following two facts were obtained in [9, Lemmas 6.2.4, 6.2.5], their proofs consist in manipulations with the Hook formula and the Stirling approximation. If λ and μ are partitions then $\mu \le \lambda$ if $D_\mu \subseteq D_\lambda$, that is for each i one has $\mu_i \le \lambda_i$.

Lemma 1

1. *If $\lambda \vdash n$ and $\mu \vdash n'$ satisfy $\mu \le \lambda$, and if $n - n' \le c$ for some constant c then $d_\mu \le d_\lambda \le n^c d_\mu$.*
2. *Fix d and l two nonnegative integers, then there exist constants C and r such that $\sum d_\lambda \le Cn^r(d + l)^n$. Here the sum runs over all $\lambda \vdash n$ with $\lambda \in H(d, l)$. In particular, whenever $l = 0$, that is $\lambda \in H(d, 0)$, one has $d_\lambda \le Cn^r d^n$.*

As we already discussed above, we have a structure of an S_n-module on $P_n(A, L) = P_n/(P_n \cap Id(A, L))$. The same holds for the T-ideals $Id(A)$ of associative algebras A, and also for the T-ideals of Lie algebra $Id(L)$. In the latter case though one considers $\mathcal{L}(X) \subseteq K\langle X\rangle^{(-)}$.

We define the nth weak codimension of the pair (A, L) in analogy with the ordinary case as $c_n(A, L) = \dim P_n(A, L)$. The character of the S_n-module $P_n(A, L)$ is $\chi_n(A, L)$, it is the nth cocharacter of the pair (A, L). Hence $\chi_n(A, L) = \sum_{\lambda\vdash n} m_\lambda(A, L)\chi_\lambda$, here the $m_\lambda(A, L)$ are the multiplicities of the corresponding irreducible modules. If $\mathcal{V}$ is an associative or Lie variety with corresponding T-ideal $Id(\mathcal{V})$ then we define $c_n(\mathcal{V})$, $\chi_n(\mathcal{V})$ and $m_\lambda(\mathcal{V})$ as in the case of pairs. The following theorem can be found in [9, Theorem 2.4.5, p. 55], stated and proved for the case of T-ideals of associative algebras. Its proof holds word by word for the case of associative–Lie pairs.

Theorem 1 *Suppose $\chi_n(A, L) = \sum_{\lambda\vdash n} m_\lambda(A, L)\chi_\lambda$ is the nth cocharacter of the pair (A, L). Then $m_\lambda(A, L) = 0$ for some λ if and only if for each Young tableau T_λ corresponding to the diagram D_λ, and for each $f \in P_n$ one has $e_{T_\lambda} f = 0$ in the pair (A, L).*

An easy argument shows that if $\dim L = k < \infty$ then for each $\lambda = (\lambda_1, \lambda_2, \ldots)$ with $\lambda_{k+1} \geq 1$ (that is λ has at least $k+1$ rows) then $m_\lambda(A, L) = 0$.

Now let V_n denote the vector space $P_n \cap \mathcal{L}(X)$, that is the Lie multilinear elements of degree n, and let (A, L) be an associative–Lie pair. We shall use the notation $P_n(A) = P_n/(P_n \cap Id(A))$, $V_n(L) = V_n/(V_n \cap Id(L))$; these are S_n-modules. Their dimensions are the codimensions $c_n(A)$ and $c_n(L)$, respectively. We write $\chi_n(A) = \sum_{\lambda \vdash n} m_\lambda(A)\chi_\lambda$, and $\chi_n(L) = \sum_{\lambda \vdash n} m_\lambda(L)\chi_\lambda$ for the corresponding cocharacters. Put $\mathcal{V}_1 = var(A)$, $\mathcal{V}_2 = var(L)$, then denote $c_n(\mathcal{V}_1) = c_n(A)$, $\chi_n(\mathcal{V}_1) = \chi_n(A)$, and analogously for $\mathcal{V}_2$ and L.

Lemma 2 *For every $\lambda \vdash n$ the inequalities $m_\lambda(L) \leq m_\lambda(A, L) \leq m_\lambda(A)$ hold. It follows $c_n(L) \leq c_n(A, L) \leq c_n(A)$ and $\chi_n(L) \subseteq \chi_n(A, L) \subseteq \chi_n(A)$.*

Proof If $m_\lambda(L) = k$ then there exist k independent Lie polynomials in $V_n(L)$ that generate irreducible S_n-modules corresponding to λ. Using the equality $\mathcal{L}(X) \cap Id(A, L) = Id(L)$ one obtains that these k polynomials are independent modulo $Id(A, L)$, thus $m_\lambda(A, L) \geq k$. The inequality $m_\lambda(A, L) \leq m_\lambda(A)$ follows from a similar argument by using the inclusion $Id(A) \subseteq Id(A, L)$. □

2.4 The Exponent

If (A, L) is a pair such that $c_n(A, L)$ is exponentially bounded we denote

$$\underline{\exp}(A, L) = \liminf(c_n(A, L)^{1/n}), \qquad \overline{\exp}(A, L) = \limsup(c_n(A, L)^{1/n}),$$

and call these the lower and upper exponent of (A, L). If $\underline{\exp}(A, L) = \overline{\exp}(A, L)$ then this limit is called the (weak) exponent $\exp(A, L)$ of the pair. In the same way one defines the exponent $\exp(B)$ of an associative or Lie algebra B, and also the exponent of a variety of algebras or pairs, if the exponent exist. If the pair is of the form $(A, \rho(L))$ where ρ is a representation of L then we denote $\exp(A, \rho(L)) = \exp(\rho)$ whenever the exponent exist.

The celebrated theorem of Regev [26] implies that if A is associative and PI then $c_n(A)$ is exponentially bounded. We recall below three important theorems describing the exponent. All of them hold over a field of characteristic 0.

1. If A is an associative PI algebra then $\exp(A)$ exists and is an integer, see for example [9, Chapter 6].
2. If L is a finite dimensional Lie algebra then $\exp(L)$ exists and is an integer, see [28].
3. If $\rho\colon L \to \mathfrak{gl}(V)$ is a finite dimensional representation of a Lie algebra L then $\exp(\rho)$ exists and is an integer, see [12].

It follows from Lemma 2 that for a pair (A, L) one has $\exp(L) \leq \exp(A, L) \leq \exp(A)$ whenever the exponents exist.

The following easy fact will be used throughout without explicit mention.

Remark 1 Let A be an associative algebra and form the pair $(A, A^{(-)})$. Then $Id(A) = Id(A, A^{(-)})$. Therefore $c_n(A) = c_n(A, A^{(-)})$ and $\chi_n(A) = \chi_n(A, A^{(-)})$. If $Id(A, A^{(-)}) \neq \{0\}$ ($Id(A) \neq \{0\}$) it follows that both associative and weak exponents exist by Remark 1 above. In this case, we have that $\exp(A) = \exp(A, A^{(-)})$.

We draw the reader's attention that a basis (that is a generating set) of $Id(A)$ is not necessarily one for $Id(A, A^{(-)})$ (the rules for taking consequences in the pair are "weaker" than those in the algebra).

If $\mathcal{V}$ is a variety (of associative or Lie algebras, or of pairs) then the lower and upper exponents of $\mathcal{V}$, denoted by $\underline{\exp}(\mathcal{V})$ and $\overline{\exp}(\mathcal{V})$ respectively, are similarly defined considering $c_n(\mathcal{V})$. If they are equal we have the exponent of $\mathcal{V}$ denoted by $\exp(\mathcal{V})$.

Definition 1 A sequence $\{a_n\}_{n\in\mathbb{N}}$ has polynomial growth if there exist constants C and r such that $a_n \leq Cn^r$ for every $n \in \mathbb{N}$.

Let $\mathcal{V}$ be a variety (of associative or Lie algebras or of pairs). The growth of $\mathcal{V}$ is defined to be the growth of the sequence $\{c_n(\mathcal{V})\}_{n\in\mathbb{N}}$ of its codimensions. We say that $\mathcal{V}$ has *almost polynomial growth* if $\mathcal{V}$ is not of polynomial growth but any proper subvariety of $\mathcal{V}$ has polynomial growth.

We give some examples that are important for the next sections.

Example 3 ([19, 24]) For the infinite dimensional Grassmann algebra E the following conditions hold:

1. The T-ideal of identities $Id(E)$ is generated by the polynomial $[x_1, x_2, x_3] \equiv 0$;
2. $c_n(E) = 2^{n-1}$ and hence $\exp(E) = 2$;
3. $\chi_n(E) = \sum_{\lambda\in H(1,1)} \chi_\lambda$.

Here and in what follows we assume that long commutators without inner brackets are left normed, that is $[a, b, c] = [[a, b], c]$ and so on.

Example 4 ([20]) For the algebra $UT_2(K)$ of 2×2 upper triangular matrices over the field K the following conditions hold:

1. The T-ideal of identities $Id(UT_2)$ is generated by the polynomial $[x_1, x_2][x_3, x_4] \equiv 0$;
2. $c_n(UT_2) = 2^{n-1}(n-2) + 2$ and hence $\exp(UT_2) = 2$;
3. The cocharacter is given by

$$\chi_n(UT_2) = \sum_{\lambda\vdash n} m_\lambda(UT_2)\chi_\lambda$$

where $m_\lambda(UT_2) = q + 1$ if either

$$\lambda = (p+q, p),\ p \geq 1, q \geq 0 \quad \text{or} \quad \lambda = (p+q, p, 1),\ p \geq 1, q \geq 0.$$

In all remaining cases $m_\lambda(UT_2) = 0$, except for the case $m_{(n)}(UT_2) = 1$.

2.5 The Action of the General Linear Group

In the next sections we also work with the representation theory of the general linear group which is closely related to that of the symmetric group. To this end we introduce the vector space of homogeneous polynomials in a given set of variables. Let $U = span_K\{x_1, \ldots, x_m\}$ and let $K_m\langle X\rangle = K\langle x_1, \ldots, x_m\rangle$ be the free associative algebra in m variables. The group $GL(U) \simeq GL_m$ acts naturally on the left on the vector space U, as linear transformations, and we can extend this action diagonally to get an action of $GL(U)$ on $K_m\langle X\rangle$.

The vector space $K_m\langle X\rangle \cap Id(A, L)$ is invariant under this action, hence

$$K_m(A, L) = \frac{K_m\langle X\rangle}{K_m\langle X\rangle \cap Id(A, L)}$$

has a structure of a left GL_m-module. Let K_m^n be the vector space of homogeneous polynomials of degree n in the variables $x_1, \ldots, x_m$. Then

$$K_m^n(A, L) = \frac{K_m^n}{K_m^n \cap Id(A, L)}$$

is canonically isomorphic to a GL_m-submodule of $K_m(A, L)$ and we denote its character by $\psi_n(A, L)$. Write

$$\psi_n(A, L) = \sum_{\lambda\vdash n} \bar{m}_\lambda(A, L)\psi_\lambda$$

where ψ_λ is the irreducible GL_m-character associated to the partition λ and $\bar{m}_\lambda(A, L)$ is the corresponding multiplicity. Analogously as in [3] and [4], if the nth cocharacter of (A, L) has the decomposition $\chi_n(A, L) = \sum_{\lambda\vdash n} m_\lambda(A, L)\chi_\lambda$ then $m_\lambda(A, L) = \bar{m}_\lambda(A, L)$ for every $\lambda \vdash n$ whose corresponding diagram has height at most m (that is it has at most m rows).

It is also well known (see for instance [5, Theorem 12.4.12]) that any irreducible submodule of $K_m^n(A, L)$ corresponding to λ is generated by a non-zero polynomial f_λ, called highest weight vector, of the form

$$f_\lambda = f_\lambda(x_1, \ldots, x_q) = \prod_{i=1}^{\lambda_1} St_{h_i(\lambda)}(x_1, \ldots, x_{h_i(\lambda)}) \sum_{\sigma\in S_n} \alpha_\sigma \sigma \tag{1}$$

where $\alpha_\sigma \in K$, the right action of S_n on $K_m^n(A, L)$ is defined by place permutation, and $h_i(\lambda)$ is the height of the ith column of the diagram of λ. Here $St_p(x_1, \ldots, x_p) = \sum_{\sigma\in S_p}(-1)^\sigma x_{\sigma(1)} \cdots x_{\sigma(p)}$ is the standard polynomial of degree p: the alternating sum of monomials $x_{\sigma(1)} \cdots x_{\sigma(p)}$ where $\sigma \in S_p$, the symmetric group permuting $\{1, 2, \ldots, p\}$. Recall that f_λ is unique up to a multiplicative constant.

For a Young tableau T_λ, denote by f_{T_λ} the highest weight vector obtained from (1) by considering the only permutation $\sigma \in S_n$ such that the integers $\sigma(1), \ldots, \sigma(h_1(\lambda))$, in this order, fill in from top to bottom the first column of T_λ, $\sigma(h_1(\lambda)+1), \ldots, \sigma(h_1(\lambda)+h_2(\lambda))$ the second column of T_λ, and so on.

We also have the following equality (see for instance [5, Proposition 12.4.14]).

Remark 2 Suppose that

$$\psi_n(A, L) = \sum_{\lambda \vdash n} \bar{m}_\lambda \psi_\lambda$$

is the GL_m-character of $K_m^n(A, L)$. Then $\bar{m}_\lambda \neq 0$ if and only if there exists a tableau T_λ such that the corresponding highest weight vector f_{T_λ} is not a weak identity for (A, L). Moreover $\bar{m}_\lambda$ is equal to the maximal number of linearly independent highest weight vectors f_{T_λ} in $K_m^n(A, L)$.

2.6 Special Pairs and Pairs of Associative Type

Definition 2 An associative–Lie pair (A, L) is special whenever A is a PI algebra. A variety of pairs is special if it is generated by a special pair.

An associative–Lie pair (A, L) is of associative type if there exist nonnegative integers d and l such that $\chi_n(A, L) \subseteq H(d, l)$ for every $n \geq 1$. A variety of pairs is of associative type if it is generated by a pair of associative type.

Lemma 3

1. *If (A, L) is a special pair then it is of the associative type.*
2. *Let (A, L) be a special pair and let $\chi_n(A, L) = \sum_{\lambda \vdash n} m_\lambda(A, L)\chi_\lambda$ be its cocharacter. Assume that there exist a constant $C > 0$ and a positive integer k such that $m_\lambda(A, L) \neq 0$ whenever $n-(\lambda_1+\cdots+\lambda_k) \leq C$. Then $c_n(A, L) \leq n^t k^n$ for some $t > 0$, and for every n.*

Proof The first statement is immediate. As for the second, since (A, L) is special we have A is a PI algebra. Hence $\chi_n(A) = \sum_{\lambda \vdash n} m_\lambda(A)\chi_\lambda$ and there exist $C > 0$ and a positive integer k with $\sum_{\lambda \vdash n} m_\lambda(A) \leq Cn^t$, see [9, Theorem 4.9.3]. Since $m_\lambda(A, L) \leq m_\lambda(A)$ it follows $\sum_{\lambda \vdash n} m_\lambda(A, L)$ is polynomially bounded. By applying Lemma 1 and the Hook formula one gets that if $m_\lambda(A, L) \neq 0$ then $d_\lambda = \deg \chi_\lambda \leq n^t k^n$ for some $t > 0$, and the proof follows. □

3 Polynomial Growth of the Codimensions

3.1 Slow Growth of the Codimensions

In this section we shall see that some properties of polynomial growth in the Lie case do not hold in the associative–Lie case.

Let $\mathcal{A}$ be the variety of abelian Lie algebras and let $\mathcal{N}_t$ be the variety of nilpotent Lie algebras (of index of nilpotency $t+1$). We denote by $\mathcal{N}_t\mathcal{A}$ the variety of Lie algebras consisting of algebras with nilpotent commutator subalgebra with index of nilpotency $t+1$. This variety is the product of the varieties $\mathcal{A}$ and $\mathcal{N}_t$ (see [1]). Moreover, the varieties of this type are related to almost polynomial growth of the Lie codimensions. The following theorem is due to Mishchenko.

Theorem 2 ([23, Theorem 2.2, p. 33]) *Let $\mathcal{V}$ be a variety of Lie algebras. The following conditions are equivalent:*

1. *$\mathcal{V}$ has polynomial growth;*
2. *For some $s \in \mathbb{N}$, $\mathcal{N}_2\mathcal{A} \not\subseteq \mathcal{V} \subseteq \mathcal{N}_s\mathcal{A}$;*
3. *There exists a constant q such that*

$$\chi_n(\mathcal{V}) = \sum_{\substack{\lambda \vdash n \\ |\lambda|-\lambda_1 \le q}} m_\lambda(\mathcal{V})\chi_\lambda,$$

for every $n \geq 1$.

Take $t \in \mathbb{N}$. Analogously to the Lie case, we define $\mathcal{W}_t$, the variety of associative–Lie pairs defined by the identity

$$[[x_1, x_2], \ldots, [x_{2t+1}, x_{2t+2}]] \equiv 0.$$

It is the variety consisting of the pairs (A, L) where L has nilpotent commutator subalgebra of index $t+1$ ($L \in \mathcal{N}_t\mathcal{A}$). We shall see that the conditions (1) and (2) in the Theorem 2 are not equivalent for varieties of associative–Lie pairs.

Given $f = f(x_1, \ldots, x_n) \in K\langle X\rangle$ and L a Lie algebra, let

$$L_f = \{z \in L \mid zf(\operatorname{ad} y_1, \ldots, \operatorname{ad} y_n) = 0, \quad \text{for every} \quad y_1, \ldots, y_n \in L\}$$

where $y \operatorname{ad} x = [y, x]$ is the action corresponding to the adjoint representation.

Lemma 4 *Take $f = f(x_1, \ldots, x_n) \in K\langle X\rangle$ a multihomogeneous polynomial of degree at most 2 in each variable x_i, $i = 1, \ldots, n$. Then L_f is an ideal of L.*

Proposition 1 ([21]) *Let L be a Lie algebra and suppose that L satisfies the identity $[x_1, \ldots, x_n, y, y] \equiv 0$, for some $n \in \mathbb{N}$. Then L is nilpotent.*

Theorem 3 *Let $\mathcal{V}$ be a variety of associative–Lie pairs. Suppose there exists an integer $n \in \mathbb{N}$ such that $c_n(\mathcal{V}) < 2^{[\frac{n-1}{2}]}$ where $[a]$ denotes, as usual, the integer part of the real number a. Then*

$$\mathcal{V} \subseteq \mathcal{W}_c, \quad \textit{for some} \quad c \in \mathbb{N}. \tag{2}$$

Proof Given $k, r \in \mathbb{N}$ with $r \leq k$, consider the transpositions $\delta_i = (2i-1, 2i) \in S_{2k}$, $i = 1, \ldots, r$. Let $H_r^k = \langle \delta_1, \ldots, \delta_r \rangle$ be the subgroup of S_{2k} generated by δ_1, $\ldots$, δ_r. Note that H_r^k is an abelian 2-group.

Fix $n \in \mathbb{N}$ such that $c_n(\mathcal{V}) < 2^{[\frac{n-1}{2}]}$ and take $k = [\frac{n-1}{2}]$. We can view H_k^k as a subgroup of S_{n-1}. Consider the elements of $\mathcal{L}(X)$

$$[x_n, x_{\sigma(n-1)}, \ldots, x_{\sigma(1)}], \quad \sigma \in H_k^k.$$

We have $|H_k^k| = 2^k$. Hence, since $c_n(\mathcal{V}) < 2^k$, there exist $\lambda_\sigma \in K, \sigma \in H_k^k$, not all of them zero, such that

$$\sum_{\sigma \in H_k^k} \lambda_\sigma [x_n, x_{\sigma(n-1)}, \ldots, x_{\sigma(1)}] \equiv 0 \pmod{Id(\mathcal{V})}.$$

If n is odd, we have that $2k = n-1$ and we replace $x_n = [x, t]$. If n is even, we have that $2k+1 = n-1$ and $\sigma(2k+1) = 2k+1$ for all $\sigma \in H_k^k$. In this case we replace $x_n = x$ and $x_{\sigma(n-1)} = t$. In both cases, we obtain

$$f = \sum_{\sigma \in H_k^k} \lambda_\sigma [x, t, x_{\sigma(2k)}, \ldots, x_{\sigma(1)}] \equiv 0 \pmod{Id(\mathcal{V})}. \tag{3}$$

We distinguish two cases:

Case 1 Suppose $\lambda_\sigma = -\lambda_{\sigma\delta_k}$ for all $\sigma \in H_k^k$. Then, the Jacobi identity implies

$$[z_1, z_2, z_3, z_4] - [z_1, z_2, z_4, z_3] = [[z_1, z_2], [z_3, z_4]],$$

and we obtain from f that

$$\sum_{\sigma \in H_{k-1}^k} \lambda'_\sigma [[[[x, t], [x_{2k}, x_{2k-1}]], x_{\sigma(2k-2)}], \ldots, x_{\sigma(1)}] \equiv 0 \pmod{Id(\mathcal{V})},$$

for some $\lambda'_\sigma \in K$. Here we joined the terms in order to eliminate the permutations of H_k^k where δ_k appears. Hence we have $\lambda'_\sigma = \lambda_\sigma$ for all $\sigma \in H_{k-1}^k$.

Case 2 Suppose $\lambda_\sigma \neq -\lambda_{\sigma\delta_k}$ for some $\sigma \in H_k^k$. We have that two transpositions δ_i and δ_j, $i \neq j$, do not exchange the same integers. Hence, taking $x_{\sigma(2k)} = x_{\sigma(2k-1)} = y_1$ in each permutation $\sigma \in H_k^k$ of (3), we obtain that

$$\sum_{\sigma \in H_{k-1}^k} \lambda'_\sigma [[[x,t](\mathrm{ad}\, y_1)^2, x_{\sigma(2k-2)}], \ldots, x_{\sigma(1)}] \equiv 0 \pmod{Id(\mathcal{V})},$$

for some $\lambda'_\sigma \in K$, not all of them zero, because $\lambda'_\sigma = \lambda_\sigma + \lambda_{\sigma\delta_k}$ for at least one $\sigma \in H_{k-1}^k$.

Repeating the argument with $\delta_{k-1}, \ldots, \delta_1$, we obtain that $\mathcal{V}$ satisfies an identity of the form $[x,t]\bar{g}$, where $\bar{g}$ is an associative monomial g in ad $[x_{i+1}, x_i]$ or $(\mathrm{ad}\, y_i)^2$ of degree at most 2 in each variable.

Now we apply induction on the number of variables of degree 2 in g. Suppose there does not exist variable of degree 2, then

$$[x,t]\bar{g} = [[x,t], [x_{2k}, x_{2k-1}], \ldots, [x_2, x_1]] \equiv 0$$

and we obtain (2).

Suppose there exists at least one variable of degree 2, then we write $g = f_1 f_2$ with $f_1 = hz^2$ where z is variable of degree 2, and h has no variable of degree 2. Denote by $\bar{f_1}$ and $\bar{f_2}$ the corresponding evaluation of f_1 and f_2 by ad $[x_{i+1}, x_i]$ and/or $(\mathrm{ad}\, y_i)^2$, then we have $\bar{g} = \bar{f_1}\bar{f_2}$.

Take $(A, L) \in \mathcal{V}$ an arbitrary pair. Each variable in f_2 corresponding to a substitution by ad $[x_{i+1}, x_i] = [\mathrm{ad}\, x_{i+1}, \mathrm{ad}\, x_i]$ is replaced by a commutator $[z_{i+1}, z_i]$. Then we obtain, from f_2, an associative polynomial q in the variables z_i. Let $\bar{q}$ be the evaluation of q of the form $z_i \to \mathrm{ad}\, x_i$. We have that $\bar{q} = \bar{f_2}$. Hence $[x,t]\bar{f_1}\bar{q} = [x,t]\bar{f_1}\bar{f_2} \equiv 0$ in L and it follows that $\bar{L} = L/L_q$ satisfies the identity $[x,t]\bar{f_1} \equiv 0$, that is it satisfies

$$[x,t]\mathrm{ad}\,[x_{2k}, x_{2k-1}] \cdots \mathrm{ad}\,[x_p, x_{p-1}](\mathrm{ad}\, y_1)^2 \equiv 0$$

for some $p \in \mathbb{N}$. Then, by Proposition 1, it follows that $\bar{L}$ has nilpotent commutator subalgebra. Therefore there exists $r \in \mathbb{N}$ such that L satisfies

$$[x,t]\mathrm{ad}\,[z_1, z_2] \cdots \mathrm{ad}\,[x_{2r-1}, x_{2r}]\bar{q} \equiv 0,$$

and then

$$[x,t]\mathrm{ad}\,[z_1, z_2] \cdots \mathrm{ad}\,[x_{2r-1}, x_{2r}]\bar{f_2} \equiv 0,$$

which is a polynomial with fewer variables of degree 2. Applying induction the result follows. □

Corollary 1 *If $\mathcal{V}$ is a variety of associative–Lie pairs of polynomial growth then*

$$\mathcal{W}_2 \not\subseteq \mathcal{V} \subseteq \mathcal{W}_c, \quad \textit{for some} \quad c \in \mathbb{N}. \tag{4}$$

Here we observe that it is well known that in the case of varieties of associative PI algebras one has $\exp \mathcal{V} \leq 1$ if and only if $\mathcal{V}$ has polynomial growth. By repeating word by word the proof of [9, Theorem 7.2.2] one obtains that the statement holds for the case of varieties of associative–Lie pairs.

Example 5 Consider $\mathcal{V} = var(E, E^{(-)})$. We have that $[[x_1, x_2], [x_3, x_4], [x_5, x_6]] \in Id(\mathcal{V})$, then $Id(\mathcal{W}_2) \subseteq Id(\mathcal{V})$, i.e., $\mathcal{V} \subseteq \mathcal{W}_2$. Moreover, if $\mathcal{W}_2 \subseteq \mathcal{V}$, then $(UT_2, UT_2^{(-)}) \in \mathcal{V}$ and $(UT_2, UT_2^{(-)})$ satisfies the identity $[x_1, x_2, x_3] \equiv 0$, a contradiction. Hence, $\mathcal{V}$ satisfies (4) with $c = 2$. On the other hand, $\exp(\mathcal{V}) = \exp(E, E^{(-)}) = \exp(E) = 2$ and then $\mathcal{V}$ has no polynomial growth.

3.2 Characterizing Varieties of Pairs of Polynomial Growth

In this section we shall give a characterization of the varieties of polynomial growth through the behavior of their sequences of cocharacters.

Consider $I = Id(E, E^{(-)})$ and $J = \langle [x, y, z], [xy, y, z] \rangle^W$. The pair $(E, E^{(-)})$ satisfies the identities $[x, y, z] \equiv 0$ and $[xy, y, z] \equiv 0$, then $J \subseteq I$. Using analogous arguments of the associative case [9, Theorem 4.1.8], we obtain that $I = J$.

The next lemma is similar to the associative case.

Lemma 5 *The variety $\mathcal{V}$ of associative–Lie pairs satisfies a standard identity if and only if $(E, E^{(-)}) \notin \mathcal{V}$.*

Given an irreducible KS_n-module M and $\lambda = (\lambda_1, \ldots, \lambda_s) \vdash n$ a partition associated to this module, we define $H(\lambda) = H(M) = n - \lambda_1$, $h(\lambda) = h(M) = s$, and $H^-(\lambda) = H^-(M) = n - h(\lambda)$. In other words in the diagram D_λ, $H(\lambda)$ is the number of boxes below the first row, $h(\lambda)$ is the length of the first column and $H^-(\lambda)$ is the number of boxes outside the first column.

Let $\mathcal{V}$ be a variety of associative–Lie pairs. If $n \in \mathbb{N}$ is fixed, we have finitely many irreducible submodules in the decomposition of the KS_n-module $P_n/(P_n \cap Id(\mathcal{V}))$. Thus we can obtain a set $\mathbb{N}_0 \subseteq \mathbb{N}$ such that $\Gamma_\mathcal{V} = \{M_n \mid n \in \mathbb{N}_0\}$ is the set of all irreducible modules ($\neq 0$) in each decomposition for each $n \in \mathbb{N}$. Notice that if $\mathbb{N}_0$ is finite and then $\Gamma_\mathcal{V}$ is finite, we have $c_n(\mathcal{V}) = 0$ for n large enough.

Therefore we shall consider only varieties $\mathcal{V}$ such that $\mathbb{N}_0$ is infinite.

Lemma 6 *Let $\mathcal{V}$ be a variety of associative–Lie pairs and let $\chi_n(\mathcal{V}) = \sum_{\lambda \vdash n} m_\lambda \chi_\lambda$ be its cocharacter. Suppose $St_m \in Id(\mathcal{V})$ for some $m \in \mathbb{N}$. Given $n > m$ and $\lambda \vdash n$ we have that $m_\lambda = 0$ if $H^-(\lambda) = n - h(\lambda) < \frac{n}{m} - 1$.*

Proof Let M_λ be a non-zero irreducible S_n-module associated to some partition $\lambda = (\lambda_1, \ldots, \lambda_r)$ in the decomposition of $\chi_n(\mathcal{V})$. Then M_λ is generated by a multilinear polynomial g that alternates in λ_1 sets of variables. Consider C the set consisting of $h(\lambda) = s$ variables. The elements of C are the variables corresponding to the first column of λ. Hence they must alternate in g.

We shall show that every monomial of g contains at least one submonomial consisting of m variables in C. As the variables in C alternate, then g belongs to the weak ideal generated by St_m and the lemma will follow.

Suppose the latter assertion fails. Then g has at least one monomial of type

$$m_1 w_1 m_2 w_2 m_3 \cdots w_{k-1} m_k w_k m_{k+1}$$

where each w_i is a monomial in the variables of C of length at most $m-1$, $i = 1, \ldots, k$, and the m_j's are monomials in the variables outside of C. Moreover, if $j \in \{2, \ldots, k\}$, then m_j cannot be the empty word. Hence we obtain that

$$n - s \geq k - 1 \quad \text{and} \quad s \leq k(m-1). \tag{5}$$

By hypothesis, we have the inequalities

$$n > m(n - s + 1) \quad \text{and} \quad s > \frac{(m-1)n}{m} + 1 > \frac{(m-1)n}{m}. \tag{6}$$

Combining (5) and (6), we obtain

$$n > m(k - 1 + 1) = mk \quad \text{and} \quad \frac{(m-1)n}{m} < k(m-1)$$

and then $mk > n > mk$, a contradiction. □

The previous lemma says that if a variety satisfies a standard identity, then in the decomposition of its cocharacter the number of boxes outside the first column cannot be bounded. On the other hand, under some assumptions, we shall see that the same does not apply to the number of boxes below the first row.

Lemma 7 *Let $\mathcal{V}$ be a variety of associative–Lie pairs and consider the set $\Gamma_{\mathcal{V}} = \{M_n \mid n \in \mathbb{N}_0\}$ where $\mathbb{N}_0 \subseteq \mathbb{N}$ is an infinite set of positive integers. If $n - h(M_n) \to \infty$, $M_n \in \Gamma_{\mathcal{V}}$, and there exists $k \in \mathbb{N}$ such that* $\dim M_n < n^k$, *for every $M_n \in \Gamma_{\mathcal{V}}$, then the set $H = \{H(M_n) \mid M_n \in \Gamma_{\mathcal{V}}\}$ is finite.*

Proof Given $n \in \mathbb{N}_0$ let $\lambda = (\lambda_1, \ldots, \lambda_s) \vdash n$ be the partition associated to $M_n \in \Gamma_{\mathcal{V}}$. We put $a_n = \lambda_1$ and $b_n = h(\lambda) = s$.

Suppose that H is infinite. Then $H(M_n) \to \infty$, $M_n \in \Gamma_{\mathcal{V}}$. We shall consider two cases:

Case 1 Suppose that $\min\{a_n, b_n\}$ is not bounded. This means that the first row and the first column of the tableaux associated to the modules $M_n \in \Gamma_{\mathcal{V}}$ grow indefinitely when $n \in \mathbb{N}_0$.

For each $n \in \mathbb{N}_0$ and each $\lambda \vdash n$, consider the partition $\mu = (a_n, 1^{b_n-1}) \vdash p = a_n + b_n - 1$ and M_p the irreducible S_p-module associated with μ. By the hook formula [13, Theorem 20.1, p. 77], it follows that $\dim M_p = d_\mu \le d_\lambda = \dim M_n$ and

$$d_\mu = \frac{(a_n + b_n - 1)!}{(a_n + b_n - 1)(a_n - 1)!(b_n - 1)!} = \frac{(a_n + b_n - 2)!}{(a_n - 1)!(b_n - 1)!} = \binom{a_n + b_n - 2}{a_n - 1}.$$

Moreover, by the arithmetic–geometric means inequality, we have

$$(a_n + b_n - 2)^2 = ((a_n - 1) + (b_n - 1))^2 \ge 2(a_n - 1)(b_n - 1) > n.$$

Here the last inequality holds for a_n and b_n large enough. For these values of a_n and b_n, let $c = a_n - 1$ or $c = b_n - 1$ according as $\min\{a_n, b_n\} = a_n$ or $\min\{a_n, b_n\} = b_n$, respectively. Then

$$\dim M_n \ge \dim M_p = d_\mu \ge \frac{[\sqrt{n}]!}{(c!)^2}$$

which contradicts $\dim M_n < n^k$.

Case 2 Suppose there exists $t \in \mathbb{N}$ such that $\min\{a_n, b_n\} < t$ for all $n \in \mathbb{N}_0$. Without loss of generality we can assume that there exists an infinite subset $\mathbb{N}_0' \subseteq \mathbb{N}_0$ such that $b_n < t$, $n \in \mathbb{N}_0'$. Let $n \in \mathbb{N}_0'$ and let $\lambda \vdash n$ be a partition associated to the module $M_n \in \Gamma_\mathcal{V}$. We put $c_n = \lambda_2$ and we take $\mu = (a_n, c_n) \vdash q = a_n + c_n$ with associated irreducible module M_q. Notice that we can suppose $c_n \neq 0$, since H was supposed to be infinite. As in the previous case, we have that $\dim M_q = d_\mu \le d_\lambda = \dim M_n$ and $q \le n$. Moreover, once again by the hook formula,

$$d_\mu = \frac{(a_n + c_n)!}{\frac{(a_n+1)!c_n!}{a_n - c_n + 1}} = \frac{(a_n + c_n)!(a_n - c_n + 1)}{(a_n + 1)a_n!c_n!} \ge \frac{(a_n + c_n)!}{na_n!c_n!} = \binom{a_n + c_n}{c_n}\frac{1}{n}$$

Hence if $c_n^2 \ge \frac{n}{2}$ for some $n \in \mathbb{N}_0'$, then

$$\dim M_n = d_\lambda \ge d_\mu = \dim M_q > 2^{c_n - 1}\frac{1}{n} \ge \frac{1}{n}2^{\sqrt{\frac{n}{2}} - 1}$$

and we reach a contradiction.

Suppose finally $c_n^2 < \frac{n}{2}$ for every $n \in \mathbb{N}_0'$. As $b_n < t$, that is the number of rows of the tableaux associated with $M_n, n \in \mathbb{N}_0'$, is bounded by t, it follows $n - h(M_n) \to \infty$, $n \in \mathbb{N}_0'$. Hence as H is infinite, we must have $c_n \to \infty$. Then take $n \in \mathbb{N}_0'$ large enough such that $c_n > b_n$. We have that $a_n + c_n^2 > a_n + b_n c_n > n$ and then

$$a_n + c_n > a_n > n - c_n^2 > n - n/2 = n/2.$$

Consider $n \in \mathbb{N}_0'$ large enough such that $c_n > k+2$, then

$$\dim M_q \geq \binom{a_n + c_n}{c_n}\frac{1}{n} > \binom{[\frac{n}{2}]}{k+2}\frac{1}{n} > n^k,$$

and this is a contradiction once again.

In all cases we reach a contradiction, thus H must be finite. □

Theorem 4 *Let $\mathcal{V}$ be a variety of associative–Lie pairs. The following conditions are equivalent:*

1. *$\mathcal{V}$ has polynomial growth;*
2. *There exists a constant q such that*

$$\chi_n(\mathcal{V}) = \sum_{\substack{\lambda \vdash n \\ |\lambda| - \lambda_1 \leq q}} m_\lambda(\mathcal{V})\chi_\lambda,$$

for every $n \geq 1$.

Proof Suppose $\mathcal{V}$ has polynomial growth and $c_n(\mathcal{V}) < n^k$ for every $n \geq 1$ and for some fixed $k \in \mathbb{N}$. Consider the set $\Gamma_{\mathcal{V}}$ introduced above. We have that $\overline{\exp}(\mathcal{V}) \leq 1$. Then $(E, E^{(-)}) \notin \mathcal{V}$ since $\exp(E, E^{(-)}) = \exp(E) = 2$. By Lemma 5, it follows that $\mathcal{V}$ satisfies a standard identity.

Thus, by Lemma 6, if

$$\chi_n(\mathcal{V}) = \sum_{\lambda \vdash n} m_\lambda(\mathcal{V})\chi_\lambda \tag{7}$$

is the cocharacter of $\mathcal{V}$, then $n - h(M_n) \to \infty$, $M_n \in \Gamma_{\mathcal{V}}$. In other words the number of boxes outside the first column of the diagram of M_n is not bounded in the decomposition (7). Moreover, for each $M_n \in \Gamma_{\mathcal{V}}$ we have that $\dim M_n \leq c_n(\mathcal{V}) < n^k$.

By Lemma 7, we have that $H = \{H(M_n) \mid M_n \in \Gamma_{\mathcal{V}}\}$ is finite. This means that the number of boxes below the first row is bounded and then it is enough to take $q = \max H$.

The opposite implication follows a standard argument, see for example [9, Theorem 7.2.2, p. 169]. □

3.3 Almost Polynomial Growth

In this section we consider the pairs $(E, E^{(-)})$, $(UT_2, UT_2^{(-)})$ and (M_2, sl_2). As we saw in Examples 1 and 2, these pairs are related to certain natural representations of Lie algebras. In this section we obtain that these pairs generate varieties of almost polynomial growth. Hence the principal result of this section is the following theorem.

Theorem 5 *The associative–Lie pairs* $(E, E^{(-)})$, $(UT_2, UT_2^{(-)})$ *and* (M_2, sl_2) *generate varieties of pairs with almost polynomial growth of the codimensions.*

3.3.1 The Pair $(UT_2, UT_2^{(-)})$

We start with the pair $(UT_2, UT_2^{(-)})$. By Example 4, we have that $\exp(UT_2, UT_2^{(-)}) = \exp(UT_2) = 2$ and

$$\chi_n(UT_2, UT_2^{(-)}) = \sum_{\lambda \vdash n} m_\lambda(UT_2, UT_2^{(-)})\chi_\lambda$$

where $m_\lambda(UT_2, UT_2^{(-)}) = q + 1$ if either

1. $\lambda = (p + q, p)$, for all $p \geq 1, q \geq 0$, or
2. $\lambda = (p + q, p, 1)$, for all $p \geq 1, q \geq 0$.

In all remaining cases $m_\lambda(UT_2, UT_2^{(-)}) = 0$, with the exception of $m_{(n)}(UT_2, UT_2^{(-)}) = 1$.

Let $I = Id(UT_2, UT_2^{(-)})$ and $J = \langle [x_1, x_2][x_3, x_4] \rangle^W$. The pair $(UT_2, UT_2^{(-)})$ satisfies the identity $[x_1, x_2][x_3, x_4] \equiv 0$ and then $J \subseteq I$. Using analogous arguments of the associative case [9, Theorem 4.1.5], we obtain that $I = J$.

Fix $n \in \mathbb{N}$ and consider the S_n-module $P_n(UT_2, UT_2^{(-)})$. Using the decomposition of the cocharacter $\chi_n(UT_2, UT_2^{(-)})$, we shall determine the generators of the irreducible modules associated to partitions whose multiplicities are non zero. It turns out more convenient here to work with modules over the general linear group rather than with those over the symmetric group. (As one has at most three rows in the corresponding diagrams then passing to modules over the general linear group one can work with at most three variables though of higher degrees.) If $\lambda = (n)$, the corresponding highest weight vector $f_{T_\lambda} = x^n$ is not an identity of $(UT_2, UT_2^{(-)})$ since $f_{T_\lambda}(E_{11}) = E_{11} \neq 0$. Then x^n is the generator corresponding to $\lambda = (n)$.

Given $p \geq 1$ and $q \geq 0$, consider $\lambda = (p + q, p)$ and T_λ^i, $i = 0, \ldots, q$, the tableau

$i+1$	$i+2$	$\cdots$	$i+p-1$	$i+p+1$	1	$\cdots$	i	$i+2p+1$	$\cdots$	n
$i+p+2$	$i+p+3$	$\cdots$	$i+2p$	$i+p$						

We associate to T_λ^i the polynomial

$$a_{p,q}^{(i)}(y_1, y_2) = y_1^i \underbrace{\bar{y}_1 \cdots \tilde{y}_1}_{p-1}[y_2, y_1] \underbrace{\bar{y}_2 \cdots \tilde{y}_2}_{p-1} y_1^{q-i} \tag{8}$$

where "$-$" and "$\sim$" stand for alternation on the corresponding "labeled" variables. We shall prove that the $q+1$ polynomials $a_{p,q}^{(i)}(y_1, y_2)$, $i = 0, \ldots, q$, are linearly independent modulo I. Suppose they are linearly dependent, then there exist $\alpha_0, \ldots, \alpha_q \in K$, not all of them zero, such that

$$\sum_{i=0}^{q} \alpha_i a_{p,q}^{(i)} \equiv 0 \pmod{I}.$$

Take $t = \max\{i \mid \alpha_i \neq 0\}$, then

$$\alpha_t a_{p,q}^{(t)} + \sum_{i<t} \alpha_i a_{p,q}^{(i)} \equiv 0 \pmod{I}.$$

If we now substitute y_1 with $y_1 + y_3$, we obtain

$$\alpha_t (y_1+y_3)^t \underbrace{\overline{(y_1+y_3)} \cdots \widetilde{(y_1+y_3)}}_{p-1}[y_2, y_1+y_3] \underbrace{\bar{y}_2 \cdots \tilde{y}_2}_{p-1}(y_1+y_3)^{q-t} +$$

$$+ \sum_{i<t} \alpha_i (y_1+y_3)^i \underbrace{\overline{(y_1+y_3)} \cdots \widetilde{(y_1+y_3)}}_{p-1}[y_2, y_1+y_3] \underbrace{\bar{y}_2 \cdots \tilde{y}_2}_{p-1}(y_1+y_3)^{q-i}.$$

modulo I. Since K is an infinite field, it follows that all homogeneous components are still identities for $(UT_2, UT_2^{(-)})$. Let us consider the homogeneous component g of degree $t+p$ in y_1 and of degree $q-t$ in y_3. Making the substitution

$$y_1 = E_{11}, \; y_2 = E_{12} + E_{22}, \; y_3 = E_{22}$$

we obtain

$$0 = \alpha_t E_{11}(-E_{12})(E_{12}+E_{22})E_{22} = -\alpha_t E_{12},$$

which means $\alpha_t = 0$, a contradiction. Since $m_{(p+q,p)}(UT_2, UT_2^{(-)}) = q+1$, it follows that the $q+1$ polynomials of type (8) generate all distinct copies of the same irreducible module associated to $\lambda = (p+q, p)$ in the decomposition of $P_n(UT_2, UT_2^{(-)})$.

Fix $p \geq 1$ and $q \geq 0$, and consider $\lambda = (p+q, p, 1)$, and T_λ^i, $i = 0, \ldots, q$, the tableau

$i+p$	$i+1$	$\cdots$	$i+p-1$	1	$\cdots$	i	$i+2p+2$	$\cdots$	n
$i+p+1$	$i+p+3$	$\cdots$	$i+2p+1$						
$i+p+2$									

We associate to T_λ^i the polynomial

$$b_{p,q}^{(i)}(y_1, y_2, y_3) = y_1^i \underbrace{\hat{y}_1 \cdots \tilde{y}_1}_{p-1} \bar{y}_1 \bar{y}_2 \bar{y}_3 \underbrace{\hat{y}_2 \cdots \tilde{y}_2}_{p-1} y_1^{q-i} \tag{9}$$

where "$\wedge$", "$-$" and "$\sim$" stand for alternation on the corresponding "labeled" variables. Using the same arguments as in the previous case, we obtain that the $q+1$ polynomials $b_{p,q}^{(i)}(y_1, y_2, y_3)$, $i = 0, \ldots, q$, are linearly independent modulo I. Since $m_{(p+q,p,1)}(UT_2, UT_2^{(-)}) = q+1$, it follows that the $q+1$ polynomials of type (9) generate all distinct copies of the same irreducible module associated to $\lambda = (p+q, p, 1)$, in the decomposition of $P_n(UT_2, UT_2^{(-)})$.

Theorem 6 *Let $\mathcal{V}$ be a variety of associative–Lie pairs and suppose that $\mathcal{V} \subsetneq var(UT_2, UT_2^{(-)})$. Then there exists a constant N such that for every $n \in \mathbb{N}$ and $\lambda \vdash n$ we have that $m_\lambda(\mathcal{V}) \leq N$. Moreover $\mathcal{V}$ has polynomial growth.*

Proof Let $I = Id(UT_2, UT_2^{(-)})$. Since $\mathcal{V} \subsetneq var(UT_2, UT_2^{(-)})$, there exists $\lambda \vdash n$ such that $m_\lambda(\mathcal{V}) < m_\lambda(UT_2, UT_2^{(-)})$. Let $a_{p,q}^{(i)}$, $b_{p,q}^{(i)}$, $i = 0, \ldots, q$, be the polynomials constructed above. Then either

$$\sum_{i=0}^{q} \alpha_i a_{p,q}^{(i)} \equiv 0 \pmod{I}, \text{ and } \alpha_i \neq 0 \text{ for some } i; \tag{10}$$

or

$$\sum_{i=0}^{q} \beta_i b_{p,q}^{(i)} \equiv 0 \pmod{I}, \text{ and } \beta_i \neq 0 \text{ for some } i. \tag{11}$$

By the identity $[x_1, x_2][x_3, x_4]$ and replacing y_3 with $[y_2, y_1]$, from (11) we obtain the relation (10).

Suppose that

$$\sum_{i=0}^{q} \alpha_i y_1^i \bar{y}_1 \cdots \tilde{y}_1 [y_2, y_1] \bar{y}_2 \cdots \tilde{y}_2 y_1^{q-i} \equiv 0 \pmod{I}.$$

Let $J = Id(UT_2)$ be the associative T-ideal of UT_2. Since $J = I$, we have that

$$\sum_{i=0}^{q} \alpha_i y_1^i \bar{y}_1 \cdots \tilde{y}_1 [y_2, y_1] \bar{y}_2 \cdots \tilde{y}_2 y_1^{q-i} \equiv 0 \pmod{J}.$$

In what follows we shall use associative consequences, that is consequences of identities of associative algebras. We draw the readers' attention that these are admissible when we consider the identities modulo the T-ideal J.

Take $t = \max\{i \mid \alpha_i \neq 0\}$. By substituting y_2 with $y_3 + y_4$, the above polynomial becomes

$$\alpha_t y_1^t \bar{y}_1 \cdots \tilde{y}_1 [y_3 + y_4, y_1] \overline{(y_3 + y_4)} \cdots (\widetilde{y_3 + y_4}) y_1^{q-t}$$
$$+ \sum_{i<t} \alpha_i y_1^i \bar{y}_1 \cdots \tilde{y}_1 [y_3 + y_4, y_1] \overline{(y_3 + y_4)} \cdots (\widetilde{y_3 + y_4}) y_1^{q-i} \equiv 0 \pmod{J}.$$

We consider the homogeneous component g of degree 1 in y_4, we substitute y_3 with y_1^2 and y_4 with y_2 in g. Thus we obtain

$$f = f(y_1, y_2) = \alpha_t y_1^t \underbrace{\bar{y}_1 \cdots \tilde{y}_1}_{p-1} [y_2, y_1] \underbrace{\overline{y_1^2} \cdots \widetilde{y_1^2}}_{p-1} y_1^{q-t}$$
$$+ \sum_{i<t} \alpha_i y_1^i \underbrace{\bar{y}_1 \cdots \tilde{y}_1}_{p-1} [y_2, y_1] \underbrace{\overline{y_1^2} \cdots \widetilde{y_1^2}}_{p-1} y_1^{q-i} \equiv 0 \pmod{J}.$$

Let $N = \deg f = 3p + q - 1$. Expanding the alternators in the above polynomial, identifying $y_2 = [z, y_1]$ and using the equality $[[z, y_1], y_1] = zy_1^2 - 2y_1 z y_1 + y_1^2 z$, it follows that

$$\alpha_t y_1^{t+2p} z y_1^{N-t-2p} \equiv \sum_{i<t+2p} \gamma_i y_1^i z y_1^{N-i} \pmod{J}$$

for some coefficients $\gamma_i \in K$. Let $M = t + 2p$. Recall that $\alpha_t \neq 0$, then we rewrite the above equivalence as follows

$$y_1^M z y_1^{N-M} \equiv \sum_{i<M} \delta_i y_1^i z y_1^{N-i} \pmod{J}, \delta_i \in K. \tag{12}$$

We shall prove that $m_\lambda(\mathcal{V}) \leq N$, for every partition λ. By the cocharacter of $(UT_2, UT_2^{(-)})$, it is enough to consider the two cases $\lambda = (p + q, p)$ and $\lambda = (p + q, p, 1)$. Consider $\lambda = (p + q, p, 1)$. If $q < N$, we have nothing to prove, since $m_\lambda(UT_2, UT_2^{(-)}) = q + 1$. Suppose $q \geq N$. Then we can replace z with

$$\underbrace{\hat{y}_1 \cdots \check{y}_1}_{p-1} \bar{y}_1 \bar{y}_2 \bar{y}_3 \underbrace{\hat{y}_2 \cdots \check{y}_2}_{p-1}$$

and apply relation (12) to every polynomial $b_{p,q}^{(i)}(y_1, y_2, y_3)$ such that $i \geq M$. We obtain that

$$b_{p,q}^{(i)} \equiv \sum_{j<M} \delta_j b_{p,q}^{(j)} \pmod{J}$$

and, recalling that $J = I$, it follows that

$$b_{p,q}^{(i)} \equiv \sum_{j<M} \delta_j b_{p,q}^{(j)} \pmod{I}.$$

Therefore $m_\lambda(\mathcal{V}) \leq M - 1 \leq N$. The case $\lambda = (p+q, p)$ is analogous.

Finally we prove that the variety $\mathcal{V}$ has polynomial growth. Linearizing (12), we obtain

$$\sum_{\sigma \in S_N} y_{1\sigma(1)} \cdots y_{1\sigma(M)} z y_{1\sigma(M+1)} \cdots y_{1\sigma(N)} \equiv$$

$$\equiv \sum_{i<M} \sum_{\sigma \in S_N} \delta_i y_{1\sigma(1)} \cdots y_{1\sigma(i)} z y_{1\sigma(i+1)} \cdots y_{1\sigma(N)} \pmod{J}. \qquad (13)$$

We identify $z = [y_3, y_4]$, multiply (13) on the right by $y_{21} \cdots y_{2M}$ and alternate y_{1i} with y_{2i} for $i = 1, \ldots, M$. It follows that

$$\bar{y}_{11} \hat{y}_{12} \cdots \bar{y}_{1M} [y_3, y_4] \bar{y}_{21} \hat{y}_{22} \cdots \bar{y}_{2M} y_{1M+1} \cdots y_{1N} \equiv 0 \pmod{J}.$$

If we multiply on the left by $y_{2M+1} \cdots y_{2N}$ and alternate y_{1j} with y_{2j} for $j = M+1, \ldots, N$ we obtain

$$\bar{y}_{11} \hat{y}_{12} \cdots \bar{y}_{1N} [y_3, y_4] \bar{y}_{21} \hat{y}_{22} \cdots \bar{y}_{2N} \equiv 0 \pmod{J}.$$

Since $J = I$, it follows that

$$\bar{y}_{11} \hat{y}_{12} \cdots \bar{y}_{1N} [y_3, y_4] \bar{y}_{21} \hat{y}_{22} \cdots \bar{y}_{2N} \equiv 0 \pmod{I}.$$

This proves that if $\lambda = ((N+1)^2)$ then $m_\lambda(\mathcal{V}) = 0$.

Similarly, if we identify $z = \bar{y}_3 \bar{y}_4 \bar{y}_5$ in (13), by using the identity

$$\bar{y}_3 \bar{y}_4 \bar{y}_5 [z_1, z_2] \equiv 0,$$

we obtain the identity

$$\check{y}_{11} \hat{y}_{12} \cdots \bar{y}_{1N} \bar{y}_3 \bar{y}_4 \bar{y}_5 \check{y}_{21} \hat{y}_{22} \cdots \bar{y}_{2N} \equiv 0 \pmod{I}.$$

This proves that $\lambda = ((N+1)^2, 1)$, and thus $m_\lambda(\mathcal{V}) = 0$.

It follows that if λ is a partition of n such that $\lambda_2 \geq N+1$ then $m_\lambda(\mathcal{V}) = 0$. Therefore

$$\chi_n(\mathcal{V}) = \sum_{\substack{\lambda \vdash n \\ |\lambda|-\lambda_1 \leq N}} m_\lambda(\mathcal{V})\chi_\lambda.$$

and, by Theorem 4, $\mathcal{V}$ has polynomial growth. □

3.3.2 The Pair $(E, E^{(-)})$

We consider the pair $(E, E^{(-)})$ where E is the infinite dimensional Grassmann algebra. As we saw at the beginning of Sect. 3.2 we have that $Id(E, E^{(-)}) = \langle [x, y, z], [xy, y, z] \rangle^W$ and, by Example 3, $\exp(E, E^{(-)}) = \exp(E) = 2$ and $\chi_n(E, E^{(-)}) = \sum_{\lambda \in H(1,1)} \chi_\lambda$.

We obtain generators of the irreducible modules in the decomposition of the cocharacter of $(E, E^{(-)})$. Given $k \in \mathbb{N}$, take $\lambda = (k, 1^{n-k})$ a partition of n such that $m_\lambda(E, E^{(-)}) = 1 \neq 0$ in the decomposition of $\chi_n(E, E^{(-)})$. The highest weight vector

$$f_k = f_k(x_1, x_2, \ldots, x_{n-k+1}) = x_1^{k-1} St_{n-k+1}(x_1, x_2, \ldots, x_{n-k+1}). \tag{14}$$

is not an identity of $(E, E^{(-)})$ and generates the irreducible module associated to $\lambda = (k, 1^{n-k})$, for every $k \in \mathbb{N}$.

Theorem 7 *Let $\mathcal{V}$ be a variety of associative–Lie pairs and suppose $\mathcal{V} \subsetneq var(E, E^{(-)})$. Then $\mathcal{V}$ has polynomial growth.*

Proof Since $\mathcal{V}$ is a proper subvariety, it follows that $(E, E^{(-)}) \notin \mathcal{V}$. By Lemma 5, there exists $m \in \mathbb{N}$ such that $St_m \equiv 0$ is identity of $\mathcal{V}$. Thus we have that $St_{m+l} \in Id(\mathcal{V})$ for each $l \geq 1$.

By the form of the generators f_k in (14), we obtain $m_\lambda(\mathcal{V}) = 0$ for every partition $\lambda = (1^p)$ with $p \geq m$. Moreover, by multiplying each highest weight vector f_k associated to this partitions by a power of the variable x_1, it follows that if $\lambda = (k, 1^{n-k})$ then $m_\lambda(\mathcal{V}) = 0$ for $k \geq 1$ and $n-k \geq m$. Therefore the number of boxes below the first row in the decomposition of $\chi_n(\mathcal{V})$ is bounded by m. By applying Theorem 4 we obtain the statement. □

3.3.3 The Pair (M_2, sl_2)

We recall that the identity $[x \circ y, z] \equiv 0$ generates all weak identities of (M_2, sl_2). Moreover $m_\lambda(M_2, sl_2) = 1$ and $\lambda_4 = 0$ for every partition λ of n in the decomposition of the cocharacter of (M_2, sl_2), see [7]. In other words

$$\chi_n(M_2, sl_2) = \sum_{\lambda=(\lambda_1,\lambda_2,\lambda_3)\vdash n} \chi_\lambda. \tag{15}$$

We deduce properties of the decomposition of the cocharacter and of the codimension sequence of the pair (M_2, sl_2).

The next two lemmas can be found in [9, Sections 1.11 and 1.12] and [1, Section 6.4.2].

Lemma 8 *Let A be a semisimple associative PI-algebra over an algebraically closed field K and suppose that A satisfies a polynomial identity of degree d. Then A is a subdirect product of matrix algebras over the field K with order bounded by $d/2$.*

Lemma 9 *Let L be a Lie algebra over a field of characteristic zero K and suppose there exists a faithful and irreducible L-module V. Then L is abelian or contains some subalgebra isomorphic to sl_2.*

The next lemma is a natural property of the codimensions of (M_2, sl_2). This is valid for example for the Lie identities of sl_2.

Lemma 10 *For the pair (M_2, sl_2) we have $c_{n+1}(M_2, sl_2) \geq c_n(M_2, sl_2)$, for every $n \geq 1$.*

Proof We give a sketch of the proof. Let $I = Id(M_2, sl_2)$ and let $f_1, \dots, f_k$ be multilinear polynomials in $x_1, \dots, x_n$. Suppose $f_1, \dots, f_k$ are linearly independent modulo I, it suffices to prove that $f_1 x_{n+1}, \dots, f_k x_{n+1}$ are independent modulo I. Form a linear combination of the latter polynomials; put $x_{n+1} = h \in sl_2$, that is the diagonal matrix with entries 1 and -1 on the diagonal. Since h is invertible we can cancel it and obtain a linear combination for the f_i. The f_i were chosen linearly independent hence our linear combination is trivial. □

Lemma 11 *The exponent of (M_2, sl_2) exists. More precisely* $\exp(M_2, sl_2) = 3$.

Proof The algebra M_2 satisfies the standard identity St_4. Therefore the pair (M_2, sl_2) is special and $\chi_n(M_2, sl_2) \subseteq H(3, 0)$, $n \geq 1$. Hence by Lemma 3 we have that

$$c_n(M_2, sl_2) \leq n^t 3^n, \ n \geq 1, \tag{16}$$

for some $t > 0$.

On the other hand, the polynomial

$$g_k = St_3(x_1^1, x_2^1, x_3^1) \cdots St_3(x_1^k, x_2^k, x_3^k)$$

of degree $3k$ is not a weak identity for (M_2, sl_2) for any $k \geq 1$. By considering the action of S_{3k} on P_{3k}, we obtain that g_k generates an irreducible S_{3k}-module with cocharacter χ_λ, $\lambda = (k, k, k)$. According to [9, Lemma 5.10.1, p. 139], one has the inequality $d_\lambda \geq 3^{3k}/(3k)^3$. Therefore $c_n(M_2, sl_2) \geq 3^n/n^3$ for every $n \geq 1$, by Lemma 10.

Combining this inequality with (16) we complete the proof of the lemma. □

Let $\lambda = (p+q+r, p+q, p) \vdash n = 3p+2q+r$ be a partition in (15). The polynomial

$$f_{T_\lambda}(x_1, x_2, x_3) = St_3(x_1, x_2, x_3)^p St_2(x_1, x_2)^q x_1^r = \\ = \underbrace{\bar{x}_1\bar{x}_2\bar{x}_3 \cdots \bar{x}_1\bar{x}_2\bar{x}_3}_{p} \underbrace{[x_1, x_2] \cdots [x_1, x_2]}_{q} x_1^r \tag{17}$$

generates an irreducible module associated to λ (see [7]).

Codimensions of representations do not change upon an extension of the base field. The proof is analogous to the cases of codimensions of associative [9, Theorem 4.1.9] and Lie algebras [28, Section 2]. Thus without loss of generality we may assume K to be algebraically closed.

Example 6 ([27]) Let $\rho : sl_2 \to \mathfrak{gl}(V)$ be a representation of the Lie algebra sl_2. Suppose that ρ is faithful, irreducible and of finite dimensional n. Then the polynomial

$$\delta x_4 St_3(\text{ad } x_1, \text{ad } x_2, \text{ad } x_3) - x_4 St_3(x_1, x_2, x_3) \equiv 0 \tag{18}$$

is an identity for ρ, where $\delta = \frac{n^2-1}{8}$ and *ad* denotes the adjoint representation. In particular, for the pair (M_2, sl_2) we have the identity

$$\frac{3}{8} x_4 St_3(\text{ad } x_1, \text{ad } x_2, \text{ad } x_3) - x_4 St_3(x_1, x_2, x_3) \equiv 0. \tag{19}$$

Lemma 12 *Let $\rho : sl_2 \to \mathfrak{gl}(V)$ be a finite dimensional representation of the Lie algebra sl_2. Consider the pair (A, sl_2) corresponding to ρ and $V = V_1 \oplus \cdots \oplus V_l$ the decomposition of V in a direct sum of irreducibles. If $Id(M_2, sl_2) \subseteq Id(A, sl_2)$, then each V_i, $i = 1, \ldots, l$, corresponds to a representation of dimension 2 and $Id(A, sl_2) = Id(M_2, sl_2)$.*

Proof Let ρ_i be the irreducible representation corresponding to V_i, $i = 1, \ldots, n$. Since sl_2 is simple, ρ_i is faithful, $i = 1, \ldots, n$. By identities (18) and (19), it follows that faithful irreducible representations of sl_2 of dimension greater than 2 can not satisfy all identities of (M_2, sl_2). Moreover, $Id(A, sl_2) \subseteq Id(\rho_i)$, $i = 1, \ldots, l$. Hence if dim $V_i \geq 3$ for some $i \in \{1, \ldots, r\}$ we obtain a contradiction, as in this case

$$Id(M_2, sl_2) \subseteq Id(A, sl_2) \subseteq Id(\rho_i).$$

Therefore V decomposes into a sum of irreducible representations of dimension 2. Consequently $Id(M_2, sl_2) = Id(A, sl_2)$. □

Theorem 8 *Let $\mathcal{V}$ be a variety of associative–Lie pairs and suppose that $\mathcal{V} \subsetneq var(M_2, sl_2)$. Then*

$$St_3(x, y, z)^M = \underbrace{\bar{x}\bar{y}\bar{z} \cdots \tilde{x}\tilde{y}\tilde{z}}_{M\ triples} \equiv 0$$

and

$$\underbrace{[x, y] \cdots [x, y]}_{2M\ commutators} \equiv 0$$

are identities of $\mathcal{V}$, for some $M \in \mathbb{N}$.

Proof Let $I = Id(\mathcal{V})$ be the ideal of weak identities of $\mathcal{V}$. Consider

$$(A, L) = (K\langle X\rangle/I, \mathcal{L}(X)/\mathcal{L}(X) \cap I)$$

the relatively free pair in the variety $\mathcal{V}$. We have that

$$Id(M_2) \subseteq Id(M_2, sl_2) \subsetneq I.$$

Since M_2 is an associative PI-algebra, there exists a non zero polynomial $f = f(x_1, \ldots, x_n) \in K\langle X\rangle$ such that $f(g_1, \ldots, g_n) \in Id(M_2) \subseteq I$, for every $g_1, \ldots, g_n \in K\langle X\rangle$. Therefore, $A = K\langle X\rangle/I$ is an associative PI-algebra.

Since in the decomposition of the cocharacter of (M_2, sl_2) we have at most three rows in each Young tableau, let us consider the subpair (A_1, L_1) of (A, L) generated by three elements x, y, z. In other words, (A_1, L_1) is the relatively free pair in $\mathcal{V}$ of rank equal to 3. Since A_1 is a finitely generated associative PI-algebra, it follows that its Jacobson radical $J = J(A_1)$ is a nilpotent ideal. Moreover, $A_2 = A_1/J$ is semisimple and, by Lemma 8, is a subdirect product of matrix algebras $M_{n_\gamma}(K)$, $\gamma \in \Gamma$, over the field K. Furthermore the sizes of these matrix algebras are bounded. For each $\gamma \in \Gamma$, let V_γ be the vector space such that $End_K(V_\gamma) \simeq M_{n_\gamma}(K)$. Notice that $M_{n_\gamma}(K)$ acts irreducibly and faithfully in V_γ. Consider the quotient $L_2 = L_1/(L_1 \cap J)$. Thus, given $\gamma \in \Gamma$, the image of L_2 (by the projection π_γ of the subdirect product) in each M_{n_γ} acts irreducibly and faithfully in V_γ, since A_2 is generated by L_2. By Lemma 9, this image is abelian or contains a subalgebra isomorphic to sl_2. In the second case, we obtain a subpair (R, sl_2) of (A_1, L_1) corresponding to a finite dimensional representation of sl_2, then

$$Id(M_2, sl_2) \subseteq Id(\mathcal{V}) = Id(A_1, L_1) \subseteq Id(R, sl_2)$$

By Lemma 12, we must have equality in the above inclusions, a contradiction, since $\mathcal{V}$ is a proper subvariety of $var(M_2, sl_2)$. Therefore, the image of L_2 in each component M_{n_γ} is abelian, $\gamma \in \Gamma$, and it follows that the derived algebra L_1' of L_1 is contained in J.

Now it is enough to notice that $\bar{x}\bar{y}\bar{z}$ belongs to the associative ideal of A_1 generated by L_1' and then $(\bar{x}\bar{y}\bar{z})^M = 0$, where $M \in \mathbb{N}$ is such that $J^M = 0$. Therefore,

$$St_3(x, y, z)^M = \underbrace{\bar{x}\bar{y}\bar{z} \cdots \bar{x}\bar{y}\bar{z}}_{M\ triples} \equiv 0 \tag{20}$$

is an identity for the relatively free pair (A, L) and, consequently, is an identity of $\mathcal{V}$.

For the second identity notice that $[x^2, y] \equiv 0$ is an identity for the pair (M_2, sl_2) and then

$$x[x, y] + [x, y]x = x^2y - xyx + xyx - yx^2 \equiv 0 \tag{21}$$

$$y[x, y] + [x, y]y = yxy - y^2x + xy^2 - yxy \equiv 0 \tag{22}$$

are also identities for (M_2, sl_2). By (21) and (22), it follows that

$$\begin{aligned} \bar{x}\bar{y}\overline{[x, y]} &= \bar{x}\bar{y}[x, y] - \bar{x}[x, y]\bar{y} + [x, y]\bar{x}\bar{y} = \\ &= [x, y][x, y] - x[x, y]y + y[x, y]x + [x, y][x, y] \equiv \\ &\equiv [x, y][x, y] + [x, y]xy - [x, y]yx + [x, y][x, y] = 3[x, y][x, y] \end{aligned} \tag{23}$$

modulo $Id(M_2, sl_2) \subseteq Id(\mathcal{V})$. Taking $z = [x, y]$ and using (20) and (23), we obtain

$$0 \equiv \underbrace{\bar{x}\bar{y}\overline{[x, y]} \cdots \bar{x}\bar{y}\overline{[x, y]}}_{M\ triples} \equiv 3^M \underbrace{[x, y][x, y] \cdots [x, y][x, y]}_{2M\ commutators}.$$

modulo $Id(\mathcal{V})$ and the result follows.

□

Corollary 2 *Let $\mathcal{V}$ be a variety of associative–Lie pairs and suppose that $\mathcal{V}$ is a proper subvariety of $var(M_2, sl_2)$. Then $\mathcal{V}$ is of polynomial growth.*

Proof By Theorem 8, there exists $M \in \mathbb{N}$ such that

$$St_3(x, y, z)^M = \underbrace{\bar{x}\bar{y}\bar{z} \cdots \bar{x}\bar{y}\bar{z}}_{M\ triples} \equiv 0$$

and

$$St_2(x, y)^{2M} = \underbrace{[x, y] \cdots [x, y]}_{2M\ commutators} \equiv 0$$

are identities of $\mathcal{V}$. By the form of generators of the irreducible modules in (17), these identities imply that if $\lambda = (\lambda_1, \lambda_2, \lambda_3)$ is a partition such that $\lambda_2 - \lambda_3 \geq 2M$ or $\lambda_3 \geq M$, then $m_\lambda(\mathcal{V}) = 0$ in the decomposition of the cocharacter of $\mathcal{V}$. Thus the multiplicities $m_\lambda(\mathcal{V}) \neq 0$ correspond to partitions such that $\lambda_3 < M$ and $\lambda_2 - \lambda_3 < 2M$. Therefore

$$\chi_n(\mathcal{V}) = \sum_{\substack{\lambda \vdash n \\ |\lambda| - \lambda_1 \leq C}} m_\lambda(\mathcal{V}) \chi_\lambda$$

where $C = 4M$. The result follows from Theorem 4. □

3.3.4 More Examples

We construct here two examples of pairs in which the Lie algebras of these pairs have almost polynomial growth, but the pairs do not have the same property.

The Lie algebra of the first example appears in [9, Example 12.3.13, p. 318]. We use the same notation as in [9]. Let $L = span_K\{h, e\}$ be the soluble nonabelian Lie algebra of dimension 2 with multiplication $[h, e] = e$. Consider the left L-action on the polynomial ring $K[t]$ defined by

$$h(f) = tf', \qquad e(f) = tf \tag{24}$$

where f' is the usual derivation in t. Then the vector space

$$B = L + K[t] = span_K\{h, e, 1, t, t^2, \dots\}$$

is an infinite dimensional Lie algebra if we define the multiplication as follows:

$$[\lambda h + \mu e + f, \alpha h + \beta e + g] = (\lambda\beta - \mu\alpha)e + \lambda h(g) + \mu e(g) - \alpha h(f) - \beta e(f).$$

In particular, $K[t]$ is abelian ideal of B of codimension 2 and B is soluble. Note also that $B' = span_K\{e\} + K[t]$ is a non-nilpotent Lie algebra.

Theorem 9

1. *(see [23, Theorem 5.4] or else [9, Theorem 12.4.4, p. 324]) The exponent of B exists and we have* $\exp(B) = 2$. *Moreover B generates a Lie variety of almost polynomial growth.*
2. *Let (A, B) be an associative–Lie pair where B is the Lie algebra just constructed. Then $\mathcal{V} = var(A, B)$ has a proper subvariety $\mathcal{W}$ of non-polynomial growth.*

Our second example is based on the Lie algebra given in [9, Example 12.3.14]. Let $H = span_K\{x, y, z\}$ be the Heisenberg algebra, that is H is the three-dimensional Lie algebra with basis x, y, z, and multiplication defined by $[x, y] = z$, all other commutators among the basis elements are zero. Consider the left H-action on $K[t]$ given by

$$x(f) = f',\ y(f) = tf,\ z(f) = f$$

and define the multiplication on

$$C = H + K[t] = span_K\{x, y, z, 1, t, t^2, \ldots\}$$

as follows

$$[\alpha x + \beta y + \gamma z + f, \lambda x + \mu y + \nu z + g] =$$
$$(\alpha\mu - \beta\lambda)z + \alpha x(g) + \beta y(g) + z(g) - \lambda x(f) - \mu y(f) - z(f) =$$
$$(\alpha\mu - \beta\lambda)z + \alpha g' - \lambda f' + \beta tg - \mu tf + f - g.$$

Then C becomes a soluble infinite dimensional Lie algebra. Moreover, $C' = span_K\{z\} + K[t]$ is not a nilpotent algebra.

Theorem 10

1. *(see [23, Theorem 5.4], see also [9, Theorem 12.4.4, p. 324]) The exponent of C exists and we have* $\exp(C) = 3$. *Moreover, C generates a Lie variety with almost polynomial growth.*
2. *Let (A, C) an associative–Lie pair where C is the above Lie algebra. Then $\mathcal{V} = var(A, C)$ has a proper subvariety $\mathcal{W}$ of non-polynomial growth.*

4 Graded Pairs and Amitsur's Conjecture

4.1 Weak Graded Polynomial Identities

If G is a group and A an algebra (not necessarily associative) then A is G-graded whenever $A = \oplus_{g\in G} A_g$. Here A_g are vector subspaces of A such that $A_g A_h \subseteq A_{gh}$ for every g, $h \in G$. We shall need only the case $G = \mathbb{Z}_2$ where $\mathbb{Z}_2$ denotes the cyclic group of order 2, and accordingly we stick to the additive notation. Then $A = A_0 \oplus A_1$; the elements of $A_0 \cup A_1$ are homogeneous. We call the elements of A_0 even and those of A_1 odd elements. The G-degree of a homogeneous element a will be denoted by $|a| \in G$. A vector subspace (subalgebra, ideal) B of A is homogeneous (or graded) if $B = (B \cap A_0) \oplus (B \cap A_1)$.

The free associative and Lie algebras have a natural $\mathbb{Z}_2$-grading. We write $X = Y \cup Z$, a disjoint union of infinite sets, and declares the variables from Y of degree 0, and those from Z of degree 1. This is extended to the monomials in $K\langle X\rangle$ and thus the latter becomes the free $\mathbb{Z}_2$-graded associative algebra. This construction transfers to $\mathcal{L}(X)$. In this case we also denote $K\langle X\rangle$ by $K\langle Y, Z\rangle$ and $\mathcal{L}(X)$ by $\mathcal{L}(Y, Z)$. The Grassmann algebra is one of the most widely used $\mathbb{Z}_2$-graded algebra. The natural grading on it is defined as $E = E_0 \oplus E_1$. Here E_i is the span of all basic monomials of length $k \equiv i \pmod 2$, $i = 0, 1$. It is clear that E_0 is the centre of E and E_1 is the "anti-commuting" part of E.

Assume that $A = A_0 \oplus A_1$ is a (not necessarily associative) $\mathbb{Z}_2$-graded algebra, then the Grassmann envelope $G(A)$ of A is the algebra $G(A) = (A_0 \otimes E_0) \oplus (A_1 \otimes E_1)$. This leads immediately to the notion of a superalgebra. Assume $\mathcal{V}$ is a variety of (not necessarily associative) algebras. A $\mathbb{Z}_2$-graded algebra A is called a $\mathcal{V}$-superalgebra if $G(A) \in \mathcal{V}$. Pay attention that one does not require $A \in \mathcal{V}$. If $\mathcal{V}$ is the variety of all associative algebras then an associative superalgebra is just a $\mathbb{Z}_2$-graded associative algebra. When $\mathcal{V}$ is the variety of all Lie algebras, if L is $\mathbb{Z}_2$-graded, $L = L_0 \oplus L_1$ then $G(L) \in \mathcal{V}$ if and only if L satisfies the super-forms of the anticommutativity $[a, b] = (-1)^{|a||b|}[b, a]$, and of the Jacobi identity:

$$(-1)^{|a||c|}[a, [b, c]] + (-1)^{|a||b|}[b, [c, a]] + (-1)^{|b||c|}[c, [a, b]] = 0,$$

for every homogeneous elements $a, b, c \in L$. Thus L_0 is a Lie algebra and L_1 is a module over L_0.

The notions of subalgebra, ideal and factor algebra of a superalgebra are defined in the natural way. Also solubility and nilpotence for superalgebras are defined in the canonical way. If $A = A_0 \oplus A_1$ is an associative superalgebra then defining the super-bracket $[a, b] = ab - (-1)^{|a||b|}ba$ on the homogeneous elements of A yields a Lie superalgebra denoted by $A^{(\sim)}$. If (A, L) is a pair (not necessarily associative–Lie) then (A, L) is said to be $\mathbb{Z}_2$-graded whenever A is an associative $\mathbb{Z}_2$-graded algebra and L is a homogeneous vector subspace of A which generates A as an associative algebra.

The free object in the class of associative–Lie $\mathbb{Z}_2$-graded pairs is the pair $(K\langle X\rangle, \mathcal{L}(X))$ equipped with the $\mathbb{Z}_2$-grading commented above. As in the case of $\mathbb{Z}_2$-graded associative (Lie) algebras one defines a graded identity for a $\mathbb{Z}_2$-graded associative–Lie pair. Let $(A, L) = (A_0 \oplus A_1, L_0 \oplus L_1)$ be a $\mathbb{Z}_2$-graded associative–Lie pair. A polynomial $f(x_1, \ldots, x_n) \in K\langle X\rangle$ is a $\mathbb{Z}_2$-graded (or simply graded) weak identity for (A, L) if $f(a_1, \ldots, a_n) = 0$ in A for every $a_i \in L_0 \cup L_1$ where $a_i \in L_{|x_i|}$, $i = 1, \ldots, n$. In other words f vanishes on (A, L) when one makes substitutions respecting the grading. We denote by $Id_2(A, L)$ the ideal of graded weak identities for the associative–Lie $\mathbb{Z}_2$-graded pair (A, L).

The $\mathbb{Z}_2$-graded pair $(A, L) = (A_0 \oplus A_1, L_0 \oplus L_1)$ is a superpair if L is a sub(super)algebra of the Lie superalgebra $A^{(\sim)}$. The notions of associative–Lie homogeneous (or $\mathbb{Z}_2$-graded) subpair and sub(super)pair are defined in the natural way. With certain abuse of notation we shall use the terms subalgebra and subpair, omitting "super" when this causes no confusion.

As in the ordinary case one defines a free object in the case of superpairs. Let $K\langle X\rangle$ be the free associative superalgebra and let $\mathcal{L}(X)$ be the Lie superalgebra of $K\langle X\rangle^{(\sim)}$ generated by X. Note that we are using the same notation for the $\mathbb{Z}_2$-graded Lie algebra and the superalgebra generated by the set X. Here we assume

$X = Y \cup Z$, a disjoint union of infinite countable sets as above where Y are the even and Z the odd variables, respectively. As in the case of associative–Lie $\mathbb{Z}_2$-graded pairs one defines weak identities for the superpair (A, L), also called graded weak identities; we shall also denote $Id_2(A, L)$ the ideal of weak identities for the superpair. A variety of associative–Lie ($\mathbb{Z}_2$-graded) pairs and a supervariety of superpairs are defined exactly in the same way as in the ordinary case.

Let $\mathcal{V}$ be a supervariety of superpairs and put $I = Id_2(\mathcal{V})$ the ideal of the weak identities for all superpairs in $\mathcal{V}$. Then I is an ideal of graded weak identities and it is closed under endomorphisms of $K\langle X\rangle$ which respect the superstructures of $K\langle X\rangle$ and also of $\mathcal{L}(X)$. The superpair $\mathcal{F}_{Y,Z}(\mathcal{V}) = (K\langle Y, Z\rangle/I, \mathcal{L}(Y, Z)/\mathcal{L}(Y, Z) \cap I)$ is free in $\mathcal{V}$, it is the relatively free pair in $\mathcal{V}$. Here we consider $\mathcal{L}(Y, Z)$ as the Lie superalgebra in $K\langle Y, Z\rangle^{(\sim)}$ generated by $Y \cup Z$. This construction is analogous to the case of a variety of $\mathbb{Z}_2$-graded associative–Lie pairs.

Now let $E = E_0 \oplus E_1$ be the Grassmann algebra with its canonical $\mathbb{Z}_2$-grading. Let $A = A_0 \oplus A_1$ and $L = L_0 \oplus L_1$ be an associative and a Lie superalgebra, respectively. The Grassmann envelope $G(A) = (A_0 \otimes E_0) \oplus (A_1 \otimes E_1)$ is an associative algebra while $G(L) = (L_0 \otimes E_0) \oplus (L_1 \otimes E_1)$ is a Lie algebra. We stress that $G(L)$ is a Lie algebra. Suppose that (A, L) is a superpair and denote $G_A(L)$ the associative subalgebra of $A \otimes E$ generated by $G(L)$. This implies $G_A(L)$ is spanned by products of elements $(l_1 \otimes x_1) \cdots (l_k \otimes x_k) = l_1 \cdots l_k \otimes x_1 \cdots x_k$. Here the $x_j \in E_0 \cup E_1$ are homogeneous elements in E and $l_j \in L_0 \cup L_1$, moreover $|l_j| = |x_j|$ for every j. Hence $|l_1 \cdots l_k| = |x_1 \cdots x_k|$. Now the spanning set of $G_A(L)$ can be split into two subsets $G_A(L)_i \subseteq A_i \otimes E_i$, $i = 0, 1$, and it follows $G_A(L) = G_A(L)_0 \oplus G_A(L)_1$. Thus $G_A(L)$ is a homogeneous (associative) subalgebra of $G(A)$, and $G(L)$ becomes a homogeneous subspace of $G_A(L)$. A direct verification shows that $G(L)$ is a Lie subalgebra of $G_A(L)^{(-)}$. It follows $(G_A(L), G(L))$ is an associative–Lie pair which is $\mathbb{Z}_2$-graded. We call it the Grassmann envelope of the superpair (A, L) and we denote it by $G(A, L)$.

In the opposite direction, let (A, L) be an associative–Lie pair, and put $B = (A \otimes E_0) \oplus (A \otimes E_1)$, $M = (L \otimes E_0) \oplus (L \otimes E_1)$. Then one writes down the products in B and in M and obtains that (B, M) is a superpair.

4.2 Finitely Generated Superpairs

We denote $P_{k,m}$ the vector space of the polynomials in $K\langle Y, Z\rangle$ which are multilinear in the variables $y_1, \ldots, y_k \in Y$, and $z_1, \ldots, z_m \in Z$. If $f \in P_{k,m}$ then it can be written as follows:

$$f = \sum \alpha_{\sigma,W} w_0 z_{\sigma(1)} w_1 z_{\sigma(2)} \cdots w_{m-1} z_{\sigma(m)} w_m.$$

Here $W = (w_0, w_1, \ldots, w_m)$ is a sequence of monomials in the variables $y_1, \ldots, y_k$, $\alpha_{\sigma,W} \in K$, and $\sigma \in S_m$ where S_m is the symmetric group permuting $\{1, 2, \ldots, m\}$.

Let

$$\widetilde{f}=\sum(-1)^{\sigma}\alpha_{\sigma,W}w_0z_{\sigma(1)}w_1z_{\sigma(2)}\cdots w_{m-1}z_{\sigma(m)}w_m\in P_{k,m}$$

where $\sigma\in S_m$ and $(-1)^{\sigma}$ is the sign of the permutation σ. Clearly the map $f\mapsto\widetilde{f}$ defines an isomorphism on the vector space $P_{k,m}$, and moreover $\widetilde{\widetilde{f}}=f$. Recall that in the case of associative algebras this automorphism was introduced and used extensively by Kemer, see for example [17]. We follow the treatment given to it in [9, pp. 81–82, 110–112] with the adaptations to our case.

Lemma 13 *Let $f\in P_{k,m}$ and let $(A,L)=(A_0\oplus A_1,L_0\oplus L_1)$ be a superpair. Then f is a graded identity for $G(A,L)$ if and only if $\widetilde{f}$ is a graded identity for (A,L).*

If $\mathcal{V}$ is a variety of associative–Lie pairs we denote $\mathcal{V}^*$ the class of all superpairs (A,L) such that $G(A,L)\in\mathcal{V}$.

Lemma 14 *The class $\mathcal{V}^*$ is a supervariety.*

We observe that the above lemma can be obtained by applying directly the theorem of Birkhoff which is also valid in the case of pairs, that is a class of pairs is a variety if and only if it is closed under subpairs, direct products of pairs and homomorphic images.

Since $\mathcal{V}^*$ is a supervariety it has relatively free superpairs. If $\mathcal{V}$ is a variety of associative–Lie pairs we denote $\mathcal{F}=\mathcal{F}_{Y,Z}(\mathcal{V}^*)$ the relatively free superpair in the supervariety $\mathcal{V}$ freely generated by the even variables $Y=\{u_1,\ldots,u_k\}$ and the odd ones $Z=\{z_1,\ldots,z_m\}$.

Proposition 2 *Let $f=f(y_1^1,\ldots,y_1^{p_1},\ldots,y_k^1,\ldots,y_k^{p_k},z_1^1,\ldots,z_1^{q_1},\ldots,z_m^1,\ldots,z_m^{q_m})$ be a multilinear polynomial in the given variables. Suppose f is symmetric in each set of variables $\{y_i^1,\ldots,y_i^{p_i}\}$, $1\le i\le k$, and is skew-symmetric in each $\{z_j^1,\ldots,z_j^{q_j}\}$, $1\le j\le m$. Form the polynomial $\widetilde{f}$ considering the y as even and the z as odd variables. If*

$$\widetilde{f}(\underbrace{u_1,\ldots,u_1}_{p_1},\ldots,\underbrace{u_k,\ldots,u_k}_{p_k},\underbrace{w_1,\ldots,w_1}_{q_1},\ldots,\underbrace{w_m,\ldots,w_m}_{q_m})=0 \tag{25}$$

in $\mathcal{F}$ then f is a weak identity in $\mathcal{V}$.

Proof The polynomial $\widetilde{f}(y_1^1,\ldots,y_1^{p_1},\ldots,y_k^1,\ldots,y_k^{p_k},z_1^1,\ldots,z_1^{q_1},\ldots,z_m^1,\ldots,z_m^{q_m})$ is symmetric in each set of variables $\{y_i^1,\ldots,y_i^{p_i}\}$, $1\le i\le k$, and also in each set $\{z_j^1,\ldots,z_j^{q_j}\}$, $1\le j\le m$, according to the definition of $\widetilde{f}$. Since $u_1,\ldots,u_k$ and $w_1,\ldots,w_m$ are the free generators of $\mathcal{F}$, the equality (25) means that the polynomial

$$\widetilde{f}(\underbrace{y_1,\ldots,y_1}_{p_1},\ldots,\underbrace{y_k,\ldots,y_k}_{p_k},\underbrace{z_1,\ldots,z_1}_{q_1},\ldots,\underbrace{z_m,\ldots,z_m}_{q_m})$$

is a graded identity for $\mathcal{V}^*$.

Let $S = (S_A, S_L)$ be the relatively free pair in $\mathcal{V}$ freely generated by $\{\overline{y}_i^j, \overline{z}_i^j \mid i, j \in \mathbb{N}\}$. Form the superpair (B, M) where $B = B_0 \oplus B_1 = (S_A \otimes E_0) \oplus (S_A \otimes E_1)$ and $M = M_0 \oplus M_1 = (S_L \otimes E_0) \oplus (S_L \otimes E_1)$. For the Grassmann envelope $G(B, M)$ of the latter superpair we have $G(B, M) = (G_A(M), G(M))$, and

$$G(M) = (M_0 \otimes E_0) \oplus (M_1 \otimes E_1) \subseteq S_L \otimes R,$$

$$G_A(M) \subseteq G(B) = (B_0 \otimes E_0) \oplus (B_1 \otimes E_1) \subseteq S_A \otimes R.$$

Here $R = (E_0 \otimes E_0) \oplus (E_1 \otimes E_1)$ is a commutative and associative algebra. Hence if a multilinear polynomial g vanishes as an element of S_A when evaluated on S_L, then g vanishes under any substitution with elements from $S_L \otimes R$. Here we consider the latter evaluation inside $S_A \otimes R$. Therefore g is a weak identity for $G(B, M)$. In this way we have that $G(B, M)$ satisfies all weak identities from S, hence $G(B, M) \in \mathcal{V}$. This implies $(B, M) \in \mathcal{V}^*$, and thus

$$\widetilde{f}(\underbrace{c_1, \ldots, c_1}_{p_1}, \ldots, \underbrace{c_k, \ldots, c_k}_{p_k}, \underbrace{d_1, \ldots, d_1}_{q_1}, \ldots, \underbrace{d_m, \ldots, d_m}_{q_m}) = 0 \tag{26}$$

for every choice of $c_i \in M_0$, $d_i \in M_1$. We substitute

$$c_i = \overline{y}_i^1 \otimes a_i^1 + \cdots + \overline{y}_i^{p_i} \otimes a_i^{p_i}, \quad d_j = \overline{z}_j^1 \otimes b_j^1 + \cdots + \overline{z}_j^{q_j} \otimes b_j^{q_j}$$

where $1 \le i \le k$, $1 \le j \le m$, a_i^t are monomials from E_0 written on distinct generators, and $b_j^t \in E_1$ are monomials also written in distinct generators of E, so that the product of all a_i^t and all b_j^t is nonzero.

Now we compute $\widetilde{f}$ from Eq. (26). We have $(a_i^t)^2 = (b_j^t)^2 = 0$. Also $\widetilde{f}$ is symmetric in each of the sets $\{y_i^1, \ldots, y_i^{p_i}\}$ and in each of the sets $\{z_j^1, \ldots, z_j^{q_j}\}$. Thus we obtain

$$\widetilde{f}(c_1, \ldots, d_m) = p_1! \cdots q_m! \widetilde{\widetilde{f}}(\overline{y}_1^1, \ldots, \overline{y}_1^{p_1}, \ldots, \overline{z}_1^1, \ldots, \overline{z}_m^{q_m}) \otimes a_1^1 \cdots a_1^{p_1} \cdots b_1^1 \cdots b_m^{q_m}.$$

But $\widetilde{\widetilde{f}} = f$, the product of factorials is nonzero, and the rightmost product is (up to a sign) a basic element of E which is also nonzero. Therefore $f(\overline{y}_1^1, \ldots, \overline{z}_m^{q_m})$ is a weak identity for $\mathcal{F}$. Since the $\overline{y}_i^t$ and $\overline{z}_j^t$ are free generators of S, the relatively free pair in $\mathcal{V}$, it follows f is a weak identity for $\mathcal{V}$, and the proposition is proved. □

Now we have the necessary ingredients in order to obtain the following theorem.

Theorem 11 *Let $\mathcal{V}$ be a nontrivial variety of associative–Lie pairs. If $\mathcal{V}$ is of associative type then there exists a superpair $(A, L) = (A_0 \oplus A_1, L_0 \oplus L_1)$ such that $\mathcal{V} = var(G(A, L))$.*

Proof Since $\mathcal{V}$ is of associative type there exist k, m with $\chi_n(\mathcal{V}) \subseteq H(k, m)$ for every n. As in Proposition 2 we form the supervariety $\mathcal{V}^*$ and its relatively free superpair $\mathcal{F} = \mathcal{F}_{Y,Z}(\mathcal{V}^*)$. Here $Y = \{u_1, \ldots, u_k\}$ and $Z = \{w_1, \ldots, w_m\}$ are the even and odd free generators of $\mathcal{F}$, respectively. We claim that $\mathcal{V}$ is generated by the Grassmann envelope $G(\mathcal{F})$.

It follows from the definition of $\mathcal{V}^*$ that $G(\mathcal{F}) \in \mathcal{V}$ hence it suffices to prove that $\mathcal{V}$ satisfies all identities for $G(\mathcal{F})$. Suppose f is a multilinear identity of degree n for $G(\mathcal{F})$. By [9, Theorem 2.4.7], the identity f is equivalent to a collection of identities of the form $e_{T_\lambda} g = 0$. Here $g = g(x_1, \ldots, x_n)$ is multilinear, $\lambda \vdash n$, and T_λ is a tableau corresponding to λ. Thus we shall consider f of the form $e_{T_\lambda} g$.

If $\lambda \notin H(k, m)$ then $f \in Id(\mathcal{V})$ since $\chi_n(\mathcal{V}) \subseteq H(k, m)$. So we suppose $\lambda \in H(k, m)$. By [9, Lemma 2.5.6], we can take f to be symmetric in some $k' \leq k$ sets of variables $\{y_i^1, \ldots, y_i^{p_i}\}$, $1 \leq i \leq k'$, and skew-symmetric in the $m' \leq m$ sets $\{z_j^1, \ldots, z_j^{r_j}\}$, $1 \leq j \leq m'$. In order to simplify the notation we take $k' = k$ and $m' = m$. Write f as

$$f = f(y_1^1, \ldots, y_1^{p_1}, \ldots, y_k^1, \ldots, y_k^{p_k}, z_1^1, \ldots, z_1^{q_1}, \ldots, z_m^1, \ldots, z_m^{q_m}),$$

then f satisfies the statement of Proposition 2. Recall we assume the y_i^p as even variables and z_j^q as odd ones. In this way we consider f as a graded identity of $G(\mathcal{F})$. By Lemma 13 the superpair $\mathcal{F}$ satisfies the graded identity $\widetilde{f}$. This implies

$$\widetilde{f}(\underbrace{u_1, \ldots, u_1}_{p_1}, \ldots, \underbrace{u_k, \ldots, u_k}_{p_k}, \underbrace{w_1, \ldots, w_1}_{q_1}, \ldots, \underbrace{w_m, \ldots, w_m}_{q_m}) = 0.$$

Now by Proposition 2 f is an identity for $\mathcal{V}$, and the proof is complete. □

We want to describe varieties of special pairs which do not contain representations of sl_2. First we state a pair of results.

Lemma 15 *Let $\mathcal{V} = var(B, M)$ be a variety of associative–Lie pairs where (B, M) is a special pair. Then each pair in $\mathcal{V}$ is special. In particular, as $\mathcal{V} = var(G(A, L))$ for some superpair $(A, L) = (A_0 \oplus A_1, L_0 \oplus L_1)$ we have that $G_A(L)$ is a PI algebra.*

The motivation for the following proposition is a fact deduced easily by Regev's theorem concerning the tensor product of two PI algebras, see [26]. If $B = B_0 \oplus B_1$ is PI and a $\mathbb{Z}_2$-graded algebra then $B \otimes E$ is also PI (as E is PI, satisfying the identity $[x_1, x_2, x_3] = 0$ the tensor product $B \otimes E$ is also PI by Regev's theorem). It follows the Grassmann envelope $G(B)$ is also PI. Put $J = Id(G(B))$, $I = Id(B \otimes E)$, then one has $I \neq 0$ and $I \subseteq J$.

Proposition 3 *Let $(A, L) = (A_0 \oplus A_1, L_0 \oplus L_1)$ be a superpair such that $G(A, L)$ is a special pair. Then A is a PI algebra.*

Theorem 12 *Suppose the base field is algebraically closed (and of characteristic 0). Let $\mathcal{V}$ be a special variety of associative–Lie pairs such that $(R, sl_2) \notin \mathcal{V}$ for* *every pair (R, sl_2) associated to a representation of sl_2. Then $\mathcal{V}$ is a soluble variety of pairs.*

Proof As $\mathcal{V}$ is special it is of associative type, we have that $\mathcal{V} = var(G(A, L))$ where $(A, L) = (A_0 \oplus A_1, L_0 \oplus L_1)$ is a superpair of finite rank. Without loss of generality we can assume (A, L) is relatively free. Thus A is a finitely generated associative superalgebra. Then by Lemma 15 we have that $G_A(L)$ is an associative PI algebra, and Proposition 3 yields A is PI too. It is well known the Jacobson radical $J = J(A)$ of an associative finitely generated PI algebra A is a nilpotent ideal. Then $\overline{A} = A/J$ is semisimple hence $\overline{A}$ is a subdirect product of matrix algebras $M_{n_\gamma}(K)$ over the base field K. (At this point only we need K algebraically closed.) Since A is PI then the n_γ are bounded.

It follows $\overline{L} = L/(L \cap J)$ embeds into a direct product of finite dimensional superalgebras (contained in the respective $M_{n_\gamma}(K)$). The corresponding associative–Lie pairs obtained from the even components all belong to $\mathcal{V}$, and are of bounded dimensions. These components cannot contain pairs (R, sl_2) hence the Lie algebras in the components do not contain copies of sl_2. Therefore all of them are soluble.

But a Lie superalgebra $B = B_0 \oplus B_1$ is soluble if and only if B_0 is a soluble Lie algebra, see for example [14, Proposition 1.3.3]. Therefore $\overline{L}$ is soluble since each component in the embedding above is soluble of bounded index. But J is nilpotent and this implies L is soluble, and so $G(L)$ is soluble. The theorem is proved. □

4.3 Non-integral Exponent: An Example

Here we shall construct an associative–Lie pair such that its exponent is not an integer (if it exists). More precisely we shall prove that both the lower and upper exponents are contained in the open interval $(6, 7)$. According to Gordienko's theorem [12] such a pair cannot be the one obtained by a finite dimensional representation ρ of the corresponding Lie algebra.

We begin with several notions, definitions and statements. These can be found in [9, Section 10.4] in the context of algebras. Here we will need them for pairs and superpairs. The proofs of these statements follow verbatim the ones for algebras, and that is why we shall omit them.

We start with the free associative $\mathbb{Z}_2$-graded algebra $K\langle Y, Z\rangle$ in the even variables Y and odd variables Z. Let $P_{k,n-k}$ be the vector space of the multilinear polynomials in $y_1, \ldots, y_k, z_1, \ldots, z_{n-k}$. Let (A, L) be an associative–Lie $\mathbb{Z}_2$-graded pair or a superpair. The intersection $P_{k,n-k} \cap Id_2(A, L)$ consists of the multilinear graded identities for (A, L) of degree k in the even variables and of degree $n-k$ in the odd variables. The group $S_k \times S_{n-k}$ acts on $P_{k,n-k}$ in a natural way: S_k permutes the even variables while S_{n-k} permutes the odd variables. Then $P_{k,n-k}$ becomes an $S_k \times S_{n-k}$-module, $P_{k,n-k} \cap Id_2(A, L)$ is a submodule, and we denote $P_{k,n-k}(A, L) = P_{k,n-k}/(P_{k,n-k} \cap Id_2(A, L))$ the factor module of the "non-identities" for (A, L). Its cocharacter $\chi_{k,n-k}(A, L)$ is the $(k, n-k)$th graded weak cocharacter of (A, L). It can be decomposed as a sum of irreducibles as follows: $\chi_{k,n-k}(A, L) = \sum m_{\lambda,\mu}\chi_\lambda \otimes \chi_\mu$ where $\lambda \vdash k$, $\mu \vdash n-k$, and $m_{\lambda,\mu}$ is the multiplicity of the irreducible character associated to the pair of partitions (λ, μ). Clearly $\deg(\chi_\lambda \otimes \chi_\mu) = d_\lambda d_\mu$.

Let $c_{k,n-k}(A, L) = \dim P_{k,n-k}(A, L)$ be the $(k, n-k)$th weak codimension of (A, L), then $c_n^{\mathbb{Z}_2}(A, L) = \sum_{k=0}^{n}\binom{n}{k}c_{k,n-k}(A, L)$ is the nth weak graded codimension of (A, L). According to [9, Lemma 10.1.2], for an associative–Lie pair (A, L) the inequality $c_n(A, L) \le c_n^{\mathbb{Z}_2}(A, L)$ holds for each n. If B is an associative or Lie $\mathbb{Z}_2$-graded algebra one defines in a similar way the graded codimensions and cocharacters of B. The analog of Theorem 1 holds in this situation too.

Theorem 13 *Let $\chi_{k,n-k}(A, L) = \sum m_{\lambda,\mu}\chi_\lambda \otimes \chi_\mu$, $\lambda \vdash k$, $\mu \vdash n-k$ be the cocharacter of the associative–Lie graded pair (or superpair) (A, L) and let λ and μ be given. Then $m_{\lambda,\mu} = 0$ if and only if for every Young tableaux T_λ and T_μ, and for every $f = f(y_1, \ldots, y_k, z_1, \ldots, z_{n-k}) \in P_{k,n-k}$ the pair (A, L) satisfies the graded identity $e_{T_\lambda}e_{T_\mu}f = 0$.*

We shall use several ideas and constructions "borrowed" from the papers by Giambruno and Zaicev [10, 11]. In these papers the authors provided examples of a special Lie algebra and a Lie superalgebra L respectively such that $\liminf(c_n(L)^{1/n})$ and $\limsup(c_n(L)^{1/n})$ both exist and belong to the open interval $(6, 7)$, and in the superalgebra case coincide. Clearly neither of these can be integer.

Let $A = M_4(K)$ be the 4×4 matrix algebra over K. We fix the following $\mathbb{Z}_2$-grading on $A = A_0 \oplus A_1$:

$$A_0 = \left\{\begin{pmatrix} P & 0 \\ 0 & Q \end{pmatrix} \mid P, Q \in M_2(K)\right\}, \qquad A_1 = \left\{\begin{pmatrix} 0 & S \\ T & 0 \end{pmatrix} \mid S, T \in M_2(K)\right\}.$$

Form the Lie superalgebra $A^{(\sim)}$ and its homogeneous subalgebra $L = L_0 \oplus L_1$ where

$$L_0 = \left\{\begin{pmatrix} X & 0 \\ 0 & -X^t \end{pmatrix} \mid X \in M_2(K), tr(X) = 0\right\},$$

$$L_1 = \left\{\begin{pmatrix} 0 & Y \\ Z & 0 \end{pmatrix} \mid Y, Z \in M_2(K), Y^t = Y, Z^t = -Z\right\}.$$

We denote as usual $tr(X)$ the trace of X, and Y^t stands for the transpose of Y. Thus $\dim L = 7$, $\dim L_0 = 3$, and $\dim L_1 = 4$.

Let R be the associative subalgebra of A generated by L, then R is spanned by the set $\beta = \{a_1 \cdots a_k \mid a_i \in L_0 \cup L_1, k \ge 1\}$. We split $\beta = \beta_0 \cup \beta_1$ where β_i is formed by all products $a_1 \cdots a_k$ which are even ($i = 0$) or odd ($i = 1$) as elements of A. Denote R_i the span of β_i, $i = 0, 1$, then $R = R_0 \oplus R_1$, and moreover L is a homogeneous subalgebra of $R^{(\sim)}$. We form the superpair (R, L), its Grassmann envelope is $G(R, L) = (B, M) = (G_A(L), G(L)) = (B_0 \oplus B_1, M_0 \oplus M_1)$. Here $M_i = L_i \otimes E_i$, and $B_i \subseteq R_i \otimes E_i$, $i = 0, 1$.

The $(k, n-k)$th graded cocharacters of (B, M) and of (R, L) are given by

$$\chi_{k,n-k}(B, M) = \sum m_{\lambda,\mu}\chi_\lambda \otimes \chi_\mu, \quad \chi_{k,n-k}(R, L) = \sum \tilde{m}_{\lambda,\mu}\chi_\lambda \otimes \chi_\mu, \tag{27}$$

where $\lambda \vdash k$, $\mu \vdash n-k$. We observe here that if $A = A_0 \oplus A_1$ is an associative superalgebra then $(A, A^{(-)})$ is an associative–Lie graded pair. As $Id_2(A, A^{(-)}) = Id_2(A)$ we have that $m_{\lambda,\mu}(A, A^{(-)}) = m_{\lambda,\mu}(A)$ for every choice of $n \geq k \geq 0$ and $\lambda \vdash k$, $\mu \vdash n-k$. Here $Id_2(A)$ denotes the ideal of $\mathbb{Z}_2$-graded identities of the superalgebra A.

Recall that for a partition λ its conjugate is denoted by λ', it corresponds to the "transpose" diagram of λ (that is the diagram obtained by exchanging the rows and the columns in the corresponding diagram). If T_λ is a λ-tableau put $e^*_{T_\lambda} = \sum(-1)^\sigma \sigma\tau$ where $\sigma \in R_{T_\lambda}$, $\tau \in C_{T_\lambda}$. (Recall that for e_{T_λ} one alternates on the elements from the column stabilizer of T_λ while here we alternate on the row stabilizer.) The following lemma can be found in [9, Lemma 4.8.6].

Lemma 16 *Let $y_1, \ldots, y_l$ and $z_1, \ldots, z_m$ be even and odd variables, respectively, and let f and g be two multilinear polynomials in these variables. Suppose S_m permutes the variables $z_1, \ldots, z_m$, $\mu \vdash m$, T_μ is a μ-tableau, and for the element $e_{T_\mu} \in KS_m$ we have $f = e_{T_\mu}h$. Then $\widetilde{f} = \pm e^*_{T_{\mu'}}h$. (Recall the linear transformation $f \mapsto \widetilde{f}$ was defined at the beginning of Sect. 4.2.)*

Lemma 17 *If (A, L) is an associative–Lie $\mathbb{Z}_2$-graded pair and (A_1, L_1) is a homogeneous (associative–Lie) subpair then $m_{\lambda,\mu}(A_1, L_1) \leq m_{\lambda,\mu}(A, L)$ for every $\lambda \vdash k$, $\mu \vdash n-k$.*

Lemma 18 *In the decomposition of the cocharacter of (B, M) in Eq. (27), there exist constants C and r which do not depend on n and such that $\sum m_{\lambda,\mu} \leq Cn^r$, $\lambda \vdash k$, $\mu \vdash n-k$.*

Lemma 19 *If $\lambda \vdash k$, $\mu \vdash n-k$ then $\tilde{m}_{\lambda,\mu} \neq 0$ if and only if $m_{\lambda,\mu'} \neq 0$. (Recall the multiplicities $\tilde{m}_{\lambda,\mu}$ were defined in Eq. (27).)*

Proof If $\tilde{m}_{\lambda,\mu} \neq 0$ let $g = g(y_1, \ldots, y_k, z_1, \ldots, z_{n-k}) \in P_{k,n-k} \setminus Id_2(R, L)$. Suppose $K(S_k \times S_{n-k})g$ is an irreducible $S_k \times S_{n-k}$-module in $P_{k,n-k}$ with a character $\chi_\lambda \otimes \chi_\mu$, then $f = e_{T_\lambda}e_{T_\mu}g = e_{T_\mu}e_{T_\lambda}g$ is not an identity for (R, L). If $h = e_{T_\lambda}g$ then $f = e_{T_\mu}h$. The linear map $f \mapsto \widetilde{f}$ fixes the variables $y_1, \ldots, y_k$, therefore by Lemma 16 we get $\widetilde{f} = \pm e^*_{T_{\mu'}}e_{T_\lambda}g = e_{T_\lambda}e^*_{T_{\mu'}}g$. On the other hand $\widetilde{f}$ generates an irreducible $S_k \times S_{n-k}$-module in $P_{k,n-k}$ whose character is $\chi_\lambda \otimes \chi_{\mu'}$. By Lemma 13 it follows $\widetilde{f}$ is not a graded identity for (B, M) and thus $m_{\lambda,\mu'} \neq 0$.

If $m_{\lambda,\mu'} \neq 0$ by using the above argument and $(\mu')' = \mu$ one has $\tilde{m}_{\lambda,\mu} \neq 0$ and we are done. □

Now we find an upper bound for $\overline{\exp}(B, M)$. If $\mu = (\mu_1, \dots, \mu_t) \vdash m$ following [11] we define the weight of μ, $wt(\mu) = -\mu_1 + \mu_2 + \cdots + \mu_t = m - 2\mu_1$. The following lemma is quite similar to Lemma 5 of [11].

Lemma 20 *Let λ and μ be partitions such that $\tilde{m}_{\lambda,\mu} \neq 0$ in Eq. (27). Then*

1. $\lambda_4 = 0$ *and* $\mu_5 = 0$.
2. $wt(\mu) \le 1$, *that is* $\mu_1 + 1 \ge \mu_2 + \mu_3 + \mu_4$.
3. *There exist constants* $\alpha_1, \alpha_2, q_1, q_2$ *which do not depend on* k *and* $n - k$ *and such that* $d_\lambda \le \alpha_1 n^{q_1} 3^k$, *and* $d_\mu \le \alpha_2 n^{q_2} (2\sqrt{3})^{n-k}$.

Corollary 3 *There exist constants* α_3, q_3 *which do not depend on* n *and such that* $c_n^{\mathbb{Z}_2}(B, M) \le \alpha_3 n^{q_3} (3 + 2\sqrt{3})^n$ *for each* n. *Thus* $\overline{\exp}(B, M) \le 3 + 2\sqrt{3}$.

Proof The first statement follows as in [11, Lemma 7]. Since $c_n(B, M) \le c_n^{\mathbb{Z}_2}(B, M)$ we get $\overline{\exp}(B, M) = \limsup(c_n(B, M)^{1/n}) \le 3 + 2\sqrt{3}$. □

In the remainder we deduce a lower bound for the codimensions of (B, M). We follow ideas from [11, Section 4]. Recall that (B, M) is the Grassmann envelope $G(R, L)$. We fix a basis of L_0 as follows. Let $e = e_{12} - e_{43}$, $f = e_{21} - e_{34}$, $h = e_{11} - e_{22} - e_{33} + e_{44}$ where e_{ij} is the matrix unit with 1 at position (ij) and 0 elsewhere. Then clearly $he = -eh = e$, $fh = -hf = f$, and $ef = e_{11} + e_{44}$, $fe = e_{22} + e_{33}$. An easy manipulation shows then that the standard polynomial $s_3(e, f, h) = 3(ef + fe) = 3I \in R_0$ where I stands for the identity matrix. Hence if $v \in R$ one obtains $s_3(e, f, h)^q v = 3^q v$ for every $q \ge 1$.

Consider $L^{(1)}$ and $L^{(-1)}$ the upper right and the lower left block 2×2 of L_1, respectively. (Recall $L = L_0 \oplus L_1$ was defined after Theorem 13.) Take linearly independent elements $a, b, c \in L^{(1)}$ (these exist since $\dim L^{(1)} = 3$). Let $0 \neq d \in L^{(-1)}$ then d has in its lower left corner a skew symmetric matrix D of order 2. It is immediate that if $x \in L^{(1)}$ has the symmetric matrix X in its upper right corner then

$$[x, d] = \left[\begin{pmatrix} 0 & X \\ 0 & 0 \end{pmatrix}, \begin{pmatrix} 0 & 0 \\ D & 0 \end{pmatrix} \right] = \begin{pmatrix} XD & 0 \\ 0 & DX \end{pmatrix}.$$

Therefore $u_1 = [a, d]$, $u_2 = [b, d]$, $u_3 = [c, d]$ are linearly independent. As $\dim L_0 = 3$ they form a basis of L_0. Writing down the u_i as linear combinations of e, f, g, one obtains $s_3(u_1, u_2, u_3) = \alpha I$, $0 \neq \alpha \in K$. It follows $s_3(u_1, u_2, u_3)v = \alpha v$ for every $v \in R$. Taking a scalar multiple of d we can suppose $\alpha = 1$. This means $s_3(u_1, u_2, u_3)v = [\overline{a}, d][\overline{b}, d][\overline{c}, d]v = v$. Here and in what follows we use the bar and symbols like $\widetilde{a}$ and/or $\hat{a}$ to indicate alternating sets of variables.

Now we iterate and apply $s_3(u_1, u_2, u_3)^{q+1}$ to v, and we write it as

$$[\overline{a}, d][\overline{b}, d][\overline{c}, d][\hat{a}, d][\hat{b}, d][\hat{c}, d] \cdots [\tilde{a}, d][\tilde{b}, d][\tilde{c}, d]v = v.$$

Since $[d, d] = 0$ we can rewrite the latter equality as

$$[\overline{a}, \hat{d}][\overline{b}, d][\overline{c}, d][\hat{a}, \tilde{d}][\hat{b}, d][\hat{c}, d] \cdots [\tilde{a}, d][\tilde{b}, d][\tilde{c}, d]v = v.$$

In other words the latter expression is alternating on q sets of variables, each of them $\{a, b, c, d\}$, and on one alternating set $\{a, b, c\}$. Moreover there are $2q + 3$ additional variables d (which do not alternate) and one v. The left-hand side yields a polynomial of degree $4q + 3 + 2q + 3 + 1 = 6q + 7$, and this polynomial depends on odd variables only. Linearizing (polarizing) it we obtain a multilinear polynomial of degree $6q + 7$, in odd variables. Denote the latter multilinear polynomial by

$$f_q = f_q(t_1^1, t_2^1, t_3^1, z_1^1, z_2^1, z_3^1, t_1^2, t_2^2, t_3^2, z_1^2, z_2^2, z_3^2, \ldots, t_1^{q+1}, t_2^{q+1}, t_3^{q+1}, z_1^{q+1}, z_2^{q+1}, z_3^{q+1}, z).$$

Thus f_q is alternating on each one of the sets $\{t_1^i, t_2^i, t_3^i, z_1^{i+1}\}$, $1 \le i \le q$, and also on the set $\{t_1^{q+1}, t_2^{q+1}, t_3^{q+1}\}$.

If we specialize the variables as follows: $t_1^i \mapsto a$, $t_2^i \mapsto b$, $t_3^i \mapsto c$, $z_j^i \mapsto d$, for all i and j, and $z \mapsto v$ where $0 \ne v \in L_1$ is arbitrary then clearly f_q does not vanish.

Denote by g_q the symmetrization (restitution) of f_q in the four sets of variables

$$\{t_1^1, \ldots, t_1^q\}, \quad \{t_2^1, \ldots, t_2^q\}, \quad \{t_3^1, \ldots, t_3^q\}, \quad \{z_2^1, z_3^1, z_1^2, z_2^2, z_3^2, \ldots, z_1^q, z_2^q, z_3^q, z_1^{q+1}\}$$

containing q, q, q, $3q$ variables, respectively. The above specialization shows that g_q is not an identity for the pair (B, L). As we work in characteristic 0 we can linearize g_q and obtain a multilinear element. Let $P_{0,6q+7}$ be the vector space of the multilinear polynomials in the odd variables appearing in the complete linearization of g_q. If $m = 6q$ then the symmetric group S_m permutes the variables from the above four sets, and $P_{0,6q+7}$ becomes an S_m-module. The linearization of g_q generates an irreducible S_m-module (this follows from the form of the polynomial) corresponding to the partition $\nu = (3q, q^3) = (3q, q, q, q)$. The above specialization of g_q falls in R_1 since $\deg g_q$ is odd and there are only odd variables in the polynomial; the same holds for its linearization.

We already saw $s_3(e, f, h)^q v = 3^q v$ for every $v \in R$. It follows that for $v = g_q$ we have that the polynomial $h_q = s_3(y_1^1, y_2^1, y_3^1) \cdots s_3(y_1^q, y_2^q, y_3^q) g_q$ does not vanish on the superpair (R, L) whatever the even variables y_j^i are. Indeed it is sufficient to substitute all y_1^i for e, all y_2^i for f and y_3^i for h. Then as above we symmetrize first on the three sets $\{y_1^i\}$, $\{y_2^i\}$, $\{y_3^i\}$, then linearize and obtain a multilinear polynomial p_q of degree $3q$ in the even variables, and of degree $6q + 7$ in the odd variables. Clearly p_q generates an irreducible $S_{3q} \times S_{6q}$-submodule in $P_{3q,6q+7}$ which corresponds to a pair of partitions (λ, ν), where $\lambda = (q, q, q)$ and $\nu = (3q, q, q, q)$. But then one takes the induced representation for $S_{3q} \times S_{6q+7}$ and chooses an irreducible component in it. We summarize the above considerations in the following proposition.

Proposition 4 (Cf. [11, Lemma 9]) *If $q \ge 1$ then there is a multilinear polynomial p_q in $3q$ even and $6k+7$ odd variables which generates an irreducible $S_{3q} \times S_{6q+7}$-module corresponding to the pair of partitions (λ, μ). The polynomial p_q does not* *vanish on* (R, L). *Moreover* $\lambda = (q^3) = (q, q, q)$, *and if* $\nu = (3q, q^3)$ *then* $\nu \le \mu$. *Moreover the character of this module is* $\chi_\lambda \otimes \chi_\mu$

Now we follow fairly close the exposition of [11].

Lemma 21 *If* $n = 9q + 7$ *then* $c_n(B, M) \ge \alpha_4 n^{q_4} (3^{5/3})^n$ *where* $\alpha_4 > 0$ *and* q_4 *are constants.*

Proof According to the above Proposition 4 there exist $\lambda = (q^3)$, $\mu \vdash 6q + 7$, and $\nu = (3q, q^3) \le \mu$ such that $\tilde{m}_{\lambda,\mu} \ne 0$ in the decomposition given in Eq. (27). This means, once again by (27) that $m_{\lambda,\mu'} \ne 0$. Here μ' is the conjugate of μ, hence $\nu' = (4^q, 1^{2q}) \le \mu'$. Therefore there is an irreducible $S_{3q} \times S_{6q+7}$-module N in $P_{3q,6q+7}$ whose character is $\chi_\lambda \otimes \chi_{\mu'}$. Moreover N is generated by a multilinear polynomial f in $3q$ even and $6q+7$ odd variables and f is not a graded identity for (B, M). Then f cannot be an ordinary (nongraded) identity for (B, M).

Applying the Stirling formula exactly in the same way as in [11, Lemma 10] one gets the conclusion of the Lemma. □

Lemma 22 *The inequality* $c_n(B, M) \le c_{n+1}(B, M)$ *holds for every n.*

Corollary 4 *One has* $\underline{\exp}(B, M) \ge 3^{5/3}$.

Theorem 14 (Cf. [11, Theorem 1]) *Let* (R, L) *be the superpair defined in the beginning of this section. Then its Grassmann envelope* $G(R, L) = (B, M)$ *is an associative–Lie pair, and its exponent, if it exists, is not an integer. More precisely the following inequalities hold:*

$$6.24 \approx 3^{5/3} \le \underline{\exp}(B, M) \le \overline{\exp}(B, M) \le 3 + 2\sqrt{3} \approx 6.46.$$

The proof of the theorem is contained in the statements preceding it. We note that we do not know whether $\underline{\exp}(B, M) < \overline{\exp}(B, M)$ (that is the exponent does not exist) or $\underline{\exp}(B, M) = \overline{\exp}(B, M)$ (that is the exponent exists but is not an integer).

Acknowledgments We thank the Referee whose remarks to the point were taken into account.

Plamen Koshlukov was partially supported by FAPESP grant No. 2018/23690-6 and by CNPq grant No. 302238/2019-0. David Levi da Silva Macêdo was financed in part by the Coordenação de Aperfeiçoamento de Pessoal de Nível Superior - Brasil (CAPES) – Finance Code 001, and in part by CNPq.

References

1. Bahturin, Yu.A.: Identical Relations in Lie Algebras. VNU Science Press, Utrecht (1987)
2. Benediktovich, I.I., Zalesskii, A.E.: T-ideals of free Lie algebras with polynomial growth of the sequence of codimensions (Russian). Vestsi Akad. Navuk BSSR Ser. Fiz.-Mat Navuk **3**(139), 5–10 (1980)
3. Berele, A.: Homogeneous polynomial identities. Israel J. Math. **42**, 258–272 (1982)

4. Drensky, V.: Representations of the symmetric group and varieties of linear algebras (Russian). Mat. Sb. **115**, 98–115 (1980). English translation: Math. USSR-Sb. **43**(1), 85–100 (1982)
5. Drensky, V.: Free Algebras and PI-Algebras. Springer, Singapore (2000)
6. Drensky, V., Koshlukov, P.: Weak polynomial identities for a vector space with a symmetric bilinear form. Mathematics and mathematical education (Bulgarian) (Sunny Beach (Slanchev Bryag), 1987), 213–219, Publ. House Bulgar. Acad. Sci., Sofia 1987. Also available at https://arxiv.org/abs/1905.08351
7. Drensky, V., Vladimirova, L.: Varieties of pairs of algebras with a distributive lattice of subvarieties. Serdica **12**(2), 166–170 (1986)
8. Giambruno, A., Zaicev, M.: Exponential codimension growth of PI-algebras: an exact estimate. Adv. Math. **142**, 221–243 (1999)
9. Giambruno, A., Zaicev, M.: Polynomial Identities and Asymptotic Methods. Mathematical Surveys Monographs, vol. 122. AMS, Providence (2005)
10. Giambruno, A., Zaicev, M.: On codimension growth of finite-dimensional Lie superalgebras. J. Lond. Math. Soc. **85**, 534–548 (2012)
11. Giambruno, A., Zaicev, M.: Non-integrality of the PI-exponent of special Lie algebras. Adv. Appl. Math. **51**(5), 619–634 (2013)
12. Gordienko, A.S.: Codimensions of polynomial identities of representations of Lie algebras. Proc. Amer. Math. Soc. **141**(10), 3369–3382 (2013)
13. James, G.D.: Representation Theory of the Symmetric Groups. Springer Lecture Notes in Mathematics, vol. 692. Springer, Berlin (1978)
14. Kac, V.: Lie superalgebras. Adv. Math. **26**, 8–96 (1977)
15. Kemer, A.: T-ideals with power growth of the codimensions are Specht. Sibirsk. Mat. Zh. **19**, 54–69 (1978) (Russian). English translation: Siberian Math. J. **19**, 37–48 (1978)
16. Kemer, A.: Varieties of finite rank (Russian). In: Proceedings of the 15-th All the Union Algebraic Conference, Krasnoyarsk, vol. 2, p. 73 (1979)
17. Kemer, A.: Ideals of Identities of Associative Algebras. Translations of Mathematical Monographs, vol. 87. American Mathematical Society, Providence (1988)
18. Koshlukov, P.: Weak polynomial identities for the matrix algebra of order two. J. Algebra **188**(2), 610–625 (1997)
19. Krakowski, D., Regev, A.: The polynomial identities of the Grassmann algebra. Trans. Amer. Math. Soc. **181**, 429–438 (1973)
20. Malcev, Yu.N.: A basis for the identities of the algebra of upper triangular matrices (Russian). Algebra Logika **10**, 393–400 (1971). English translation: Algebra Logic **10**, 242–247 (1973)
21. Mishchenko, S.P.: Varieties of Lie algebras with weak growth of the sequence of codimensions (Russian). Vestnik Moskov. Univ. Ser. I Mat. Mekh. **5**, 63–66 (1982). English translation: Mosc. Univ. Math. Bull. **37**, No. 5, 78–82 (1982)
22. Mishchenko, S.P.: On varieties of polynomial growth of Lie algebras over a field of characteristic zero (Russian). Mat. Zametki **40**, 713–721 (1986). English translation: Math. Notes **40**, 5–6, 901–905 (1986)
23. Mishchenko, S.P.: Growth of varieties of Lie algebras (Russian). Uspekhi Mat. Nauk **45**, 25–45 (1990). English translation: Russian Math. Surveys **45**, 27–52, (1990)
24. Olson, J.B., Regev, A.: An application of representation theory to PI-algebras. Proc. Amer. Math. Soc. **55**, 253–257 (1976)
25. Razmyslov, Yu.P.: Finite basing of the identities of a matrix algebra of second order over a field of characteristic 0 (Russian). Algebra Logika **12**, 83–113 (1973). English translation: Algebra Logic **12**, 47–63 (1974)
26. Regev, A.: Existence of identities in $A \otimes B$. Israel J. Math. **11**, 131–152 (1972)
27. Trishin, I.M.: Identities of irreducible representations of a three-dimensional simple Lie algebra (Russian). Algebra i Logika **22**(3), 316–342 (1983). English translation: Algebra Logic **22**(3), 229–250 (1983)
28. Zaicev, M.: Integrality of exponents of growth of identities of finite-dimensional Lie algebras (Russian). Izv. Ross. Akad. Nauk Ser. Mat. **66**(3), 23–48 (2002). English translation: Izv. Math. **66**, 463–478 (2002)

On Codimensions of Algebras with Involution

Daniela La Mattina

Dedicated to my dear colleague Antonio Giambruno on the occasion of his anniversary.

Abstract Let A be an associative algebra with involution $*$ over a field F of characteristic zero. One associates to A, in a natural way, a numerical sequence $c_n^*(A)$, $n = 1, 2, \ldots$, called the sequence of $*$-codimensions of A which is the main tool for the quantitative investigation of the polynomial identities satisfied by A. In this paper we focus our attention on $c_n^*(A)$, $n = 1, 2, \ldots$, by presenting some recent results about it.

Keywords *-identities · *-codimensions · Growth

1 Introduction

Let A be an algebra with involution $*$ over a field F of characteristic zero. Recall that one can attach to A a numerical sequence $c_n^*(A)$, $n = 1, 2, \ldots$, called the sequence of $*$-codimensions of A. Such sequence is built out of the dimensions of the multilinear $*$-polynomial identities of degree $n \geq 1$ satisfied by the algebra A. Such sequence has been extensively studied (see [9, 12–16]) but it turns out that it can be explicitly computed only in very few cases. In case A is a PI-algebra, i.e., it satisfies a non trivial polynomial identity, it was proved in [6] that, as in the ordinary case, $c_n^*(A)$, $n = 1, 2, \ldots$, is exponentially bounded. As a consequence in the last years the interest has been focused in the computation of its asymptotics since they represent an invariant of the T*-ideal of the $*$-polynomial identities satisfied by A. The exponential rate of growth of the sequence of $*$-codimensions was computed

D. La Mattina (✉)
Dipartimento di Matematica e Informatica, Università di Palermo, Palermo, Italy
e-mail: daniela.lamattina@unipa.it

O. M. Di Vincenzo, A. Giambruno (eds.), *Polynomial Identities in Algebras*,
Springer INdAM Series 44, https://doi.org/10.1007/978-3-030-63111-6_13

for finite dimensional algebras in [7] and for general PI-algebras in [9] and it turns out to be a non-negative integer called the $*$-exponent of the algebra.

In this paper we present some results proved recently on $c_n^*(A)$, $n = 1, 2, \ldots$. First we shall point out that if A is any algebra with involution satisfying a non trivial polynomial identity, then its sequence of $*$-codimensions is eventually non decreasing. Then, starting with the well-known inequality for PI-algebras given in [9]:

$$C_1 n^t \exp^*(A)^n \leq c_n^*(A) \leq C_2 n^s \exp^*(A)^n \tag{1}$$

with $C_1 > 0, C_2, t, s$ constants, we shall see that, for finite dimensional algebras [11] and, as a consequence [17] for finitely generated algebras, $t = s \in \frac{1}{2}\mathbb{Z}$. In this way we get a second invariant $\lim_{n\to\infty} \log_n \frac{c_n^*(A)}{exp^*(A)^n}$ of a T*-ideal, after the $*$-exponent.

Such result is accomplished by studying especial class of algebras, the so-called $*$-fundamental algebras. These are finite dimensional algebras that can be defined in terms of some multialternating polynomials and for such algebras the polynomial factor t in (1) is related to the structure of the algebra and can be determined explicitly.

Finally, we shall give a characterization of the varieties of algebras with involution whose exponential growth is bounded by 2.

2 $*$-Codimensions and $*$-Fundamental Algebras

Throughout this paper F will denote a field of characteristic zero, A an associative F-algebra with involution $*$ and $F\langle X, *\rangle = F\langle x_1, x_1^*, x_2, x_2^*, \ldots\rangle$ the free associative algebra with involution on a countable set $X = \{x_1, x_1^*, x_2, x_2^*, \ldots\}$ over F.

Recall that a $*$-polynomial identity (or simply a $*$-identity) of A is a $*$-polynomial $f(x_1, x_1^*, \ldots, x_n, x_n^*) \in F\langle X, *\rangle$ such that $f(a_1, \ldots, a_n) = 0$, for all $a_1, a_1^*, \ldots, a_n, a_n^* \in A$.

Obviously, any ordinary polynomial identity can be viewed as a $*$-identity and, so, if an algebra is PI, i.e., it satisfies a non-trivial ordinary identity then it also satisfies a non-trivial $*$-identity. The converse is also true for a well-known result of Amitsur [2], i.e., if A satisfies a non-trivial $*$-identity, then A satisfies an ordinary identity. This result gives a close relation between identities and $*$-identities. Moreover, an explicit bound related to the ordinary identities of the algebra A was found in [3].

As for the ordinary case, we have a positive answer to the Specht problem [1]: every proper T*-ideal of $F\langle X, *\rangle$ is finitely generated as a T*-ideal. Here T*-ideal refers to an ideal of $F\langle X, *\rangle$ invariant under all endomorphisms of the free algebra commuting with $*$. Nevertheless, the T*-ideals of the free algebra are quite obscure objects, since finding a finite set of generators is not at all simple. So, in order to get information about the $*$-identities satisfied by an algebra, one associates to an algebra numerical invariants. One of the most important numerical invariants of a PI-algebra is its codimension sequence $c_n^*(A), n = 1, 2, \ldots$. It is well known that in characteristic zero, every $*$-identity is equivalent to a system of multilinear $*$-identities. We denote by

$$P_n^* = \text{span}_F\{w_{\sigma(1)} \cdots w_{\sigma(n)} \mid \sigma \in S_n, w_i = x_i \text{ or } w_i = x_i^*, 1 \le i \le n\}$$

the space of multilinear $*$-polynomials of degree n in $x_1, \ldots, x_n$, i.e., for every $i = 1, \ldots, n$, either x_i or x_i^* appears in every monomial of P_n^* at degree 1 (but not both).

So, if we denote by $\text{Id}^*(A)$ the T*-ideal of all $*$-identities satisfied by A, its study is equivalent to the study of $P_n^* \cap \text{Id}^*(A)$, for all $n \ge 1$ and we denote by

$$c_n^*(A) = \dim_F \frac{P_n^*}{P_n^* \cap \text{Id}^*(A)}, n \ge 1$$

the n-th $*$-codimension of A. Recently, it was proved in [4] that such a sequence is eventually non-decreasing.

Theorem 1 ([4]) *Let A be a PI-algebra with involution $*$. Then the sequence of $*$-codimensions $c_n^*(A), n = 1, 2, \ldots$, is eventually non-decreasing, that is, $c_{n+1}^*(A) \ge c_n^*(A)$, for n large enough.*

Despite its importance the exact computation of the $*$-codimensions of an algebra is extremely difficult, and it has been done for very few algebras. That is why one is led to study the asymptotic behaviour of the sequence of $*$-codimensions. Such a sequence is bounded from above by the dimension of P_n^* which is $2^n n!$ but, in case A is a PI-algebra, it was proved in [6] that, as in the ordinary case, $c_n^*(A), n = 1, 2, \ldots$, is exponentially bounded. The exponential rate of growth of $c_n^*(A), n = 1, 2, \ldots$ was computed and shown to be an integer for finite dimensional algebras in [7] and for general PI-algebras in [9].

Theorem 2 ([9]) *Let A be a PI-algebra with involution $*$ over a field of characteristic zero. Then there exist constants $C_1 > 0, C_2, t_1, t_2$ such that*

$$C_1 n^{t_1} d^n \le c_n^*(A) \le C_2 n^{t_2} d^n. \tag{2}$$

Hence $\lim_{n\to\infty} \sqrt[n]{c_n^*(A)} = \exp^*(A)$, *the $*$-exponent of A, exists and is an integer.*

As a consequence of the above theorem we have that the sequence of $*$-codimensions $c_n^*(A)$, $n = 1, 2, \ldots$, is either polynomially bounded or grows as an exponential function d^n with $d \ge 2$.

In case of polynomial growth, if A is an algebra with 1, in [14] it was proved that

$$c_n^*(A) = qn^k + O(n^{k-1})$$

is a polynomial with rational coefficients. Moreover its leading term satisfies the inequalities

$$\frac{1}{k!} \leq q \leq \sum_{i=0}^{k} 2^{k-i} \frac{(-1)^i}{i!}.$$

Let us write down the disequality given in (2) keeping in mind that $d = \exp^*(A)$:

$$C_1 n^{t_1} \exp^*(A)^n \leq c_n^*(A) \leq C_2 n^{t_2} \exp^*(A)^n. \tag{3}$$

Now one can ask if the polynomial factor in (3) is uniquely determined, i.e., $t_1 = t_2$, giving in this way a second invariant of a T*-ideal, after the $*$-exponent. The answer is positive for finite dimensional algebras with involution [11] and, as a consequence, by the main result in [17], for finitely generated algebras.

Theorem 3 ([11]) *Let A be a finitely generated $*$-algebra over a field F of characteristic zero. If A satisfies a polynomial identity then*

$$C_1 n^t exp^*(A)^n \leq c_n^*(A) \leq C_2 n^t exp^*(A)^n,$$

where $t \in \frac{1}{2}\mathbb{Z}$, for some constants $C_1 > 0, C_2$. Hence $\lim_{n\to\infty} \log_n \frac{c_n^*(A)}{exp^*(A)^n}$ exists and is a half integer.

Now, a more concrete question would be the following: can one compute such polynomial factor for a certain class of algebras relating it to the structure of the algebra itself? The answer is positive for the class of $*$-fundamental algebras defined in [11].

Let us recall the definition of $*$-fundamental algebra.

We recall that a $*$-polynomial $f(x_1, \ldots, x_n, Y)$ linear in the variables $x_1, \ldots, x_n$ (and in some other set of variables Y) is alternating in $x_1, \ldots, x_n$ if f vanishes whenever we identify any two of these variables This is equivalent to say that the polynomial changes sign whenever we exchange any two of these variables (here we exchange the indices of the two variables).

Now assume that $A = \bar{A} + J$ is a finite dimensional $*$-algebra over an algebraically closed field, where $\bar{A}$ is a semisimple subalgebra of A and $J = J(A)$ is the Jacobson radical. We recall that the (t, s)-index of A is $Ind_{t,s}(A) = (\dim \bar{A}, s_A)$ where $s_A \geq 0$ is the smallest integer such that $J^{s_A+1} = 0$.

Next we define the Kemer $*$-index of A.

Let $\Gamma \subseteq F\langle X, *\rangle$ be the ideal of $*$-identities of A. Then $\beta(\Gamma)$ is defined as the greatest integer t such that for every $\mu \geq 1$, there exists a multilinear $*$-polynomial $f(X_1, \ldots, X_\mu, Y) \notin \Gamma$ alternating in the μ sets X_i with $|X_i| = t$. Moreover $\gamma(\Gamma)$ is defined as the greatest integer s for which there exists for all $\mu \geq 1$, a multilinear $*$-polynomial $f(X_1, \ldots, X_\mu, Z_1, \ldots, Z_s, Y) \notin \Gamma$ alternating in the μ sets X_i with $|X_i| = \beta(\Gamma)$ and in the s sets Z_j with $|Z_j| = \beta(\Gamma) + 1$.

Then $Ind_K^*(\Gamma) = (\beta(\Gamma), \gamma(\Gamma))$ is called the Kemer $*$-index of Γ.

Since $\Gamma = \mathrm{Id}^*(A)$, we also say that $(\beta(\Gamma), \gamma(\Gamma)) = (\beta(A), \gamma(A)) = Ind_K^*(A)$ is the Kemer $*$-index of A.

In general we have that $Ind_K^*(A) \leq Ind_{t,s}(A)$ in the left lexicographic order.

Next we give a definition of $*$-fundamental algebra through the Kemer $*$-index.

Definition 1 ([11, Theorem 6.1]) A finite dimensional $*$-algebra A is $*$-fundamental if and only if $Ind_K^*(A) = Ind_{t,s}(A)$.

The main feature of these algebras is that any finite dimensional algebras satisfies the same $*$-identities as a finite direct sum of $*$-fundamental algebras.

Theorem 4 ([11]) *Let $A = \bar{A} + J$ be a $*$-fundamental algebra over an algebraically closed field F of characteristic zero and let $s \geq 0$ be the least integer such that $J^{s+1} = 0$. Write $\bar{A} = A_1 \oplus \cdots A_r \oplus A_{r+1} \oplus \cdots \oplus A_q$, a direct sum of $*$-simple algebras with $A_1, \ldots, A_r$ not simple algebras, then*

$$C_1 n^{-\frac{1}{2}(\dim(\bar{A})^- - r)+s} (\dim \bar{A})^n \leq c_n^*(A) \leq C_2 n^{-\frac{1}{2}(\dim(\bar{A})^- - r)+s} (\dim \bar{A})^n,$$

for some constants $C_1 > 0, C_2$, where $(\bar{A})^- = \{a \in \bar{A} \mid a^* = -a\}$ is the Lie algebra of skew elements of $\bar{A}$. Hence

$$\lim_{n\to\infty} \log_n \frac{c_n^*(A)}{exp^*(A)^n} = -\frac{1}{2}(\dim(\bar{A})^- - r) + s.$$

3 Low Exponential Growth

A structure theorem for PI-algebras with involutions proved in [1] is fundamental in proving Theorem 2; it asserts that any PI-algebra with involution A over a field of characteristic zero, satisfies the same $*$-identities as the Grassmann envelope $G(B)$ of a finite dimensional superalgebra with superinvolution B, i.e.,

$$\mathrm{Id}^*(A) = \mathrm{Id}^*(G(B)). \tag{4}$$

Now let us recall the basic definitions in order to see some applications of such a result.

Let $B = B_0 \oplus B_1$ be an associative superalgebra over F endowed with a superinvolution $\sharp$. Recall that a superinvolution on B is a graded linear map $\sharp : B \longrightarrow B$ such that $(a^\sharp)^\sharp = a$ for all $a \in B$ and $(ab)^\sharp = (-1)^{(\deg a)(\deg b)} b^\sharp a^\sharp$ for any homogeneous elements $a, b \in B$. Here $\deg c$ denotes the homogeneous degree of $c \in B_0 \cup B_1$.

Since char $F = 0$, we can write $B = B_0^+ \oplus B_0^- \oplus B_1^+ \oplus B_1^-$, where for $i = 0, 1$, $B_i^+ = \{a \in B_i \mid a^* = a\}$ and $B_i^- = \{a \in B_i \mid a^* = -a\}$ denote the sets of symmetric and skew elements of B_i, respectively.

In a natural way one defines the free algebra with superinvolution $F\langle X, \sharp\rangle$, the ideal of identities with superinvolution $\mathrm{Id}^{\sharp}(B)$, etc.

Let G be the infinite dimensional Grassmann algebra over F, i.e., the algebra generated by the elements $1, e_1, e_2, \ldots$ subject to the relations $e_ie_j = -e_je_i$, for all $i, j \geq 1$. Recall that G has a natural $\mathbb{Z}_2$-grading $G = G_0 \oplus G_1$, where G_0 and G_1 are the spans of the monomials in the e_i's of even and odd length, respectively. One defines a superinvolution $\sharp$ on the Grassmann algebra $G = G_0 \oplus G_1$ by requiring that $e_i^{\sharp} = -e_i$, for $i \geq 1$. Hence $G^+ = G_0$ and $G^- = G_1$.

Now if $B = B_0 \oplus B_1$ is a superalgebra endowed with a superinvolution $\sharp$, it was proved in [1] that the Grassmann envelope of B, $G(B) = B_0 \otimes G_0 \oplus B_1 \otimes G_1$ has an induced involution $*$ by requiring that $(a \otimes g)^* = a^{\sharp} \otimes g^{\sharp}$, on all homogeneous elements $g \in G$ and $a \in B$. The main property of such a Grassmann envelope is the one we have seen above: if A is a PI-algebra with involution over a field of characteristic zero, then there exists a finite dimensional superalgebra with superinvolution B, such that $\mathrm{Id}^*(A) = \mathrm{Id}^*(G(B))$.

As a consequence we have that $c_n^*(A) = c_n^*(G(B))$, for all $n \geq 1$.

Such a result allowed the authors in [9] to determine the exponential rate of growth of the $*$-codimensions of $G(B)$, and consequently of A. They also proved that the $*$-exponent, when the field is algebraically closed, is just the dimension of a suitable subalgebra of B.

In order to state this result we make the following definition. Let F be an algebraically closed field of characteristic zero and let B be a finite dimensional superalgebra with superinvolution. Then by Giambruno et al. [8] $B = \bar{B} + J$, where $\bar{B}$ is a maximal semisimple subalgebra with induced superinvolution and $J = J^{\sharp}$ is the Jacobson radical of B. Let $\bar{B} = B_1 \oplus \cdots \oplus B_q$ be a direct sum of simple superalgebras with superinvolution. Then a subalgebra $C = C_1 \oplus \cdots \oplus C_t$ of B, where $C_1, \ldots, C_t$ are distinct subalgebras from the set $\{B_1, \ldots, B_q\}$ is called admissible if

$$C_1JC_2J \cdots JC_t \neq 0.$$

The subalgebra $C + J$ with induced superinvolution will be called reduced.

The result in [9] reads as follows. If $B = B_1 \oplus \cdots \oplus B_q + J$ is a finite dimensional superalgebra with superinvolution defined as above, i.e., $\mathrm{Id}^*(A) = \mathrm{Id}^*(G(B))$, then there exist constants $C_1 > 0, C_2, t_1, t_2$ such that

$$C_1n^{t_1}d^n \leq c_n^*(G(B)) \leq C_2n^{t_2}d^n, \tag{5}$$

where d is the maximal dimension of an admissible subalgebra of B.

Since the codimensions do not change by extending the base field, by putting together the results in (4) and (5) we get the result in Theorem 2.

Hence, in order to characterize the varieties of $*$-algebras of a given $*$-exponent t, a starting point is the study of the varieties of algebras with superinvolution generated by reduced finite dimensional algebras whose semisimple part is of dimension t. We recall that the $*$-exponent of a variety is the $*$-exponent of a generating algebra.

Next we give a characterization of the varieties of algebras with involution whose $*$-exponent is bounded by 2. To this end we list nine algebras that will play a basic role in what follows.

Next we denote by $UT_n = UT_n(F)$ the algebra of $n \times n$ upper triangular matrices over the field F. We consider the following two algebras with involution:

1. $F \oplus F$, a two dimensional algebra endowed with the exchange involution $(a, b)^* = (b, a)$;
2. $M = F(e_{11}+e_{44}) \oplus F(e_{22}+e_{33}) \oplus Fe_{12} \oplus Fe_{34}$, the subalgebra of UT_4 endowed with the reflection involution. Here the e_{ij}s are the usual matrix units.

Such algebras were extensively studied in [5] and [16].

Now we consider some algebras with involution introduced in [4] in order to characterize the varieties of $*$-exponent ≤ 2 :

1. $D_1 = Fe_{11} \oplus F(e_{22} + e_{33}) \oplus Fe_{44} \oplus Fe_{12} \oplus Fe_{34} \subseteq UT_4$ with reflection involution $*$;
2. $D_2 = G_0e_{11} \oplus G_0(e_{22}+e_{33}) \oplus G_0e_{44} \oplus G_1e_{12} \oplus G_1e_{34} \subseteq UT_4(G)$ is the algebra with involution defined on a basis by

$$(ge_{ij})^\circ = \begin{cases} -ge_{ij}^* & \text{if } (i, j) \in \{(1, 2), (3, 4)\} \\ ge_{ij}^* & \text{otherwise} \end{cases},$$

   where * denotes the reflection involution on $UT_4(G)$;
3. $D_3 = F(e_{11} + e_{66}) \oplus F(e_{22} + e_{55}) \oplus F(e_{33} + e_{44}) \oplus Fe_{12} \oplus Fe_{13} \oplus Fe_{23} \oplus Fe_{45} \oplus Fe_{46} \oplus Fe_{56} \subseteq UT_6$ with reflection involution $*$;
4. $D_4 = G_0(e_{11}+e_{66}) \oplus G_0(e_{22}+e_{55}) \oplus G_0(e_{33}+e_{44}) \oplus G_0e_{12} \oplus G_1e_{13} \oplus G_1e_{23} \oplus G_1e_{45} \oplus G_1e_{46} \oplus G_0e_{56} \subseteq UT_6(G)$ is the algebra with involution defined on a basis by

$$(ge_{ij})^\circ = \begin{cases} -ge_{ij}^* & \text{if } (i, j) \in \{(1, 3), (2, 3), (4, 5), (4, 6)\} \\ ge_{ij}^* & \text{otherwise} \end{cases},$$

   where * denotes the reflection involution on $UT_6(G)$;
5. $D_5 = G_0(e_{11}+e_{66}) \oplus G_0(e_{22}+e_{55}) \oplus G_0(e_{33}+e_{44}) \oplus G_1e_{12} \oplus G_0e_{13} \oplus G_1e_{23} \oplus G_1e_{45} \oplus G_0e_{46} \oplus G_1e_{56} \subseteq UT_6(G)$ is the algebra with involution defined on a basis by

$$(ge_{ij})^\circ = \begin{cases} -ge_{ij}^* & \text{if } (i, j) \in \{(2, 3), (4, 5)\} \\ ge_{ij}^* & \text{otherwise} \end{cases},$$

   where * denotes the reflection involution on $UT_6(G)$;

6. $D_6 = (M_2(F), s)$ is the algebra of 2×2 matrices over F with symplectic involution;
7. $D_7 = (M_2(F), t)$ is the algebra of 2×2 matrices over F with transpose involution;
8. $D_8 = M_{1,1}(G) = \begin{pmatrix} G_0 & G_1 \\ G_1 & G_0 \end{pmatrix}$ with involution: $\begin{pmatrix} g_0 & g_1 \\ g_1' & g_o' \end{pmatrix}^* = \begin{pmatrix} g_o' & g_1 \\ -g_1' & g_o \end{pmatrix}$;
9. $D_9 = G \oplus G^{op}$ with involution $(a, b)^* = ((-1)^{degb} b, (-1)^{dega} a)$, for $a, b \in G_0 \cup G_1$.

In what follows we denote by $\mathrm{var}^*(A)$ the variety of algebras with involution generated by A.

Theorem 5 ([4]) *Let A be a PI-algebra with involution $*$ over a field F of characteristic zero. Then*

(a) $\exp^*(A) \leq 2$ *if and only if $D_i \notin var^*(A)$, for any $i \in \{1, \ldots, 9\}$.*
(b) $\exp^*(A) = 2$ *if and only if $D_i \notin var^*(A)$, for any $i \in \{1, \ldots, 9\}$ and either $F \oplus F$ or $M \in var^*(A)$.*
(c) $\exp^*(A) \leq 1$ *if and only if $F \oplus F$, $M \notin var^*(A)$.*

Notice that last item in the previous theorem is equivalent to the following result, which was obtained in [5] without using the $*$-exponent.

Theorem 6 ([4, 5]) *Let A be a PI-algebra with involution $*$ over a field F of characteristic zero. Then the sequence $c_n^*(A), n = 1, 2, \ldots,$ is polynomially bounded if and only if M, $F \oplus F \notin var^*(A)$.*

We recall that two algebras are T^*-equivalent if they satisfy the same $*$-identities. As a consequence of the above theorem we get the only two $*$-algebras, up to T^*-equivalence, generating varieties of almost polynomial growth, i.e., such that they grow exponentially but any proper subvariety grows polynomially. Here the growth of a variety is the growth of the $*$-codimension sequence of a generating algebra.

Corollary 1 *The algebras M and $F \oplus F$ are the only algebras, up to T^*-equivalence, generating varieties of almost polynomial growth.*

Much interest was put into the study of the varieties of almost polynomial growth. In [12] all subvarieties of the varieties generated by M and $F \oplus F$ were completely classified and a complete list of finite dimensional $*$-algebras generating them was given.

Moreover the authors classified all the minimal subvarieties of polynomial growth of the varieties generated by M and $F \oplus F$; minimal varieties of polynomial growth are varieties $\mathcal{V}^*$ satisfying the property: $c_n^*(\mathcal{V}^*) \approx qn^k$ for some $k \geq 1, q > 0$, and for any proper subvariety $\mathcal{U}^* \subsetneq \mathcal{V}^*$, $c_n^*(\mathcal{U}^*) \approx q'n^t$ with $t < k$.

Now, we recall that a variety $\mathcal{V}^*$ of algebras with involution is minimal with respect to the $*$-exponent if for any proper subvariety $\mathcal{U}^*$ we have that $\exp^*(\mathcal{V}^*) > \exp^*(\mathcal{U}^*)$. Here the $*$-exponent (resp. the nth $*$-codimension) of a variety is the $*$-exponent (resp. the nth $*$-codimension) of a generating algebra.

By using this definition we can say that the algebras M and $F \oplus F$ are the only algebras, up to T^*-equivalence, generating minimal varieties of $*$-exponent 2.

We finish this section by giving an equivalent formulation of Theorem 6 through another numerical sequence.

Let H_n be the hyperoctahedral group of degree n, i.e., $H_n = \mathbb{Z}_2 \wr S_n$, the wreath product of the multiplicative group of order two with S_n. The space P_n^* has a natural left H_n-module structure induced by defining for $h = (a_1, \ldots, a_n; \sigma) \in H_n$, $hy_i = y_{\sigma(i)}$, $hz_i = z_{\sigma(i)}^{a_{\sigma(i)}} = \pm z_{\sigma(i)}$.

Since $P_n^* \cap \mathrm{Id}^*(A)$ is invariant under this H_n-action, the space $P_n^*/(P_n^* \cap \mathrm{Id}^*(A))$ has the structure of a left H_n-module and its character $\chi_n^*(A)$, called the nth $*$-cocharacter of A, decomposes as

$$\chi_n^*(A) = \sum_{|\lambda|+|\mu|=n} m_{\lambda,\mu}\chi_{\lambda,\mu}, \tag{6}$$

where $\lambda \vdash r$, $\mu \vdash n-r$, $r = 0, 1, \ldots, n$ and $m_{\lambda,\mu} \geq 0$ is the multiplicity of the irreducible H_n-character $\chi_{\lambda,\mu}$ associated to the pair (λ, μ).

Also

$$l_n^*(A) = \sum_{|\lambda|+|\mu|=n} m_{\lambda,\mu}$$

is called the nth $*$-colength of A.

Theorem 7 ([15]) *Let A be an algebra with involution over a field F of characteristic zero. Then $c_n^*(A)$, $n = 1, 2, \ldots$, is polynomially bounded if and only if $l_n^*(A) \leq k$, for some constant k and for all $n \geq 1$.*

Notice that if $l_n^*(A) \leq 3$, then for n large enough, $l_n^*(A)$ is always constant.

Acknowledgment Partially supported by GNSAGA of INDAM.

References

1. Aljadeff, E., Giambruno, A., Karasik, Y.: Polynomial identities with involution, superinvolutions and the Grassmann envelope. Proc. Am. Math. Soc. **145**(5), 1843–1857 (2017)
2. Amitsur, S.A.: Identities in rings with involutions. Isr. J. Math. **7**, 63–68 (1969)
3. Bahturin, Y., Giambruno, A., Zaicev, M.: G-identities on associative algebras. Proc. Am. Math. Soc. **127**(1), 63–69 (1999)
4. Giambruno, A., La Mattina, D.: Codimensions of star-algebras and low exponential growth. Isr. J. Math. (2020). https://doi.org/10.1007/s11856-020-2023-y
5. Giambruno, A., Mishchenko, S.: On star-varieties with almost polynomial growth. Algebra Colloq. **8**(1), 33–42 (2001)
6. Giambruno, A., Regev, A.: Wreath products and P.I. algebras. J. Pure Appl. Algebra **35**(2), 133–149 (1985)

7. Giambruno, A., Zaicev, M.: Involution codimensions of finite-dimensional algebras and exponential growth. J. Algebra **222**(2), 471–484 (1999)
8. Giambruno, A., Ioppolo, A., La Mattina, D.: Varieties of algebras with superinvolution of almost polynomial growth. Algebra Represent. Theory **19**(3), 599–611 (2016)
9. Giambruno, A., Polcino Milies, C., Valenti, A.: Star-polynomial identities: computing the exponential growth of the codimensions. J. Algebra **469**, 302–322 (2017)
10. Giambruno, A., La Mattina, D., Polcino Milies, C.: Understanding star-fundamental algebras. Proc. Amer. Math. Soc. (in press)
11. Giambruno, A., La Mattina, D., Polcino Milies, C.: Star-fundamental algebras: polynomial identities and asymptotics. Trans. Am. Math. Soc. **373**(11), 7869–7899 (2020). https://doi.org/10.1090/tran/8182
12. La Mattina, D., Martino, F.: Polynomial growth and star-varieties. J. Pure Appl. Algebra **220**(1), 246–262 (2016)
13. La Mattina, D., Misso, P.: Algebras with involution with linear codimension growth. J. Algebra **305**(1), 270–291 (2006)
14. La Mattina, D., Mauceri, S., Misso, P.: Polynomial growth and identities of superalgebras and star-algebras. J. Pure Appl. Algebra **213**, 2087–2094 (2009)
15. La Mattina, D., do Nascimento, T.S., Vieira, A.C.: Minimal star-varieties of polynomial growth and bounded colength. J. Pure Appl. Algebra **222**(7), 1765–1785 (2018)
16. Mishchenko, S., Valenti, A.: A star-variety with almost polynomial growth. J. Algebra **223**(1), 66–84 (2000)
17. Sviridova, I.: Finitely generated algebras with involution and their identities. J. Algebra **383**, 144–167 (2013)

Context-Free Languages and Associative Algebras with Algebraic Hilbert Series

Roberto La Scala and Dmitri Piontkovski

Abstract In this paper, homological methods together with the theory of formal languages of theoretical computer science are proved to be effective tools to determine the growth and the Hilbert series of an associative algebra. Namely, we construct a class of finitely presented associative algebras related to a family of context-free languages. This allows us to connect the Hilbert series of these algebras with the generating functions of such languages. In particular, we obtain a class of finitely presented graded algebras with non-rational algebraic Hilbert series.

Keywords Algebraic generating functions · Associative algebras · Graded homology

1 Introduction

The Hilbert series (or growth series) of graded and filtered structures is one of the most important invariants for infinite dimensional algebraic objects. In particular for associative algebras, such series is the most natural tool for finding the growth. For a number of important classes of algebras, Hilbert series are of special form, so that they are useful to characterize Koszul algebras, Noetherian algebras, some classes of PI algebras, algebras of small homological dimensions (such as noncommutative complete intersection) and many other algebras (see, for instance [16, 22]).

For many important classes, the Hilbert series is a rational function. It was Hilbert himself who proved this property for (finitely generated) commutative algebras. Govorov [6] proved in 1972 that finitely presented monomial algebras

R. L. Scala (✉)
Dipartimento di Matematica, Università di Bari, Bari, Italy
e-mail: roberto.lascala@uniba.it

D. Piontkovski
Department of Mathematics for Economics, National Research University Higher School of Economics, Moscow, Russia
e-mail: dpiontkovski@hse.ru

O. M. Di Vincenzo, A. Giambruno (eds.), *Polynomial Identities in Algebras*, Springer INdAM Series 44, https://doi.org/10.1007/978-3-030-63111-6_14

have rational Hilbert series. After that, the rationality of Hilbert series has been proved for a number of classes of associative algebras, such as, for example, prime PI algebras [3] and relatively free PI algebras [4].

Moreover, Ufnarovski have introduced [21] a general class of algebras with rational Hilbert series by connecting the theory of algebras with the theory of formal languages of theoretical computer science. The regular languages are the ones that are recognized by finite-state automata and it is well-known that such languages have rational generating functions (see, e.g., [17]). A finitely generated algebra defined by a monomial set of relations is called automaton (or Ufnarovski automaton) if the set of relations forms a regular language, see details in Sect. 2.2. Moreover, the set of normal words of such an algebra is also a regular language. Since regular languages are known to have rational generating functions, the Hilbert series of automaton algebras are always rational. Moreover, the finitely presented monomial algebras are automaton, so that the Govorov's rationality theorem follows from this result as a particular case. Optimal algorithms to compute the rational (univariate and multivariate) Hilbert series of an automaton algebra are due to La Scala and Tiwari [10, 11].

Govorov had conjectured in 1972 that all finitely presented graded algebras have rational series. However, a couple of counterexamples were found by Shearer [20] in 1980 and Kobayashi [13] in 1981. We remark that the corresponding non-rational Hilbert series were algebraic functions, that is, roots of polynomials with coefficients in the rational function field. At the same time, classes of finitely presented universal enveloping algebras with intermediate growth (having hence transcendental Hilbert series) were also discovered in [22]. Examples of such algebras have been recently introduced also by Koçak [14, 15]. Other examples of finitely presented algebras with transcendental Hilbert series are recently given in [8, 19]. Note that for important classes of algebras (such as Koszul algebras or graded Noetherian algebras) the question about rationality of Hilbert series is still open.

Whereas a number of algebras with either rational or transcendental Hilbert series are known, there are only few examples of algebras with non-rational *algebraic* Hilbert series. We have therefore defined the class of *homologically unambiguous* algebras [12]. These are the monomial algebras such that their relations together with the monomial bases of their homologies are unambiguous context-free languages (see details in Sect. 2.3). If such an algebra has finite homological dimension, then its Hilbert series is algebraic. An example of a finitely presented graded algebra with non-rational algebraic Hilbert series is constructed in [12]. This is in fact an algebra such that the associated monomial algebra is homologically unambigous.

In this paper, we give a general method to construct a finitely presented algebra of finite homological dimension (in fact, we provide three of them) such that for the associated monomial algebra, both the set of relations and the monomial bases of the homologies are context-free languages. Namely, for each context-free language L which is a homomorphic image of the Dyck language over a finite alphabet, we construct a finitely presented algebra such that its Hilbert series is calculated in terms of the generating function $H_L(z)$ of the language L.

The paper is organized as follows. In Sect. 2, we briefly recall the notions of Hilbert series of associative algebras, monomial algebras, and context-free languages. In Sect. 3, we discuss a construction that assigns a finitely presented algebra to such a language and give explicit formulae for the homologies and the Hilbert series of the algebra. Finally, we describe several examples of such algebras with particular choices of L. These are new examples of finitely presented graded algebras with non-rational algebraic Hilbert series.

2 Preliminaries

2.1 Associative Algebras and Their Hilbert Series

The free monoid generated by a set X is denoted by X^*. Following theoretical computer science, we call the elements of X^* *words* and the subsets of X^* *languages*.

Let A be a unital associative algebra over a field k generated by a finite subset X. The words and languages then correspond to elements and subsets of A which we denote by the same symbols.

Let us define a degree function on A by putting $\deg x_i = d_i \in \mathbb{Z}_{>0}$ for all $x_i \in X$. Then, we put $\deg w = d_{i_1} + \cdots + d_{i_s}$ for a word $w = x_{i_1} \cdots x_{i_s} \in X^*$ and $\deg a = \max\{\deg w_i\}$ for an element $0 \neq a = \sum_i c_i w_i \in A$ $(c_i \in k)$. This gives an ascending filtration $F_0 = k1$, $F_d = k\{a \mid \deg a \leq d\}$ on A.

The Hilbert series of A is then defined as the formal power series

$$H_A(z) = \sum_{n \geq 0} z^n \dim(F_n / F_{n-1}).$$

The algebra A is *graded* if it is equal to the direct sum $A_0 \oplus A_1 \oplus A_2 \oplus \dots$, where $A_0 = k1$ and $A_d = k\{w \mid \deg w = d\}$. In this case, we have $H_A(z) = \sum_{n \geq 0} z^n \dim A_n$.

We assume now that the reader is familiar with the theory of noncommutative Gröbner bases which are also called Gröbner-Shirshov bases (see, for instance, [5, 18, 22]). We recall here some basic foundations. Let I be a two-sided ideal of the free associative algebra $F = k\langle X \rangle$ such that $A = F/I$. Suppose we have a multiplicative well-ordering $\prec$ on X^*. This gives a monomial ordering on F. Then, let $0 \neq f = \sum_{i=1}^{s} c_i w_i \in F$ with $0 \neq c_i \in k$, $w_i \in X^*$ and $w_1 \succ w_2 \succ \dots \succ w_s$. The word $\mathrm{lm}(f) = w_1$ is called the *leading monomial of* f. A (possibly infinite) subset $U \subset I$ is called a *Gröbner-Shirshov basis*, briefly a *GS-basis of* I, if $\mathrm{lm}(U) = \{\mathrm{lm}(f) \mid 0 \neq f \in U\} \subset X^*$ is a monomial basis of the monomial ideal

$$\mathrm{LM}(I) = \langle \mathrm{lm}(f) \mid 0 \neq f \in I \rangle \subset F.$$

The GS-basis U is called *minimal* if the monomial basis $\mathrm{lm}(U)$ is such. We call $\mathrm{LM}(I)$ the *leading monomial ideal of* I. If $J = \mathrm{LM}(I)$, one defines the corresponding monomial algebra $B = F/J$.

If the algebra A is graded, then it is easy to prove that $\mathrm{HS}(A) = \mathrm{HS}(B)$. Moreover, the same is true for non-graded algebras if the ordering is degree-compatible, that is, $w_1 \prec w_2$ provided that $\deg w_1 < \deg w_2$. So, to compute the Hilbert series of the general algebra A it is sufficient to calculate it for the associated monomial algebra B.

2.2 The Homology and Hilbert Series of Monomial Algebras

In this sections, we recall some basic facts about monomial algebras and their homology. We have discussed this topic in details in [12]. For a complete introduction, we refer the reader to [9, 22].

Let $A = k\langle X\rangle/\langle L\rangle$ be an associative algebra generated by a finite set $X = \{x_1, \dots, x_n\}$ subject to a monomial set of relations $L_1 \subset X^*$. We assume that this set of relations is minimal, that is, the language L_1 is subword-free.

The homology $\mathrm{Tor}^A_\cdot(k, k)$ of monomial algebras are calculated via so-called chains [1, 2]. More precisely, we have $\mathrm{Tor}^A_0(k, k) = k$, while for $n \geq 0$ the graded vector space $\mathrm{Tor}^A_{n+1}(k, k)$ is isomorphic to the span of a language L_n called *the language of n-chains* of the monomial algebra A. Denote $L = X^* L_1 X^*$ and $X^+ = X^* \setminus \{1\}$. Then, for all $t \geq 1$ it holds that

$$L_{2t} = (X^+ L^t \cap L^t X^+) \setminus (X^+ L^t X^+ \cup L^{t+1}),$$
$$L_{2t-1} = (X^+ L^{t-1} X^+ \cap L^t) \setminus (X^+ L^t \cup L^t X^+).$$

In particular, $L_0 = X$, and for $t = 1$ we get L_1 both on the left- and right-hand sides.

Note that the classical definition of chains is recursive and the above one is based on the Govorov's formulae for homologies of associative algebras [7]. We discuss these definitions in [12].

Given a degree function on X^*, one can define a generating function of any language $W \subset X^*$ by $H_W(z) = \sum_{w \in W} z^{\deg w}$. Then, the Hilbert series of the graded vector space $\mathrm{Tor}^A_{n+1}(k, k)$ is equal to $H_{L_n}(z)$. From the exact sequence corresponding to the minimal free resolution of the A-module k, one gets the following formula for its Hilbert series:

$$H_A(z) = \left(1 - \sum_{i=0}^{k} (-1)^i H_{L_i}(z)\right)^{-1}. \tag{1}$$

2.3 Automaton Algebras and Homologically Unambiguous Algebras

The automaton algebras were introduced by Ufnarovski [21]. In our terms, one can defined them as follows.

Definition 1 A monomial algebra is called automaton, if the following equivalent conditions hold:

(i) the set S of nonzero words in A is a regular language;
(ii) the set of relations L_1 is a regular language;
(iii) the chain languages L_n are regular, for all n.

We refer the reader to [10, 11, 21, 22] for details on automaton algebras. In particular, the Hilbert series of each automaton algebra is a rational function. It can be calculated using the methods of formal language theory as the generating function of the regular language S (see [10–12]).

If an algebra A has a non-rational Hilbert series, it cannot be automaton. So, for such algebras the condition that the corresponding languages are automaton should be weakened. A more general class of languages is the class of context-free (c-f for short) languages. Unfortunately, there does not exist an algorithm to calculate the generating function of any c-f language. However, for some classes of c-f languages such algorithms do exist. The most important result in this direction is a theorem by Chomsky and Schützenberger stating that the generating function of each *unambiguous* c-f language is an algebraic function. Moreover, the theorem provides a way to construct a system of algebraic equations which defines the generating function. For a detailed description of these methods, see [12, 17].

A monomial algebra is called *(homologically) unambiguous* if all chain languages L_n are unambiguous c-f. We refer the reader to [12] for the discussion and examples of such algebras. Note that unlike the automaton case, it is not sufficient to check this condition for L_1 [12, Example 5.4]. If the unambiguous algebra A has finite homological dimension, then it follows from (1) that the Hilbert series $H_A(z)$ is an algebraic function.

Suppose that a finitely generated algebra A is not monomial. If the associated monomial algebra $\widehat{A}$ is unambiguous, then $H_A(z)$ is algebraic. In [12], we have provided an example of a finitely presented algebra of that kind such that its Hilbert series is not rational.

3 Finitely Presented Algebras Associated to Context-Free Languages

3.1 The General Construction

Let us describe a class of finitely presented associative algebras. Each algebra of this class is related to an arbitrary context free language L which is a homomorphic image of a Dyck language $D_n = D_n(a_1, b_1, \dots, a_n, b_n)$. Recall that D_n consists of the words with balanced parentheses of n possible kinds, where a_i is the opening parenthesis and b_i is the closing parenthesis of the i-th pair. Note that a classical theorem by Chomsky and Schützenberger provides that each context-free language is an intersection of such a language L with a regular one, so that this class of languages is quite general.

Suppose $L = \phi(D_n) \subset T^*$ with $T = \{t_1, \dots, t_m\}$. Let A be an algebra generated by the set X of variables $a_i, b_i, a_i^j, b_i^j, x, e, y, t_k$, where i, j run in $\{1, \dots, n\}$ and k runs in $\{1, \dots, m\}$. The relations are defined for each $i, j, l \in \{1, \dots, n\}$ as follows:

(i) $a_i^i x - x a_i^i, b_1^1 x - xe$;
(ii) $a_i^j a_l - a_i a_l^j, a_i^j b_l - a_i b_l^j, b_i^j a_l - b_i a_l^j, b_i^j b_l - b_i b_l^j, a_i^j e - a_i b_j, b_i^j e - b_i b_j$;
(iii) $a_i y - y\phi(a_i), b_i y - y\phi(b_i), a_i^j y, b_i^j y$;
(iv) xye.

We denote by R the set of the above relations.

Denote $d = \max\{\deg \phi(a_1), \deg \phi(b_1), \dots, \deg \phi(a_n), \deg \phi(b_n)\}$. We introduce a new degree function on the variables as follows: $|a_i| = |b_i| = |a_i^j| = |b_i^j| = |x| = |y| = |e| = D$ and $|t_k| = 1$ for all possible i, j, k. Set a deglex ordering over the words on all variables by putting $a_{\bullet}^{\bullet} > b_{\bullet}^{\bullet} > a_{\bullet} > b_{\bullet} > e > x > y > t_{\bullet}$.

Theorem 1 *The minimal Gröbner basis of the ideal generated by R is the disjoint union of the set of the relations (i)–(iii) with the set of monomials*

$$xP_n yLe,$$

where $P_n = (D_n e)^*$ is the language consisting of the empty word and the words of the form

$$e^{i_0} w_1 e^{i_1} \dots w_s e^{i_s}$$

for all $s, i_0 \geq 0$, $i_1, \dots, i_s > 0$ and $w_1, \dots, w_s \in D_n$.

Proof First, note that the relations (i)–(iii) form the minimal Gröbner basis of the ideal generated by them. Indeed, the first term of each relation is equal to its leading monomial. The only overlapping of them are between the ones of the first four types of relations (ii) and the leading monomials to the first two types of relations (iii). In all cases, the resulting s-polynomials are reduced to zero. For example, the intersection of the leading monomials of $a_i^j b_l - a_i b_l^j$ and $b_l y - y\phi(b_l)$ gives an s-polynomial

$$(a_i^j b_l - a_i b_l^j)y - a_i^j(b_l y - y\phi(b_l)) = a_i^j y\phi(b_l) - a_i b_l^j y,$$

which is immediately reduced to zero by the last two elements of (iii). All other cases of overlapping are similar.

Now, we are ready to prove that the Gröbner basis mentioned in the theorem consists of the relations (i)–(iv) and a subset of $xP_n yLe$. We proceed by the induction on the length d of the leading monomial of a Gröbner basis element which we denote by g. The induction base $d = 2$ follows immediately.

Let $d \geq 3$. The element g is obtained as a complete reduction of an s-polynomial based on the elements having leading monomials with lower lengths. By the induction, the only possible (new) overlappings of such leading monomials are between the relations of type (i) and some element $g' \in xP_n yLe$ of the Gröbner basis. Let $g' = xpyev$ (with $p \in P_n, v \in L$). Then, g is the complete reduction of one of the s-polynomials $s_1 = a_i^i g' - (a_i^i x - xa_i^i)pyve = xa_i^i pyve$ or $s_2 = b_1^1 g' - (b_1^1 x - xe)pyve = xepyve$. Here s_2 belongs to $xP_n yLe$, so that we can assume that g is the complete reduction of s_1.

If $p = 1$, then $s_1 = xa_i^i yve$ is reduced to 0 by (iii).

Suppose that $p \in D_n e$; then either $p = e$ or $p = \tilde{p}b_j e$ for some subword $\tilde{p}$ and some j. In the first case, $s_1 = xa_i^i eyve$ is reduced to $g = xy\phi(a_i b_i)ve \in xyLe \subset xP_n yLe$. In the second case, $s_1 = xa_i^i \tilde{p}b_j eyve$ is reduced by (ii) to the monomial $xa_i \tilde{p}b_j b_i yve$. This monomial is reduced by (iii) to the monomial $g = xy\phi(w)ve$, where $w = a_i \tilde{p}b_j b_i \in D_n$. We see that in this case again $g \in xyLe$.

Now, let $p \neq 1$ and $p \in P_n \setminus D_n e$. We have $p = wep'$ for some $w \in D_n$, $p' \in P_n e$. Then, the complete reduction $g = xa_i wb_i p'yve$ of $s_1 = xa_i^i wep'yve$ belongs to $xP_n eyL$.

Now, it remains to prove that all elements of the set $xP_n yLe$ belong to the minimal Gröbner basis. Indeed, let us define a homomorphism $\alpha : \{a_i, b_i, e \mid i = 1, \ldots, n\}^* \to \{a_i^i, b_1^1, e \mid i = 1, \ldots, n\}^*$ by putting $\alpha : a_i \mapsto a_i^i, b_i \mapsto b_1^1, e \mapsto b_1^1$. Then, for each two words $v \in D_n, w \in P_n$ the element $xwyve$ is the complete reduction of the word $\alpha(wv)xye$. Since this word $\alpha(wv)xye$ is divisible by the monomial (iv), it belongs to the ideal $\langle R \rangle$, so that the (normal) element $xwy\phi(v)e$ belongs to the minimal Gröbner basis, for all $v \in D_n, w \in P_n$. □

Note that the generating series of the languages D_n and P_n (with the standard degree functions) are

$$H_{D_n}(q) = \frac{1 - \sqrt{1 - 4nq^2}}{2nq^2}$$

and

$$H_{P_n}(q) = H_{(D_n e)^*}(q) = \frac{1}{1 - qH_{D_n}(q)} = \frac{2nq}{2nq - 1 + \sqrt{1 - 4nq^2}}.$$

Corollary 1

(*a*) *For the associated monomial algebra $\widehat{A}$ to the algebra A above, the graded vectors spaces $Tor_i^{\widehat{A}}(k,k)$ are spanned, respectively, by the following sets L_{i-1}*

- *X (for $i = 1$),*
- *the disjoint union of the set of the first terms of (i)–(iii) with the set $xP_n yL$ (for $i = 2$),*
- *the set $\{a_1^1, a_2^2, \dots, a_n^n, b^1\}xP_n yL$ (for $i = 3$),*
- *and the empty set for each $i \geq 4$.*

In particular, all these languages L_{i-1} are c-f. Moreover, they are unambiguous c-f if the language L is unambiguous c-f.

(*b*) *The generating functions of these graded vectors spaces with respect to the degree function $|\cdot|$ are*

$$(2n^2 + 2n + 3)z^d + mz \text{ for } i = 1,$$

$$z^{2d}\left(4n^3 + 4n^2 + 3n + 1\right) + z^{3d} H_{P_n}(z^d) H_L(z) \text{ for } i = 2,$$

$$(n+1)z^{4d} H_{P_n}(z^d) H_L(z) \text{ for } i = 3,$$

$$\text{and } 0 \text{ for } i \geq 4.$$

(*c*) *Both algebras A and $\widehat{A}$ have global dimension 3.*

(*d*) *The Hilbert series of both algebras A and $\widehat{A}$ with respect to the degree function $|\cdot|$ is*

$$\Big(1 - mz - z^d(2n^2 + 2n + 3) + z^{2d}(4n^3 + 4n^2 + 3n + 1) + z^{3d}(1 - (n+1)z^d) H_{P_n}(z^d) H_L(z)\Big)^{-1}.$$

In particular, both algebras have exponential growth.

Remark 1 The graded vector space $\mathrm{Tor}_3^A(k,k)$ has Hilbert series

$$z^{3d} - z^{3d}\left(1 - (n+1)z^d\right) H_{P_n}(z^d) H_L(z).$$

This formula follows from the equality $H_A(z)^{-1} = H_{\widehat{A}}(z)^{-1}$, where for each of the two 3-dimensional algebras A and $\widehat{A}$ the inverse of the Hilbert series is equal to the Euler characteristic $1 - H_{\mathrm{Tor}_1^{\cdot}(k,k)}(z) + H_{\mathrm{Tor}_2^{\cdot}(k,k)}(z) - H_{\mathrm{Tor}_3^{\cdot}(k,k)}(z)$.

3.2 *Graded Algebras Examples*

In this subsection, we give new examples of graded finitely presented algebras with non-rational algebraic Hilbert series. These examples are based on the general construction described above.

Example 1 With the above notations, let L be the Dyck langauge on the alphabet $T = \{t_1, \ldots, t_{2n}\}$, so that ϕ is the obvious isomorphism and $d = 1$. Then, the algebra A is graded with the standard degree function with Hilbert series

$$\Big(1 - z(2n^2 + 4n + 3) + z^2(4n^3 + 4n^2 + 3n + 1) \\ +z^3(1 - (n+1)z)H_{P_n}(z)H_{D_n}(z)\Big)^{-1}.$$

Note that here

$$H_{P_n}(z)H_{D_n}(z) = \frac{1 - \sqrt{1 - 4nz^2}}{z\left(2nz - 1 + \sqrt{1 - 4nz^2}\right)} = \frac{1 - 2z - \sqrt{-4nz^2 + 1}}{2z\,(nz + z - 1)},$$

so that the Hilbert series

$$H_A(z) = \left(1 - z(2n^2 + 4n + 3) + z^2(4n^3 + 4n^2 + 3n + \frac{1}{2}) + z^3 \right. \\ \left. + \frac{z^2\sqrt{1 - 4nz^2}}{2}\right)^{-1}$$

is not rational. This algebra is homologically unambiguous.

Example 2 Now, let L be the language consisting of all words on t_1 and t_2 with the same number of t_1-s and t_2-s. It is the image of D_2 under the homomorphism $\phi : a_1 \mapsto t_1, b_1 \mapsto t_2, a_2 \mapsto t_2, b_2 \mapsto t_1$. So, here $d = 1$ (so that the algebra A is graded with the standard grading) and $n = m = 2$. Then, the Hilbert series of A is

$$\left(1 - 17z + 55z^2 + z^3(1 - 3z)H_{P_2}(z)H_L(z)\right)^{-1},$$

where $H_L(z) = \sum_{n\geq 0} \binom{2n}{n} z^{2n} = 1/\sqrt{1 - 4z^2}$.

Example 3 Finally, let L be the language over the 26 capital letter alphabet consisting of all words with balanced pairs of the words "BEG", "END" and of the words "FOR, END". It is the image of D_2 under the homomorphism $\phi : a_1 \mapsto BEG, b_1 \mapsto END, a_2 \mapsto FOR, b_2 \mapsto END$. Then, we have $n = 2, m = 26, d = 3$. Since all generators of L have the same length 3, the algebra A is graded with the Hilbert series

$$\left(1 - 26z - 15z^3 + 55z^6 + z^9(1 - 3z^3)H_{P_2}(z^3)H_L(z)\right)^{-1},$$

where

$$H_L(z) = D_2(z^3) = \frac{1 - \sqrt{1 - 8z^6}}{4z^6}.$$

In the explicit form, we have therefore

$$H_A(z) = \left(1 - 26z - 15z^3 + \frac{109}{2}z^6 + z^9 + z^6\frac{\sqrt{1 - 8z^6}}{2}\right)^{-1}.$$

Acknowledgments We acknowledge the support of the University of Bari. The research of the author D.P. is supported by RFBR project 18-01-00908.

References

1. Anick, D.: On the homology of associative algebras. Trans. Amer. Math. Soc. **296**(2), 641–659 (1986)
2. Backelin, J.: La série de Poincaré-Betti d'une algèbre graduée de type fini à une relation est rationnelle. C. R. Acad. Sei. Paris. Sér. A **287**, 843–846 (1978)
3. Bell, J.P.: The Hilbert series of prime PI rings. Israel J. Math. **139**, 1–10 (2004)
4. Belov, A.Y.: On the rationality of Hilbert series of relatively free algebras. Russ. Math. Surv. **52**, 394 (1997)
5. Bokut, L.A., Chen, Y.: Gröbner-Shirshov bases and their calculation. Bull. Math. Sci. **4**, 325–395 (2014)
6. Govorov, V.E.: Graded algebras. (Russian) Mat. Zametki **12**, 197–204 (1972); Translation in Math. Notes **12**, 552–556 (1972)
7. Govorov, V.E.: The global dimension of algebras (Russian). Mat. Zametki **14**, 399–406 (1973); Translation in Math. Notes **14**, 789–792 (1973)
8. Iyudu, N., Shkarin, S.: Automaton algebras and intermediate growth. J. Combin. Algebra **2**, 147–167 (2018)
9. La Scala, R.: Computing minimal free resolutions of right modules over noncommutative algebras. J. Algebra **478**, 458–483 (2017)
10. La Scala, R.: Monomial right ideals and the Hilbert series of noncommutative modules. J. Symb. Comp. **80**, 403–415 (2017)
11. La Scala, R., Tiwari, S.K.: Multigraded Hilbert series of noncommutative modules. J. Algebra **516**, 514–544 (2018)

12. La Scala, R., Piontkovski, D., Tiwari, S.K.: Noncommutative algebras, context-free grammars and algebraic Hilbert series. J. Symb. Comp. **101**, 28–50 (2020). https://doi.org/10.1016/j.jsc.2019.06.005
13. Kobayashi, Y.: Another graded algebra with a nonrational Hilbert series. Proc. Amer. Math. Soc. **81**, 19–22 (1981)
14. Koçak, D.: Finitely presented quadratic algebras of intermediate growth. Algebra Discrete Math. **20**, 69–88 (2015)
15. Koçak, D.: Intermediate growth in finitely presented algebras. Internat. J. Algebra Comput. **27**, 391–401 (2017)
16. Krause, G.R., Lenegan, T.H.: Growth of Algebras and Gelfand-Kirillov Dimension. Revised Edition. Graduate Studies in Mathematics, vol. 22. American Mathematical Society, Providence (2000)
17. Kuich, W., Salomaa, A.: Semirings, automata, languages. EATCS Monographs on Theoretical Computer Science, vol. 5. Springer, Berlin (1986)
18. Mora, T.:An introduction to commutative and noncommutative Gröbner bases, second international colloquium on words, languages and combinatorics (Kyoto, 1992). Theoret. Comput. Sci. **134**, 131–173 (1994)
19. Piontkovski, D.: Algebras of linear growth and the dynamical Mordell-Lang conjecture. Adv. Math. **343**, 141–156 (2019)
20. Shearer, J.: A graded algebra with a nonrational Hilbert series. J. Algebra **62**, 228–231 (1980)
21. Ufnarovski, V.A.: On the use of graphs for calculating the basis, growth and Hilbert series of associative algebras (Russian). Mat. Sb. **180**, 1548–1560 (1989); Translation in Math. USSR-Sb. **68**, 417–428 (1991)
22. Ufnarovski, V.A.: Combinatorial and Asymptotic Methods in Algebra, Algebra VI. Encyclopaedia of Mathematical Sciences, vol. 57, pp. 1–196. Springer, Berlin (1995)

On Almost Nilpotent Varieties of Linear Algebras

Sergey P. Mishchenko and Angela Valenti

To Antonio Giambruno for his 70th birthday

Abstract A variety $\mathcal{V}$ is almost nilpotent if it is not nilpotent but all proper subvarieties are nilpotent. Here we present the results obtained in recent years about almost nilpotent varieties and their classification.

Keywords Varieties · Almost nilpotent · Codimension growth

1 Introduction

Let F be a field of characteristic zero and let $F\{X\}$ be the free non associative algebra on a countable set X over F. If A is a non necessarily associative algebra over F we denote by $Id(A)$ the T-ideal of polynomial identities of A. In the study of $Id(A)$ an important role is played by the sequence $\{c_n(A)\}_{n\geq 1}$ of the codimensions of A. In fact a general strategy in the study of $Id(A)$ is that of studying the space of multilinear polynomials in n fixed variables modulo the identities of the algebra A through the representation theory of the symmetric group S_n on n symbols. Then one attaches to $Id(A)$ a sequence of S_n-modules, $n = 1, 2, \ldots$, and studies the corresponding sequence of characters.

More precisely, for every $n \geq 1$, we consider the space P_n of multilinear polynomials of $F\{X\}$ in the first n variables $x_1, \ldots, x_n$. Since char$F = 0$, the sequence of spaces $P_n \cap Id(A)$, $n = 1, 2, \ldots$, carries all informations about $Id(A)$. The symmetric group S_n acts on P_n by permuting the variables: if $\sigma \in S_n$,

S. P. Mishchenko
Department of Applied Mathematics, Ulyanovsk State University, Ulyanovsk, Russia
e-mail: mishchenkosp@mail.ru

A. Valenti (✉)
Dipartimento di Ingegneria, Università di Palermo, Palermo, Italy
e-mail: angela.valenti@unipa.it

O. M. Di Vincenzo, A. Giambruno (eds.), *Polynomial Identities in Algebras*,
Springer INdAM Series 44, https://doi.org/10.1007/978-3-030-63111-6_15

$f(x_1, \ldots, x_n) \in P_n$,

$$\sigma f(x_1, \ldots, x_n) = f(x_{\sigma(1)}, \ldots, x_{\sigma(n)}).$$

The space $P_n \cap Id(A)$ is invariant under this action and one studies the structure of $P_n(A) = P_n/(P_n \cap Id(A))$ as an S_n-module. The S_n-character of $P_n(A)$, denoted $\chi_n(A)$, is the n-th cocharacter of A. By complete reducibility one writes

$$\chi_n(A) = \sum_{\lambda \vdash n} m_\lambda \chi_\lambda,$$

where χ_λ is the irreducible S_n-character corresponding to the partition λ of n and $m_\lambda \geq 0$ is the corresponding multiplicity (see for example [8] for the representation theory of the symmetric group).

The number of irreducible summands in $\chi_n(A)$, i.e., the length of the S_n-module $P_n(A)$, is called the n-th colength of A, and is denoted $l_n(A)$. Its degree $c_n(A) = \dim P_n(A)$ is the n-th codimension of A and gives a quantitative estimate of the polynomial identities satisfied by A. Then clearly

$$l_n(A) = \sum_{\lambda \vdash n} m_\lambda$$

and

$$c_n(A) = \sum_{\lambda \vdash n} m_\lambda d_\lambda,$$

where $d_\lambda = \deg \chi_\lambda$ is the degree of the irreducible character χ_λ.

In the language of varieties if $\mathcal{V} = var(A)$ is the variety generated by an algebra A, then we write $Id(\mathcal{V}) = Id(A)$, $\chi_n(\mathcal{V}) = \chi_n(A)$, $l_n(\mathcal{V}) = l_n(A)$ and $c_n(\mathcal{V}) = c_n(A)$. The growth of $\mathcal{V}$ is the growth of the sequence of codimensions of A.

The first result on the asymptotic behavior of $c_n(\mathcal{V})$ is due to Regev [23]. He proved that if $\mathcal{V}$ is a non-trivial variety of associative algebras, then the sequence of codimensions is exponentially bounded, i.e., there exist constants $\alpha, a > 0$ such that $c_n(\mathcal{V}) \leq \alpha a^n$, for all n. In case $\mathcal{V}$ is a variety of non associative algebras, such sequence has a much more involved behavior and can have overexponential growth (see [3, 6, 22, 25]).

If the sequence of codimensions $c_n(\mathcal{V})$ is exponentially bounded then one naturally defines, Exp($\mathcal{V}$), the exponent of the variety. Let

$$\overline{\mathrm{Exp}(\mathcal{V})} = \lim\sup_{n\to\infty} \sqrt[n]{c_n(\mathcal{V})}, \quad \underline{\mathrm{Exp}(\mathcal{V})} = \lim\inf_{n\to\infty} \sqrt[n]{c_n(\mathcal{V})}$$

the upper and lower exponent respectively of the variety $\mathcal{V}$. If $\overline{\mathrm{Exp}(\mathcal{V}} = \underline{\mathrm{Exp}(\mathcal{V})}$ then

$$\mathrm{Exp}(\mathcal{V}) = \overline{\mathrm{Exp}(\mathcal{V})} = \underline{\mathrm{Exp}(\mathcal{V})}.$$

Let us recall that a variety $\mathcal{V}$ has polynomial growth if there exist α, t such that $c_n(\mathcal{V}) = \alpha n^t$. Moreover $\mathcal{V}$ has intermediate growth if for any $k > 0$, $a > 1$ there exist constants C_1, C_2, such that, for any n, takes place the inequality

$$C_1 n^k < c_n(\mathcal{V}) < C_2 a^n.$$

We say that a variety $\mathcal{V}$ has subexponential growth if for any constant $B > 1$ there exists n_0 such that for all $n > n_0$, $c_n(\mathcal{V}) < B^n$. Clearly varieties with polynomial growth or intermediate growth have subexponential growth and it can be shown that varieties realizing each growth can be constructed. For instance a class of varieties of intermediate growth was constructed in [7].

Recall that an algebra A is nilpotent if, for some $k \geq 1$, any product of k elements of A (with all possible arrangements of the brackets) is zero. Clearly if A is a nilpotent algebra then $c_n(A) = 0$, for n large. Accordingly we say that a variety is nilpotent if it is generated by a nilpotent algebra. So we say that a variety $\mathcal{V}$ is almost nilpotent if it is not nilpotent but all proper subvarieties are nilpotent.

Almost nilpotent varieties exist and not only those having linear growth, which is not surprising, but also almost nilpotent varieties with exponential growth. In fact in [16] an almost nilpotent variety of exponent two was constructed and, in [15], this example was extended to prove the existence of almost nilpotent varieties with any integral exponent.

The aim of this paper is to review on the results obtained in recent years about almost nilpotent varieties and their classification.

We present nine almost nilpotent varieties in different classes of algebras and we characterize those having subexponential growth. Moreover we recall some results about infinite series of almost nilpotent varieties with polynomial growth and we describe almost nilpotent varieties of exponential growth.

We next recall some basic properties of the representation theory of the symmetric group that we shall use in the sequel. Let $\lambda \vdash n$ be a partition of n and let T_λ be a Young tableau of shape $\lambda \vdash n$. We denote by e_{T_λ} the corresponding essential idempotent, i.e., $e_{T_\lambda}^2 = \alpha e_{T_\lambda}$, $0 \neq \alpha \in F$, of the group algebra FS_n. Recall that $e_{T_\lambda} = R_{T_\lambda}^+ C_{T_\lambda}^-$ where $R_{T_\lambda}^+ = \sum_{\sigma \in R_{T_\lambda}} \sigma$, and $C_{T_\lambda}^- = \sum_{\tau \in C_{T_\lambda}} (\mathrm{sgn}\tau)\tau$ and R_{T_λ}, C_{T_λ} are the groups of row and column stabilizers of T_λ, respectively. Recall that if M_λ is an irreducible S_n-submodule of $P_n(A)$ corresponding to λ there exists a polynomial $f(x_1, \ldots, x_n) \in P_n$ and a tableau T_λ such that $e_{T_\lambda} f(x_1, \ldots, x_n) \notin Id(A)$. Let $e'_{T_\lambda} = C_{T_\lambda}^- R_{T_\lambda}^+ C_{T_\lambda}^-$. Since $R_{T_\lambda}^+ C_{T_\lambda}^- R_{T_\lambda}^+ C_{T_\lambda}^- \neq 0$ then e'_{T_λ} is a nonzero essential idempotent that generates the same irreducible module and so also $e'_{T_\lambda} f(x_1, \ldots, x_n) \notin Id(A)$.

In what follows we shall also denote by $g(\lambda)$ the polynomial obtained from the essential idempotent corresponding to a tableau of shape λ by identifying the elements in each row. Recall that $g(\lambda)$ is an highest weight vector of the general linear group $GL_k(F)$ where k is the number of distinct part of λ (see [4]).

We recall that for an algebra A, R_a denotes the operator of right multiplication by $a \in A$. From now on we shall write $X_i = R_{x_i}$ for the right multiplication in $F\{X\}$ by the variable x_i. In order to simplify the notation we shall also use the following convention: a monomial M in which some variables are overlined by the same sign (say $\bar{}$, $\tilde{}$ etc), must be read as the polynomial in which those variables are alternated. Moreover through the paper we shall omit the parenthesis in left normed monomials, i.e. $xyz = (xy)z$.

2 Almost Nilpotent Varieties Generated by One or Two Dimensional Algebra

It is well known that a not nilpotent one dimensional algebra A_0 is isomorphic to the basic field F. Clearly the variety generated by this algebra $\mathcal{V}_0 = \text{var}(A_0) = \text{var}(F)$ is an almost nilpotent variety and $\chi_n(\mathcal{V}_0) = \chi_{(n)}$, where $(n) \vdash n$ is a partition of n.

Let's now consider four two-dimensional algebras A_1, A_2, A_3, A_4 with basis $\{a, b\}$ and with the following tables of multiplication:

A_1	a	b
a	0	a
b	0	0

A_2	a	b
a	0	0
b	a	0

A_3	a	b
a	0	a
b	a	0

A_4	a	b
a	0	a
b	$-a$	0

The purpose of this section is to prove that these two-dimensional algebras generate almost nilpotent varieties.

Definition 1 Let $\mathcal{V}_1$ be the variety of algebras satisfying the following identities:

(1) $x(yz) \equiv 0$;
(2) $(xy)z \equiv (xz)y + x(yz)$.

Let ${}_2\mathcal{N}$ denote the variety of left nilpotent algebras of index two, that is the variety of algebras satisfying the identity $x(yz) \equiv 0$. In others words $\mathcal{V}_1$ is the variety of Leibniz algebras contained in ${}_2\mathcal{N}$. From (1) and (2) it follows that

$$(3)\ xyz \equiv xzy$$

holds in $\mathcal{V}_1$.

Let A_1 be the algebra with basis $\{a, b\}$ and such that $ab = a$ and $a^2 = b^2 = ba = 0$. Clearly $A_1 \in \mathcal{V}_1$. We have the following (see [17])

Proposition 1 *$\mathcal{V}_1 = var(A_1)$ and $\chi_n(\mathcal{V}_1) = \chi_{(n)} + \chi_{(n-1,1)}$.*

Proof Let P_n^i be the space of left normed multilinear polynomials in $x_1, \ldots, x_n$ starting with x_i. Then clearly

$$P_n(\mathcal{V}_1) = P_n^1(\mathcal{V}_1) + \ldots + P_n^n(\mathcal{V}_1).$$

Since by (3) dim $P_n^i(\mathcal{V}_1) = 1$ it follows that $c_n(\mathcal{V}_1) \leq n$.

Let $\chi_n(A_1) = \sum m_\lambda \chi_\lambda$. Clearly the polynomial $g((n)) = x_1 X_1^{n-1}$ is not an identity of A_1 therefore $m_{(n)} \geq 1$. Notice that modulo the identities of A_1 all non-zero monomials of the free algebra are left normed then $m_{(n)}(A_1) = d_{(n)} = 1$.

Consider now the partition $\lambda = (n-1, 1)$ and let $g((n-1, 1)) = \bar{x_1}\bar{x_2}X_1^{n-2}$. It is easy to see that $g((n-1, 1))$ is not an identity of A_1 therefore we obtain that $m_{(n-1,1)} \geq 1$. Since $d_{(n)} = 1$ and $d_{(n-1,1)} = n-1$, it follows that

$$n \geq c_n(\mathcal{V}_1) \geq c_n(A_1) \geq d_{(n)} + d_{(n-1,1)} = n.$$

Hence $c_n(\mathcal{V}_1) = c_n(A_1) = n$, $\mathcal{V}_1 = \text{var}(A_1)$, $\chi_n(\mathcal{V}_1) = \chi_{(n)} + \chi_{(n-1,1)}$ and we are done.

□

A characterization of the algebra A_1 is given in the following

Proposition 2 (Mishchenko and Valenti [17]) *Let $\mathcal{V}$ be a variety of algebras satisfying the identity $x(yz) \equiv 0$. Then $A_1 \notin \mathcal{V}$ if and only if $x_0 X_1^m \equiv 0$ holds in $\mathcal{V}$, for some $m \geq 1$.*

Proof Consider the basis $\{a, b\}$ of A_1. Since $aR_b^m \neq 0$, then $x_0 X_1^m \notin Id(A_1)$, for all $m \geq 1$. Hence if, for some m, $x_0 X_1^m \equiv 0$ is an identity of $\mathcal{V}$ we have that $A_1 \notin \mathcal{V}$.

Viceversa if $A_1 \notin \mathcal{V}$, then for some n, there exists an irreducible S_n-character χ_λ appearing with multiplicity m'_λ in $\chi_n(A_1)$ and m_λ in $\chi_n(\mathcal{V})$ with $m'_\lambda > m_\lambda$.

Since $\chi_n(A_1) = \chi_{(n)} + \chi_{(n-1,1)}$ it follows that either $g((n)) = x_1 X_1^{n-1} \equiv 0$ holds in $\mathcal{V}$ or $g((n-1, 1)) = \sum_{s=0}^{n-2} \alpha_s \bar{x_1} X_1^s \bar{x_2} X_1^{n-s-2} \equiv 0$ holds in $\mathcal{V}$, for some (not all zero) coefficients $\alpha_s \in F$. Notice that here $g((n-1, 1))$ is a highest weight vector of $GL_2(F)$ written as a linear combination of highest weight vectors corresponding to standard tableaux of shape $(n-1, 1)$.

If $g((n)) = x_1 X_1^{n-1} \equiv 0$ in $\mathcal{V}$ we make the substitution $x_1 = x_1 + x_0 x_1$ and we get $x_0 X_1^n \equiv 0$.

If $g((n-1, 1)) = \sum_{s=0}^{n-2} \alpha_s \bar{x_1} X_1^s \bar{x_2} X_1^{n-s-2} \equiv 0$ after the substitution $x_2 = x_1^2$ we get $(\sum_{s=0}^{n-2} \alpha_s) x_1^2 X_1^{n-2} \equiv 0$. If $(\sum_{s=0}^{n-2} \alpha_s) = 0$ then $g((n-1, 1))$ is an identity also for A_1, and this is a contradiction. Then $(\sum_{s=0}^{n-2} \alpha_s) \neq 0$ and this implies that $x_1^2 X_1^{n-2} \equiv 0$ holds in $\mathcal{V}$. If we now make the substitution $x_1 = x_1 + x_0 x_1$ we obtain that $x_0 X_1^n \equiv 0$ is an identity of $\mathcal{V}$ and we are done. □

Remark 1 The variety $\mathcal{V}_1$ is generated by the free Liebniz algebra of rank 1.

Note that $x_0 X_1^n \not\equiv 0$ is not an identity of $\mathcal{V}_1$ so $\mathcal{V}_1$ is not a nilpotent variety. Let's prove that

Theorem 1 *$\mathcal{V}_1$ is an almost nilpotent variety.*

Proof Let $\mathcal{W} \subsetneq \mathcal{V}_1$ be a proper subvariety of $\mathcal{V}_1$, we want to prove that $\mathcal{W}$ is nilpotent. If $A_1 \in \mathcal{W}$ then $\mathcal{W} = \mathcal{V}_1$. So $A_1 \notin \mathcal{W}$ and, by Proposition 2, we have that, for some m, $x_0 X_1^m \equiv 0$ is an identity of $\mathcal{W}$. Clearly $g((m+1)) \equiv 0$ and we claim that $g((m+1,1)) \equiv 0$. In fact let's make the substitution $x_0 = x_2 x_1$ or $x_0 = x_1 x_2$. We obtain that $x_2 X_1^{m+1} \equiv 0$ and $x_1 x_2 X_1^m \equiv 0$. Combining these two identities we obtain that $g((m+1,1)) = \bar{x_1}\bar{x_2} X_1^m \equiv 0$ as claimed.

Therefore $\chi_n(\mathcal{W}) = 0$ for all $n \geq m+2$ and $\mathcal{W}$ is a nilpotent variety. □

Let's now consider the algebra A_2 with basis $\{a, b\}$ such that $ba = a$ and $a^2 = b^2 = ab = 0$.

Remark 2 The algebra A_2 is right nilpotent and satisfies the identities $(xy)z \equiv 0$, $x(yz) \equiv x(zy)$. All properties of A_2 coincide with the properties of the previous algebra A_1 in fact we have only to consider right-normed polynomials instead of left-normed polynomials. As before we have that $\mathcal{V}_2 = \text{var}(A_2)$ is an almost nilpotent variety.

Let's now look at the algebra A_3 with basis $\{a, b\}$ and the following table of multiplication $ab = ba = a$ and $a^2 = b^2 = 0$. The algebra A_3 is commutative and metabelian hence

Remark 3 $\text{var}(A_3)$ is a commutative metabelian variety.

Definition 2 Let $\mathcal{V}_3$ be the variety satisfying the following identities

$$xy \equiv yx, \tag{1}$$

$$(xy)(zt) \equiv 0, \tag{2}$$

$$xyzt \equiv xytz, \tag{3}$$

$$\bar{x}\tilde{y}\bar{z}\tilde{t} = xyzt - zyxt - xtzy + ztxy \equiv 0. \tag{4}$$

The following results are proved in [20].

Proposition 3 *$\mathcal{V}_3 = var(A_3)$ and $\chi_n(\mathcal{V}_3) = \chi_{(n)} + \chi_{(n-1,1)}$.*

Proof First we prove that $A_3 \in \mathcal{V}_3$. Since the identities defining $\mathcal{V}_3$ are multilinear, it is sufficient to verify them on the basis elements of the algebra A_3. By definition, the algebra A_3 is a commutative metabelian algebra, so, the identities (1) and (2) hold in A_3. The identity (4) is skew-symmetric on the pairs of generators x, z and y, t, hence if we substitute the same elements of A_3 in at least one pair of alternating elements we obtain zero. If we substitute different elements in each alternating pairs then we get at least two elements of A_3^2 and so the result of the substitution is also equal to zero. Therefore the identity (4) holds in A_3. Direct calculations show that also the identity (3) holds in A_3 hence $A_3 \in \mathcal{V}_3$.

Now we show that, modulo the identities of the variety $\mathcal{V}_3$, any multilinear monomial of degree $n \geq 3$ is equal to a linear combination of n monomials of the following type:

$$x_{n-1}x_{n-2}x_nx_{n-3}\dots x_1, \tag{5}$$

$$x_nx_ix_{i_1}x_{i_2}\dots x_{i_{n-2}}, \quad i = 1,\dots,n-1,\ i_1 > i_2 > \dots > i_{n-2}. \tag{6}$$

If $n = 3$ from the commutativity we have that $P_3(\mathcal{V}_3)$ is the linear span of the monomials $x_3x_1x_2, x_3x_2x_1, x_2x_1x_3$. It follows that $c_3(\mathcal{V}_3) \leq 3$.

If $n > 3$, the identity (3) allows us to sort the indices of the generators of each monomial starting from the third position in decreasing order, therefore, taking into account the commutativity of the multiplication, we have to consider monomials of the following type $x_ix_jx_{i_1}x_{i_2}\dots x_{i_{n-2}}, i > j, i_1 > i_2 > \dots > i_{n-2}$.

If one of these monomials is different from the monomials (5), (6), then it can be one of the following three types:

$$x_ix_jx_nx_{n-1}x_{n-2}\dots,\quad 1 \leq j < i < n-2; \tag{7}$$

$$x_{n-1}x_jx_nx_{n-2}\dots,\quad 1 \leq j < n-2; \tag{8}$$

$$x_{n-2}x_jx_nx_{n-1}\dots,\quad 1 \leq j < n-2. \tag{9}$$

If we apply the identity (4) to monomials of the first type we have that

$$x_ix_jx_nx_{n-1}x_{n-2}\dots \equiv x_ix_{n-1}x_nx_j\dots + x_nx_jx_ix_{n-1}\dots - x_nx_{n-1}x_ix_j\dots.$$

In the sum obtained the second and third term are identically equal to monomials of type (6) and the first term has the second form.

So let's apply the identity (4) to monomials of type (8), we obtain

$$x_{n-1}x_jx_nx_{n-2}\dots \equiv x_{n-1}x_{n-2}x_nx_j\dots + x_nx_jx_{n-1}x_{n-2}\dots - x_nx_{n-2}x_{n-1}x_j\dots.$$

On the right side of the equality, the first term is identically equal to the monomial (5), the second and third term refer to monomials of the form (6).

It remains to consider monomials of type (9). As before, applying the identity (4), we obtain that

$$x_{n-2}x_jx_nx_{n-1}\dots \equiv x_{n-2}x_{n-1}x_nx_j\dots + x_nx_jx_{n-2}x_{n-1}\dots - x_nx_{n-1}x_{n-2}x_j\dots.$$

It follows that any multilinear monomial of degree n is a linear combination of monomials of type (5) and (6) hence for $n \geq 3$, $c_n(\mathcal{V}_3) \leq n$.

Let's now estimate a lower bound of $c_n(A_3)$. The one-dimensional irreducible submodule of the module $P_n(A_3)$ corresponding to the partition $(n) \vdash n$ is generated by the complete linearization of the monomial $g_{(n)}(x) = xX^{n-1}$. Notice that

$$g_{(n)}(a+b) = \underbrace{(a+b)\dots(a+b)}_{n} = 2a \neq 0.$$

Since modulo the identities of A_3 all non-zero monomials of the free algebra can be rewritten as left normed monomials it follows that $m_{(n)}(A_3) = d_{(n)} = 1$.

The irreducible submodule of the module $P_n(A_3)$ corresponding to the partition $(n-1, 1) \vdash n$ is generated by the complete linearization of the polyhomogeneous polynomial

$$g_{(n-1,1)}(x, y) = \bar{x}x\bar{y}X^{n-3} = xxyX^{n-3} - yX^{n-1}.$$

This polynomial is not identically zero in the algebra A_3, in fact

$$g_{(n-1,1)}(b, a) = bba\,\underbrace{b\dots b}_{n-3} - a\,\underbrace{b\dots b}_{n-1} = -a.$$

It follows that $m_{(n-1,1)}(A_3) \geq 1$. Now, by the hook formula, $d_{(n)} = 1$ and $d_{(n-1,1)} = n-1$ hence $c_n(A_3) \geq n$.

Since $A_3 \in \mathcal{V}_3$ we have that $m_\lambda(A_3) \leq m_\lambda(\mathcal{V}_3)$ hence $c_n(A_3) \leq c_n(\mathcal{V}_3)$ and so

$$n \leq c_n(A_3) \leq c_n(\mathcal{V}_3) \leq n.$$

It follows that $\mathcal{V}_3 = \mathrm{var}(A_3)$ and $\chi_n(\mathcal{V}_3) = \chi_{(n)} + \chi_{(n-1,1)}$. □

A characterization of the algebra A_3 is given in the following

Proposition 4 (Mishchenko et al. [20]) *The algebra A_3 does not belong to a commutative metabelian variety $\mathcal{V}$ if and only if in the variety $\mathcal{V}$, for some $k \geq 1$, the identity $x_0X^k \equiv 0$ holds.*

Proof Clearly if the identity $x_0X^k \equiv 0$ holds in the variety $\mathcal{V}$ then the algebra A_3 does not belong to $\mathcal{V}$.

If the algebra $A_3 \notin \mathcal{V}$ then, by Proposition 3, for some $n \geq 3$, the identities corresponding to the partition (n) or $(n-1, 1)$ hold in the variety $\mathcal{V}$ but do not hold in the algebra A_3. Note that the proof in the cases $n = 1, 2$ is obvious.

Suppose that for the partition (n) the identity $xX^{n-1} \equiv 0$ holds in the variety $\mathcal{V}$, then we substitute the sum $x_0x + x$ instead of x_0 and, by virtue of the commutative and metabelian identities, we get that $x_0X^n \in Id(\mathcal{V})$.

Let's now consider the polyhomogeneous elements corresponding to the standard tableaux of shape $\lambda = (n-1, 1)$ whose linearizations generate irreducible submodules of the module $P_n(\mathcal{V})$. They are polynomials of the form $\bar{x}X^{s-2}\bar{y}X^{n-s}$, $s = 2, 3, \dots, n$, therefore in the variety $\mathcal{V}$ holds the following identity

$$\sum_{s=3}^{n} \alpha_s \bar{x}X^{s-2}\bar{y}X^{n-s} \equiv 0. \tag{10}$$

The summation begins with $s = 3$, since for $s = 2$ the identity $\overline{x}\,\overline{y}X^{n-2} \equiv 0$ holds in any commutative variety. If in (10) the sum $\sum_{s=3}^{n} \alpha_s = 0$ then it is an identity in A_3 since by (3), for $3 \le s \le n$, $\overline{x}X^{s-2}\overline{y}X^{n-s} \equiv \overline{x}x\overline{y}X^{n-3}$.

Therefore, suppose that $\sum_{s=3}^{n} \alpha_s \ne 0$, and in the identity (10) we substitute y with $x_0 x$. We get the identity

$$\left(\sum_{s=3}^{n} \alpha_s\right) x_0 X^n \equiv 0,$$

and as consequence we obtain the desired identity $x_0 X^n \equiv 0$. The proposition is proved. □

Now we are able to prove

Theorem 2 *$\mathcal{V}_3$ is an almost nilpotent variety.*

Proof Let $\mathcal{W}$ be a proper subvariety of $\mathcal{V}_3$ then, by Proposition 4, in the variety $\mathcal{W}$ the identity $x_0 X^k \equiv 0$ holds for some $k \ge 1$. We replace x_0 by the product $x_1 x_2$ and x by $x_3 + x_4 + \ldots + x_{k+2}$. Taking the multilinear part and using (3) we get that $x_1 x_2 x_3 \ldots x_{k+2} \equiv 0$. □

Let A_4 be the algebra having two basic element a, b and defining relations:

1. $ab = -ba = a$,
2. $a^2 = b^2 = 0$.

A_4 is a two dimensional not nilpotent metabelian Lie algebra and it is well known that the variety generated by A_4, $\mathcal{V}_4 = \mathrm{var}(A_4)$, is the variety of all metabelian Lie algebras.

Proposition 5 ([1, p. 186]) *The variety $\mathcal{V}_4$ has the following numerical characteristics*

$$c_1(\mathcal{V}_4) = 1, \quad \chi_1(\mathcal{V}_4) = \chi_{(1)}, \quad l_1(\mathcal{V}_4) = 1,$$

$$c_n(\mathcal{V}_4) = n - 1, \quad \chi_n(\mathcal{V}_4) = \chi_{(n-1,1)}, \quad l_n(\mathcal{V}_4) = 1, \quad n = 2, 3, \ldots. \tag{11}$$

Moreover $\mathcal{V}_4$ is one of three varieties whose colength is identically equal to 1 (see [10]).

We denote by $\mathcal{MA}$ the variety of all anticommutative metabelian algebras.

Proposition 6 (Shulezhko and Panov [24]) *The variety $\mathcal{V}_4$ is not a subvariety of $\mathcal{V} \subset \mathcal{MA}$ if and only if, for some $k \ge 1$, the identity $x_0 X^k \equiv 0$ holds in $\mathcal{V}$.*

Proof By the definition of the algebra A_4, for any $k \ge 1$, we have that $aR_b^k = a$, therefore the identity $x_0 X^k \equiv 0$ does not hold in the variety $\mathcal{V}_4$ and it remains to prove the necessary condition.

Let's assume that $\mathcal{V}_4$ is not a subvariety of $\mathcal{V}$, then by (11) the identities corresponding to the Young diagram associated to $\lambda = (n-1, 1)$, $n \geq 2$, hold in the variety $\mathcal{V}$ but do not hold in the variety $\mathcal{V}_4$. The different standard tableaux of this diagram correspond to the polyhomogeneous polynomials $\overline{x}_1 X_1^i \overline{x}_2 X_1^{n-2-i}$, $i = 0, \ldots, n-2$, and in the variety $\mathcal{V}$ we have that

$$\sum_{i=0}^{n-2} \alpha_i \overline{x}_1 X_1^i \overline{x}_2 X_1^{n-2-i} \equiv 0, \quad \alpha_i \in F. \tag{12}$$

Using the anticommutativity we obtain that

$$\left(2\alpha_0 + \sum_{i=1}^{n-2} \alpha_i\right) x_2 X_1^{n-1} \equiv 0.$$

Since, by assumption, (12) does not hold in the variety $\mathcal{V}_4$, we have that $2\alpha_0 + \sum_{i=1}^{n-2} \alpha_i \neq 0$. Thus, for some $k \geq 1$, the identity $x_0 X_1^k \equiv 0$ holds in the variety $\mathcal{V}$. □

The following result is well known (see for example [1]).

Theorem 3 *$\mathcal{V}_4$ is an almost nilpotent variety.*

3 Almost Nilpotent Varieties and Skew Symmetric Polynomials

In this section we consider almost nilpotent varieties of subexponential growth for which $x_0\bar{x}_1 \cdots \bar{x}_n$ is not identity (see [17]).

Definition 3 Let $\mathcal{V}_5$ be the variety of algebras satisfying the following identities:

(1) $x(yz) \equiv 0$.
(2) $xyz \equiv -xzy$.

Clearly $\mathcal{V}_5$ is a variety contained in ${}_2\mathcal{N}$. Next we find an algebra A_5 generating the variety $\mathcal{V}_5$.

Definition 4 Let A_5 be the algebra over F generated by the countable set of elements $e_1, e_2, \ldots$ satisfying the following relations

(1) $ue_ie_j = -ue_je_i$ for any nonempty word u in $e_1, e_2, \ldots$,
(2) $uv = 0$, for any non empty words u, v in $e_1, e_2, \ldots$, with $|v| \geq 2$.

From the definition it follows that $x(yz) \equiv 0$ and $xyz \equiv -xzy$ are polynomial identities of A_5. Hence $\mathcal{V}_5 \supseteq var(A_5)$.

We have the following

Proposition 7 (Mishchenko and Valenti [17]) $\mathcal{V}_5 = var(A_5)$ *and* $\chi_n(\mathcal{V}_5) = \chi_{(1^n)} + \chi_{(2,1^{n-2})}$.

Proof Let P_n^i be the space of left normed multilinear polynomials in $x_1, \ldots x_n$ starting with x_i. Then

$$P_n(\mathcal{V}_5) = P_n^1(\mathcal{V}_5) + \ldots + P_n^n(\mathcal{V}_5).$$

Since $xyz \equiv -xzy$, dim $P_n^i(\mathcal{V}_5) = 1$ and this implies that $c_n(\mathcal{V}_5) \leq n$.

Let $\chi_n(A_5) = \sum m_\lambda \chi_\lambda$. Consider the partition (1^n) and the polynomial $g((1^n)) = \bar{x}_1\bar{x}_2\cdots\bar{x}_n$. We claim that $g((1^n))$ is not an identity of A_5. In fact if we make the substitutions $x_1 = e_1^2$ and $x_i = e_i$ for $i \neq 1$, then, modulo the relations (1) and (2) of A_5, we obtain that $g((1^n)) = (n-1)!e_1^2e_2\cdots e_n \neq 0$.

Since $d_{(1^n)} = 1$ and only left normed monomials can have a nonzero evaluation on A_5 then it follows that $m_{(1^n)} = d_{(1^n)} = 1$.

Consider now the partition $(2, 1^{n-2})$ and let $g((2, 1^{n-2})) = \bar{x}_1x_1\bar{x}_2\cdots\bar{x}_{n-1}$. We claim that $g((2, 1^{n-2}))$ is not an identity of A_5. In fact, from the relations satisfied by A_5, if we substitute $x_i = e_i$, for $i \geq 1$ we obtain $g((2, 1^{n-2})) = (n-1)!e_1^2e_2\cdots e_{n-1} \neq 0$. Since $d_{(2,1^{n-2})} = n-1$, it follows that $n \geq c_n(\mathcal{V}_5) \geq c_n(A_5) \geq d_{(1^n)} + d_{(2,1^{n-2})} = n$. Hence $c_n(\mathcal{V}_5) = c_n(A_5) = n$, $\mathcal{V}_5 = \text{var}(A_5)$ and $\chi_n(\mathcal{V}_5) = \chi_{(1^n)} + \chi_{(2,1^{n-2})}$. □

Our aim is to prove that $\mathcal{V}_5$ is an almost nilpotent variety. Let $\mathcal{W} \subsetneq \mathcal{V}_5$ be a proper subvariety of $\mathcal{V}_5$ then there exists $n \geq 1$ and an irreducible S_n-character χ_λ appearing with multiplicity m'_λ in $\chi_n(\mathcal{W})$ and m_λ in $\chi_n(V_5)$ with $m'_\lambda < m_\lambda$. Since

$$\chi_n(\mathcal{V}_5) = \chi_{(1^n)} + \chi_{(2,1^{n-2})}$$

it follows that either $g((1^n)) \equiv 0$ or $g((2, 1^{n-2})) \equiv 0$ holds in $\mathcal{W}$. Notice that, modulo the identities (1) and (2), any highest weight vector of $GL_{n-1}(F)$ can be reduced to the form $g((2, 1^{n-1}))$. We claim that $g((1^{n+1})) \equiv 0$ and $g((2, 1^{n-1})) \equiv 0$.

In fact suppose first that $g((1^n)) = \bar{x}_1\cdots\bar{x}_n \equiv 0$. Then $\bar{x}_1\cdots\bar{x}_nR_{x_1} \equiv 0$ and this implies that $g((2, 1^{n-1})) \equiv 0$. Moreover $\bar{x}_1\cdots\bar{x}_nR_{x_{n+1}} \equiv 0$. Hence if we apply the operator of alternation, taking into account (1) and (2), we obtain

$$\sum_{\sigma\in S_{n+1}} (sgn\sigma)\bar{x}_{\sigma(1)}\cdots\bar{x}_{\sigma(n)}x_{\sigma(n+1)} \equiv 0.$$

This implies that $\bar{x}_{\sigma(1)}\cdots\bar{x}_{\sigma(n+1)} = g((1^{n+1})) \equiv 0$ and we are done in this case.

Now suppose that $g((2, 1^{n-2})) = \bar{x}_1x_1\bar{x}_2\cdots\bar{x}_{n-1} \equiv 0$. Then if we make the substitution $x_1 = x_1 + x_nx_{n+1}$, since $x(yz) \equiv 0$, we obtain that

$$x_nx_{n+1}x_1\bar{x}_2\cdots\bar{x}_{n-1} \equiv 0$$

and, so,

$$\sum_{\sigma\in S_{n+1}} (sgn\sigma)x_{\sigma(n)}x_{\sigma(n+1)}x_{\sigma(1)}\bar{x}_{\sigma(2)}\cdots\bar{x}_{\sigma(n-1)} \equiv 0.$$

This implies that $\bar{x}_1\bar{x}_2\cdots\bar{x}_{n+1} \equiv 0$.

Moreover from $g((2, 1^{n-2})) \equiv 0$ we also obtain $\bar{x}_1x_1\bar{x}_2\cdots\bar{x}_{n-1}x_n \equiv 0$ and, so, by applying the operator of alternation,

$$\sum_{\sigma\in S_n} (sgn\sigma)\bar{x}_{\sigma(1)}x_1\bar{x}_{\sigma(2)}\cdots\bar{x}_{\sigma(n)} \equiv 0.$$

This says that $g((2, 1^{n-1})) \equiv 0$ and the claim is proved.

It follows (see [17])

Proposition 8 *Any proper subvariety of $\mathcal{V}_5$ is nilpotent.*

We complete this section with a characterization of the algebra A_5.

Theorem 4 (Mishchenko and Valenti [17]) *Let $\mathcal{V}$ be a variety such that $x(yz) \equiv 0$. Then $A_5 \notin \mathcal{V}$, if and only if $x_0\bar{x}_1\cdots\bar{x}_m \equiv 0$ in $\mathcal{V}$, for some $m \geq 1$.*

Proof Clearly if $x_0\bar{x}_1\cdots\bar{x}_m \equiv 0$ holds in $\mathcal{V}$ then $A_5 \notin \mathcal{V}$. Conversely suppose that $A_5 \notin \mathcal{V}$. Since $\chi_n(A_5) = \chi_{(1^n)} + \chi_{(2,1^{n-2})}$ it follows that

$$g((1^n)) = \bar{x}_1\cdots\bar{x}_n \equiv 0$$

or

$$g((2, 1^{n-2})) = \sum_{s=0}^{n-1} \alpha_s\bar{x}_1\cdots\bar{x}_sx_1\bar{x}_{s+1}\cdots\bar{x}_{n-s} \equiv 0$$

hold in $\mathcal{V}$ but not in A_5, for some $n \geq 1$.

Suppose that $g((1^n)) = \bar{x}_1\cdots\bar{x}_n \equiv 0$ holds in $\mathcal{V}$. If we make the substitution $x_1 = x_0x_1$, then we get that $x_0x_1\bar{x}_2\cdots\bar{x}_n \equiv 0$ holds in $\mathcal{V}$. This implies that

$$\sum_{\sigma\in S_n} (\mathrm{sgn}\,\sigma)x_0x_{\sigma(1)}\bar{x}_{\sigma(2)}\cdots\bar{x}_{\sigma(n)} \equiv 0$$

and so $(n-1)!x_0\bar{x}_1\cdots\bar{x}_n \equiv 0$ holds in $\mathcal{V}$.

Now suppose that $g((2, 1^{n-2})) \equiv 0$. Let

$$g_\lambda^1 = x_1\bar{x}_1\bar{x}_2\cdots\bar{x}_{n-1}$$

$$g_\lambda^2 = \bar{x}_1x_1\bar{x}_2\cdots\bar{x}_{n-1}$$

$$\vdots$$

$$g_\lambda^n = \bar{x}_1\bar{x}_2\cdots\bar{x}_{n-1}x_1$$

be the polynomials obtained from the essential idempotents corresponding to the standard tableaux of shape $\lambda = (2, 1^{n-2})$. Notice that, since $x_1\bar{x}_1\bar{x}_2 = \bar{x}_1x_1\bar{x}_2 - \bar{x}_1\bar{x}_2x_1$, g_λ^1 is a linear combination of g_λ^i, $i = 2, \ldots, n$. It follows that

$$g((2, 1^{n-2})) = \sum_{i=2}^{n} \alpha_i g_\lambda^i = \sum_{i=2}^{n} \alpha_i \bar{x}_1\bar{x}_2\cdots\bar{x}_{i-1}x_1x_i\cdots\bar{x}_{n-1}.$$

If we now make the substitutions $x_1 = x_0x_1 + x_2$ and, for $i = 2, \ldots, n-1$, $x_i = x_{i+1}$ we obtain

$$\left(\sum_{i=2}^{n} \alpha_i\right) x_0x_1x_2\bar{x}_3\cdots\bar{x}_n \equiv 0.$$

If $\sum_{i=2}^{n} \alpha_i = 0$ then $g((2, 1^{n-2}))$ is an identity for A_5, a contradiction. It follows that $\sum_{i=2}^{n} \alpha_i \neq 0$ and by alternating $x_1, x_2, \ldots, x_n$ we obtain

$$\sum_{\sigma\in S_n} (\operatorname{sgn}\sigma) \left(\sum_{i=2}^{n} \alpha_i\right) x_0x_{\sigma(1)}x_{\sigma(2)}\bar{x}_{\sigma(3)}\cdots\bar{x}_{\sigma(n)} \equiv 0.$$

Therefore $(n-2)!(\sum_{i=2}^{n} \alpha_i)x_0\bar{x}_1\cdots\bar{x}_n \equiv 0$ holds in $\mathcal{V}$ and we are done. □

Remark 4 Let's now consider A_6 the opposite algebra of the algebra A_5, that is the algebra with the same elements of A_5 and the same addition operation but with the multiplication performed in the reverse order. Clearly A_6 has the same properties of A_5 but instead of left-normed polynomials we have to consider right-normed polynomials.

Let $\mathcal{V}_6$ be the variety defined by the following identities

1. $(zy)x \equiv 0$.
2. $z(yx) \equiv -y(zx)$.

Then $\mathcal{V}_6 = \operatorname{var}(A_6)$ and we have another example of almost nilpotent variety.

4 A Commutative Metabelian Algebra with Skew Symmetric Polynomials (a Jordan Algebra)

In this section the almost nilpotent variety considered will be generated by a commutative metabelian Jordan algebra.

Definition 5 Let A_7 be the algebra with generators $b, e_i, i = 1, 2, \ldots$, and defining relations

(1) $ue_ie_j = -ue_je_i$,
(2) $e_ie_j = 0$,
(3) $ue_i = e_iu$,
(4) $uv = 0, \ \deg_b u + \deg_b v \geq 2$,

where $u, v \in A_7$ are monomials.

Since char $F = 0$ from the first relation we have that, for $i = j, ue_ie_i = 0$. Moreover from (2) (3), and (4) it follows that A_7 is a commutative metabelian algebra.

A basis of the algebra A_7 is given by the left-normed monomials

$$b, \ e_i, \quad i = 1, 2, \ldots,$$

$$be_{j_1}e_{j_2}\ldots e_{j_n}, \quad j_1 < j_2 < \ldots < j_n, \ n = 1, 2, \ldots.$$

Proposition 9 (Mishchenko et al. [20]) *In the commutative metabelian algebra A_7 the following identities hold:*

$$xyzt \equiv -xytz, \tag{13}$$

$$x^3 \equiv 0, \tag{14}$$

$$xyz + yzx + zxy \equiv 0. \tag{15}$$

Proof Let's verify the multilinear identity (13) by substituting basis elements of the algebra A_7. Notice that if we substitute the variables z or t with the element b or with a monomial u then both part of the identity (13) vanish and we are done. Therefore we substitute z with e_i and t with e_j and also in this case, by the defining relations (1), for any substitution of x, y we obtain an equality of A_7.

Consider now the identity (14). From the commutative identity we obtain that $x(xx) \equiv (xx)x$. Let's consider the evaluation ϕ such that

$$\phi(x) = \alpha b + \sum_i \alpha_i e_i + u, \quad \alpha, \alpha_i \in F,$$

where the sum over the index i is finite, and the element $u \in A_7^2$ is a linear combination of a finite number of basis elements. By the definition of the algebra

A_7 we have that $u^2 = b^2 = ub = bu = 0$, and so

$$\phi(x^3) = \left(\alpha b + u + \sum_i \alpha_i e_i\right)\left(\alpha b + u + \sum_i \alpha_i e_i\right)\left(\alpha b + u + \sum_i \alpha_i e_i\right) =$$

$$= \left((\alpha b + u)\left(\sum_i \alpha_i e_i\right) + \left(\sum_i \alpha_i e_i\right)(\alpha b + u)\right)\left(\sum_i \alpha_i e_i\right) =$$

$$= 2\alpha b \left(\sum_i \alpha_i e_i\right)\left(\sum_i \alpha_i e_i\right) + 2u\left(\sum_i \alpha_i e_i\right)\left(\sum_i \alpha_i e_i\right) =$$

$$= 2\alpha \sum_{\substack{i,j \\ i<j}} \alpha_i \alpha_j (be_i e_j + be_j e_i) + 2\sum_{\substack{i,j \\ i<j}} \alpha_i \alpha_j (ue_i e_j + ue_j e_i) = 0.$$

The Jacobi identity (15) follows directly from the complete linearization of the identity (14) and the commutativity identity. □

The algebra A_7 is a Jordan algebra, i.e., it satisfies the identities

$$xy \equiv yx, \quad (x^2 y)x \equiv x^2(yx).$$

In fact from the metabelian identity, it suffices to show that $(x^2 y)x \equiv 0$. This follows from (13) and (14) indeed $(x^2 y)x \equiv -((xx)x)y \equiv 0$.

Proposition 10 (Mishchenko et al. [20]) *For the algebra A_7 the following equalities hold:*

$$c_1(A_7) = 1, \quad \chi_1(A_7) = \chi_{(1)}, \quad l_1(A_7) = 1,$$

$$c_n(A_7) = n - 1, \quad \chi_n(A_7) = \chi_{(2,1^{n-2})}, \quad l_n(A_7) = 1, \quad n = 2, 3, \ldots.$$

Let's now consider the variety $\mathcal{V}_7 = \mathrm{var}(A_7)$ generated by A_7.

We want to show that $\mathcal{V}_7$ is also generated by the algebra defined by Zhevlakov (see [26, p. 86] and [27, p. 103, Example 1]) in the following way.

Let $Z = \{z_1, z_2, \ldots\}$ be a countable set. We say that a word in the alphabet Z, $z_{i_1} z_{i_2} \ldots z_{i_n}$, is correct if it satisfies the condition $i_1 < i_2 < \ldots < i_n$. On the set of correct words, a lexicographic linear order is given as follows. Let $d(u)$ denote the length of the word u. If $i < j$ then $z_i > z_j$ and if $u = vv_1$ with $d(v_1) \geq 1$, then $u > v$. For example we have the inequalities

$$z_1 z_2 z_3 > z_1 z_2 > z_1 z_3 > z_1 > z_2 z_3 > z_2 > z_3.$$

Definition 6 Let A_Z be the algebra over a field F of characteristic zero, with multiplication $*$, having as its basis the set of all regular words in the alphabet Z, and with the following table of multiplication: for any u, v regular words, then

(1) $u * v = v * u$,
(2) $u * v = 0$, $d(u) > 1$ and $d(v) > 1$,
(3) $u * z_i = 0$, if z_i is contained in u,
(4) $u * z_i = 0$, if $d(u) > 1$ and $z_i > u$,
(5) $z_i * z_j = z_i z_j$, if $i < j$,
(6) $u * z_j = (-1)^{\theta} v$, if $d(u) > 1$, $z_j < u$, the word u does not contain z_j, v is a correct word consisting of all elements of the word u and z_j, and θ is the number of inversions in the permutation $\theta = (i_1, \ldots, i_n, j)$.

We have that

Proposition 11 $\mathcal{V}_7 = var(A_7) = var(A_Z)$.

Proof By the defining relations (1), (2), the algebra A_Z is a commutative metabelian algebra. Moreover in [27] it was proved that the Jacobi identity (15) holds in the algebra A_Z.

Let us show that the algebra A_Z also satisfies the identity (13)

$$xyzt \equiv -xytz.$$

Denote by ϕ an evaluation on the algebra A_Z. By the definition of the algebra A_Z, if both sides of this identity after the evaluation are not equal to zero then the following conditions must be simultaneously satisfied: $\phi(xy) \neq 0$, $d(\phi(z)) = d(\phi(t)) = 1$, the elements $\phi(z)$ and $\phi(t)$ are different and are not contained in $\phi(xy)$, and $\phi(z), \phi(t) < \phi(xy)$. In this case, by the defining relation (6), we obtain an equality and we are done.

Since the identities (13) and (15) hold in the commutative metabelian algebra A_Z, as in the proof of the Proposition 10, we obtain that $c_n(A_Z) \leq n - 1, n = 3, 4, \ldots$. Moreover, for any $n \geq 2$ and any permutation $\sigma \in S_n$, such that $\sigma(1) = 1$ we have that

$$z_1 * z_{\sigma(2)} * \ldots * z_{\sigma(n)} = (\text{sgn}\sigma) z_1 z_2 \ldots z_n.$$

Therefore, the polynomial $x_n \bar{x}_1 \ldots \bar{x}_{n-1}$, $n \geq 2$, is not identically equal to zero in the algebra A_Z and, as for the algebra A_7, for any $n \geq 3$ we obtain

$$m_{(2,1^{n-2})}(A_Z) = m_{(2,1^{n-2})}(A_7) = 1.$$

So, for any $n \geq 2$, the irreducible submodules of the modules $P_n(A_Z)$ and $P_n(A_7)$ corresponding to the partition $(2, 1^{n-2})$ coincide.

Thus, similarly to what proved in Proposition 10, we obtain that:

$$c_n(A_Z) = n-1, \quad \chi_n(A_Z) = \chi_{(2,1^{n-2})}, \quad l_n(A_Z) = 1, \quad n = 2, 3, \ldots,$$

$$c_1(A_Z) = 1, \quad \chi_1(A_Z) = \chi_{(1)}, \quad l_1(A_Z) = 1.$$

We conclude that $P_n(A_7) = P_n(A_Z)$ and $\mathcal{V}_7 = \text{var}(A_Z)$, the proposition is proved. □

Now we show that the algebra A_7 can be defined using the algebra A_Z.

Let's denote by I_Z the ideal of the algebra A_Z generated by the regular words u, which simultaneously satisfy two conditions: $d(u) > 1$ and $u < z_1$. By the defining relation (2) of the algebra A_Z, the product of any two elements of I_Z is equal to zero. Moreover:

1. $z_1 * u = z_1 v * u = 0, \quad u, v \in I_Z, d(v) \geq 1,$
2. $z_i * z_j \in I_Z, \quad i, j = 2, 3, \ldots.$

Since the algebra A_Z is commutative I_Z is a two-sided ideal.

Let $\overline{z_i} = z_i + I_Z \in A_Z/I_Z$. Then a basis of the algebra A_Z/I_Z is composed by the elements

$$\overline{z_i}, \; i = 1, 2, \ldots,$$

$$\overline{z_1 z_{j_1} \ldots z_{j_n}}, \quad 1 < j_1 < j_2 < \ldots < j_n, \; n = 1, 2, \ldots.$$

Let $\phi : \{\overline{z_1}, \overline{z_2}, \ldots\} \to A_7$ the evaluation such that

$$\phi(\overline{z_1}) = b,$$

$$\phi(\overline{z_i}) = e_{i-1}, \quad i = 2, 3, \ldots.$$

By the defining relations of both algebras A_7 and A_Z and by the definition of the ideal I_Z, the mapping ϕ uniquely extends to an isomorphism of algebras $A_Z/I_Z \cong A_7$.

Let's now consider the algebra defined by Shestakov (see [27, p. 104, Example 2]) as follows.

Let $\wedge(M)$ be the outer algebra of the vector space M with basis $\{x_1, x_2, \ldots\}$, and $\wedge^0(M)$ be the subalgebra of $\wedge(M)$ generated by the set M. Let $A_J = \wedge^0(M) \oplus M$ be the algebra with multiplication

$$(u + x)(v + y) = v \wedge x + u \wedge y,$$

where $u, v \in \wedge^0(M)$, $x, y \in M$.

Let $\mathcal{J} = \mathrm{var}(A_J)$ be the variety generated by the algebra A_J. If we consider the variety of all associative commutative algebras $\mathcal{V}_0$ and the variety of all metabelian Lie algebras $\mathcal{V}_4$, then the main result of [10] is the following

Theorem 5 *Let $\mathcal{V}$ be the variety of linear algebras over a field of characteristic zero. If $l_n(\mathcal{V}) = 1$, for $n = 1, 2, \ldots$, then the variety $\mathcal{V}$ coincides with one of the following three varieties: $\mathcal{V}_0$, $\mathcal{V}_4$ or $\mathcal{J}$.*

By Proposition 10, it follows that, for $n \geq 1, l_n(A_7) = 1$, therefore by the previous theorem we have

Proposition 12 $\mathcal{V}_7 = \mathcal{J}$.

5 Anti-Commutative Metabelian Algebra with Skew Symmetric Polynomials

In this section we consider the almost nilpotent variety $\mathcal{V}_8$ generated by the following anticommutative metabelian algebra A_8.

Definition 7 Let A_8 be the algebra with generators $e_1, e_2, \ldots$ and with the following defining relations:

(1) $ue_i = -e_i u, \quad u \in A_8,$
(2) $u_1 u_2 = 0, \quad u_1, u_2 \in A_8^2,$
(3) $e_{i_{\sigma(1)}} e_{i_{\sigma(2)}} \ldots e_{i_{\sigma(n)}} = (\mathrm{sgn}\sigma) e_{i_1} e_{i_2} \ldots e_{i_n}, \quad \sigma \in S_n, n \geq 3.$

In the algebra A_8, by the defining relation (3), any monomial of degree ≥ 2 in only one generator is equal to zero. By definition a basis of the algebra A_8 consists of the following elements

$$e_i, \quad i = 1, 2, \ldots,$$

$$e_{j_1} e_{j_2} \ldots e_{j_n}, \quad j_1 < j_2 < \ldots < j_n, \ n = 2, 3, \ldots.$$

Note that A_8^2, by the defining relation (2), is an algebra with zero multiplication.

Let $\mathcal{V}_8 = \mathrm{var}(A_8)$, we have the following

Proposition 13 (Shulezhko and Panov [24]) *In the variety $\mathcal{V}_8$ the following identities hold*

$$xyzt \equiv -xytz, \tag{16}$$

$$xyzt + zyxt + xtzy + ztxy \equiv 0. \tag{17}$$

Proof Let's consider various substitutions of the variables of the identities (16) and (17) with basis elements of the algebra A_8.

If in (16) we substitute z or t by an element of the algebra A_8^2, then by the defining relation (2), both parts of the identity will be equal to zero. Instead if we replace z and t by $e_i, i = 1, 2, \ldots$, then by the defining relation (3) we obtain the correct equality.

In the identity (17), let's substitute one of the free generators with the basis element $b \in A_8^2$ and the remaining variables with the elements e_i, $i \geq 1$. By the defining relations (1), (2), as a result of the substitution, there remains a pair of terms of the same sign, in each of which the element b is in the first position, and the permutations of the generators e_i differ by one transposition. Thus, by the defining relation (3), we obtain an equality.

Since the terms $zyxt$, $xtzy$ are obtained from the monomial $xyzt$ by one transposition of generators, and the monomial $ztxy$ from two transpositions, by the defining relation (3) the identity (17) turns into the correct equality when we substitute all variables with the elements e_i. The Proposition is proved. □

Proposition 14 (Shulezhko and Panov [24]) *For the variety $\mathcal{V}_8$ we have:*

$$c_1(\mathcal{V}_8) = 1, \quad \chi_1(\mathcal{V}_8) = \chi_{(1)}, \quad l_1(\mathcal{V}_8) = 1,$$

$$c_2(\mathcal{V}_8) = 1, \quad \chi_2(\mathcal{V}_8) = \chi_{(1,1)}, \quad l_2(\mathcal{V}_8) = 1,$$

$$c_n(\mathcal{V}_8) = n, \quad \chi_n(\mathcal{V}_8) = \chi_{(2,1^{n-2})} + \chi_{(1^n)}, \quad l_n(\mathcal{V}_8) = 2, \quad n = 3, 4, \ldots.$$

Proof The proof of the equalities for $n = 1, 2$ is obvious, therefore, we further estimate the values of $c_n(\mathcal{V}_8)$ for $n \geq 3$. First we determine an upper bound for $c_n(\mathcal{V}_8)$. If $n = 3$ then by the anticommutativity it follows that there are no more than three linearly independent multilinear monomials. Therefore, $c_3(\mathcal{V}_8) \leq 3$.

If $n \geq 4$, we show that any multilinear monomial, modulo the identities of the variety $\mathcal{V}_8$, is equal to a linear combination of the n monomials

$$x_{n-1}x_{n-2}x_n x_{n-3} \ldots x_1, \tag{18}$$

$$x_n x_i x_{j_1} x_{j_2} \ldots x_{j_{n-2}}, \quad i = 1, \ldots, n-1, \; j_1 > j_2 > \cdots > j_{n-2}. \tag{19}$$

Since $n \geq 4$, by virtue of the identity (16) any monomial different from the monomials (18) and (19), is identically equal to one of the monomials of the following three types:

(1) $x_{n-1}x_i x_n x_{n-2} \ldots$, $\quad 1 \leq i < n-2$,
(2) $x_{n-2}x_i x_n x_{n-1} \ldots$, $\quad 1 \leq i < n-2$,
(3) $x_i x_j x_n x_{n-1} \ldots$, $\quad 1 \leq j < i < n-2$.

If we apply the identity (17) to these monomials we obtain:

$$x_{n-1}x_ix_nx_{n-2}\ldots \equiv -x_nx_ix_{n-1}x_{n-2}\ldots - x_{n-1}x_{n-2}x_nx_i\ldots - x_nx_{n-2}x_{n-1}x_i\ldots,$$

$$x_{n-2}x_ix_nx_{n-1}\ldots \equiv -x_nx_ix_{n-2}x_{n-1}\ldots - x_{n-2}x_{n-1}x_nx_i\ldots - x_nx_{n-1}x_{n-2}x_i\ldots,$$

$$x_ix_jx_nx_{n-1}\ldots \equiv -x_nx_jx_ix_{n-1}\ldots - x_ix_{n-1}x_nx_j\ldots - x_nx_{n-1}x_ix_j\ldots.$$

On the right side of the last identity, the second term, up to a sign, is identically equal to a monomial of the first kind. All other monomials, up to a sign, are identically equal to the monomials (18) or (19). Thus we obtain that $c_n(\mathcal{V}_8) \leq n, n = 3, 4, \ldots$.

Let's now determine a lower bound for $c_n(\mathcal{V}_8)$, $n = 3, 4, \ldots$. Consider the polynomial

$$g_{(2,1^{n-2})}(x_1, x_2, \ldots, x_{n-1}) = \overline{x}_1\overline{x}_2\ldots\overline{x}_{n-1}x_1$$

corresponding to the standard Young tableau of shape $\lambda = (2, 1^{n-2})$ where the first line contains $1, n$. Let $b \in A_8^2$ a non-zero monomial such that $\deg_{e_i} b = 0$, for $i = 1, \ldots, n-1$, then

$$g_{(2,1^{n-2})}(b+e_1, e_2, \ldots, e_{n-1}) = \overline{(b+e_1)}\overline{e}_2\ldots\overline{e}_{n-1}(b+e_1) =$$

$$= b\overline{e}_2\ldots\overline{e}_{n-1}e_1 - \overline{e}_2b\ldots\overline{e}_{n-1}e_1 = 2b\overline{e}_2\ldots\overline{e}_{n-1}e_1 = 2(n-2)!be_2\ldots e_{n-1}e_1 \neq 0.$$

Therefore, for $n \geq 3$, $m_{(2,1^{n-2})}(\mathcal{V}_8) \geq 1$.

Let $g_{(1^n)}(x_1, \ldots, x_n) = \overline{x}_1\overline{x}_2\ldots\overline{x}_n$ the polynomial corresponding to the unique standard Young tableau of shape (1^n). By the defining relation (3) of the algebra A_8 we have that $g_{(1^n)}(e_1, \ldots, e_n) = n!e_1e_2\ldots e_n$. Therefore, for $n \geq 3$, $m_{(1^n)}(\mathcal{V}_8) \geq 1$.

Thus, for $n \geq 1$,

$$c_n(\mathcal{V}_8) \geq m_{(2,1^{n-2})}(\mathcal{V}_8)d_{(2,1^{n-2})} + m_{(1^n)}(\mathcal{V}_8)d_{(1^n)} = (n-1)m_{(2,1^{n-2})}(\mathcal{V}_8) + m_{(1^n)}(\mathcal{V}_8) \geq n.$$

Since $c_n(\mathcal{V}_8) \leq n$, then for $n \geq 3$, $c_n(\mathcal{V}_8) = n$, $m_{(2,1^{n-2})}(\mathcal{V}_8) = m_{(1^n)}(\mathcal{V}_8) = 1$ and we are done. □

Proposition 15 (Shulezhko and Panov [24]) *Let $\mathcal{V} \subset$* **MA**. *The variety $\mathcal{V}_8$ is not a subvariety of $\mathcal{V}$ if and only if, for some $l \geq 1$, $x_0\overline{x}_1\overline{x}_2\ldots\overline{x}_l \in Id(\mathcal{V})$.*

Proof Let $f(x_0, x_1, \ldots, x_k) = x_0\overline{x}_1\overline{x}_2\ldots\overline{x}_k$, $k \geq 1$, then

$$f(e_1, e_2, \ldots e_{k+1}) = k!e_1\ldots e_{k+1}, \quad e_i \in A_8.$$

This implies that $f \notin Id(\mathcal{V}_8)$ and it remains to prove the necessary condition.

If $\mathcal{V}_8 \not\subset \mathcal{V}$ then, by Proposition 14, the identities corresponding to the partition (1^n) or $(2, 1^{n-2})$, $n \geq 3$, hold in $\mathcal{V}$ but not in $\mathcal{V}_8$. Note that the proof in the cases $n = 1, 2$ is obvious.

Fix $n \geq 3$. Suppose first that $\overline{x}_1\overline{x}_2\ldots\overline{x}_n \in Id(\mathcal{V})$. Since the variety $\mathcal{V}$ is anticommutative and metabelian by replacing x_1 with x_0x_1 we obtain that $x_0x_1\overline{x}_2\ldots\overline{x}_n \equiv 0$. We rewrite it in the form

$$\sum_{\sigma\in H_n}(\mathrm{sgn}\sigma)x_0x_{\sigma(1)}x_{\sigma(2)}\ldots x_{\sigma(n)} \equiv 0,$$

where H_n is the group of all permutations $\sigma \in S_n$ such that $\sigma(1) = 1$. Let's write out the result of the skew-symmetrization of the last identity with respect to the indices $1,\ldots,n$,

$$\sum_{\sigma\in H_n}(\mathrm{sgn}\sigma)x_0\overline{x}_{\sigma(1)}\overline{x}_{\sigma(2)}\ldots\overline{x}_{\sigma(n)} \equiv 0.$$

We use the relation

$$\overline{x}_1\overline{x}_2\ldots\overline{x}_n = (\mathrm{sgn}\sigma)\overline{x}_{\sigma(1)}\overline{x}_{\sigma(2)}\ldots\overline{x}_{\sigma(n)},\ \sigma \in S_n,$$

and we obtain that $(n-1)!x_0\overline{x}_1\overline{x}_2\ldots\overline{x}_n \equiv 0$ is an identity of $\mathcal{V}$.

The identity corresponding to the partition $(2, 1^{n-2})$ has the form

$$g(x_1,\ldots,x_{n-1}) = \sum_{i=1}^{n-1}\alpha_i\overline{x}_1\ldots\overline{x}_i x_1\overline{x}_{i+1}\ldots\overline{x}_{n-1} \equiv 0, \tag{20}$$

where for $i = n-1$ we have the term $\alpha_{n-1}\overline{x}_1\ldots\overline{x}_{n-1}x_1$.

Using the anticommutative identity, we represent the term with coefficient α_1 in the form

$$\overline{x}_1x_1\overline{x}_2\ldots\overline{x}_{n-1} \equiv -x_1\overline{x}_1\overline{x}_2\ldots\overline{x}_{n-1} \equiv \sum_{j=2}^{n-1}(-1)^j x_1\overline{x}_2\ldots\overline{x}_j x_1\overline{x}_{j+1}\ldots\overline{x}_{n-1}. \tag{21}$$

In the original identity (20), instead of x_1 let's substitute the sum $x_0x_1 + x_n$ and, by using the identities (20) and (21), we obtain

$$\alpha_1\sum_{i=2}^{n-1}(-1)^i(x_0x_1+x_n)\overline{x}_2\ldots\overline{x}_i(x_0x_1+x_n)\overline{x}_{i+1}\ldots\overline{x}_{n-1}+$$

$$+\sum_{i=2}^{n-1}\alpha_i\overline{(x_0x_1+x_n)}\overline{x}_2\ldots\overline{x}_i(x_0x_1+x_n)\overline{x}_{i+1}\ldots\overline{x}_{n-1} \equiv$$

$$\equiv\sum_{i=2}^{n-1}\Big((-1)^i\alpha_1+2\alpha_i\Big)x_0x_1\overline{x}_2\ldots\overline{x}_i x_n\overline{x}_{i+1}\ldots\overline{x}_{n-1} \equiv 0.$$

We write out the skew-symmetrization of this identity with respect to the indices $1, \dots, n$. Let's denote by $\widetilde{H}_n$ the subgroup of S_n of all permutations that leave 1 and n fixed and consider the result of skew symmetrization of the term with coefficient $(-1)^i\alpha_1 + 2\alpha_i$, for $2 \le i \le n-1$,

$$\sum_{\sigma\in\widetilde{H}_n} \operatorname{sgn}\sigma\, x_0\overline{x}_{\sigma(1)}\overline{x}_{\sigma(2)}\dots\overline{x}_{\sigma(i)}\overline{x}_{\sigma(n)}\overline{x}_{\sigma(i+1)}\dots\overline{x}_{\sigma(n-1)} =$$

$$= \sum_{\sigma\in\widetilde{H}_n} x_0\overline{x}_1\overline{x}_2\dots\overline{x}_i\overline{x}_n\overline{x}_{i+1}\dots\overline{x}_{n-1} = (-1)^{n-1-i}(n-2)!x_0\overline{x}_1\overline{x}_2\dots\overline{x}_n.$$

Thus,

$$\left((-1)^{n-1}(n-2)\alpha_1 + \sum_{i=2}^{n-1}(-1)^{n-1-i}2\alpha_i\right)x_0\overline{x}_1\overline{x}_2\dots\overline{x}_n \equiv 0 \tag{22}$$

is an identity of $\mathcal{V}$. We want to show that the sum of the coefficients in parentheses is nonzero.

Consider the polynomial (20) modulo the identities of the variety $\mathcal{V}_8$. By virtue of (16) we have that

$$\overline{x}_1\dots\overline{x}_i x_1\overline{x}_{i+1}\dots\overline{x}_{n-1} \equiv (-1)^i\overline{x}_1\overline{x}_2x_1\overline{x}_3\dots\overline{x}_{n-1}, \quad i = 2, \dots, n-1,$$

where for $n = 3$ we have the polynomial $\overline{x}_1\overline{x}_2x_1$. So

$$g(x_1, \dots, x_{n-1}) \equiv \alpha_1\overline{x}_1x_1\overline{x}_2\dots\overline{x}_{n-1} + \left(\sum_{i=2}^{n-1}(-1)^i\alpha_i\right)\overline{x}_1\overline{x}_2x_1\overline{x}_3\dots\overline{x}_{n-1}. \tag{23}$$

We write the first term in the form

$$\overline{x}_1x_1\overline{x}_2\dots\overline{x}_{n-1} = x_1x_1\overline{x}_2\dots\overline{x}_{n-1} - \sum_{i=2}^{n-1}x_ix_1\underset{2}{\overline{x}_2}\dots\underset{i}{\overline{x}_1}\dots\underset{n-1}{\overline{x}_{n-1}} \equiv$$

$$\equiv (n-2)!\sum_{i=2}^{n-1}x_1x_i\underset{2}{x_2}\dots\underset{i}{x_1}\dots\underset{n-1}{x_{n-1}} \equiv (n-2)!\sum_{i=2}^{n-1}(-1)^ix_1x_ix_1\dots\widehat{x_i}\dots x_{n-1},$$

where for $n = 3$ we get the monomial $x_1x_2x_1$.

Note that a monomial of the form $xy\dots z\dots z\dots$, by virtue of (16), is an identity and we transform the second term of (23) as follows,

$$\bar{x}_1\bar{x}_2x_1\bar{x}_3\dots\bar{x}_{n-1} \equiv x_1\bar{x}_2x_1\bar{x}_3\dots\bar{x}_{n-1} - \bar{x}_2x_1x_1\bar{x}_3\dots\bar{x}_{n-1} \equiv 2x_1\bar{x}_2x_1\bar{x}_3\dots\bar{x}_{n-1} \equiv$$

$$\equiv 2(n-3)!x_1x_2x_1x_3\dots x_{n-1} - 2(n-3)!\sum_{i=3}^{n-1} x_1x_ix_1\underset{3}{x_3}\dots\underset{i}{x_2}\dots\underset{n-1}{x_{n-1}} \equiv$$

$$\equiv 2(n-3)!\sum_{i=2}^{n-1}(-1)^i x_1x_ix_1\dots\widehat{x_i}\dots x_{n-1},$$

where for $n=3$ we get the monomial $2x_1x_2x_1$. Adding the resulting polynomials we obtain,

$$g(x_1,\dots,x_{n-1}) \equiv$$

$$\equiv (n-3)!\left((n-2)\alpha_1 + \sum_{i=2}^{n-1}(-1)^i 2\alpha_i\right)\sum_{i=2}^{n-1}(-1)^i x_1x_ix_1\dots\widehat{x_i}\dots x_{n-1}.$$

Thus, if in the identity (22) the sum of the coefficients in parentheses is zero, then the polynomial $g(x_1,\dots,x_{n-1})$ is identically equal to zero in the variety $\mathcal{V}_8$, a contradiction. Therefore, for some $l \geq 1$, $x_0\bar{x}_1\bar{x}_2\dots\bar{x}_l \equiv 0$ is an identity of $\mathcal{V}$ and the proposition is proved. □

Proposition 16 (Shulezhko and Panov [24]) *The variety $\mathcal{V}_8$ is almost nilpotent.*

Proof By the previous proposition for any proper subvariety $\mathcal{V}$ of the variety $\mathcal{V}_8$ there exists $l \geq 1$ such that $x_0\bar{x}_1\bar{x}_2\dots\bar{x}_l \equiv 0$ is an identity of $\mathcal{V}$. If we substitute x_0 with the product x_0y_0 and we use the identity (16) then we get the identity $x_0y_0x_1\dots x_l \equiv 0$. Therefore, any proper subvariety of the variety $\mathcal{V}_8$ is nilpotent, and the proposition is proved. □

6 A Characterization of Almost Nilpotent Varieties in Different Classes of Algebras

If we consider varieties of associative algebras it is not hard to prove that the only almost nilpotent variety is the variety $\mathcal{V}_0$ of commutative algebras satisfying the identities:

$$xy \equiv yx, \qquad (xy)z \equiv x(yz).$$

For the sequence of codimensions we have that $c_n(\mathcal{V}_0) = 1, n \geq 1$.

Also in case of varieties of Lie algebras it is well known that there is only one almost nilpotent variety, the variety $\mathcal{V}_4$ of metabelian Lie algebras satisfying the identities:

$$xy \equiv -yx, \qquad (xy)z \equiv (xz)y + x(yz), \qquad (x_1x_2)(x_3x_4) \equiv 0.$$

For this variety we have that $c_n(\mathcal{V}_4) = n - 1, n \geq 1$.

In [5] it was proved that there exist only two almost nilpotent varieties of Leibniz algebras, the varieties $\mathcal{V}_1$ and $\mathcal{V}_4$, and both varieties have at most linear growth.

What about varieties with subexponential growth in the class ${}_2\mathcal{N}$ of left nilpotent algebras of index two? In [17] it was proved that there exist only two almost nilpotent varieties having subexponential growth. More precisely

Theorem 6 (Mishchenko and Valenti [17]) *Let $\mathcal{V}$ be a subvariety of ${}_2\mathcal{N}$. If $\mathcal{V}$ has subexponential growth then either $\mathcal{V}_1 \subseteq \mathcal{V}$ or $\mathcal{V}_5 \subseteq \mathcal{V}$ or $\mathcal{V}$ is nilpotent.*

Corollary 3 *Let $\mathcal{V}$ be an almost nilpotent subvariety of ${}_2\mathcal{N}$ with subexponential growth, then either $\mathcal{V} = \mathcal{V}_1$, or $\mathcal{V} = \mathcal{V}_5$.*

For commutative or anticommutative metabelian algebras similar results were obtained in [20] and [14].

Theorem 7 (Mishchenko et al. [20]) *Let $\mathcal{V}$ be a variety of commutative metabelian algebras whose growth is not higher than subexponential, then either $\mathcal{V}_3 \subseteq \mathcal{V}$, or $\mathcal{V}_7 \subseteq \mathcal{V}$, or the variety $\mathcal{V}$ is nilpotent.*

Corollary 4 *Let $\mathcal{V}$ be an almost nilpotent variety of commutative metabelian algebras with subexponential growth, then either $\mathcal{V} = \mathcal{V}_3$, or $\mathcal{V} = \mathcal{V}_7$.*

In [24] it was proved the existence of only two almost nilpotent anti-commutative metabelian varieties with subexponential growth.

Theorem 8 *Let $\mathcal{V}$ be an almost nilpotent subexponential growth variety of anti-commutative metabelian algebras, then either $\mathcal{V} = \mathcal{V}_4$, or $\mathcal{V} = \mathcal{V}_8$.*

7 An Infinite Series of Almost Nilpotent Metabelian Varieties with Polynomial Growth

In this section we recall some results about the existence of two families of almost nilpotent varieties. The first is a countable family of varieties of at most linear growth and the second is an uncountable family of at most quadratic growth (see [12, 13, 18]).

Throughout A will be the algebra generated by one element a such that every word in A containing two or more subwords equal to a^2 must be zero.

Note that in particular the algebra A is metabelian, i.e., it satisfies the identity

$$(x_1x_2)(x_3x_3) \equiv 0.$$

Next for every real number between 0 and 1 we shall construct a quotient algebras of A.

We need to recall that a Sturmian word is an infinite word such that for every $n = 0, 1, \ldots$ admits exactly $n + 1$ different subwords of length n.

Moreover a lower mechanical word is the word $\underline{w}^{\alpha,\rho}$ with parameters $\alpha \in [0, 1]$ and $\rho \geq 0$, in which the letter $\underline{w}_n^{\alpha,\rho}$ appearing in the position $n = 0, 1, 2, \ldots$, is given by

$$\underline{w}_n^{\alpha,\rho} = \lfloor \alpha(n+1) + \rho \rfloor - \lfloor \alpha n + \rho \rfloor,$$

where $\lfloor x \rfloor$ is the largest integer less than or equal to $x.\alpha$ is the slope of the word.

Also, an infinite word $w = w_1w_2\cdots$ is *periodic* with period T if $w_i = w_{i+T}$ for $i = 1, 2, \ldots$.

We are going to associate to every finite word in the alphabet $\{0, 1\}$ a monomial in $End(A)$ in left and right multiplications: if $u(0, 1)$ is such a word we associate to u the monomial $u(L_a, R_a)$ obtained by substituting 0 with L_a and 1 with R_a.

Let α be a real number, $0 < \alpha < 1$, and let w_α be a Sturmian or periodic infinite word in the alphabet $\{0, 1\}$ whose slope is $\pi(w_\alpha) = \alpha$.

Let I_α be the ideal of the algebra A generated by the elements $a^2u(L_a, R_a)$ where $u(0, 1)$ is not a subword of the word w_α.

Let $A_\alpha = A/I_\alpha$ denote the corresponding quotient algebra and let $\mathcal{V}_\alpha$ be the variety generated by the algebra A_α.

We have the following

Theorem 9 *For any real number α, $0 < \alpha < 1$, the variety $\mathcal{V}_\alpha$ is almost nilpotent and has linear or quadratic growth according as w_α is a periodic or a Sturmian word. Moreover for words with different slopes the varieties are different.*

8 Almost Nilpotent Varieties with Exponentional Growth

The existence of almost nilpotent variety with exponential growth was proved in [15, 16, 21].

Let B_m be the algebra with generator $z, a_1, \ldots, a_m$ and relations

$$a_iu = 0, \qquad u \in B_m, \ 1 \leq i \leq m,$$

$$\big(zw(R_{a_1}, \ldots, R_{a_m})\big)\big(zw'(R_{a_1}, \ldots, R_{a_m})\big) = 0, \quad \deg w, \deg w' \geq 0,$$

$$z(R_{a_1}\dots R_{a_m})^k R_{a_{i_1}}\dots R_{a_{i_s}} R_{a_{i_{s+1}}}\dots R_{a_{i_t}} +$$

$$z(R_{a_1}\dots R_{a_m})^k R_{a_{i_1}}\dots R_{a_{i_{s+1}}} R_{a_{i_s}}\dots R_{a_{i_t}} = 0, \quad k \geq 0,$$

where $1 \leq s < t \leq m$, $1 \leq i_1, \dots, i_t \leq m$, and R_a is the operator of right multiplication on a.

A basis of B_m is given by the monomials:

$$a_1, \dots, a_m, \quad z(R_{a_1}\dots R_{a_m})^k, \quad z(R_{a_1}\dots R_{a_m})^k R_{a_{i_1}} R_{a_{i_2}}\dots R_{a_{i_t}},$$

$$k \geq 0, \qquad 1 \leq t < m, \qquad 1 \leq i_1 < i_2 < \dots < i_t \leq m.$$

For the variety generated by B_m we have the following

Theorem 10 *For any $m \geq 2$ $var(B_m)$ is an almost nilpotent variety of exponent m.*

The existence of almost nilpotent varieties with non-integer exponent was proved in [11]

Theorem 11 *There is an almost nilpotent variety with upper and lower growth exponents belonging to the interval (1, 2).*

We finish by suggesting the following

Problem Construct almost nilpotent varieties having over exponential growth or prove that such varieties do not exist.

Acknowledgement A. Valenti was partially supported by INDAM-GNSAGA of Italy.

References

1. Bakhturin, Yu.A. Identities in Lie Algebras. Nauka, Moscow, p. 448 (1985)
2. Chang, N.T.K., Frolova, Yu.Yu.: Commutative metabelian almost nilpotent varieties growth not higher than exponential (Russian). In: International Conference Mal'tsev Readings: 10–13 November 2014, p. 119. Novosibirsk, (2014). http://www.math.nsc.ru/conference/malmeet/14/Malmeet2014.pdf
3. Drensky, V.: Polynomial identities for the Jordan algebra of a symmetric bilinear form. J. Algebra **108**, 66–87 (1987)
4. Drensky, V.: Free Algebras and PI-Algebras. Graduate Course in Algebra. Springer, Singapore (2000)
5. Frolova, Y., Shulezhko, O.V.: Almost nilpotent varieties of Leibniz algebras (Russian). Prikladnaya Discretnaya Mat. **2**(28), 30–36 (2015)
6. Giambruno, A., Zelmanov, E.: On growth of codimensions of Jordan algebras. In: Groups, Algebras and Applications. Contemporary Mathematics, vol. 537, pp. 205–210. American Mathematical Society, Providence (2011)
7. Giambruno, A., Mishchenko, S., Zaicev, M.: Algebras with intermediate growth of the codimensions. Adv. Applied Math. **37** , 360–377 (2006)

8. James, G., Kerber, A.: The representation theory of the symmetric group. In: Encyclopedia of Mathematics and Its Applications, vol. 16. Addison-Wesley, London (1981)
9. Mishchenko, S.P.: Lower bounds on the dimensions of irreducible representations of symmetric groups and of the exponents of the exponential of varieties of Lie algebras (Russian) Mat. Sb. **187**, 83–94 (1996); translation in Sb. Math. **187**, 81–92 (1996)
10. Mishchenko, S.P.: Varieties of linear algebras with colength one (Russian). Vestn. Mosk. Univ. Ser. I N 1, 25–30 (2010); translation in Mosc. Univ. Math. Bull. **65**, 23–27 (2010)
11. Mishchenko, S.P.: Almost nilpotent varieties with non-integer exponents do exist. Mosc. Univ. Math. Bull. **71**(3), 115–118 (2016)
12. Mishchenko, S.P.: Infinite periodic words and almost nilpotent varieties. Mosc. Univ. Math. Bull. **72**(4), 173–176 (2017)
13. Mishchenko, S.P., Panov, N.P.: Sturmian words and uncountable set of almost nilpotent varieties of quadratic growth. Mosc. Univ. Math. Bull. **72**(6), 251–254 (2017)
14. Mishchenko, S.P., Shulezhko, O.V.: Description of almost nilpotent anticommutative metabelian varieties with subexponential growth (Russian). In: International Conference Mal'tsev Readings: 10–13 November 2014. Novosibirsk, 2014, p. 104. http://www.math.nsc.ru/conference/malmeet/14/Malmeet2014.pdf
15. Mishchenko, S.P., Shulezhko, O.V.: Almost nilpotent varieties of arbitrary integer exponent (Russian). Vestn. Moskov. Univ. Ser. I Mat. Mekh. N 2, 53–57 (2015); translation in Mosc. Univ. Math. Bull. **70**(2), 92–95 (2015)
16. Mishchenko, S.P., Valenti, A.: An almost nilpotent variety of exponent 2. Israel J. Math. **199**, 241–258 (2014)
17. Mishchenko, S., Valenti, A.: On almost nilpotent varieties of subexponential growth. J. Algebra **423**, 902–915 (2015)
18. Mishchenko, S., Valenti, A.: An uncountable family of almost nilpotent varieties of polynomial growth. J. Pure Appl. Algebra **222**(7), 1758–1764 (2018)
19. Mishchenko, S.P., Zaicev, M.: An example of a variety of linear algebras with the fractional polynomial growth. Vestn. Mosc. State Univ. Math. Mech. **1**, 25–31 (2008)
20. Mishchenko, S., Panov, N., Frolova, Y., Trang, N.: On the varieties of commutative metabelian algebras. Fundam. Prikl. Mat. **21**(1), 165–180 (2016). Russian; translation in J. Math. Sci. **233**(5), 713–723
21. Panov, N.P.: Almost nilpotent varieties with integer exponent. Izv. Saratov Univ., Ser. Math. Mech. Inform. **17**(3), 331–343 (2017)
22. Petrogradsky, V.M.: Growth of polynilpotent varieties of Lie algebras, and rapidly increasing entire functions, (Russian). Mat. Sb. **188**(6), 119–138 (1997); translation in Sb. Math. **188**(6), 913–931 (1997)
23. Regev, A.: Existence of identities in $A \otimes B$. Israel J. Math. **11**, 131–152 (1972)
24. Shulezhko, O.V., Panov, N.P.: On almost nilpotent varieties of anticommutative metabelian algebras. Appl. Discrete Math. **38**, 35–48 (2017)
25. Volicenko, I.B.: Varieties of Lie algebras with identity $[[X_1, X_2, X_3], [X_4, X_5, X_6]] = 0$ over a field of characteristic zero. Sibirsk. Math. Zh **25**(3), 370–382 (1984)
26. Zhevlakov, K.A., Slin'ko, A.M., Shestakov, I.P., Shirshov, A.I.: Jordan algebras [in Russian]. Novosibirsk (1976)
27. Zhevlakov, K.A., Slinko, A.M., Shestakov, I.P., Shirshov, A.I.: Rings that are nearly associative. Nauka, Moscow (1978). English Translation: Academic Press, New York (1982)

(δ, ε)-Differential Identities of $UT_m(F)$

Vincenzo C. Nardozza

Abstract Let δ and ε be the inner derivations of $UT_m(F)$ induced by the unit matrices e_{1m} and e_{mm} respectively. We study the differential polynomial identities of the algebra $UT_m(F)$ under the coupled action of δ and ε. We produce a basis of the differential identities, then we determine the S_n-structure of their proper multilinear spaces and, for the minimal cases $m = 2, 3$, their exact differential codimension sequence.

Keywords Differential polynomial identities · Upper triangular matrices · Differential codimensions · Lie algebras · Derivation action

1 Introduction

Differential polynomial identities are certainly not a brand new topic in PI-theory. Significant contributions to this subject may be dated back to the late 1970s, due to a series of fundamental papers by Kharchenko involving both derivations and automorphisms, but in fact a vast literature on this topics is available (a good source is [1] and its bibliography to this and related topics). In present days, however, new interest is flowing into this subject, mainly because of a new unifying approach towards the several areas related to PI-theory.

As a consequence of the evolution of classic (so to say) PI-theory, almost every special PI-theory has developed a suitable set of tools, techniques and results modeled on those available for the classic case. So, for instance, when dealing with algebras with involutions, superalgebras, or with more general graded algebras, one can properly define universal objects, identities, cocharacters, codimensions and so on, resembling what happens in the ordinary case. It turns out that several results holding for the classic case can be restated for the special ones, although

V. C. Nardozza (✉)
Dipartimento di Matematica, Università degli Studi di Bari, Bari, Italia
e-mail: vincenzo.nardozza@uniba.it

O. M. Di Vincenzo, A. Giambruno (eds.), *Polynomial Identities in Algebras*, Springer INdAM Series 44, https://doi.org/10.1007/978-3-030-63111-6_16

under some suitable, reasonable, assumptions. This is the case, for instance, for the growth behaviour of codimensions of superalgebras, algebras with involution, graded algebras, algebras with derivations. On the contrary, other central results of classical PI-Theory are more eluding. Among them, the most prominent ones are Kemer's Representability Theorem and the finite basis property holding in classic PI-Theory [2]. A generalization of these results has been obtained for algebras graded by a finite group in [3].

Starting from a sparkling intuition of Berele in his influential paper [4] (more precisely, the Remark at page 878), an effort to a unifying approach started, turning around the notion of Hopf-algebra action. This is for instance the case of the relatively recent paper by Gordienko [5]. The Specht property and the Representability theorem have been recently faced within this framework in [6].

In the case of algebras under the derivation action of a Lie algebra L, the involved Hopf-algebra is the universal enveloping algebra $U(L)$ of L, a more than natural connection. Of course, even when a final unifying theory should be established, it could conceal but not cancel the differences among concrete situations, so differential identities, as well as other types of identities, will still have to be treated and studied according to their specific features, although empowered perhaps with some new profitable idea coming from some other specific situation.

The present paper is based on a very recent joint work with Di Vincenzo [7] which, at the moment, has been just submitted for publication, so it partially serves as an announcement for the results contained in [7] and concerning the description of the differential polynomial identities satisfied by algebras $UT_m(F)$ under the derivation actions of the two-dimensional non abelian Lie algebra, a problem which was in some sense inspired by the recent paper of Giambruno and Rizzo [8].

At the same time, I wanted this paper to be an expository one, hoping to convey the reader the same pleasure I sensed working on this problem. So in writing these notes I chose to present the material in a maybe rather unusual way, that is from an operative point of view rather than from a formal one. The basic definition of differential polynomial identity is therefore given in the next section within this perspective, together with all the necessary background and tools needed to quickly understand the problem and how to attack it, in the same spirit. The subsequent sections are devoted to give the answers, in case the Lie algebra acts faithfully (coupled actions of δ, ε) on $UT_m(F)$ or not (separate actions). Due to the nature of this paper, technical details and proofs have been omitted, but I tried to at least address the reader to the main ideas involved in them. By the way, a couple of statements which were missing on the original paper have been added here, and their proofs are therefore provided within these pages. The last section is instead devoted to present the general topics within a more theoretical setup, in order to confer the objects and tools presented in the preceding sections a more sound and deeper sense.

2 The Problem

Let F denote a field of characteristic zero and let A be an associative F-algebra. An F-linear map $d : A \to A$ is called an *F-derivation* if the usual Leibniz rule $(ab)^d = a^d b + ab^d$ holds for all $a, b \in A$. Throughout the paper, we will adopt the exponential notation for derivations, hence derivations will compose from left to right. It is easy to produce concrete derivations on A: for any $a \in A$, just consider the map $[\cdot, a]$ sending $x \in A$ into the Lie product $[x, a] = xa - ax$. It is called the *inner derivation induced by a*. For some relevant algebras, these derivations are actually the only ones available: this is the case for the full matrix algebra $M_m(F)$ and its subalgebra $UT_m(F)$ of upper triangular matrices [9]. In this paper, we are going to deal with the latter one. More precisely, let δ and ε be the inner derivations of $UT_m(F)$ induced by the unit matrices e_{1m} and e_{mm} respectively, that is $\delta = [\cdot, e_{1m}]$ and $\varepsilon = [\cdot, e_{mm}]$. The algebra $UT_m(F)$ is enriched with these derivation actions on it, and we denote U_m this structure, to distinguish it from the simpler algebra structure $UT_m(F)$. Then the identity relations among the elements of $UT_m(F)$ are still valid in U_m, but they are just a part of those holding in U_m: new relations, involving both elements of $UT_m(F)$ and derivations of elements of $UT_m(F)$, appear. For instance, for any $a, b, c \in UT_m(F)$, it holds $a^\varepsilon bc = a^\varepsilon cb$. These more general identity relations are called *differential* identities, and the basic problem we are going to face is the following:

Determine and describe the differential identities holding in U_m.

Some clarifications are in order: first of all, we need to be more precise on what a differential identity is. Then, we have to agree on what the verbs *determine* and *describe* should mean. About the first point, we are going to pursue a very intuitive approach. It will fit perfectly the operative aspects of our investigations, though is a bit too naïve to be fully satisfying. A more sound and solid approach will be postponed to the last section.

Let us start with a countable set of indeterminates X, and define a new one, namely

$$X^D := \{x^w \mid x \in X,\ w \text{ word in } \delta, \varepsilon\}.$$

So, for instance, $x^{\delta\varepsilon\delta} \in X^D$ for all $x \in X$. We will call *letters* the elements of X^D. More precisely, if w is not the empty word, we call x^w a *differential* letter; the letters in X^D which are not differential are substantially indistinguishable from their parent indeterminate, so we identify the letter $x \in X^D$ (corresponding to the empty word) with the indeterminate $x \in X$, and call it an *ordinary* letter. Hence we write $X \subseteq X^D$.

The free associative unitary F-algebra $F\langle X^D\rangle$ generated by X^D inherits a formal derivation action of δ, ε: just define $(x^w)^\alpha := x^{w\alpha}$, for $\alpha \in \{\delta, \varepsilon\}$, on the letters $x^w \in X^D$, and then extend this (right) action to the whole $F\langle X^D\rangle$ by F-linearity and the Leibniz rule. The elements of $F\langle X^D\rangle$ are called differential polynomials; in case

$f \in F\langle X^D \rangle$ involves ordinary letters only, it will be called an ordinary polynomial. The natural inclusion $F\langle X \rangle \subseteq F\langle X^D \rangle$ then follows from the definition.

Definition 1 An element $f(x_1, \ldots, x_n) \in F\langle X^D \rangle$ is a *differential polynomial identity* of U_m if $f(a_1, \ldots, a_n) = 0$ for all $a_1, \ldots, a_n \in U_m$. The set of all differential polynomial identities of U_m is denoted by $T_D(U_m)$.

Notice that the writing $f(x_1, \ldots, x_n)$ points just to the parent indeterminates of the letters occurring in the polynomial: this is legitimate for the following fact

Lemma 2 *Let A be an F-algebra with derivation actions of δ, ε. Any map $\varphi_0 : X \to A$ uniquely extends to an algebra homomorphism $\varphi : F\langle X^D \rangle \to A$ commuting with the derivation action of δ, ε.*

As a first sign of the naïve nature of our definitions notice that, at the moment, just two algebras with derivation action of δ and ε are available: U_m, from which δ and ε have been constructed, and $F\langle X^D \rangle$ itself. Hence what a generic algebra with (δ, ε)-derivation actions should be is left too vague and subject to criticism, not to mention the definition of $T_D(A)$ for a generic algebra A. Nevertheless, we shall pursue our intuitive perpsective for the moment, and interprete the first statement of the Lemma as a shortcut to mean $A \in \{F\langle X^D \rangle, U_m\}$, while we merely focus on $T_D(U_m)$. Also, a homomorphism $\varphi : F\langle X^D \rangle \to A$ commuting with δ and ε will be called a *D-homomorphism*, for short.

It is worth noticing that the set $T(UT_m(F))$ of usual polynomial identities satisfied by the algebra $UT_m(F)$ coincides with the set of *ordinary* polynomial identities of U_m and is a subset of $T_D(U_m)$, as a consequence of our definitions; this is consistent with the idea that we are adding more general identity relations to the ones related to the mere algebra structure.

The set $T_D(U_m)$ is clearly a two-sided ideal of $F\langle X^D \rangle$, but actually is more than this: it is invariant under all D-endomorphism of $F\langle X^D \rangle$, and is called a T_D-ideal. If $\mathscr{I} \subseteq F\langle X^D \rangle$, it makes sense to consider the least T_D-ideal containing $\mathscr{I}$, and call it the T_D-ideal *generated* by $\mathscr{I}$. So a possible, acceptable answer for *determine the differential polynomial identities of* U_m, is to exhibit a few differential polynomial identities of U_m generating the whole $T_D(U_m)$ as a T_D-ideal.

This also gives a first, very rough, sense to what we may mean by *describing* the T_D-ideal $T_D(U_m)$: in fact, even when a generating set $\mathscr{I}$ is given, it is extremely hard to decide if a random polynomial $f \in F\langle X^D \rangle$ follows from $\mathscr{I}$ (that is if f belongs to the T_D-ideal generated by $\mathscr{I}$). Since F has characteristic zero, a more refined description of $T_D(U_m)$ is provided by its multilinear spaces:

Definition 3 Define $P_n^D := \operatorname{span}_F\langle x_{\sigma(1)}^{w_1} \cdots x_{\sigma(n)}^{w_n} \mid \sigma \in S_n,\ w_i \text{ word in } \delta, \varepsilon \rangle$ for all $n \geqslant 1$, and let $P_n^D(U_m)$ denote the factor space $P_n^D / \big(P_n^D \cap T_D(U_m)\big)$.

Any element of P_n^D is called a *multilinear differential polynomial of degree n*, and those in $P_n^D \cap T_D(U_m)$ are named multilinear differential polynomial identities of U_m of degree n. From here on, we shall abbreviate it in *multilinear D-PI*.

The slices $P_n^D \cap T_D(U_m)$ completely characterize $T_D(U_m)$, because their union generates $T_D(U_m)$, by standard Vandermonde argument and multilinearization process. Moreover, each $P_n^D \cap T_D(U_m)$ is actually a submodule of the S_n-module P_n^D, where the (left) action of S_n on the multilinear differential polynomials is the natural one; namely, the one defined on the letters by $\sigma \bullet x_i^w := x_{\sigma(i)}^w$ for all $\sigma \in S_n$. Hence $P_n^D(U_m)$ is a left S_n-module as well, and its S_n-character $\chi_n^D(U_m)$, called the n-th *D-cocharacter* of U_m, indirectly gives a picture of the S_n-structure of $P_n^D \cap T_D(U_m)$. Moreover, the dimension $c_n^D(U_m) := \dim_F P_n^D(U_m)$, called the *n-th codimension of* U_m, gives a quantitative measure on how big the slice $P_n^D \cap T_D(U_m)$ is: the greater $c_n^D(U_m)$, the smaller is the space of multilinear D-PI's $P_n^D \cap T_D(U_m)$. A word of caution is due to this proposal: there is no reason, at the moment, to believe that $P_n^D(A)$ is finite dimensional. Indeed, by definition, P_n^D is infinite-dimensional: for instance the set $\{x_1^{\delta^i} x_2 \dots x_n \mid i \in \mathbb{N}\}$ is an infinite set of F-independent elements of P_n^D.

Since the multilinear spaces do provide so many useful information on $T_D(U_m)$, both of qualitative and quantitative nature, it is more than agreeable to accept the S_n-cocharacter sequence of U_m as a description of the D-PI of U_m. So our tasks are now operatively clear: to answer the problem, we have to find a small set of D-PI generating $T_D(U_m)$, and give the decomposition of the n-th D-cocharacter $\chi_n^D(U_m)$ into irreducible S_n-characters. The D-codimension sequence $c_n^D(U_m)$, once computed, will give the quantitative description on how big $T_D(U_m)$ is.

These tasks can be made easier if we exploit the fact that U_m is a unitary algebra. In this case, all features of $P_n^D(U_m)$ are encoded in smaller multilinear spaces, consisting of the so-called *proper* multilinear polynomials. There are several ways of presenting the notion of proper polynomials, and the easiest is the following

Definition 4 A polynomial $f \in F\langle X^D \rangle$ is called a *proper* polynomial if $\dfrac{\partial f}{\partial x} = 0$ for all ordinary letters $x \in X$.

A word of caution is needed, also with this definition: we are pointing to the elements of X^D as free generators, so there is no interaction among the formal derivations δ, ε and the usual formal partial derivatives $\frac{\partial}{\partial x}$. Explicitly, $\frac{\partial x^\delta}{\partial x} = 0$, and the same holds for x^ε: x, x^δ and x^ε are different elements among those freely generating $F\langle X^D \rangle$. Actually, talking of *X-proper polynomials* would be more precise. So, in particular, x^δ and x^ε are among the proper polynomials. What is the form of a generic proper polynomial? We need the following

Definition 5 Let $n \geqslant 2$ and let $z_1, \dots, z_n \in X^D$. The *higher commutator* $[z_1, \dots, z_n]$ is defined recursively by $[z_1, z_2] = z_1 z_2 - z_2 z_1$ and, for $n \geqslant 3$, $[z_1, z_2, \dots, z_n] := \big[[z_1, \dots, z_{n-1}], z_n\big]$. The number n is the *length* of the commutator.

Higher commutators are therefore a (left-normed) generalization of the Lie product among letters of X^D. We may extend the notion to commutators of lengths 0 and 1: a commutator of length 0 is simply an element $a \in F$, while by commutator of length

1 we mean any differential letter. Actually, the proper polynomials are precisely the elements of the unitary subalgebra B^D of $F\langle X^D\rangle$ generated by the commutators of any length. Hence, they are F-linear combinations of products of commutators.

Much more could be said on proper polynomials and their general properties. We address the interested reader to the book [10] for the basic definitions and results. Actually, Drensky re-discovered and gave new life to this class of polynomials, turning them into an amazing and powerful tool employed in several papers. By the way, we just need to focus on a specific type of proper polynomials:

Definition 6 Define $\Gamma_0^D := F$, and for $n \geqslant 1$ let $\Gamma_n^D := P_n^D \cap B^D$. The elements of the set $\Gamma^D := \bigcup_{n\in\mathbb{N}} \Gamma_n^D$ are called *proper multilinear polynomials.*

Proper multilinear polynomials share the same basic property of multilinear polynomials in our settings:

Lemma 7 *$T_D(U_m)$ is generated, as T_D-ideal, by $\Gamma^D \cap T_D(U_m)$.*

Therefore the proper multilinear polynomials in $T_D(U_m)$ completely determine the T_D-ideal, as the multilinear polynomials do. Moreover, since Γ_n^D is an S_n-module as well, the factor space $\Gamma_n^D(U_m) = \Gamma_n^D/\big(\Gamma_n^D \cap T_D(U_m)\big)$ is an S_n-module. Let $\xi_n^D(U_m)$ be its S_n-character (the n-th *proper* differential cocharacter of U_m), and $\gamma_n^D(U_m)$ be its dimension (the n-th *proper* codimension of U_m). Then the cocharacter sequence $\chi_n^D(U_m)$ is simply the so-called Young-derived sequence of $(\xi_n^D(U_m))_{n\in\mathbb{N}}$, that is $\chi_n^D(U_m)$ is obtained from the cocharacters $\xi_0^D(U_m), \dots, \xi_n^D(U_m)$ via the Young–Pieri rule (see [11], but also the most comprehensive exposition in [10] illustrating the interplay between proper and ordinary polynomials not only in the multilinear case, but in the more general case of multi-homogeneous one, involving the action of the general linear groups); hence the codimension sequence can be computed from the proper codimension sequence by the simple relation

$$c_n^D(U_m) = \sum_{k=0}^{n} \binom{n}{k} \gamma_k^D(U_m).$$

Therefore, a significatively simpler meaning for the verb *describe* is made available: in order to describe $T_D(U_m)$, it is sufficient to get the decomposition of the proper cocharacters of U_m for all $n \in \mathbb{N}$; the quantitative information on $T_D(U_m)$ are carried from the proper codimension sequence.

A further, last simplification towards this description is possible, by selecting a particular basis for the vector spaces Γ_n^D (see [12]. Proper polynomials with respect to a distinct set of indeterminates were first presented in [13]). Let us fix a total order $\leqslant$ on X^D, such that ordinary letters precede the differential ones.

Definition 8 A higher commutator $[z_1, \dots, z_n]$ is *normal* if $z_2, \dots, z_n$ are ordinary letters. Moreover, the normal commutator $[z_1, \dots, z_n]$ is *standard* if $z_1 > z_2 < \dots < z_n$.

We include the commutators of lengths 0 and 1 among the normal standard commutators. Of course, if z_1 is a differential letter in the normal commutator $[z_1, z_2, \ldots, z_n]$ then just the order among $z_2, \ldots, z_n$ matters in being standard. The reason for bringing up normal standard commutators is that, as a Corollary of a stronger statement (Proposition 7 in [12]), it holds

Theorem 9 *The elements of Γ_n^D which are products of normal standard commutators constitute an F-basis of Γ_n^D.*

Proof The products of normal *semi*standard commutators (that is: normal commutators $[z_1, \ldots, z_n]$ such that $z_1 > z_2 \leqslant \cdots \leqslant z_n$) form a basis for the algebra B^D of proper polynomials. Since $\Gamma_n^D = P_n^D \cap B^D$, any polynomial in Γ_n^D is a linear combination of products of normal semistandard commutators but, being multilinear, it is actually a linear combination of products of normal standard commutators. Then, just note that normal standard commutators are in particular semistandard, to get the linear independence. □

3 The Coupled Actions of δ and ε on $UT_m(F)$

As elements of $End_F(U_m)$, the operators δ and ε satisfy the following relations:

$$\delta^2 = 0, \quad \varepsilon^2 = \varepsilon, \quad \delta\varepsilon = \delta, \quad \varepsilon\delta = 0$$

(recall that their compositions are computed from left to right in our notation). Therefore the following D-PI's are readily available, and depend just upon the selected derivations:

Lemma 10 *The polynomials $x_1^{\delta^2}$, $x_1^{\varepsilon^2} - x_1^{\varepsilon}$, $x_1^{\delta\varepsilon} - x_1^{\delta}$, $x_1^{\varepsilon\delta}$ are in $T_D(U_m)$.*

Notice that they are all in Γ_1^D. Moreover, they cause any differential letter x^w related to a word w of length $\geqslant 2$ to be congruent, modulo $T_D(U_m)$, either to 0 or to a single differential letter x^{ε}, x^{δ}. Therefore just ordinary letters or the differential letters x^{δ}, x^{ε} need to be considered in the sequel.

The following monomial identities also belong to $T_D(U_m)$:

Corollary 11 *The monomials $x_1^{\delta}x_2^{\varepsilon}$, $x_1^{\varepsilon}x_2^{\delta}$, $x_1^{\delta}x_2^{\delta}$, $x_1^{\varepsilon}x_2^{\varepsilon}$ follow from the identities listed in the previous Lemma. In particular, they are all in $T_D(U_m)$.*

Proof Let I be the T_D-ideal generated by the polynomials listed in Lemma 10. Then $(x_1x_2)^{\varepsilon\delta} \in I$. Explicitly, one has

$$(x_1x_2)^{\varepsilon\delta} = x_1^{\varepsilon\delta}x_2 + x_1^{\varepsilon}x_2^{\delta} + x_1^{\delta}x_2^{\varepsilon} + x_1x_2^{\varepsilon\delta} \in I. \text{ Hence } x_1^{\varepsilon}x_2^{\delta} + x_1^{\delta}x_2^{\varepsilon} \in I.$$

Replacing x_1 by x_1^{ε} yields $x_1^{\varepsilon^2}x_2^{\delta} + x_1^{\varepsilon\delta}x_2^{\varepsilon} \in I$, so $x_1^{\varepsilon^2}x_2^{\delta} \in I$. By the way, since $x^{\varepsilon^2} \equiv x^{\varepsilon} \pmod{I}$, it follows $x_1^{\varepsilon}x_2^{\delta} \in I$.

The other identities follow easily. Then, since $I \subseteq T_D(U_m)$, they are in particular differential polynomial identities of U_m. □

Hence, any nonvanishing multilinear monomial will involve at most a single differential letter.

There are other basic identities, not depending just on the selected inner derivations but more properly on the interplay of δ and ε with the algebra structure. They are listed in the following

Lemma 12 *Let x, y, x_i, y_i denote distinct elements of X. The following polynomials are all in $T_D(U_m)$:*

(1) $[x_1, x_2]^\delta$;

(2) $x^\delta[x_1, x_2]$, $[x_1, x_2]x^\delta$, $x^\varepsilon[x_1, x_2]$;

(3) $[x_1, y_1] \dots [x_{m-1}, y_{m-1}]x^\varepsilon$;

(4) $[x_1, y_1] \dots [x_m, y_m]$;

(5) $[x_1, y_1] \dots [x_{m-2}, y_{m-2}]\Big([x, y]^\varepsilon - [x, y]\Big)$.

A different, maybe better, way to write the identity (1) is $[x_1^\delta, x_2] + [x_1, x_2^\delta]$. The identity (3) may be considered, in some sense, the ε-analogous of $[x_1, x_2]x^\delta$. The identity (4) is the one generating the whole T-ideal of ordinary polynomial identities of the algebra $UT_m(F)$, as proved in [14] . The last identity of the list is undoubtedly the most remarkable one, and the most difficult to find.

Collecting together the polynomials of Lemmas 10 and 12 we get all the necessary identities we need to generate the whole $T_D(U_m)$. Precisely, it holds

Theorem 13 *Let I be the T_D-ideal generated by the following differential polynomials*

(1) x^{δ^2}, $x^{\varepsilon^2} - x^\varepsilon$, $x^{\delta\varepsilon} - x^\delta$, $x^{\varepsilon\delta}$

(2) $[x_1, x_2]^\delta$

(3) $x^\delta[x_1, x_2]$, $[x_1, x_2]x^\delta$, $x^\varepsilon[x_1, x_2]$

(4) $[x_1, y_1] \dots [x_{m-1}, y_{m-1}]x^\varepsilon$

(5) $[x_1, y_1] \dots [x_m, y_m]$

(6) $[x_1, y_1] \dots [x_{m-2}, y_{m-2}]\Big([x, y]^\varepsilon - [x, y]\Big)$,

where all indeterminates x, y, x_i, y_i belong to X. Then $T_D(U_m) = I$.

It is worth noticing that all these polynomials are multilinear proper polynomials, and are expressed as linear combinations of products of normal standard commutators. The proof of this theorem is quite direct although rather structured, and brings up some objects which turn useful in describing the multilinear spaces, so I am giving the reader a glimpse of the main ideas involved in it.

Of course, $I \subseteq T_D(U_m)$, and in order to prove the reverse inclusion it is enough to compare the proper multilinear parts of the two T_D-ideals. The first thing to do is therefore to exhibit a set of polynomials spanning Γ_n^D modulo I for each $0 < n \in \mathbb{N}$, and then prove that this spanning set is linearly independent modulo $T_D(U_m)$. From this, the fact that $T_D(U_m) \subseteq I$ follows immediately.

The spanning set $\mathscr{S}_n$ we are looking for is partitioned according to the differential letters occurring in its elements, if any. So let us separately construct the sets $\mathscr{S}_n^1, \mathscr{S}_n^\delta, \mathscr{S}_n^\varepsilon$ partitioning $\mathscr{S}_n$.

- $\mathscr{S}_n^1$: take any product $c_1 \dots c_k \in \Gamma_n^D$ of $k < m$ normal standard commutators c_i involving ordinary letters only. Of course $\mathscr{S}_1^1 = \varnothing$, while $\mathscr{S}_2^1 = \{[x_2, x_1]\}$. Notice that $\mathscr{S}_n^1$ is actually an F-basis for $\Gamma_n(UT_m(F))$.
- $\mathscr{S}_n^\delta$: it is a singleton. More precisely, $\mathscr{S}_1^\delta = \{x_1^\delta\}$ and, if $n \geqslant 2$, $\mathscr{S}_n^\delta = \{[x_n^\delta, x_1, \dots, x_{n-1}]\}$.
- $\mathscr{S}_n^\varepsilon$: take any product $c_1 \dots c_k \in \Gamma_n^D$ of $k < m$ normal standard commutators such that
  - $c_1, \dots, c_{k-1}$ involve only ordinary letters. In particular, each of them has length $\geqslant 2$;
  - if $k < m - 1$ then the last commutator c_k is any $[x^\varepsilon, y_1, \dots, y_l]$, that is any normal standard commutator involving the remaining indeterminates
  - if $k = m - 1$ then the last commutator is uniquely determined. More precisely, if $y_1 < y_2 < \cdots < y_l < x$ are the remaining indeterminates, it is $[x^\varepsilon, y_1, \dots, y_l]$.

Our candidate set is therefore $\mathscr{S}_n = \mathscr{S}_n^1 \cup \mathscr{S}_n^\delta \cup \mathscr{S}_n^\varepsilon$, and it is almost easy to see that $\mathscr{S}_n$ in fact spans Γ_n^D modulo I.

To prove that $\mathscr{S}_n$ is linearly independent modulo $T_D(U_m)$ is more tricky. Essentially, we produce a sort of elimination algorithm:

1. initialize $\mathscr{S} := \mathscr{S}_n$
2. produce a substitution $\varphi : X \to U_m$ such that $\varphi(w) = 0$ for all $w \in \mathscr{S}$ but a single element w_0
3. delete w_0 from $\mathscr{S}$ and repeat the previous step until $\mathscr{S} = \varnothing$.

We may now proceed in describing the S_n-structure of the multilinear spaces $\Gamma_n^D(U_m)$. Since we are going to work modulo $T_D(U_m)$, we will simply write f instead of $f + T_D(U_m)$ and so on, in order to keep the notation as simple as possible.

As a byproduct of the preceding proof, $\Gamma_n^D(U_m)$ is finite-dimensional, since it has $\mathscr{S}_n$ as an F-basis. Moreover, setting $\Gamma_n^\alpha(U_m) = F\mathscr{S}_n^\alpha$ for $\alpha \in \{1, \delta, \varepsilon\}$, each $\Gamma_n^\alpha(U_m)$ is an S_n-submodule of $\Gamma_n^D(U_m)$, so we get the decomposition $\Gamma_n^D(U_m) = \Gamma_n^1(U_m) \oplus \Gamma_n^\delta(U_m) \oplus \Gamma_n^\varepsilon(U_m)$ and then consider the three summands separately in order to get the S_n-proper cocharacter of U_m.

Recall that the isomorphism classes of irreducible S_n-modules are in a bijective correspondence with the integer partitions λ of n (which we express by $\lambda \vdash n$). If $\lambda = [\![\lambda_1, \dots, \lambda_k]\!] \vdash n$, the corresponding irreducible character will be denoted by

λ as well, thus committing a slight abuse of notation. The character of $\Gamma_n^{\alpha}(U_m)$ will be denoted $\xi_n^{\alpha}(U_m)$.

It is easy to see that $\Gamma_0^D(U_m) = F$ and $\Gamma_1^D(U_m) = Fx_1^{\delta} \oplus Fx_1^{\varepsilon}$. For $n \geqslant 2$, just $\xi_n^{\varepsilon}(U_m)$ needs to be investigated. Indeed,

- $\xi_n^{\delta}(U_m) = [\![n]\!]$ is clear,
- $\xi_n^{1}(U_m)$ is the proper S_n-cocharacter of the algebra $UT_m(F)$ by [15].

In order to study $\xi_n^{\varepsilon}(U_m)$, let us denote $(l_1, \ldots, l_k) \vDash n$ any weak k-composition of n, that is any sequence of integers $l_1, \ldots, l_k \geqslant 0$ such that $l_1 + \cdots + l_k = n$.

Theorem 14 *$\xi_n^{\varepsilon}(U_2) = [\![n]\!]$ and, if $m \geqslant 3$,*

$$\xi_n^{\varepsilon}(U_m) = \xi_n^{\varepsilon}(U_{m-1}) + \sum_{\substack{(\lambda_1,\ldots,\lambda_{m-2})\vDash n\\ \lambda_1,\ldots,\lambda_{m-2}\geqslant 2}} \Big([\![\lambda_1 - 1, 1]\!] \otimes \cdots \otimes [\![\lambda_{m-2} - 1, 1]\!]\Big)^{S_n}$$

$$+ \sum_{\substack{(\lambda_1,\ldots,\lambda_{m-1})\vDash n\\ \lambda_1,\ldots,\lambda_{m-2}\geqslant 2\\ \lambda_{m-1}\geqslant 1}} \Big([\![\lambda_1 - 1, 1]\!] \otimes \cdots \otimes [\![\lambda_{m-2} - 1, 1]\!] \otimes [\![\lambda_{m-1}]\!]\Big)^{S_n}$$

There is a certain amount of indetermination in both $\xi_n^{1}(U_m)$ and in $\xi_n^{\varepsilon}(U_m)$, due to the induced characters involved in their description. The Littlewood–Richardson rule would turn them into a sum of irreducible S_n-characters, but this explicit decomposition, even if possible in principle, could hardly be accepted as a better one. By the way, at least in the small cases $m = 2$ and $m = 3$, they are worth of being computed, to get at least an idea of the general case.

Recall that $\Gamma_0^D(U_m) = F$ and $\Gamma_1^D(U_m) = Fx_1^{\delta} \oplus Fx_1^{\varepsilon}$ for all $m \geqslant 2$. Then

Corollary 15 *For any $n \geqslant 2$ it holds $\xi_n^D(U_2) = [\![n - 1, 1]\!] + 2[\![n]\!]$. In particular, for all $n \in \mathbb{N}$, it holds $\gamma_n^D(U_2) = n + 1$.*

The differential cocharacter sequence $\chi_n^D(U_2)$ and the differential codimension sequence $c_n^D(U_2)$ follow easily and, of course, coincide with the results in [8]

Corollary 16 *For any $n \geqslant 1$ it holds $\chi_n^D(U_2) = \sum_{\lambda\vdash n} m_\lambda \lambda$, where*

- *$\lambda = [\![n]\!]$ has multiplicity $2n + 1$;*
- *$\lambda = [\![a + b, a]\!]$, with $a > 0$, has multiplicity $3(b + 1)$;*
- *$\lambda = [\![a + b + 1, a + 1, 1]\!]$ has multiplicity $b + 1$.*

In particular, for all $n \in \mathbb{N}$ it holds $c_n^D(U_2) = 2^{n-1}(n + 2)$.

It is interesting to notice that the effective contribution of $\xi_n^{\delta}(U_2)$ and $\xi_n^{\varepsilon}(U_2)$ to $\xi_n^D(U_2)$ is limited to the trivial S_n-character $[\![n]\!]$. This however is far from being the general situation, as evidence shows already for U_3:

Corollary 17 *The proper differential cocharacter sequence of U_3 is the following:*

- $\xi_2^D(U_3) = 2[\![1,1]\!] \oplus 2[\![2]\!]$,
- $\xi_3^D(U_3) = 2[\![3]\!] \oplus 3[\![2,1]\!] \oplus [\![1,1,1]\!]$

and, for $n \geqslant 4$, $\xi_n^D(U_3) = \sum_{\lambda \vdash n} m_\lambda \lambda$ with multiplicities m_λ determined according to the table

	$\xi_n^1(U_3)$	$\xi_n^\delta(U_3)$	$\xi_n^\varepsilon(U_3)$	$\xi_n^D(U_3)$
$[\![n]\!]$		1	1	2
$[\![n-1,1]\!]$	1		n	$n+1$
$[\![a+b,a]\!]$ (*if* $a \geqslant 2$)	$b+1$		$b+1$	$2(b+1)$
$[\![n-2,1,1]\!]$	$n-3$		$n-2$	$2n-5$
$[\![1+a+b,1+a,1]\!]$ (*if* $a \geqslant 1$)	$2(b+1)$		$b+1$	$3(b+1)$
$[\![2+a+b,2+a,2]\!]$	$b+1$			$b+1$
$[\![1+a+b,1+a,1,1]\!]$	$b+1$			$b+1$

In particular, $\gamma_0^D(U_3) = 1$, $\gamma_1^D(U_3) = 2$, $\gamma_2^D(U_3) = 4$, $\gamma_3^D(U_3) = 9$ and, for $n \geqslant 4$, $\gamma_n^D(U_3) = n(n-3)2^{n-2} + 3n$.

It is evident that, even in this small case, the main contribution to $\xi_n^D(U_3)$ comes from the ordinary proper cocharacter $\xi_n(UT_3(F))$ but the contribution due to $\xi_n^\varepsilon(U_3)$ is very close to it, while $\xi_n^\delta(U_3) = [\![n]\!]$.

The explicit decomposition of the n-th differential cocharacter of U_3 would already result in an awkward list of partitions and multiplicities, so it is hard to conceive it as a better description of $T_D(U_3)$ than the one provided through proper characters. It is however interesting to compute the differential codimension sequence:

Corollary 18 *It holds $c_0^D(U_3) = 1$ and, for $n \geqslant 1$,*

$$c_n^D(U_3) = n(n-4)3^{n-2} + 3n2^{n-1} + 1.$$

4 The Separate Actions of δ and ε on $UT_m(F)$

We are going to consider the action of the single derivations δ and ε on the identities of $UT_m(F)$. Denote U_m^δ and U_m^ε these two structures, respectively. The considerations made for U_m may be replied in each case, and we want to determine and describe the differential identities of U_m^δ and U_m^ε. It is now easy to get the following results

Theorem 19 *$T_D(U_m^\delta)$ is generated by the following differential polynomials:*

- x^{δ^2}
- $[x_1, y_1] \dots [x_m, y_m]$
- $x^\delta [x_1, x_2], [x_1, x_2] x^\delta, [x_1, x_2]^\delta$

where all indeterminates belong to X.

Theorem 20 *$T_D(U_m^\varepsilon)$ is generated by the following differential polynomials*

- $x^{\varepsilon^2} - x^\varepsilon$
- $[x_1, y_1] \dots [x_m, y_m]$
- $x^\varepsilon [x_1, x_2], [x_1, y_1] \dots [x_{m-1}, y_{m-1}] x^\varepsilon$
- $[x_1, y_1] \dots [x_{m-2}, y_{m-2}] \Big([x, y]^\varepsilon - [x, y] \Big)$

where all indeterminates belong to X.

Also the proper cocharacters and codimensions follow easily; actually, they can be read off from the proper cocharacter decomposition of U_m, and summarized in

- $\xi_n^D(U_m^\delta) = \xi_n(UT_m(F)) + [\![n]\!]$ for all $n \geqslant 1$, so a bit more than the usual proper cocharacter of $UT_m(F)$;
- $\xi_n^D(U_m^\varepsilon) = \xi_n^D(U_m) - [\![n]\!]$ for all $n \geqslant 1$, so a bit less than the other extreme, the differential proper cocharacter of U_m.

This is hardly surprising. Informally speaking, in fact, the two chosen derivations have extreme, opposite, features: δ is a nilpotent transformation of class 2 while ε is an idempotent transformation. These differences are concealed by the case $m = 2$ (the algebra is too small), but emerge already in case $m = 3$.

We record here the codimension sequences for these small cases:

Corollary 21 *The proper codimension sequence and the codimension sequence of U_2^δ and U_2^ε are the following:*

- $\gamma_n^D(U_2^\delta) = n = \gamma_n^D(U_2^\varepsilon)$ *(for $n \geqslant 1$),*
- $c_n^D(U_2^\delta) = n2^{n-1} + 1 = c_n^D(U_2^\varepsilon)$ *(for $n \in \mathbb{N}$).*

The proper codimension sequence and the codimension sequence of U_3^δ and U_3^ε are the following:

- $\gamma_n^D(U_3^\delta) = 2^{n-2}(n-1)(n-4) + 2n - 1$ *(for $n \geqslant 4$),*
- $\gamma_n^D(U_3^\varepsilon) = 2^{n-2}n(n-3) + 3n - 1$ *(for $n \geqslant 4$),*
- $c_n^D(U_3^\delta) = 3^{n-2}(n^2 - 7n + 9) + 2^n(n-1) + \frac{1}{6}(2n^3 - 3n^2 + n + 6)$ *(for $n \geqslant 1$),*
- $c_n^D(U_3^\varepsilon) = 3^{n-2}n(n-4) + 2^{n-1}(3n-2) + 2$ *(for $n \geqslant 1$).*

In particular we get back the sequences $\chi_n^D(U_2^\varepsilon)$ and $c_n^D(U_2^\varepsilon)$ computed in [8].

5 Behind the Scenes

It is high time we gave a more precise and sound grounding to the notions employed so far. Let us start by recalling that the set $Der(A)$ of all derivations on A is a Lie algebra laying inside $End_F(A)$. If L is any Lie algebra, we say that L acts on A by derivation if A is a Lie L-module. It amounts to say that there is a Lie-homomorphism from L to $Der(A)$. By a fundamental property of the universal enveloping algebra $U(L)$ of the Lie algebra L, this is equivalent to say that A is turned into an $U(L)$-module (in our settings, a *right* $U(L)$-module). We summarize these facts in the following

Definition 22 Let L be a Lie algebra over F and let A be an associative F-algebra. We say that A is an *L-algebra*, or that *L acts on A by derivations*, if A is a $U(L)$-module.

One can define a universal object in the class of L-algebras: start by a countable set of indeterminates X, and consider the F-vector space $V := FX \otimes U(L)$. Then the tensor algebra of V, denoted $F\langle X \mid L\rangle$, is an associative, unitary F-algebra, spanned by the (tensor) products of the simple tensors $x \otimes w$ for $x \in X$ and $w \in U(L)$. The regular right action of $U(L)$ on the simple tensors $x \otimes w$, defined by $(x \otimes w) \bullet u := x \otimes wu$, turns $F\langle X \mid L\rangle$ into a right $U(L)$-module, therefore induces a derivation action of L on the tensor algebra and turns it into an L-algebra. Moreover, if A is any L-algebra, any map $\varphi_0 : X \to A$ uniquely defines an algebra homomorphism from $F\langle X \mid L\rangle$ commuting with the derivation action of L (which we call an L-homomorphism), due to the general properties of the tensor algebra of a vector space and to the defining right action of $U(L)$ on it. It is therefore natural to define $T_L(A)$ as the intersection of all the kernels of L-homomorphisms from $F\langle X \mid L\rangle$ to A.

In order to keep the notation under control, it is a good idea to write x^u to denote the simple tensor $x \otimes u$ for $x \in X$ and $u \in U(L)$; if $u = 1$ (the unit element of $U(L)$) one identifies $x \otimes 1$ with x. Hence, for the "critical" (for one's own understanding) case of $V^{\otimes 2}$, the spanning tensors $(x \otimes u) \otimes (y \otimes v)$, with $x, y \in X$ and $u, v \in U(L)$, can be written in the simpler form $x^u y^v$; moreover, the action of L (which is canonically embedded in $U(L)$ by the Poincaré–Birkhoff–Witt Theorem) can be written in the usual form $(x^u y^v)^a = x^{ua} y^v + x^u y^{va}$, that is the Leibniz rule.

Example 23 Let $L = Fa$ be the one-dimensional Lie algebra, spanned by the basis element a. Then its universal enveloping algebra is the (infinite dimensional) polynomial algebra $F[a]$, and $F\langle X \mid L\rangle$ is the noncommutative associative unitary F-algebra generated by the (countable) set $\{x^{a^i} \mid x \in X, i \in \mathbb{N}\}$. Each indeterminate x^{a^i} is just a simpler writing for the simple tensor $x \otimes a^i \in FX \otimes U(L)$.

So, when we considered U_m^δ and U_m^ε, what we really did was to choose a derivation $\alpha \in Der(UT_m(F))$, and fix a Lie homomorphism $\varphi : L = F\alpha \to End_F(UT_m(F))$. This uniquely defines an algebra homomorphism $\varphi^* : U(L) \to End_F(UT_m(F))$, turning $UT_m(F)$ into a right $U(L)$-module. Moreover, the algebra we intuitively produced as $F\langle X^D \rangle$ is nothing more than the free algebra $F\langle X \mid L\rangle$, with the once formal letters x^w now becoming the generators $x \otimes w$ of the tensor algebra $F\langle X \mid L\rangle$.

A natural question, at this point, arises: it does not matter if $\alpha = \delta$ or $\alpha = \varepsilon$, because the Lie algebras $F\delta$ and $F\varepsilon$ are isomorphic, being one dimensional. Therefore in both cases $U(L)$ is the same algebra, up to isomorphisms. What makes the differences among them? It is, of course, the kernel of the action: $U(L)$ acts on $UT_m(F)$ in both cases, but the in case $\alpha = \delta$ the kernel is the two-sided ideal generated by the generator α^2, in the other case it is the ideal generated by $\alpha^2 - \alpha$. These relations affect the differential polynomial identities of U_m^δ and U_m^ε, and correspond precisely to the identities x^{δ^2} and $x^{\varepsilon^2} - x^\varepsilon$, respectively.

When we considered the coupled action of δ and ε, that is U_m, a similar process was in action: this time, the Lie algebra L acting on $UT_m(F)$ is a two-dimensional Lie algebra and, since δ and ε do not commute, it must be the two-dimensional non commutative Lie algebra (sometimes named the two-dimensional *metabelian* Lie algebra). It is well known that one can choose a linear basis $\{a, b\}$ in L such that $[ab] = a$ (this is the true Lie product in L, so we are writing it without the separating comma), and the obvious map carrying a and b in δ and ε respectively provides a faithful representation of L. Once again, this uniquely defines an algebra homomorphism from $U(L)$ to $End_F(UT_m(F))$, thus turning $UT_m(F)$ into a right $U(L)$-module, whose kernel is the two-sided ideal generated by the elements $a^2, b^2 - b, ab - a, ba$, from which the differential identities $x^{\delta^2}, x^{\varepsilon^2} - x^\varepsilon, x^{\delta\varepsilon} - x^\delta$ and $x^{\varepsilon\delta}$, respectively, arise.

This also explains why the differential identities of U_m^δ and U_m^ε resemble so closely the ones of U_m involving separately the δ- and ε-letters: any map $\varphi : L \to End_F(UT_m(F))$ such that $[\varphi(a), \varphi(b)] = \varphi(a)$ uniquely defines a Lie homomorphism of $L = Fa \oplus Fb$ in $End_F(UT_m)$. The map φ involved in forming U_m of course does the job, but the same do the maps φ_δ sending $a \to 0$ and $b \to \delta$ and φ_ε carrying a in 0 and b in ε. In this case, the differences with U_m^δ and U_m^ε are little more than formal, and depend on adding the generator x^a to the kernel of the $U(L)$ action (that is, respectively, to add the differential identity x^ε or x^δ). Of course, choosing φ as the zero map still yields a Lie homomorphisms. In this case L acts trivially on $UT_m(F)$, and the ideal of differential identities coincides with the usual T-ideal of $UT_m(F)$ (that is: formally x^ε and x^δ are among the differential identities).

Another natural question is the following: how tightly the differential polynomial identities depend upon the Lie algebra L? The answer is: very weakly. An easy example has been already provided by the one-dimensional algebra $L = Fa$. Indeed, U_m^δ and U_m^ε have very different differential identities. One can think that this is due just to the different types of δ and ε: a nilpotent versus an idempotent transformation. This is true, but it is not the only reason. Let us consider the following example: let $\eta := [\cdot, -e_{11}] = [e_{11}, \cdot]$ be the inner derivation induced on $UT_m(F)$ by the matrix $-e_{11}$, and let θ be the map defined by $\theta(a) = \eta$. This of course defines a Lie homomorphism from the one-dimensional Lie algebra $L = Fa$

in $End_F(UT_m(F))$. Since $\eta^2 = \eta$, the $U(L)$ action on $UT_m(F)$ is exactly the same as the one we got assigning $a \to \varepsilon$ (both the kernels are generated by $a^2 - a$). By the way, the differential polynomial identities satisfied by U_m^η differ from the ones of U_m^ε. For instance, the basic identity $x^a[x_1, y_1]$ in the latter (where the differential letter x^a means x^ε) is no longer holding in the former, where it is replaced by $[x_1, y_1]x^a$ (so the differential letter changes side). Thus, the only direct part played by L in determining the differential polynomial identities of $UT_m(F)$ is limited to the identities arising from the kernel of the $U(L)$-action, but the relations among the selected derivations and the algebra structure play a decisive role in determining the actual differential identities of the algebra.

References

1. Beidar, K.I., Martindale III, W.S., Mikhalev, A.V.: Rings with Generalized Identities. Marcel Dekker, New York (1996)
2. Kemer, A.R.: Varieties and $\mathbb{Z}_2$-graded algebras. Izv. Akad. Nauk SSSR Ser. Mat. **48**(5), 1042–1059 (1984)
3. Aljadeff, E., Kanel-Belov, A.: Representability and Specht problem for G-graded algebras. Adv. Math. **225**, 2391–2428 (2010)
4. Berele, A.: Cocharacter sequence for algebras with Hopf algebra actions. J. Algebra **185**, 869–885 (1996)
5. Gordienko, A.S.: Amitsur's conjecture for associative algebras with a generalized Hopf action. J. Pure Appl. Algebra **217**(8), 1395–1411 (2013)
6. Karasik, Y.: Kemer's theory for H-module algebras with application to the PI exponent. J. Algebra **457**, 194–227 (2016)
7. Di Vincenzo, O.M., Nardozza, V.: Differential Polynomial Identities of Upper Triangular Matrices Under the Action of the Two-Dimensional Metabelian Lie Algebra. Algebr Represent Theor, (2021). https://doi.org/10.1007/s10468-020-10014-w
8. Giambruno, A., Rizzo, C.: Differential identities, 2×2 upper triangular matrices and varieties of almost polynomial growth. J. Pure Appl. Algebra **223**(4), 1710–1727 (2019)
9. Coelho, S.P., Polcino Milies, C.: Derivations of upper triangular matrix rings. Linear Algebra Appl. **187**, 263–267 (1993)
10. Drensky, V.: Free algebras and PI-algebras. Graduate Course in Algebra. Springer, Singapore (2000)
11. Regev, A.: Young-derived sequences of S_n-cocharacters. Adv. Math. **106**, 169–197 (1994)
12. Di Vincenzo, O.M., Nardozza, V.: On the $*$-Polynomial identities of a class of $*$-minimal algebras. Commun. Algebra **38**, 3078–3093 (2010)
13. Drensky, V., Giambruno, A.: Cocharacters, codimensions and Hilbert series of the polynomial identities for 2×2 matrices with involution. Can. J. Math. **46**, 718–733 (1994)
14. Maltsev, Yu.N.: A basis for the identities of the algebra of upper triangular matrices. Algebra I Logika **10**, 393–400 (1971) (Russian). Translation: Algebra and Logic **10**, 242–247 (1971)
15. Drensky, V., Kasparian, A.: Polynomial identities of eighth degree for 3×3 matrices. Annuaire de l'Univ. de Sofia, Fac. de Math. et Meca., Livre 1, Math. **77**, 175–195 (1983)

Identities in Group Rings, Enveloping Algebras and Poisson Algebras

Victor Petrogradsky

Abstract This is a short survey of works on identical relations in group rings, enveloping algebras, Poisson symmetric algebras and other related algebraic structures. First, the classical work of Passman specified group rings that satisfy nontrivial identical relations. This result was an origin and motivation of close research projects. Second, Latyshev and Bahturin determined Lie algebras such that their universal enveloping algebra satisfies a non-trivial identical relation. Next, Passman and Petrogradsky solved a similar problem in case of restricted enveloping algebras. Third, Farkas started to study identical relations in Poisson algebras. On the other hand, Shestakov proved that the symmetric algebra $S(L)$ of an arbitrary Lie algebra L satisfies the identity $\{x, \{y, z\}\} \equiv 0$ if, and only if, L is abelian. Also, Giambruno and Petrogradsky determined when a truncated symmetric Poisson algebra satisfies a non-trivial multilinear Poisson identical relation. We survey further results on existence of identical relations in (truncated) symmetric Poisson algebras of Lie algebras. In particular, we report on recent results on (strong) Lie nilpotency and (strong) solvability of (truncated) symmetric Poisson algebras and related nilpotency classes. Also, we discuss constructions and methods to achieve these results.

Keywords Poisson algebras · Identical relations · Solvable Lie algebras · Nilpotent Lie algebras · Symmetric algebras · Truncated symmetric algebras · Restricted Lie algebras.

1 Introduction

Now, there is an established theory of identical relations in associative and Lie algebras [4, 13]. It has many applications to group theory such as the solution of the Restricted Burnside Problem. Also, identical relations were applied to study other algebraic structures.

V. Petrogradsky (✉)
University of Brasilia, Brasilia, DF, Brazil
e-mail: petrogradsky@rambler.ru

O. M. Di Vincenzo, A. Giambruno (eds.), *Polynomial Identities in Algebras*, Springer INdAM Series 44, https://doi.org/10.1007/978-3-030-63111-6_17

In the present review, we discuss existence of identical relations in three classes of algebras. First, the starting point is the result of Passman on existence of identical relations in group rings [37] (Theorem 1). This result caused an intensive research on different types of identical relations in group rings, such as Lie nilpotence, solvability, non-matrix identical relations, classes of Lie nilpotence, solvability lengths, etc. There are at least 50 papers published in this area.

Second, Latyshev [26] and Bahturin [2] started to study identical relations in universal enveloping algebras of Lie algebras. Passman [38] and Petrogradsky [40] specified existence of identical relations in restricted enveloping algebras (Theorem 5). There are many papers in this area studying different types of identical relations, such as Lie nilpotence, solvability, non-matrix identical relations, classes of Lie nilpotence, solvability lengths, etc. In particular, Riley and Shalev determined necessary and sufficient conditions for restricted Lie algebras under which the restricted enveloping algebra is Lie nilpotent or solvable [44]. The research was further extended to new objects, such as Lie superalgebras, color Lie superalgebras, smash products. These problems were studied in numerous papers by Bahturin, Bergen, Kochetov, Lichtman, Passman, Petrogradsky, Riley, Shalev, Siciliano, Spinelli, Usefi et al.

Poisson algebras appeared in works of Berezin [8] and Vergne [62]. Free Poisson (super)algebras were introduced by Shestakov [47]. A basic theory of identical relations for Poisson algebras was developed by Farkas [16, 17]. Identical relations of symmetric Poisson algebras of Lie (super)algebras were studied by Kostant [24], Shestakov [47], and Farkas [17]. The third starting point for our research is the result of Giambruno and Petrogradsky [18] on existence of non-trivial multilinear Poisson identical relations in truncated symmetric Poisson algebras of Lie algebras (Theorem 15). Finally, we review recent results on Lie identities of truncated symmetric Poisson algebras [32].

By K denote the ground field, as a rule of positive characteristic p. By $\langle S\rangle$ or $\langle S\rangle_K$ denote the linear span of a subset S in a K-vector space. Let L be a Lie algebra. The Lie brackets are left-normed: $[x_1,\dots,x_n]=[[x_1,\dots,x_{n-1}],x_n]$, $n\geq 1$. One defines the *lower central series*: $\gamma_1(L)=L$, $\gamma_{n+1}(L)=[\gamma_n(L),L]$, $n\geq 1$. Also, $L^2=[L,L]=\gamma_2(L)$ is the *commutator subalgebra*. By $U(L)$ denote the universal enveloping algebra and $S(L)=\mathop{\oplus}\limits_{n=0}^{\infty} U_n/U_{n-1}$ the related *symmetric algebra* [4, 6, 12]. For the basic theory of restricted Lie algebras and restricted enveloping algebras see [4, 22]. Let us note that all our Lie algebras over a field of positive characteristic need not be restricted.

2 Identical Relations of Group Rings

In this section we review results on existence of nontrivial polynomial identities in group rings. This is the origin of this research direction.

Passman obtained necessary and sufficient conditions for a group ring $K[G]$ to satisfy a nontrivial polynomial identity over a field K of arbitrary characteristic p.

A group G is said to be *p-abelian* if G is abelian in case $p = 0$ and, in case $p > 0$, G', the commutator subgroup of G, is a finite p-group.

Theorem 1 ([37]) *The group algebra $K[G]$ of a group G over a field K of characteristic $p \geq 0$ satisfies a nontrivial polynomial identity if and only if the following conditions are satisfied.*

1. *There exists a subgroup $A \subseteq G$ of finite index;*
2. *A is p-abelian.*

All our associative algebras are with unity. Recall the notions of a (strong) Lie nilpotence and (strong) solvability for associative algebras. Let A be an associative algebra, and $A^{(-)}$ the related Lie algebra. Consider its *lower central series*: $\gamma_1(A) = A$, $\gamma_{i+1}(A) = [\gamma_i(A), A]$, $i \geq 1$. The algebra A is said to be *Lie nilpotent* of class s if and only if $\gamma_{s+1}(A) = 0$ and $\gamma_s(A) \neq 0$. Also consider *upper Lie powers* defined by $A^{(0)} = A$ and $A^{(n+1)} = [A^{(n)}, A]A$, $n \geq 0$ (we use the shifted enumeration in comparison with [35, 45] because one checks that $\{A^{(n)} \mid n \geq 0\}$ is a filtration). Now, A is *strongly Lie nilpotent* of class s if and only if $A^{(s)} = 0$ and $A^{(s-1)} \neq 0$. One defines the *derived series* of A by setting $\delta_0(A) = A$, $\delta_{i+1}(A) = [\delta_i(A), \delta_i(A)]$, $i \geq 0$. The algebra A is *solvable* of length s if and only if $\delta_s(A) = 0$ and $\delta_{s-1}(A) \neq 0$. Consider also the *upper derived series*: $\tilde{\delta}_0(A) = A$, $\tilde{\delta}_{i+1}(A) = [\tilde{\delta}_i(A), \tilde{\delta}_i(A)]A$, $i \geq 0$. Now, A is *strongly solvable* of length s if and only if $\tilde{\delta}_s(A) = 0$ and $\tilde{\delta}_{s-1}(A) \neq 0$.

Passi, Passman and Sehgal characterized the Lie nilpotence and solvability of $K[G]$ [35].

Theorem 2 ([35]) *Let $K[G]$ be the group ring of a group G over a field K,* char $K = p \geq 0$. *Then*

1. *$K[G]$ is Lie nilpotent if and only if G is p-abelian and nilpotent;*
2. *$K[G]$ is solvable if and only if G is p-abelian, for $p \neq 2$;*
3. *$K[G]$ is solvable if and only if G has a 2-abelian subgroup of index at most 2, for $p = 2$.*

Using the upper Lie powers, one defines *Lie dimension subgroups* of a group (our enumeration is shifted) [34]:

$$D_{(n),K}(G) = G \cap (1 + K[G]^{(n)}), \qquad n \geq 0.$$

One also has the following description [10]:

$$D_{(n),K}(G) = \prod_{(i-1)p^k \geq n} \gamma_i(G)^{p^k}, \quad n \geq 0. \tag{1}$$

There is a formula for the Lie nilpotency class of a modular group ring.

Theorem 3 ([10]) *Let G be a group, K a field of characteristic $p > 3$ such that the group ring $K[G]$ is Lie nilpotent. Then the Lie nilpotency class of $K[G]$ coincides with its strong Lie nilpotency class and is equal to*

$$1 + (p-1) \sum_{m \geq 1} m \log_p \left| D_{(m),K}(G) : D_{(m+1),K}(G) \right|.$$

The original work caused more research projects. So called generalized polynomial identities of (twisted) group rings were studied in [36, 39]. Also, a relation with gradings of PI-algebras see in [1].

A group G is said to have the *n-rewritable property* Q_n if for all elements $g_1, g_2, \dots, g_n \in G$, there exist two distinct permutations $\sigma, \tau \in Sym_n$ such that $g_{\sigma(1)} g_{\sigma(2)} \cdots g_{\sigma(n)} = g_{\tau(1)} g_{\tau(2)} \cdots g_{\tau(n)}$ [11]. The following result is a further application of delta-sets (see definitions below).

Theorem 4 ([11, 14]) *If a group G satisfies Q_n, then G has a characteristic subgroup N such that $|G : N|$ and $|N'|$ are finite and have sizes bounded by functions of n.*

Problem 1 We suggest a problem to find an analogue of the rewritable property and characterise it in case of Lie algebras.

3 Identical Relations of Enveloping Algebras

Latyshev proved that the universal enveloping algebra of a finite dimensional Lie algebra over a field of characteristic zero satisfies a nontrivial polynomial identity if and only if the Lie algebra is abelian [26]. Bahturin noticed that the condition of a finite dimensionality is inessential (see e.g. [4]).

Bahturin settled a similar problem on the existence of a nontrivial identity for the universal enveloping algebra over a field of positive characteristic [2]. Also, Bahturin found necessary and sufficient conditions for the universal enveloping algebra of a Lie superalgebra over a field of characteristic zero to satisfy a non-trivial polynomial identity [3]. PI-subrings and algebraic elements in universal enveloping algebras and their fields of fractions were studied by Lichtman [28].

Passman [38] and Petrogradsky [40] described restricted Lie algebras L whose restricted enveloping algebra $u(L)$ satisfies a nontrivial polynomial identity.

Theorem 5 ([38, 40]) *Let L be a restricted Lie algebra over a field of characteristic $p > 0$. The restricted enveloping algebra $u(L)$ satisfies a nontrivial polynomial identity if and only if there exist restricted ideals $Q \subseteq H \subseteq L$ such that*

1. $\dim L/H < \infty$, $\dim Q < \infty$;
2. *H/Q is abelian;*
3. *Q is abelian and has a nilpotent p-mapping.*

Riley and Shalev determined when $u(L)$ is Lie nilpotent, solvable (for $p > 2$), or satisfies the Engel condition [44].

Theorem 6 ([44]) *Let $u(L)$ be the restricted enveloping algebra of a restricted Lie algebra L over a field K of characteristic $p > 0$.*

1. *$u(L)$ is Lie nilpotent if and only if L is nilpotent and L^2 is finite dimensional and p-nilpotent;*
2. *$u(L)$ is n-Engel for some n if and only if L is nilpotent, L^2 is p-nilpotent, and L has a restricted ideal A such that both L/A and A^2 are finite dimensional.*
3. *$u(L)$ is solvable if and only if L^2 is finite dimensional and p-nilpotent, for $p \neq 2$.*

Let L be a restricted Lie algebra, char $K = p > 0$. Similarly to the dimension subgroups, using upper Lie powers (see the previous section), Riley and Shalev defined *Lie dimension subalgebras* [45]:

$$D_{(n)}(L) = L \cap u(L)^{(n)}, \qquad n \geq 0.$$

(Recall that our enumeration is shifted.) They also gave the following description [45]:

$$D_{(n)}(L) = \sum_{(i-1)p^k \geq n} \gamma_i(L)^{[p^k]}, \quad n \geq 0. \tag{2}$$

Siciliano proved [49] that in case $p > 2$, the strong solvability of the restricted enveloping algebra $u(L)$ is equivalent to its solvability. Moreover, the strong solvability in case $p = 2$ is described by the same conditions of Part 3 of Theorem 6. Also, in case $p = 2$ he provided an example of the restricted enveloping algebra $u(L)$ that is solvable but not strongly solvable.

The following is an analogue of results on the Lie nilpotency classes of group rings (Theorem 3).

Theorem 7 ([45]) *Let L be a restricted Lie algebra over a field K of characteristic $p > 0$ such that $u(L)$ is Lie nilpotent. Then*

1. *The strong Lie nilpotency class of $u(L)$ is equal to*

$$1 + (p-1) \sum_{m \geq 1} m \dim(D_{(m)}(L)/D_{(m+1)}(L)).$$

2. *In case $p > 3$, the Lie nilpotency class coincides with the strong Lie nilpotency class.*

The solvability of restricted enveloping algebras in case of characteristic 2 was settled in [55]. Lie nilpotence, solvability, and other non-matrix identities for (restricted) enveloping algebras of (restricted) Lie (super)algebras are studied in [9, 41, 50, 51, 53, 54, 57, 60, 61]. For other results on derived lengths, Lie nilpotency classes for $u(L)$, or identities for skew and symmetric elements of $u(L)$, etc., see the survey [56].

More general cases of (restricted) enveloping algebras for (color) Lie p-(super)-algebras are treated in [6]. Further developments have been obtained for smash products $U(L)\#K[G]$ and $u(L)\#K[G]$, where a group G acts by automorphisms on a (restricted) Lie algebra L [5]. Identities of smash products $U(L)\#K[G]$, where L is a Lie superalgebra in characteristic zero were studied by Kotchetov [23]. The Lie structure of smash products has been investigated in [58]. The results on identities of smash products are of interest because they combine, as particular cases, both, the results on identities of group ring and enveloping algebras.

4 Poisson Algebras and Their Identities

4.1 Poisson Algebras

Poisson algebras naturally appear in different areas of algebra, topology and physics. Probably Poisson algebras were first introduced in 1967 by Berezin [8], see also Vergne [62]. Poisson algebras are used to study universal enveloping algebras of finite dimensional Lie algebras in characteristic zero [24, 33]. In particular, abelian subalgebras in symmetric Poisson algebras are used to study commutative subalgebras in universal enveloping algebras of finite-dimensional semisimple Lie algebras in characteristic zero [59, 63]. Applying Poisson algebras, Shestakov and Umirbaev managed to solve a long-standing problem: they proved that the Nagata automorphism of the polynomial ring in three variables $\mathbb{C}[x, y, z]$ is wild [48]. Related algebraic properties of free Poisson algebras were studied by Makar-Limanov, Shestakov and Umirbaev [29, 30].

The free Poisson algebras were defined by Shestakov [47]. A basic theory of identical relations for Poisson algebras was developed by Farkas [16, 17]. See further developments on the theory of identical relations of Poisson algebras, in particular, the theory of so called codimension growth in characteristic zero by Mishchenko et al. [31], and Ratseev [43].

Recall that a vector space A is a *Poisson algebra* provided that, beside the addition, A has two K-bilinear operations which are related by the Leibnitz rule. More precisely,

- A is a commutative associative algebra with unit whose multiplication is denoted by $a \cdot b$ (or ab), where $a, b \in A$;
- A is a Lie algebra whose product is traditionally denoted by the Poisson bracket $\{a, b\}$, where $a, b \in A$;
- these two operations are related by the Leibnitz rule

$$\{a \cdot b, c\} = a \cdot \{b, c\} + b \cdot \{a, c\}, \qquad a, b, c \in A.$$

4.2 Examples of Poisson Algebras

Typical examples are as follows.

Example 1 Consider the polynomial ring $\mathbf{H}_{2m} = K[X_1, \dots, X_m, Y_1, \dots, Y_m]$. Set $\{X_i, Y_j\} = \delta_{i,j}$ and extend this bracket by the Leibnitz rule. We obtain the Poisson bracket:

$$\{f, g\} = \sum_{i=1}^{m} \left(\frac{\partial f}{\partial X_i} \frac{\partial g}{\partial Y_i} - \frac{\partial f}{\partial Y_i} \frac{\partial g}{\partial X_i} \right), \qquad f, g \in \mathbf{H}_{2m}.$$

The commutative product is the natural multiplication. We obtain the *Hamiltonian* Poisson algebra $\mathbf{H}_{2m}$.

Example 2 Let L be a Lie algebra over an arbitrary field K, $\{U_n \mid n \geq 0\}$ the natural filtration of its universal enveloping algebra $U(L)$. Consider the *symmetric algebra* $S(L) = \operatorname{gr} U(L) = \bigoplus_{n=0}^{\infty} U_n/U_{n-1}$ (see [12]). Recall that $S(L)$ is identified with the polynomial ring $K[v_i \mid i \in I]$, where $\{v_i \mid i \in I\}$ is a K-basis of L. Define the Poisson bracket as follows. Set $\{v_i, v_j\} = [v_i, v_j]$ for all $i, j \in I$, and extend to the whole of $S(L)$ by linearity and using the Leibnitz rule. For example,

$$\{v_i \cdot v_j, v_k\} = v_i \cdot \{v_j, v_k\} + v_j \cdot \{v_i, v_k\}, \qquad i, j, k \in I.$$

Thus, $S(L)$ has a structure of a Poisson algebra, called the *symmetric algebra* of L.

Example 3 Let L be a Lie algebra with a K-basis $\{v_i \mid i \in I\}$, where $\operatorname{char} K = p > 0$. Consider a factor algebra of the symmetric (Poisson) algebra

$$\mathbf{s}(L) = S(L)/(v^p \mid v \in L) \cong K[v_i \mid i \in I]/(v_i^p \mid i \in I),$$

we get an algebra of truncated polynomials. Observe that

$$\{v^p, u\} = pv^{p-1}\{v, u\} = 0, \qquad v \in L,\ u \in \mathbf{s}(L).$$

So, the Poisson bracket on $S(L)$ yields a Poisson bracket on $\mathbf{s}(L)$. Thus, $\mathbf{s}(L)$ is a Poisson algebra, we call it a *truncated symmetric algebra*. Remark that the Lie algebra L **need not be restricted**.

Example 4 Let K be a field of positive characteristic p. We introduce the *truncated Hamiltonian* Poisson algebra as

$$\mathbf{h}_{2m}(K) = K[X_1, \dots, X_m, Y_1, \dots, Y_m]/(X_i^p, Y_i^p \mid i = 1, \dots, m),$$

where we define the bracket as in Example 1 using the observation of Example 3.

The Hamiltonian algebras $\mathbf{h}_{2m}(K)$ and $\mathbf{H}_{2m}(K)$ in the class of Poisson algebras play a role similar to that of the matrix algebras $\mathrm{M}_n(K)$ for associative algebras.

4.3 Poisson Identities

The objective of this subsection is to supply basic facts on polynomial identities of Poisson algebras.

Consider the free Lie algebra $L = L(X)$ generated by a set X and its symmetric algebra $F(X) = S(L(X))$. Then, $F(X)$ is a *free Poisson algebra* in X, as was shown by Shestakov [47]. For example, let $L = L(x, y)$ be the free Lie algebra of rank 2. Consider its Hall basis [4]

$$L = \langle x, y, [y, x], [[y, x], x], [[y, x], y], [[[y, x], x], x], \dots \rangle_K.$$

We obtain the free Poisson algebra $F(x, y) = S(L)$ of rank 2, which has a canonical basis as follows:

$$F(x, y) = \left\langle x^{n_1} y^{n_2} \{y, x\}^{n_3} \{\{y, x\}, x\}^{n_4} \{\{y, x\}, y\}^{n_5} \{\{\{y, x\}, x\}, x\}^{n_6} \cdots \;\middle|\; n_i \geq 0 \right\rangle_K,$$

where only finitely many n_i, $i \geq 1$, are non-zero in the monomials above.

A definition of a *Poisson PI-algebra* is standard, identities being elements of the free Poisson algebra $F(X)$ of countable rank. We assume that basic facts on identical relations of linear algebras are known to the reader (see, e.g., [4, 13]). Farkas introduced so called *customary identities* [16]:

$$\sum_{\substack{\sigma \in S_{2n} \\ \sigma(2k-1)<\sigma(2k),\ k=1,\dots,n \\ \sigma(1)<\sigma(3)<\cdots<\sigma(2n-1)}} \mu_\sigma \{x_{\sigma(1)}, x_{\sigma(2)}\} \cdots \{x_{\sigma(2n-1)}, x_{\sigma(2n)}\} \equiv 0, \qquad \mu_\sigma \in K.$$

where $\mu_e = 1$, for the identity permutation. Denote by T_{2n} the set of permutations $\tau \in S_{2n}$ appearing in the some above. The importance of customary identities is explained by the following fact.

Theorem 8 ([16]) *Suppose that $\mathcal{V}$ is a nontrivial variety of Poisson algebras over a field K of characteristic zero. Then $\mathcal{V}$ satisfies a nontrivial customary identity.*

Let us show the idea of the proof. Let a Poisson algebra R satisfy the identity $f(X, Y, Z) = \{\{X, Y\}, Z\} \equiv 0$. Then, R also satisfies the identity:

$$\begin{aligned} 0 &\equiv f(X_1 X_2, Y, Z) - X_1 f(X_2, Y, Z) - X_2 f(X_1, Y, Z) \\ &= \{\{X_1 X_2, Y\}, Z\} - X_1 \{\{X_2, Y\}, Z\} - X_2 \{\{X_1, Y\}, Z\} \\ &= \{X_1, Y\}\{X_2, Z\} + \{X_1, Z\}\{X_2, Y\}, \end{aligned}$$

which is customary. Farkas called this process a *customarization* [16], it is an analogue of the linearization process for associative algebras. The arguments of [16] actually prove the following.

Theorem 9 ([16]) *Suppose that a Poisson algebra A over an arbitrary field satisfies a nontrivial* **multilinear** *Poisson identity. Then A satisfies a nontrivial customary identity.*

Remark 1 Let us explain why we *need all polynomials to be multilinear in case of positive characteristic p*. The linearization process is simply not working for Poisson algebras in positive characteristic as it does for associative and Lie algebras. For example, the Poisson identity $\{x, y\}^p \equiv 0$ is given by a nonzero element of the free Poisson algebra $F(X)$. Observe that its full linearization is trivial:

$$\sum_{\sigma,\pi\in S_p} \{x_{\sigma(1)}, y_{\pi(1)}\}\cdots\{x_{\sigma(p)}, y_{\pi(p)}\} = p!\sum_{\pi\in S_p}\{x_1, y_{\pi(1)}\}\cdots\{x_p, y_{\pi(p)}\} = 0.$$

Moreover, let us check that any truncated symmetric algebra $\mathbf{s}(L)$ satisfies the identity $\{x, y\}^p \equiv 0$. Indeed, let $a, b \in \mathbf{s}(L)$, then $\{a, b\}$ is a truncated polynomial without constant term, its pth power is zero by the Frobenius rule $(v + w)^p = v^p + w^p$. Thus, it does not make sense to study nonlinear Poisson identities for truncated symmetric algebras.

In the theory of Poisson PI-algebras, the analogue of the standard polynomial is [16, 17]:

$$\mathrm{St}_{2n} = \mathrm{St}_{2n}(x_1, \ldots, x_{2n}) = \sum_{\sigma\in T_{2n}} (-1)^\sigma \{x_{\sigma(1)}, x_{\sigma(2)}\}\cdots\{x_{\sigma(2n-1)}, x_{\sigma(2n)}\}.$$

This is a customary polynomial, skewsymmetric in all variables [16]. One has the following fact similar to the theory of associative algebras.

Theorem 10 ([31]) *In case of zero characteristic, any Poisson PI-algebra satisfies an identity* $(\mathrm{St}_{2n})^m \equiv 0$, *for some integers* n, m.

This result was proved by establishing an analogue of Regev's theorem on codimension growth. Moreover, it was proved that so called *customary codimension growth* is exponential with an integer exponent [31].

Another important fact on the standard identity is as follows.

Lemma 1 ([16]) *Let A be a Poisson algebra over an arbitrary field K and A is k-generated as an associative algebra. Then it satisfies the standard identity* $\mathrm{St}_{2m} \equiv 0$, *whenever* $2m > k$.

5 Multilinear Identities of Symmetric Poisson Algebras

The following result is an analogue of the classical Amitsur-Levitzki theorem on identities of matrix algebras. Kostant used another terminology, but as observed by Farkas [17], this is a result on identities of symmetric Poisson algebras.

Theorem 11 ([17, 24]) *Let L be a finite dimensional Lie algebra over a field of characteristic zero. The symmetric algebra $S(L)$ satisfies the standard Poisson identity $\mathrm{St}_{2d} \equiv 0$ as soon as $2d$ exceeds the dimension of a maximal coadjoint orbit of L.*

The Lie nilpotence of class 2 of symmetric algebras $S(L)$, where L is a Lie superalgebra, was characterized by Shestakov. The next statement follows from Theorem 4 and Theorem 5 of [47].

Theorem 12 ([47]) *The symmetric algebra $S(L)$ of a Lie algebra L over a field K satisfies the identity $\{x, \{y, z\}\} \equiv 0$ if and only if L is abelian.*

Farkas proved the following statement that generalizes Kostant's Theorem 11.

Theorem 13 ([17]) *Let L be a Lie algebra over a field of characteristic zero. Then the symmetric algebra $S(L)$ satisfies a nontrivial Poisson identity if and only if L contains an abelian subalgebra of finite codimension.*

Giambruno and Petrogradsky extended this result to an arbitrary characteristic [18].

Theorem 14 ([18]) *Let L be a Lie algebra over an arbitrary field. Then the symmetric algebra $S(L)$ satisfies a nontrivial* multilinear *Poisson identity if and only if L contains an abelian subalgebra of finite codimension.*

The following result was obtained for the truncated symmetric algebra $\mathbf{s}(L)$ of a restricted Lie algebra L.

Theorem 15 ([18]) *Let L be a restricted Lie algebra. Then the truncated symmetric algebra $\mathbf{s}(L)$ satisfies a nontrivial* multilinear *Poisson identity if and only if there exists a restricted ideal $H \subseteq L$ such that*

1. $\dim L/H < \infty$;
2. $\dim H^2 < \infty$;
3. *H is nilpotent of class 2.*

6 Lie Identities of Symmetric Poisson Algebras

Now we discuss special cases of identities of symmetric Poisson algebras. Remark that the identities of the (strong) Lie nilpotence and (strong) solvability are multilinear, thus Theorem 15 can be applied for such algebras.

6.1 Lie Nilpotence of Truncated Symmetric Algebras $\mathbf{s}(L)$

Let R be a Poisson algebra. Consider the *lower central series* of R as a Lie algebra, i.e., $\gamma_1(R) = R$ and $\gamma_{n+1}(R) = \{\gamma_n(R), R\}$, $n \geq 1$. We say that R is *Lie nilpotent of class s* if and only if $\gamma_{s+1}(R) = 0$ but $\gamma_s(R) \neq 0$. Clearly, the condition $\gamma_{s+1}(R) = 0$ is equivalent to the *identity of Lie nilpotence of class s*:

$$\{\dots\{\{X_0, X_1\}, X_2\}, \dots, X_s\} \equiv 0.$$

Similarly to the associative case, one defines *upper Lie powers*. At each step we take the ideal generated by commutators, namely, put $R^{(0)} = R$ and $R^{(n)} = \{R^{(n-1)}, R\} \cdot R$ for all $n \geq 1$ (the enumeration is shifted, because $\{R^{(n)} | n \geq 0\}$ is a filtration, see also [42]). A Poisson algebra R is *strongly Lie nilpotent of class s* iff $R^{(s)} = 0$ but $R^{(s-1)} \neq 0$. The condition $R^{(s)} = 0$ is equivalent to the identical relation of the *strong Lie nilpotence of class s*:

$$\{\{\dots\{\{X_0, X_1\} \cdot Y_1, X_2\} \cdot Y_2, \dots, X_{s-1}\} \cdot Y_{s-1}, X_s\} \equiv 0.$$

Observe that

$$\gamma_n(R) \subseteq R^{(n-1)}, \quad n \geq 1. \tag{3}$$

So, the strong Lie nilpotence of class s implies the Lie nilpotence of class at most s. The Lie nilpotence of class 1 is equivalent to the strong Lie nilpotence of class 1 and equivalent that R is abelian.

Theorem 16 ([32]) *Let L be a Lie algebra over a field of positive characteristic p. Consider its truncated symmetric Poisson algebra $\mathbf{s}(L)$. The following conditions are equivalent:*

1. $\mathbf{s}(L)$ *is strongly Lie nilpotent;*
2. $\mathbf{s}(L)$ *is Lie nilpotent;*
3. L *is nilpotent and* $\dim L^2 < \infty$.

Let L be a Lie algebra over a field K ($\operatorname{char} K = p > 0$). Using the upper Lie powers, define the *Poisson dimension subalgebras* (*truncated Poisson dimension subalgebras*, respectively) of L as:

$$D^S_{(n)}(L) = L \cap (S(L))^{(n)}, \qquad n \geq 0;$$

$$D^s_{(n)}(L) = L \cap (\mathbf{s}(L))^{(n)}, \qquad n \geq 0.$$

We obtain a description of these subalgebras [32] similar to that for group rings (1) and restricted enveloping algebras (2) (compare with the first term of that products):

$$D^{S}_{(n)}(L) = \gamma_{n+1}(L), \qquad n \geq 0;$$
$$D^{s}_{(n)}(L) = \gamma_{n+1}(L), \qquad n \geq 0.$$

We compute the classes of Lie nilpotence and strong Lie nilpotence. We get an analogue of the formulas known for group rings (Theorem 3) and restricted enveloping algebras (Theorem 7). The analogy is better seen in terms of truncated Poisson dimension subalgebras.

Theorem 17 ([32]) *Let L be a Lie algebra over a field of positive characteristic $p > 3$, such that the truncated symmetric Poisson algebra $\mathbf{s}(L)$ is Lie nilpotent. The following numbers are equal:*

1. *the strong Lie nilpotency class of $\mathbf{s}(L)$;*
2. *the Lie nilpotency class of $\mathbf{s}(L)$;*
3. $$1 + (p-1)\sum_{n\geq 1} n \cdot \dim(\gamma_{n+1}(L)/\gamma_{n+2}(L)).$$

In cases $p = 2, 3$, the numbers (1) and (3) remain equal.

In case $p = 2, 3$, the number above yields an upper bound for the Lie nilpotency class. Also, we have a lower bound for the Lie nilpotency class, L being non-abelian [32]:

$$2 + (p-1)\sum_{n\geq 2}(n-1) \cdot \dim(\gamma_{n+1}(L)/\gamma_{n+2}(L)).$$

6.2 Solvability of Truncated Symmetric Algebras $\mathbf{s}(L)$

Let R be a Poisson algebra. Consider its *derived series* as a Lie algebra: $\delta_0(R) = R$, $\delta_{n+1}(R) = \{\delta_n(R), \delta_n(R)\}$, $n \geq 0$. Polynomials of *solvability* are defined as: $\delta_1(X_1, X_2) = \{X_1, X_2\}$ and

$$\delta_{n+1}(X_1, X_2, \ldots, X_{2^{n+1}}) = \left\{\delta_n(X_1, \ldots, X_{2^n}), \delta_n(X_{2^n+1}, \ldots, X_{2^{n+1}})\right\}, \quad n \geq 1.$$

A Poisson algebra R is *solvable of length s* if, and only if, $\delta_s(R) = 0$ and $\delta_{s-1}(R) \neq 0$, or equivalently, R satisfies the above identity of Lie solvability $\delta_s(\ldots) \equiv 0$, s being minimal.

Define the *upper derived series*: $\tilde{\delta}_0(R) = R$ and $\tilde{\delta}_{n+1}(R) = \{\tilde{\delta}_n(R), \tilde{\delta}_n(R)\} \cdot R$, $n \geq 0$. Define polynomials of the *strong solvability* $\tilde{\delta}_1(X_1, X_2, Y_1) = \{X_1, X_2\} \cdot Y_1$, and

$$\tilde{\delta}_{n+1}(X_1, \dots, X_{2^{n+1}}, Y_1, \dots, Y_{2^{n+1}-1}) = \Big\{\tilde{\delta}_n(X_1, \dots, X_{2^n}, Y_1, \dots, Y_{2^n-1}),$$
$$\tilde{\delta}_n(X_{2^n+1}, \dots, X_{2^{n+1}}, Y_{2^n}, \dots, Y_{2^{n+1}-2})\Big\} \cdot Y_{2^{n+1}-1}, \quad n \geq 1.$$

A Poisson algebra R is *strongly solvable of length* s iff $\tilde{\delta}_s(R) = 0$ and $\tilde{\delta}_{s-1}(R) \neq 0$, or equivalently R satisfies $\tilde{\delta}_s(\dots) \equiv 0$, s being minimal. Observe that

$$\delta_s(R) \subseteq \tilde{\delta}_s(R), \qquad s \geq 0. \tag{4}$$

So, strong solvability of length s implies solvability of length at most s. The solvability of length 1 is equivalent to the strong solvability of length 1 and equivalent that R is an abelian Lie algebra.

Theorem 18 ([32]) *Let L be a Lie algebra over a field of positive characteristic $p \geq 3$. Consider its truncated symmetric Poisson algebra* $\mathbf{s}(L)$. *The following conditions are equivalent:*

1. $\mathbf{s}(L)$ *is strongly solvable;*
2. $\mathbf{s}(L)$ *is solvable;*
3. *L is solvable and* $\dim L^2 < \infty$.

In case $p = 2$, conditions (1) and (3) remain equivalent.

Solvability length of symmetric Poisson algebras was further studied in [52]. Namely, an upper and lower bounds for the strong derived length of $s(L)$ are obtained. It is established when $s(L)$ is metabelian, that is, it has derived length 2. For a non-abelian Lie algebra, a lower bound for the derived length of $s(L)$ is obtained. Finally, necessary and sufficient conditions under which that value is attained are determined.

Observe that the description of solvable group rings in characteristic 2 looks very nice (Theorem 2). But the answer to a similar question for the restricted enveloping algebras is rather complicated and was obtained only recently [55].

The problem of solvability of $\mathbf{s}(L)$ in case $\operatorname{char} K = 2$ is open. We show that the situation is different from other characteristics. Namely, in case $\operatorname{char} K = 2$, we give two examples of truncated symmetric Poisson algebras that are solvable but not strongly solvable, see Lemmas 6 and 7. A close fact is that the Hamiltonian algebras $\mathbf{H}_2(K)$ and $\mathbf{h}_2(K)$ are solvable but not strongly solvable in case $\operatorname{char} K = 2$ (Lemma 8). This is an analogue of a well-known fact that the matrix ring $\mathrm{M}_2(K)$ of 2×2 matrices over a field K, $\operatorname{char} K = 2$, is solvable but not strongly solvable.

6.3 Nilpotency and Solvability of Symmetric Algebras $S(L)$

The following extension of a result of Shestakov [47] is proved.

Theorem 19 ([32]) *Let L be a Lie algebra over a field K, and $S(L)$ its symmetric Poisson algebra. The following conditions are equivalent:*

1. *L is abelian;*
2. *$S(L)$ is strongly Lie nilpotent;*
3. *$S(L)$ is Lie nilpotent;*
4. *$S(L)$ is strongly solvable;*
5. *$S(L)$ is solvable (here assume that* $\operatorname{char} K \neq 2$*).*

In case $\operatorname{char} K = 2$, the solvability of the symmetric Poisson algebra $S(L)$ is an open question. Two examples of Lie algebras mentioned above also yield solvable symmetric algebras which are not strongly solvable (Lemmas 9 and 10).

Remark 2 Formally, our statements on ordinary Lie nilpotency and solvability are concerned only with the *Lie structure* of the Poisson algebras $\mathbf{s}(L)$ and $S(L)$. But our proof heavily relies on Theorem 22 of [18], which in turn uses the existence of a nontrivial customary identity given by Theorem 9 (Farkas [16]). In this way, we need the *Poisson structure* of our algebras to prove our results. We do not see ways to prove them using the theory of Lie identical relations only.

6.4 Delta-Sets and Multilinear Poisson Identical Relations

Now we present an important instrument to prove our results on Poisson identities in enveloping algebras. *Delta-sets* in groups were introduced by Passman to study identities in the group rings [37]. Namely, let G be a group, then

$$\Delta_n(G) := \{a \in G \mid |a^G| \le n\}, \qquad n \ge 0;$$

$$\Delta(G) := \bigcup_{n=0}^{\infty} \Delta_n(G) = \{a \in G \mid |a^G| < \infty\}.$$

A crucial step to specify group ring with identical relations was to establish that there exist integers n, m such that $|G : \Delta_n(G)| \le m$, see [37].

In case of Lie algebras, the delta-sets were introduced by Bahturin to study identical relations of the universal enveloping algebras [2]. Let L be a Lie algebra, one defines the delta-sets as sets of elements of *finite width* as follows:

$$\Delta_n(L) := \{x \in L \mid \dim[L, x] \le n\}, \qquad n \ge 0;$$

$$\Delta(L) := \bigcup_{n=0}^{\infty} \Delta_n(L) = \{x \in L \mid \dim[L, x] < \infty\}.$$

Note that $\Delta_n(L)$, $n \geq 0$, is not a subalgebra or even a subspace in a general case. The basic properties of the delta-sets are as follows.

Lemma 2 ([6, 41]) *Let L be a (restricted) Lie algebra, $n, m \geq 0$.*

1. *$\Delta_n(L)$ is invariant under scalar multiplication;*
2. *if $x \in \Delta_n(L)$, $y \in \Delta_m(L)$, then $\alpha x + \beta y \in \Delta_{n+m}(L)$, where $\alpha, \beta \in K$;*
3. *if $x \in \Delta_n(L)$, $y \in L$, then $[x, y] \in \Delta_{2n}(L)$;*
4. *if $x \in \Delta_n(L)$ and L a restricted Lie algebra, then $x^{[p]} \in \Delta_n(L)$;*
5. *$\Delta(L)$ is a (restricted) ideal of L.*

Lemma 3 ([44]) *Let L be a Lie algebra.*

1. *if I is a finite dimensional ideal of L, then $\Delta(L/I) = (\Delta(L) + I)/I$;*
2. *if H is a subalgebra of finite codimension in L, then $\Delta(H) = \Delta(L) \cap H$.*

Suppose that W is a subset in a K-vector space V. We say that W has finite codimension in V if there exist $v_1, \ldots, v_m \in V$ such that $V = \{w + \lambda_1 v_1 + \cdots + \lambda_m v_m \mid w \in W,\ \lambda_1, \ldots, \lambda_m \in K\}$. If m is the minimum integer with such property, then we write $\dim V/W = m$. We also introduce the notation $m \cdot W = \{w_1 + \cdots + w_m \mid w_i \in W\}$, where $m \in \mathbb{N}$.

Lemma 4 ([5, Lemma 6.3]) *Let V be a K-vector space. Suppose that a subset $T \subseteq V$ is stable under multiplication by scalars and* $\dim V/T \leq n$. *Then the linear span is obtained as: $\langle T \rangle_K = 4^n \cdot T$.*

We need a result on bilinear maps.

Theorem 20 (P.M. Neumann [4]) *Let U, V, W be vector spaces over a field K and $\varphi : U \times V \to W$ a bilinear map. Suppose that for all $u \in U$ and $v \in V$,* $\dim \varphi(u, V) \leq m$ *and* $\dim \varphi(U, v) \leq l$. *Then* $\dim \langle \varphi(U, V) \rangle_K \leq ml$.

The following facts were indispensable in our approach to study Poisson identical relations in symmetric algebras of Lie algebras. Actually the following result was proved for restricted Lie algebras [18], but its proof remains valid for truncated symmetric algebras as well.

Theorem 21 ([18]) *Let L be a Lie algebra. Suppose that the symmetric algebra $S(L)$ (or the truncated symmetric algebra $\mathbf{s}(L)$) satisfies a multilinear Poisson identity. Then there exist integers n, N such that* $\dim L/\Delta_N(L) < n$.

It yields the following reduction step, which is actually contained in [18].

Theorem 22 ([18]) *Let L be a Lie algebra such that the symmetric algebra $S(L)$ (or the truncated symmetric algebra $\mathbf{s}(L)$) satisfies a multilinear Poisson identity. Let $\Delta = \Delta(L)$. Then there exist integers n, M such that*

1. *$\Delta = \Delta_M(L)$;*
2. $\dim L/\Delta < n$;
3. $\dim \Delta^2 \leq M^2$.

6.5 Products of Commutators in Poisson Algebras

Now, we supply technical results on products of commutators in Poisson algebras that were used to get a lower bound on the Lie nilpotency class of $\mathbf{s}(L)$.

Products of terms of the lower central series for *associative algebras* appear in works of many mathematicians, the results being reproved without knowing the earlier works. We do not pretend to make a complete survey here. Probably, the first observations on products of commutators in associative algebras were made by Latyshev in [27] and Volichenko in [64]. There are further works, see e.g. [7, 15, 19, 20, 25].

In case of associative algebras, Claim 1 of Theorem 23, probably, first was established by Sharma-Shrivastava in [46, Theorem 2.8]. As was remarked in [45], the proof of the associative version of Claim 2 of Theorem 23 is implicitly contained in [46], where it is proved for group rings. A weaker statement (the associative version of Lemma 5) is established by Gupta and Levin [21, Theorem 3.2].

The following statement is a Poisson version of respective results for associative algebras. The validity of it is not automatically clear and it was checked directly, following a neat approach due to Krasilnikov [25].

Theorem 23 ([32]) *Let R be a Poisson algebra over a field K,* $\operatorname{char} K \neq 2, 3$.

1. *Suppose that one of integers $n, m \geq 1$ is odd, then*

$$\gamma_n(R) \cdot \gamma_m(R) \subseteq \gamma_{n+m-1}(R)R.$$

2. *For all $x_1, \dots, x_n \in R$, $n, m \geq 1$, we have*

$$\{x_1, \dots, x_n\}^m \in \gamma_{(n-1)m+1}(R)R.$$

The following is an analogue of a result for associative algebras, see [21, Theorem 3.2]. It is weaker than Claim 1 of Theorem 23, but it is valid for an arbitrary characteristic.

Lemma 5 ([32]) *Let R be a Poisson algebra over arbitrary field K. Then*

$$\gamma_m(R)\gamma_n(R) \subseteq \gamma_{m+n-2}(R)R, \qquad n, m \geq 2.$$

6.6 Solvability of Symmetric Algebras $\mathbf{s}(L)$ and $S(L)$ in Case $\operatorname{char} K = 2$

We supply two examples of truncated symmetric algebras that are solvable but not strongly solvable.

Lemma 6 ([32]) *Let* $L = \langle x, y_i \mid [x, y_i] = y_i,\ i \in \mathbb{N}\rangle_K$, char $K = 2$, *the remaining commutators being trivial. Then*

1. $L^2 = \Delta(L) = \Delta_1(L) = \langle y_i \mid i \in \mathbb{N}\rangle$;
2. $\mathbf{s}(L)$ *is solvable of length 3;*
3. $\mathbf{s}(L)$ *is not strongly solvable.*

Lemma 7 ([32]) *Let* $L = \langle x, y_i, z_i \mid [x, y_i] = z_i,\ i \in \mathbb{N}\rangle_K$, char $K = 2$, *the remaining commutators being trivial. Then*

1. $\Delta(L) = \Delta_1(L) = \langle y_i, z_i \mid i \in \mathbb{N}\rangle$ *and* $L^2 = \langle z_i \mid i \in \mathbb{N}\rangle$;
2. $\mathbf{s}(L)$ *is solvable of length 3;*
3. $\mathbf{s}(L)$ *is not strongly solvable.*

Two examples above are closely related to the following observation.

Lemma 8 ([32]) *Consider the truncated Hamiltonian Poisson algebra* $P = \mathbf{h}_2(K)$ *(or the Hamiltonian Poisson algebra* $P = \mathbf{H}_2(K)$*),* char $K = 2$. *Then*

1. P *is solvable of length 3.*
2. P *is not strongly solvable.*

Proof Let $P = \mathbf{h}_2(K) = K[X, Y]/(X^2, Y^2) = \langle 1, x, y, xy\rangle_K$, where x, y denote the images of X, Y. We have $\delta_1(P) = \{P, P\} = \langle 1, x, y\rangle_K$, $\delta_2(P) = \langle 1\rangle_K$, and $\delta_3(P) = 0$. Also, one checks that P is not strongly solvable.

Let $P = \mathbf{H}_2(K) = K[X, Y]$. The Poisson brackets of monomials $X^n Y^m$, $n, m \geq 0$ depend on parities of n, m of multiplicands. For simplicity, denote by $X^{\bar 0}Y^{\bar 1}$ all monomials $X^\alpha Y^\beta \in K[X, Y]$ such that α is even and β odd, etc. We get non-zero products only in the cases:

$$\{X^{\bar 1}Y^{\bar 0}, X^{\bar 0}Y^{\bar 1}\} = X^{\bar 0}Y^{\bar 0};$$

$$\{X^{\bar 1}Y^{\bar 1}, X^{\bar 1}Y^{\bar 0}\} = X^{\bar 1}Y^{\bar 0};$$

$$\{X^{\bar 1}Y^{\bar 1}, X^{\bar 0}Y^{\bar 1}\} = X^{\bar 0}Y^{\bar 1}.$$

Thus, $\delta_1(P)$ is spanned by monomials of three types obtained above. Consider their commutators, the first line yields that $\delta_2(P)$ is spanned by monomials of type $Y^{\bar 0}Y^{\bar 0}$. Finally, $\delta_3(P) = 0$. □

Thus, the Poisson algebras $\mathbf{h}_2(K)$, $\mathbf{H}_2(K)$ in characteristic 2 behave similarly to the associative algebra $\mathrm{M}_2(K)$ of 2×2 matrices in characteristic 2.

The question of the solvability of the symmetric algebra $S(L)$ in case char $K = 2$ is more complicated as shown below. The algebras of Lemmas 6 and 7, also yield solvable symmetric algebras which, of course, are not strongly solvable by that Lemmas.

Lemma 9 ([32]) *Let* $L = \langle x, y_i \mid [x, y_i] = y_i, i \in \mathbb{N}\rangle_K$*, the other commutators being trivial,* char $K = 2$. *Then the symmetric Poisson algebra* $S(L)$ *is solvable of length 3 but not strongly solvable.*

Proof Put $H = \langle y_i \mid i \in \mathbb{N}\rangle_K$. For a monomial $v = y_{i_1} y_{i_2} \cdots y_{i_k} \in S(H)$ we define its length $|v| = k$. Then $v' = \{x, v\} = |v|v$. A basis of $S(L)$ is formed by $x^\alpha v$, $\alpha \geq 0$, where $v \in S(H)$ are respective basis monomials. Consider the products:

$$\{x^\alpha v, x^\beta w\} = x^{\alpha+\beta-1}(\alpha|w| + \beta|v|)vw. \tag{5}$$

These products depend on the parities of $\alpha, \beta, |v|, |w|$. For simplicity, denote by $x^{\bar{0}} v^{\bar{1}}$ all monomials $x^\alpha v \in S(L)$ such that α is even and $|v|$ is odd, etc. The only non-zero products (5) are of types:

$$\{x^{\bar{1}} v^{\bar{0}}, x^{\bar{0}} v^{\bar{1}}\} = x^{\bar{0}} v^{\bar{1}};$$

$$\{x^{\bar{1}} v^{\bar{1}}, x^{\bar{1}} v^{\bar{0}}\} = x^{\bar{1}} v^{\bar{1}};$$

$$\{x^{\bar{1}} v^{\bar{1}}, x^{\bar{0}} v^{\bar{1}}\} = x^{\bar{0}} v^{\bar{0}}.$$

Thus, $\delta_1(S(L))$ is spanned by monomials of the three types obtained above. Consider their commutators, the last line yields that $\delta_2(S(L))$ is spanned by monomials of type $x^{\bar{0}} v^{\bar{0}}$. Finally, $\delta_3(S(L)) = 0$. □

Lemma 10 ([32]) *Let* $L = \langle x, y_i, z_i \mid [x, y_i] = z_i, i \in \mathbb{N}\rangle_K$*, the other commutators being trivial,* char $K = 2$. *Then the symmetric Poisson algebra* $S(L)$ *is solvable of length 3 but not strongly solvable.*

Acknowledgment The author was partially supported by grant CNPq 309542/2016-2.

References

1. Aljadeff, E., David, O.: On group gradings on PI-algebras. J. Algebra **428**, 403–424 (2015)
2. Bahturin, Yu.: Identities in the universal envelopes of Lie algebras. J. Austral. Math. Soc. **18**, 10–21 (1974)
3. Bahturin, Yu.A.: Identities of a universal enveloping algebra for a Lie superalgebra. Mat. Sb., N. Ser. **127(169)**(3(7)), 384–397 (1985)
4. Bahturin, Yu.A.: Identical Relations in Lie Algebras. VNU Science Press, Utrecht (1987)
5. Bahturin, Yu., Petrogradsky, V.: Polynomial identities in smash products. J. Lie Theory **12**(2), 369–395 (2002)
6. Bahturin, Yu.A., Mikhalev, A.A., Petrogradsky, V.M., Zaicev M.V., Infinite Dimensional Lie Superalgebras. de Gruyter Exp. Math., vol. 7. de Gruyter, Berlin (1992)
7. Bapat, A., Jordan, D.: Lower central series of free algebras in symmetric tensor categories. J. Algebra **373**, 299–311 (2013)
8. Berezin, F.A.: Several remarks on the associative hull of a Lie algebra. Funkcional. Anal. i Priložen **1**(2), 1–14 (1967) (Russian)

9. Bergen, J., Riley, D., Usefi, H.: Lie superalgebras whose enveloping algebras satisfy a non-matrix polynomial identity. Isr. J. Math.**196**(1), 161–173 (2013)
10. Bhandari, A., Passi, I.B.S.: Lie-nilpotency indices of group algebras. Bull. Lond. Math. Soc. **24**(1), 68–70 (1992)
11. Blyth, R.D.: Rewriting products of group elements I. J. Algebra **116**, 506–521 (1988)
12. Dixmier, J.: Enveloping Algebras. AMS, Rhode Island (1996)
13. Drensky, V.: Free Algebras and PI-Algebras. Graduate Course in Algebra. Springer, Singapore (2000)
14. Elashiry, M.I., Passman, D.S.: Rewritable groups. J. Algebra **345**, 190–201 (2011)
15. Etingof, P., Kim, J., Ma, X.: On universal Lie nilpotent associative algebras. J. Algebra **321**(2), 697–703 (2009)
16. Farkas, D.R.: Poisson polynomial identities. Commun. Algebra **26**(2), 401–416 (1998)
17. Farkas, D.R.: Poisson polynomial identities. II. Arch. Math. (Basel) **72**(4), 252–260 (1999)
18. Giambruno, A., Petrogradsky, V.M.: Poisson identities of enveloping algebras. Arch. Math. **87**, 505–515 (2006)
19. Gordienko, A.S.: Codimension of commutators of length 4. Russ. Math. Surv. **62**(1), 187–189 (2007); translation from Usp. Mat. Nauk **62**(1), 191–192 (2007)
20. Grishin, A.V., Pchelintsev, S.V.: On the centers of relatively free associative algebras with the identity of Lie nilpotency. Mat. Sb. **206**(11), 113–130 (2015); translation in Sb. Math. **206**, 11–12 (2015)
21. Gupta, N., Levin, F.: On the Lie ideals of a ring. J. Algebra **81**(1), 225–231 (1983)
22. Jacobson, N.: Lie Algebras. Interscience, New York (1962)
23. Kochetov, M.V.: Identities in the smash product of the universal enveloping superalgebra of a Lie superalgebra and a group algebra. Mat. Sb. **194**(1), 87–102 (2003); translation in Sb. Math. **194**(1–2), 89–103
24. Kostant, B.: A Lie algebra generalization of the Amitsur-Levitski theorem. Adv. Math. **40**, 155–175 (1981)
25. Krasilnikov, A.: The additive group of a Lie nilpotent associative ring. J. Algebra **392**, 10–22 (2013)
26. Latyshev, V.N.: Two remarks on PI-algebras. Sibirsk. Mat. Ž. **4**, 1120–1121 (1963)
27. Latyshev, V.N.: On finite generation of a T-ideal with the element $[x_1, x_2, x_3, x_4]$. Sib. Math. J. **6**, 1432–1434 (1965)
28. Lichtman, A.I.: PI-subrings and algebraic elements in enveloping algebras and their fields of fractions. J. Algebra **121**(1), 139–154 (1989)
29. Makar-Limanov, L., Shestakov, I.: Polynomial and Poisson dependence in free Poisson algebras and free Poisson fields. J. Algebra **349**(1), 372–379 (2012)
30. Makar-Limanov, L., Umirbaev, U.: The Freiheitssatz for Poisson algebras. J. Algebra **328**(1), 495–503 (2011)
31. Mishchenko, S.P., Petrogradsky, V.M., Regev, A.: Poisson PI algebras. Trans. Am. Math. Soc. **359**(10), 4669–4694 (2007)
32. Monteiro Alves, I.Z., Petrogradsky,V.: Lie structure of truncated symmetric Poisson algebras. J. Algebra **488**, 244–281 (2017).
33. Ooms, A.I.: The Poisson center and polynomial, maximal Poisson commutative subalgebras, especially for nilpotent Lie algebras of dimension at most seven. J. Algebra **365**, 83–113 (2012)
34. Passi, I.B.S.: Group Rings and Their Augmentation Ideals. Lecture Notes in Mathematics, vol. 715. Springer, Berlin (1979)
35. Passi, I.B.S., Passman, D.S., Sehgal, S.K.: Lie solvable group rings. Can. J. Math. **25**(4), 748–757 (1973)
36. Passman, D.S.: Group rings satisfying a polynomial identity II. J. Pac. J. Math. **39**, 425–438 (1971)
37. Passman, D.S.: Group rings satisfying a polynomial identity. J. Algebra **20**, 103–117 (1972)
38. Passman, D.S.: Enveloping algebras satisfying a polynomial identity. J. Algebra **134**(2), 469–490 (1990)

39. Passman, D.S.: Twisted group rings satisfying a generalized polynomial identity. Commun. Algebra **29**(9), 3683–3710 (2001)
40. Petrogradsky, V.M.: The existence of an identity in a restricted envelope. Mat. Zametki **49**(1), 84–93 (1991); translation in Math. Notes **49**(1–2), 60–66 (1991)
41. Petrogradsky, V.M.: Identities in the enveloping algebras for modular Lie superalgebras. J. Algebra **145**(1), 1–21 (1992)
42. Petrogradsky, V.M.: Codimension growth of strong Lie nilpotent associative algebras. Commun. Algebra **39**(3), 918–928 (2011)
43. Ratseev, S.M.: Correlation of Poisson algebras and Lie algebras in the language of identities. Math. Notes **96**(3–4), 538–547 (2014); Translation of Mat. Zametki **96**(4), 567–577 (2014)
44. Riley, D.M., Shalev, A.: The Lie structure of enveloping algebras. J. Algebra **162**, 46–61 (1993)
45. Riley, D.M., Shalev, A.: Restricted Lie algebras and their envelopes. Can. J. Math. **47**, 146–164 (1995)
46. Sharma, R.K., Srivastava, J.B.: Lie ideals in group rings. J. Pure Appl. Algebra **63**, 63–80 (1990)
47. Shestakov, I.P.: Quantization of Poisson superalgebras and speciality of Jordan Poisson superalgebras. Algebra i Logika **32**(5), 571–584 (1993); English translation: Algebra Logic **32**(5), 309–317 (1993)
48. Shestakov, I., Umirbaev, U.U.: The tame and the wild automorphisms of polynomial rings in three variables. J. Am. Math. Soc. **17**(1), 197–227 (2004)
49. Siciliano, S.: Lie derived lengths of restricted universal enveloping algebras. Publ. Math. **68**(3–4), 503–513 (2006)
50. Siciliano, S.: On Lie solvable restricted enveloping algebras. J. Algebra **314**, 226–234 (2007)
51. Siciliano, S.: On the Lie algebra of skew-symmetric elements of an enveloping algebra. J. Pure Appl. Algebra **215**, 72–76 (2011)
52. Siciliano, S.: Solvable symmetric Poisson algebras and their derived lengths. J. Algebra **543**, 98–110 (2020)
53. Siciliano, S., Spinelli, E.: Lie nilpotency indices of restricted universal enveloping algebras. Commun. Algebra **34**(1), 151–157 (2006)
54. Siciliano, S., Usefi, H.: Lie identities on symmetric elements of restricted enveloping algebras. Isr. J. Math. **195**(2), 999–1012 (2013)
55. Siciliano S., Usefi H.: Lie solvable enveloping algebra of characteristic two. J. Algebra **382**, 314–331 (2013)
56. Siciliano, S., Usefi, H.: Lie properties of restricted enveloping algebras. In: Lie Algebras and Related Topics. Contemp. Math., vol. 652, pp. 141–152. Amer. Math. Soc., Providence (2015)
57. Siciliano, S., Usefi, H.: Engel condition on enveloping algebras of Lie superalgebras. J. Pure Appl. Algebra **219**(12), 5631–5636 (2015)
58. Siciliano, S., Usefi, H.: Lie structure of smash products. Isr. J. Math. **217**(1), 93–110 (2017)
59. Tarasov, A.A.: On some commutative subalgebras in the universal enveloping algebra of the Lie algebra $gl_n(\mathbb{C})$. Mat. Sb. **191**(9), 115–122 (2000); translation in Sb. Math. **191**, 9–10 (2000)
60. Usefi, H.: Lie identities on enveloping algebras of restricted Lie superalgebras. J. Algebra **393**, 120–131 (2013)
61. Usefi, H.: Non-matrix polynomial identity enveloping algebras. J. Pure Appl. Algebra **217**(11), 2050–2055 (2013)
62. Vergne, M.: La structure de Poisson sur l'algèbre symétrique d'une algèbre de Lie nilpotente. C. R. Acad. Sci. Paris Sér. A-B **269**, 950–952 (1969)
63. Vinberg, E.B.: Some commutative subalgebras of a universal enveloping algebra. Izv. Akad. Nauk SSSR Ser. Mat. **54**(1), 3–25, 221 (1990); translation in Math. USSR-Izv. **36**(1), 1–22 (1991)
64. Volichenko, I.B.: The T-ideal generated by the element $[x_1, x_2, x_3, x_4]$. Preprint no. 22, Inst. of Mathematics of the Academy of Sciences of BSSR, Minsk, 13 pp. (1978) (Russian)

Notes on the History of Identities on Group (and Loop) Algebras

C. Polcino Milies

Abstract We survey the development of the theory of polynomial and group identities in group algebras, including properties of the unit groups that imply on group identities, starting from the very beginning of the theory. We also include the history of similar results for alternative loop algebras.

Keywords Group algebra · Loop algebra · Units · Polynomial identity · Group identity

1 Introduction

In this paper we survey some aspects of the historical development of the theory of polynomial and group identities for group algebras and a brief extension of these results. Fist we describe the progress made in establishing conditions for the existence of polynomial identities and conditions relating the degree of the identity to the index of a special subgroup of the given group.

In the late 70s Brian Hartley conjectured that if G is a torsion group and K a field such that $U(KG)$ satisfies a group identity, then KG must satisfy a polynomial identity. We examine the circumstances in which this conjecture was formulated to show that it was a quite natural supposition, at that time.

Finally, we describe how these results were extended to the case of alternative loop algebras, the closest analogue to group algebras in a non associative context.

C. P. Milies (✉)
Instituto de Matemática e Estatística, Universidade de São Paulo, São Paulo, SP, Brazil

O. M. Di Vincenzo, A. Giambruno (eds.), *Polynomial Identities in Algebras*,
Springer INdAM Series 44, https://doi.org/10.1007/978-3-030-63111-6_18

2 Existence of Polynomial Identities

According to a well-known paper which covers the early history of the theory of polynomial identities due to Amitsur [2] the origins of this theory goes back to a paper of Dehn [5], written in 1922. However, he also observes that:

> The modern approach was founded by Kaplansky in 1948 [16], based upon two forerunners, namely the methods introduced by Jacobson [14] and Levitzki [20].

Interestingly enough the initial work on polynomial identities for group algebras started almost immediately, in 1949, and it is also due to Kaplanski [17].

The paper starts with the definition of what we now call the *standard identity of degree n*, and his first lemma shows that if A is an algebra of dimension n over a field K, then A satisfies the standard identity of degree $n+1$. It should be noted that he assumes that all fields considered in the paper are of characteristic 0.

Then, he defines a special class of groups; in his own words: *a group G satisfies the condition $P_n (n \geq 2)$ if the following is true: for any n elements in G the set of $n!/2$ products obtained from even permutations coincides with the $n!/2$ products obtained from odd permutations,* and proves that *a group extension of an abelian group by a finite group of order n satisfies P_{n^2+1}*. i.e., we have the following.

Theorem 2.1 *Let K be a field of characteristic 0. If G is a group containing an abelian subgroup of index n then KG satisfies the standard identity of degree n^2+1.*

The eventual converse was left as an open problem.

In 1961 Amitsur [1] showed that, under the conditions above, KG actually satisfies the standard identity of degree $2n$.

He also obtained a first step in the direction of a converse. Actually, he proved the following:

Theorem 2.2 *All the absolutely irreducible representations of a group G are of degree less than or equal to 2 if and only if one of the following conditions holds:*

(i) *G is abelian.*
(ii) *G contains an abelian subgroup of index 2.*
(iii) *If $\mathcal{Z}(\mathcal{G})$ denotes the center of G, then $G/\mathcal{Z}(\mathcal{G})$ is an Abelian 2-group of order 8.*

Clearly, this result implies that if KG satisfies and identity of degree $n \leq 4$ then G contains an abelian subgroup A such that $[G : A] \leq 2$.

A complete answer in the case when $char(K) = 0$ came only in 1964.

Theorem 2.3 (D.S. Passman and M. Isaacs, [13]) *Let K be a field of characteristic 0 and G a group.*

(i) *If G contains an Abelian subgroup of finite index n then KG satisfies the standard polynomial identity of degree 2n.*
(ii) *If KG satisfies a polynomial identity of degree n then G contains an Abelian subgroup of index bounded by a fixed function of n.*

The first result in the case when $char(K) = p > 0$ came in 1970 in Martha K. Smith's PhD thesis [35], a student of I.N. Herstein at the University of Chicago who published her results in [36]. This is a rather long paper containing many results on ideals and rings of quotients of group algebras. In section §8 she considered polynomial identities and proved the following.

Theorem 2.4 *Let K be a field and G a group such that KG is prime. If KG satisfies a polynomial identity of degree n then G has an Abelian normal subgroup of index less than or equal to $[d/2]^2$.*

Given a group G, we shall denote by $\Phi(G)$ its FG-subgroup; i.e. the set of elements of G that have a finite number of conjugates.

Theorem 2.5 *Let K be a field and G a group such that KG is semiprime. If KG satisfies a polynomial identity of degree n then $[G : \Phi(G)] \leq [d/2]^8$ and G contains and Abelian subgroup of finite index.*

The complete answer in the case when $char(K) = p \geq 0$ is due to Passman [26], in 1972. To state the result shall need the following.

Definition 2.6 Let p be a prime rational integer. A group G is called *p-abelian* if G', its commutator subgroup, is a finite p-group.

Theorem 2.7 *Let K be a field of characteristic $p > 0$ and G a group.*

- *If G contains a p-abelian subgroup of finite index n then KG satisfies a polynomial the standard polynomial identity of degree $2n|A'|$.*
- *If KG satisfies a polynomial identity of degree n then G contains a p-abelian subgroup and $[g : A'] \cdot |A'|$ is bounded by a fixed function of n.*

If we agree that 0-Abelian means Abelian, we can state the main results of this section in a very condensed way.

Theorem 2.8 *Let K be a field of characteristic $p \geq 0$ and G any group. Then KG satisfies a polynomial identity if and only if G contains a p-abelian subgroup of finite index. Details proofs of these results can be found in [27];*

3 Group Theoretical Properties of Unit Groups

In the late 60s and through the 70s abundant research was conduced in an effort to characterize group algebras whose unit groups satisfy some algebraic properties:

- Starting this trend, Bateman and Coleman characterized group algebras of finite groups whose units were nilpotent in 1968.
- This work was extended by Khripta [18] in 1972 and Motose and Tominaga [24] to the case of infinite groups.

- The study of the solvabiliy of the unit group was started independently by Motose and Tominaga [25] in 1969 and by Bateman [3] in 1971 and complemented by Motose and Ninomiya [23] in 1972.
- An alternative characterization was given by Bovdi and Khripta [4] and a nice and complete exposition by Passman [28] both in 1977. See also [6].

Similar research was conduced for unit groups of integral group rings.

- The nilpotency of the unit group of $\mathbb{Z}G$, in the case when G is finite, was studied by Polcino Milies [30] in 1976.
- For arbitrary groups, results were obtained by Sehgal and Zassenhaus [33] in 1977.
- Solvability was discussed by Sehgal [31, Theorem VI.4.8] in 1978.
- Moreover, integral group rings with FC unit groups were characterized by Sehgal and Zassenhaus [34] also in 1977.

In the early stages of the theory, Higman showed, in 1940, that if G is finite, then all torsion units on $\mathbb{Z}G$ are trivial if and only if G is either Abelian or a Hamiltonian 2-group, a result he used to prove that the so-called Isomorphism Problem has a positive solution for these families of groups; namely, if G and H are finite groups such that $\mathbb{Z}G \cong \mathbb{Z}H$ then $G \cong H$.

In the case of finite groups, all the above characterization of groups with nice algebraic properties on the unit groups of $\mathbb{Z}G$ reduce to this one.

The reason became obvious with a result of Hartley and Pickel [12] in 1980.

Theorem 3.1 *Let G be a finite group. Then, the group of units of $\mathbb{Z}G$ contains a free group on two generators if and only if G is neither Abelian or a Hamiltonian 2-group.*

In the same direction we have the following.

Theorem 3.2 (Gonçalves, [6])

- *A noncommutative division ring finite dimensional over its center contains a multiplicative free group on two generators.*
- *If G is a finite noncommutative group and K is a field such that $char(K) \nmid |G|$ then the units of KG contain a free group on two generators.*

4 Group Indentities

Definition 4.1 A *group identity* for a group U is a nontrivial reduced word $w = w(x_1, \ldots, x_n)$ of the free group on $x_1, \ldots, x_n$ vanishing in U; i.e., such that for every choice of elements $u_1, \ldots, u_n \in U$, we have $w(u_1, \ldots, u_n) = 1$.

In the late 70s Brian Hartley conjectured that if G is a torsion group and $U(FG)$ satisfies a group identity, then FG must satisfy a polynomial identity.

Obviously, if a group contains a free subgroup of rank 2, it cannot verify a group identity so the last two results on the previous section seem to suggest that unit groups of group rings will seldom verify such an identity. On the other hand, algebraic properties of groups such as nilpotency or solvability can be described in terms of identities, so all the examples above are of unit groups verifying some kind of group identity. When one checks the descriptions of the groups in question, it is easy to see that their group rings do satisfy a polynomial identity so Hartley's conjecture seems quite natural under these circumstances.

Remark

- The hypothesis of G being torsion is essential: In fact, if G is a torsion free nilpotent group, then G can be ordered and it is easy to see that $U(FG)$ has only trivial units i.e., $U(FG) = F^* \times G$. Hence $U(FG)$ satisfies a group identity
- The converse is not true. In fact, if R is any non commutative finite dimensional simple algebra over a field F, then R satisfies a polynomial identity (since it is finite dimensional). Now, by Theorem 3.2, $U(R)$ contains a free group of rank 2 and thus cannot satisfy a group identity.

Hartley's conjecture was proved in a series of papers:

- For semiprime group algebras by Giambruno et al. [7] in 1994.
- For group algebras of arbitrary groups over infinite fields by Giambruno et al. [8] in 1997.
- The case of finite fields was completed by Liu and Passman in [21] and [22] in 1999.
- Passman [29] extended this result, obtaining a complete classification of those groups G which satisfy the hypotheses of Hartley's conjecture also in 1997.

Theorem 4.2 *Let F be any field and G a torsion group. Then $U(FG)$ satisfies a group identity if and only if one of the following conditions holds.*

(i) *If char $F = 0$, $U(FG)$ satisfies a group identity if and only if G is abelian.*
(ii) *If char $F = p > 0$, $U(FG)$ satisfies a group identity if and only if G has a normal p-abelian subgroup of finite index, and*
   (a) *either G' is a p-group of bounded period (and $U(FG)$ satisfies to $(x, y)^{p^k} = 1$ for some $k \geq 0$).*
   (b) *or G has bounded period and F is finite (and $U(FG)$ satisfies $x^n = 1$ for some integer n).*

It is natural to ask whether the hypothesis of G being torsion can be removed and still obtain a characterization of groups G such that the group of units $U(FG)$ satisfies a group identity. A result in this direction was given by Giambruno et al. [9] in 2000.

Theorem 4.3 *Suppose that G is a group with an element of infinite order and let F be a field of characteristic $p \geq 0$. We have the following*

(a) *If $U(FG)$ satisfies a group identity then P is a subgroup.*
(b) *If P is of unbounded exponent and $U(FG)$ satisfies a group identity then*

  (i) *G contains a p-abelian subgroup of finite index.*
  (ii) *G' is of bounded p-power exponent.*

Conversely, if P is a subgroup and G satisfies (i) and (ii) then $U(FG)$ satisfies a group identity.

(c) *If P is of bounded exponent and $U(FG)$ satisfies a group identity then*

  (1) *P is finite or G has a p-abelian subgroup of finite index.*
  (2) *$T(G/P)$ is an abelian p'-subgroup and so T is a group.*
  (3) *Every idempotent of $F(G/P)$ is central.*

Conversely, if P is a subgroup, G satisfies (1), (2), (3) *and G/T is nilpotent then $U(FG)$ satisfies a group identity.*

All results on this conjecture, up to 2010 are discussed in detail in [32] and more recent results in a paper by E. Spinelli [37] in this same volume.

5 Loop Algebras

Roughly speaking, a loop is a group which is not necessarily associative; more precisely, we have the following.

Definition 5.1 A em loop is a set L together with a (closed) binary operation $(a, b) \mapsto ab$ for which there is a two-sided identity element 1 and such that the right and left translation maps

$$R_x : a \mapsto ax \quad \text{and} \quad L_x : a \mapsto xa$$

are bijections for all $x \in L$. This requirement implies that, for any $a, b \in L$, the equations $ax = b$ and $ya = b$ have unique solutions.

The *loop algebra* of L over an associative and commutative ring with unity R was introduced in 1944 by R.H. Bruck as a means to obtain a family of examples of nonassociative algebras.

It is defined in a way similar to that of a group algebra; i.e., as the free R-module with basis L, with a multiplication induced distributively from the operation in L.

Definition 5.2 A ring R is *alternative* if

$$x(xy) = (xx)y \text{ and } (xy)y = x(yy) \text{ for all } x, y \in R.$$

In 1983, E.G. Goodaire: [10] proved the following.

Theorem 5.3 *Let L be a loop and R an associative ring with unity, of characteristic different from 2. Then RL is alternative loop ring if and only if the following properties hold:*

(i) *If three elements associate in some order then they associate in all orders and*
(ii) *If* $g, h, k \in L$ *do not associate, then* $gh.k = g.kh = h.gk$.

It follows that if RL is alternative over one ring R as in the Theorem above, then it is also alternative over *all* such rings. Also, it can be shown that, in this case, RL is alternative also for rings of characteristic 2. However if we assume, conversely, that $char(R) = 2$ and that RL is alternative, besides loops as in the theorem, there are a few other cases. It is then natural to formulate the following.

Definition 5.4 An *RA loop* (ring alternative loop) is a loop whose loop ring RL over any ring with unity R is alternative, but not associative.

A better characterization of this family of loops can be given.

Definition 5.5 A group G, with center $\mathcal{Z}(G)$, is called an *LC group* (or, that it has *limited commutativity*) if it is not commutative and for any pair of elements $x, y \in G$ we have that $xy = yx$ if and only if either $x \in \mathcal{Z}(G)$ or $y \in \mathcal{Z}(G)$ or $xy \in \mathcal{Z}(G)$.

Theorem 5.6 *A loop L is RA if and only if it is not commutative and, for any two elements a and b of L which do not commute, the subloop of L generated by its center together with a and b is a group G such that*

(i) *for any* $u \notin G$, $L = G \cup Gu$ *is the disjoint union of G and the coset Gu;*
(ii) *G is an LC group.*
(iii) *G has a unique nonidentity commutator s, which is necessarily central and of order 2.*
(iv) *the map*

$$g \mapsto g^* = \begin{cases} g & \text{if } g \text{ is central} \\ sg & \text{otherwise} \end{cases}$$

is an involution of G (i.e., an antiautomorphism of order 2);
(v) *multiplication in L is defined by*

$$g(hu) = (hg)u$$

$$(gu)h = gh^*u$$

$$(gu)(hu) = g_0 h^* g$$

where $g, h \in G$ *and* $g_0 = u^2$ *is a central element of G.*

Definition 5.7 A group G is called an *SLC group* if it is LC and contains a unique non-trivial commutator s.

Proposition 5.8 *A group G, with center $\mathcal{Z}(G)$, is an SLC group if and only if $G/\mathcal{Z}(G) \cong C_2 \times C_2$.*

Theorem 5.9 (Leal–Polcino Milies [19]) *A group G is SLC if and only if G can be written in the form $G = D \times A$, where A is abelian and D is an indecomposable 2-group generated by its center and two elements x and y which satisfy*

(i) $\mathcal{Z}(D) = C_{2^{m_1}} \times C_{2^{m_2}} \times C_{2^{m_3}}$, *where $C_{2^{m_i}}$ is cyclic of order 2^{m_i} for $i = 1, 2, 3$, $m_1 \geq 1$ and $m_2, m_3 \geq 0$;*
(ii) $(x, y) \in C_{2^{m_1}}$;
(iii) $x^2 \in C_{2^{m_1}} \times C_{2^{m_2}}$ *and* $y^2 \in C_{2^{m_1}} \times C_{2^{m_2}} \times C_{2^{m_3}}$.

Indecomposable groups as above can be classified.

Theorem 5.10 (Jespers et al. [15]) *Let G be a finite group. Then $G/\mathcal{Z}(G) \cong C_2 \times C_2$ if and only if G can be written in the form $G = D \times A$, where A is abelian and $D = \langle \mathcal{Z}(D), x, y \rangle$ is of one of the following five types of indecomposable 2-groups:*

Type	$\mathcal{Z}(D)$	D
D_1	$\langle t_1 \rangle$	$\langle x, y, t_1 \mid (x, y) = t_1^{2^{m_1-1}}, x^2 = y^2 = t_1^{2^{m_1}} \rangle$
D_2	$\langle t_1 \rangle$	$\langle x, y, t_1 \mid (x, y) = t_1^{2^{m_1-1}}, x^2 = y^2 = t_1, t^{2^{m_1}} = 1 \rangle$
D_3	$\langle t_1 \rangle \times \langle t_2 \rangle$	$\langle x, y, t_1, t_2 \mid (x, y) = t_1^{2^{m_1-1}}, x^2 = t_1^{2^{m_1}} = t_2^{2^{m_2}} = 1, y^2 = t_2 \rangle$
D_4	$\langle t_1 \rangle \times \langle t_2 \rangle$	$\langle x, y, t_1, t_2 \mid (x, y) = t_1^{2^{m_1-1}}, x^2 = t_1, y^2 = t_2, t_1^{2^{m_1}} = t_2^{2^{m_2}} = 1 \rangle$
D_5	$\langle t_1 \rangle \times \langle t_2 \rangle \times \langle t_3 \rangle$	$\langle x, y, t_1, t_2, t_3 \mid (x, y) = t_1^{2^{m_1-1}}, x^2 = t_2, y^2 = t_3, t_1^{2^{m_1}} = t_2^{2^{m_2}} = t_3^{2^{m_3}} = 1 \rangle$

Using these results, all indecomposable finite RA loops were classified in the same paper.

Because of the particular nature of RA loops, alternative loop algebras always satisfy a polynomial identity.

Theorem 5.11 (Goodaire–Polcino Milies [11]) *Over any (commutative, associative) coefficient ring R (with 1), the loop ring of an RA loop satisfies the polynomial identity* $[(XY - YX)^2, Z] = 0$.

Just as the set of units of an associative ring is a group, so the units of an alternative ring form a Moufang Loop.

A Moufang loop is *diassociative*—that is, the subloop generated by any two elements is associative.

Since a free group on n generators can always be embedded in a free group on just two generators, we can always assume that if the units of an alternative loop algebra satisfy a group identity, then it is a word on only two variables.

Because of diassociativity, the notion of a group identity extends to Moufang loops:

A Moufang loop L satisfies a group identity if and only if there is a nonempty reduced word $w = w(x_1, x_2)$ in the free group on two variables such that $w(\ell_1, \ell_2) = 1$ for all $\ell_1, \ell_2 \in L$.

Theorem 5.12 (Goodaire and Polcino Milies [11]) *Let L be a torsion RA loop and let F be a field of characteristic $p \geq 0$. Then $U(FL)$ satisfies a group identity if and only if $p = 2$, in which case $U(FL)$ satisfies $(u, v)^2 = 1$.*

It might be interesting to note that the polynomial identity given above is Wagner's identity, the first polynomial identity obtained for the ring of 2×2 matrices over a field in 1937 [38].

Theorem 5.13 ([11]) *Let F be a field of characteristic $p \geq 0$ and let L be an RA loop with torsion subloop T that is different from L. Then $U(FL)$ satisfies a group identity if and only if either $p = 2$, or T is an abelian group and every idempotent of FT is central in FL. In the latter case, $U(FL)$ satisfies the identity $((u_1, u_2), (u_3, u_4)) = 1$.*

The statement of the theorem above contains the condition "every idempotent of FT is central in FL," where T is the torsion subloop of an RA loop L. This might seem somewhat unnatural, but it is equivalent to conditions on F and L that are well established.

References

1. Amitsur, S.A.: Groups wuth representations of bounded degree II. Illinois J. Math. **5**, 198–205 (1961)
2. Amitsur, S.A.: Polynomial Identities. Israel J. Math. **19**, 183–189 (1974)
3. Bateman, J.M.: On the solvability of unit groups of group algebras. Trans. Amer. Math. Soc. **157**, 73–86 (1971)
4. Bovdi, A.A., Khripta, I.I.: Group algebras of periodic groups with solvable multiplicative group. (Russian) Mat. Zametki **22**, 421–432 (1977)
5. Dehn, M.: Über die Grundlagen der projektiven Geometrie und allgemeine Zahlsysteme. Math. Ann. **85**, 184–193 (1922)
6. Fisher, J.L., Parmenter, M.M., Sehgal, S.K.: Group rings with solvable n-Engel unit groups. Proc. Amer. Math. Soc. **59**, 195–200 (1976)
7. Giambruno, A., Jespers, E., Valenti, A.: Group identities on units of rings. Arch. Math. **63**, 291–296 (1994)
8. Giambruno, A., Sehgal, S.K., Valenti, A.: Group algebras whose units satisfy a group identity. Proc. Amer. Math. Soc. **125**, 629–634 (1997)
9. Giambruno, A., Sehgal, S.K., Valenti, A.: Group identities on units of group algebras. J. Algebra **226**, 488–504 (2000)
10. Goodaire, E.G.: Alternative loop rings. Publ. Math. Debrecen **30**, 31–38 (1983)
11. Goodaire, E.G., Polcino Milies, C.: Polynomial and group identities in alternative loop algebras. J. Algebra Appl. **7**(5), 593–599 (2008)
12. Hartley, B., Pickel, P.F.: Free subgroups in the unit groups of integral group rings. Canad. J. Math. **32**, 1342–1352 (1980)
13. Isaacs, M., Passman, D.S.: Groups with representations of bounded degree. Canad. J. Math. **16**, 299–309 (1964)

14. Jacobson, N.: Structure theory for algebraic algebras of bounded degree. Ann. Math. **46**, 695–707 (1945)
15. Jespers, E., Leal, G., Polcino Milies, C.: Classifying indecomposable loops, R.A. J. Algebra **176**, 569–584 (1995)
16. Kaplansky, I.: Rings with a polynomial identity. Bull. Amer. Math. Soc. **54**, 575–580 (1948)
17. Kaplansky, I.: Groups with representations of bounded degree. Canad. J. Math. **1**, 105–112 (1949)
18. Khripta, I.I.: The nilpotence of the multiplicative group of a group ring. Mat. Zametki **11**, 191–200 (1972); Math. Notas **11**, 119–124 (1972)
19. Leal, G., Polcino Milies, C.: Isomorphic group (and loop) algebras. J. Algebra **155**, 195–210 (1993)
20. Levitski, J.: On a problem of Kurosh. Bull. Amer. Math. Soc. **52**, 1033–1035 (1946)
21. Liu, C.H.: Group algebras with units satisfying a group identity. Proc. Amer. Math. Soc. **127**, 327–336 (1999)
22. Liu, C.H., Passman, D.S.: Group algebras whose units satisfy a group identity II. Proc. Amer. Math. Soc. **127**, 337–341 (1999)
23. Motose, K., Ninomiya, Y.: On the solvability of the unit groups of group rings. Math. J. Okayama Univ. **15**, 209–214 (1972)
24. Motose, K., Tominaga, H.: Group rings with nilpotent unit groups. Math. J. Okayama Univ. **14**, 43–46 (1969)
25. Motose, K., Tominaga, H.: Group rings with solvable unit groups. Math. J. Okayama Univ. **15**, 37–40 (1971)
26. Passman, D.S.: Group rings satisfying a polynomial identity. J. Algebra **20**, 103–117 (1972)
27. Passman, D.S.: The Algebraic Structure of Group Rings. Wiley-Interscience, New York (1977)
28. Passman, D.S.; Observations in group rings. Comm. Algebra **6**, 1119–1162 (1977)
29. Passman, D.S.: Group algebras whose units satisfy a group identity. Proc. Amer. Math. Soc. **125**, 657–662 (1997)
30. Polcino Milies, C.: Integral group rings with nilpotent unit groups. Canad. J. Math. **28**(5), 954–960 (1976)
31. Sehgal, S.K.: Topics in Group Rings. M. Dekker, New York (1978)
32. Sehgal, S.K.: A conjecture of brian hartley and developments arising. Note Mat. **30**, 73–91 (2010)
33. Sehgal, S.K., Zassenhaus, H.J.: Integral group rings with nilpotent unit groups. Comm. Algebra **5**, 101–111 (1977)
34. Sehgal, S.K., Zassenhaus, H.J.: Group rings whose units form an FC group. Math. Z. **153**, 29–35 (1977)
35. Smith, M.K.: Group Algebras, PhD Thesis, University of Chicago, 1970
36. Smith, M.K.: Group algebras. J. Algebra **18**, 477–499 (1971)
37. Spinelli, E.: Group and polynomial identities in group rings, in this same volume
38. Wagner, W.: Über die Grundlagen der projektiven geometrie and allgemeine Zahlsysteme. Math. Z. **113**, 528–567 (1936–1937).

Cayley Hamilton Algebras

Claudio Procesi

Abstract In this paper we first review the main ideas of Cayley Hamilton algebras and then, in Theorem 3.18, we give a different approach and formulation of the Theorem of Zubkov on the relations among invariants of matrices. In this approach the relations appear as those of a *universal Cayley Hamilton algebra*.

Keywords Cayley–Hamilton identity · Invariants · T–ideals

1 Introduction

All algebras are associative and over a commutative ring A, sometimes we use F when A is a field. A basic fact for an $n \times n$ matrix a with entries in a commutative ring A is the construction of its characteristic polynomial $\chi_a(t) := \det(t - a) = t^n + \sum_{i=1}^{n}(-1)^i \sigma_i(a)t^{n-i}$ and the Cayley Hamilton theorem $\chi_a(a) = 0$. If F is an algebraically closed field then the elements $\sigma_i(a)$ are the elementary symmetric functions in the eigenvalues of a.

The notion of Cayley Hamilton algebra (CH algebras for short) was introduced in 1987 by Procesi [13], Definition 2.5, as an axiomatic treatment of the Cayley Hamilton theorem. This was done in order to clarify the Theory of n-dimensional representations, of an associative and in general noncommutative algebra R (from now on just called *algebra*).

The theory was developed only in characteristic 0, for two reasons, the first being that at that time it was not clear to the author if the characteristic free results of Donkin [6] and Zubkov [30] were sufficient to found the theory in general. The second reason was mostly because it looked not likely that the *main Theorem* 3.9 could possibly hold in general.

C. Procesi (✉)
Dipartimento di Matematica, Università di Roma “La Sapienza”, Roma, Italy
e-mail: procesi@mat.uniroma1.it

O. M. Di Vincenzo, A. Giambruno (eds.), *Polynomial Identities in Algebras*,
Springer INdAM Series 44, https://doi.org/10.1007/978-3-030-63111-6_19

The first concern can now be considered to have a positive solution due to the contributions of several people and we may take the book [5] as reference. As for the second, that is the main theorem in positive characteristic, the issue remains unsettled. The present author feels that it should not be true in general but has no counterexamples.

Independently, studying deformations of representations of Galois groups, see Mazur [9] or [14], the theory of *pseudocharacters* or *pseudorepresentations*, Definition 2.7, was developed by several authors, see Wiles [29], Taylor [24], Nyssen Louise [10] and R. Rouquier [22]. The strict connection between these two concepts is given in Proposition 2.8.

2 The Cayley–Hamilton Identity

Let us quickly recall some basic facts, for the proofs the reader is referred to the books [1, 5] and [15].

Let us denote by $\det(t - X) = t^n + \sum_{i=1}^{n}(-1)^i\sigma_i(X)t^{n-i}$ the characteristic polynomial of an $n \times n$ matrix X. The Cayley–Hamilton identity is thus

$$\det(t - X) = t^n + \sum_{i=1}^{n}(-1)^i\sigma_i(X)t^{n-i}, \quad X^n + \sum_{i=1}^{n}(-1)^i\sigma_i(X)X^{n-i} = 0. \tag{1}$$

A general fact on matrices is that (F an infinite field):

Proposition 2.1 *The algebra of polynomial functions on $M_n(F)$ invariant under conjugation restricted to diagonal matrices is isomorphic to the algebra of symmetric polynomials.*

Under this isomorphism the function $\sigma_i(X)$ restricts to the elementary symmetric function $e_i(x_1, \ldots, x_n)$. While the function $tr(X^k)$ restricts to the Newton powers sums $\psi_i = \psi_i(x_1, \ldots, x_n) = x_1^i + \ldots + x_n^i$.

Over $\mathbb{Q}$ one may express the elementary symmetric function $e_i(x_1, \ldots, x_m)$ in terms of the powers sums $\psi_i = \psi_i(x_1, \ldots, x_m) = x_1^i + \ldots + x_m^i$. Using the Taylor expansion for $\log(1 + y) = \sum_{j=1}^{\infty}(-1)^{j+1}\frac{y^j}{j}$, we get

$$\sum_{i=0}^{m}\sigma_i(x_1, \ldots, x_m)y^i = \prod_{r=1}^{m}(1 + x_r y) = \exp\Big(\sum_{j=1}^{\infty}(-1)^{j+1}\frac{\psi_j(x_1, \ldots, x_m)}{j}y^j\Big).$$

One then has the following formula for a matrix X in characteristic 0:

$$\sigma_r(X) = \sum_{\substack{h_1+2h_2+\cdots rh_r=r\\ h_1\geq 0,\ldots,h_r\geq 0}} \prod_{j=1}^{r} \frac{((-1)^{j+1}tr(X^j))^{h_j}}{h_j!\,j^{h_j}}. \tag{2}$$

Finally we need

Proposition 2.2 *For each j one has a universal expression of $\sigma_i(X^j)$ as polynomial in the $\sigma_k(X)$ obtained from the theory of symmetric functions.*

Proof One considers the symmetric function $e_i(x_1^j, \ldots, x_n^j),\ n \geq i \cdot j$.

This is given by a polynomial $P_{i,j}(e_1, \ldots, e_{i \cdot j})$ independent of n, by the basic theorem on symmetric functions. Then:

$$\sigma_i(X^j) = P_{i,j}(\sigma_1(X), \ldots, \sigma_{i \cdot j}(X)). \tag{3}$$

□

$$\text{Example} \qquad \sigma_2(X^2) = \sigma_2(X)^2 - 2\sigma_1(X)\sigma_3(X) - 2\sigma_4(X). \tag{4}$$

In characteristic 0 the invariants $tr(X^i),\ i = 1, \ldots, n$ generate the algebra of all invariants.

It is convenient to use also the *multilinear* form of the Cayley–Hamilton identity and of the symmetric functions $\sigma_i(X)$ which can be obtained by *full polarization*, cf. [15]. For this, given a permutation $\sigma \in S_m$ (the symmetric group), we decompose $\sigma = (i_1 i_2 \ldots i_h) \ldots (j_1 j_2 \ldots j_\ell)(s_1 s_2 \ldots s_t)$ in cycles and set:

$$T_\sigma(X_1, X_2, \ldots, X_m) \tag{5}$$

$$:= tr(X_{i_1} X_{i_2} \ldots X_{i_h}) \ldots tr(X_{j_1} X_{j_2} \ldots X_{j_\ell}) tr(X_{s_1} X_{s_2} \ldots X_{s_t}). \tag{6}$$

In the basic invariants T_σ, of Formula (5), take $m = k + 1$. We may assume that the last cycle ends with $s_t = k + 1$ so the last factor is of the form $tr((X_{s_1} X_{s_2} \ldots X_{s_{t-1}}) X_{k+1})$. Hence we have that

$$T_\sigma(X_1, X_2, \ldots, X_{k+1}) = tr(\psi_\sigma(X_1, X_2, \ldots, X_k) X_{k+1}) \tag{7}$$

where $\psi_\sigma(X_1, X_2, \ldots, X_k)$ is the equivariant map given by the formula

$$\begin{aligned} &\psi_\sigma(X_1, X_2, \ldots, X_k) \\ &= tr(X_{i_1} X_{i_2} \ldots X_{i_h}) \ldots tr(X_{j_1} X_{j_2} \ldots X_{j_\ell}) X_{s_1} X_{s_2} \ldots X_{s_{t-1}}. \end{aligned} \tag{8}$$

Then we have the (see also Lew [8]):

Proposition 2.3 *For each $k \leq n$ the polarized form of $\sigma_k(X)$ is the expression*

$$T_k(x_1, \ldots, x_k) = \sum_{\sigma \in S_k} \epsilon_\sigma T_\sigma(x_1, \ldots, x_k). \tag{9}$$

The polarized form of $CH_n(X)$ is

$$CH_n(x_1,\ldots,x_n) = (-1)^n \sum_{\sigma\in S_{n+1}} \epsilon_\sigma \psi_\sigma(x_1,x_2,\ldots,x_n). \tag{10}$$

Here ϵ_σ denotes the sign of σ.

Example $n = 2$ (**polarize** $CH_2(x)$)

$$CH_2(x) = x^2 - t(x)x + det(x) = x^2 - t(x)x + \frac{1}{2}(t(x)^2 - t(x^2)). \tag{11}$$

$$x_1x_2 + x_2x_1 - t(x_1)x_2 - t(x_2)x_1 - t(x_1x_2) + t(x_1)t(x_2)$$

$$= CH_2(x_1 + x_2) - CH_2(x_1) - CH_2(x_2).$$

Also from the decomposition into cosets $S_{n+1} = S_n \bigcup_{i=1}^n S_n(i, n+1)$ one has the recursive formula

$$T_{n+1}(x_1,\ldots,x_{n+1}) = T_n(x_1,\ldots,x_n)tr(x_{n+1}) - \sum_{i=1}^n T_n(x_1,\ldots,x_ix_{n+1},\ldots,x_n) \tag{12}$$

One then axiomatizes the idea of trace.

Definition 2.4 An associative algebra with trace, over a commutative ring A is an associative A algebra R with a 1-ary operation

$$t : R \to R$$

which is assumed to satisfy the following axioms:

1. t is A-linear.
2. $t(a)b = b\,t(a), \quad \forall a, b \in R.$
3. $t(ab) = t(ba), \quad \forall a, b \in R.$
4. $t(t(a)b) = t(a)t(b), \quad \forall a, b \in R.$

From these axioms it follows that the image $t(R)$ of R under t is an algebra in the center of R, called the *trace algebra* and t is $t(R)$ linear.

We then have the two definitions

Definition 2.5 An algebra R over $\mathbb{Q}$ with a trace t is an n-Cayley Hamilton algebra if it satisfies the trace identity (10) and $t(1) = 1$.

In fact, since we are assuming that the algebra is over $\mathbb{Q}$, by the classical method of polarization and restitution the algebra satisfies (10) if and only if each of its elements satisfy its associated n characteristic polynomial, defined abstractly from the formal trace t as in Formula (1) using Formula (2).

Remark 2.6 Definition 2.5 can be made also if algebra R is not over $\mathbb{Q}$. Then one cannot define the n characteristic polynomial using Formula (2) since in this formula there are integers in the denominator.

So in general this definition is not useful and we need a different approach through determinants, see Definition 3.5.

Definition 2.7 A pseudocharacter (or pseudorepresentation) of a group G, of degree n with coefficients in a commutative ring A, is a map $t : G \to A$ satisfying the following three properties:

1. $t(1) = n$.
2. $t(ab) = t(ba), \ \forall a, b \in G$.
3. $T_{n+1}(g_1, \dots, g_{n+1}) = 0, \ \forall g_i \in G$ (Formula (9)).

Frobenius [7], discovered already that this is a property of an n-dimensional character.

The connection between the two definitions is the following. One considers the group algebra $A[G]$ and then extends the map t to a trace. Next one considers the *Kernel* of the trace, that is

$$K := \{a \in A[G] \mid t(ab) = 0, \ \forall b \in A[G]\}.$$

It is then an easy fact to see that, if t is a pseudocharacter of G of degree n, then $A[G]/K$ is a n-Cayley Hamilton algebra. In particular if $A \supset \mathbb{Q}$ one can apply Theorem 3.9. In general we have:

Proposition 2.8 *If R is an algebra with trace and the trace satisfies*

$$T_{n+1}(x_1, \dots, x_{n+1}) = \sum_{\sigma \in S_{n+1}} \epsilon_\sigma T_\sigma(x_1, \dots, x_{n+1}) = 0$$

then $CH_n(x_1, \dots, x_n) \in K$ and R/K is an n-Cayley Hamilton algebra.

Then, as a consequence of the strong embedding Theorem 3.9 one has that R/K embeds into $n \times n$ matrices over some commutative ring and one has that the pseudocharacter is in fact a true character.

3 The First and Second Fundamental Theorem

3.1 The Free Trace Algebra

The free trace algebra over a set X of variables will be denoted by $\mathcal{F}_T\langle X\rangle$, it can be described as follows. Start from the usual free algebra $F\langle X\rangle$, then consider the classes of cyclic equivalence of monomials M, which we formally denote $t(M)$. The algebra $\mathcal{F}_T\langle X\rangle = F\langle x_i\rangle_{i\in I}[t(M)]$ is the polynomial ring in the infinitely many commuting variables $t(M)$ over the free algebra $F\langle X\rangle$. Its trace algebra is the polynomial ring $F[t(M)]$ in the infinitely many commuting variables $t(M)$. The map $t : M \mapsto t(M)$ is the formal trace.

Each element of $\mathcal{F}_T\langle X\rangle$ can be evaluated in $n \times n$ matrices giving rise to an equivariant polynomial map from I-tuples of $n \times n$ matrices to $n \times n$ matrices. The elements of $\mathcal{F}_T\langle X\rangle$ which vanish when evaluated in $n \times n$ matrices are the *trace identities* of $n \times n$ matrices. They form a T-ideal, that is an ideal of $\mathcal{F}_T\langle X\rangle$ closed under all trace compatible endomorphisms, i.e. substitutions of variables $x_i \mapsto f_i \in \mathcal{F}_T\langle X\rangle$.

The first and second fundamental Theorem for matrix invariants for algebras over $\mathbb{Q}$, see §12.1 of [1], may be stated as:

Theorem 3.2 *The algebra $F_{T,n}\langle X\rangle$ of equivariant polynomial maps from I-tuples of $n \times n$ matrices to $n \times n$ matrices, is the free algebra with trace modulo the T-ideal generated by the nth Cayley Hamilton polynomial and $t(1) = n$.*

$$\boxed{F_{T,n}\langle X\rangle := \mathcal{F}_T\langle X\rangle / \langle CH_n(x),\ t(1) = n\rangle}. \tag{13}$$

3.3 A Characteristic Free Approach

In order to develop the Theory over a field of any characteristic or over the integers one needs to start from some axiomatization of the determinant rather than of the trace. Recall that, in [20] and [21], Roby defines:

Definition 3.4 A *polynomial law* between two F modules M, N is a natural transformation of the two set valued functors on the category C_F of commutative F algebras:

$$f_B : M \otimes_F B \to N \otimes_F B, \quad B \in C_F. \tag{14}$$

Such a law is homogeneous of degree n if:

$$f_B(ba) = b^n f_B(a),\ \forall b \in B,\ \forall a \in M \otimes_F B,\ \forall B \in C_F.$$

If we have two algebras R, S we have the notion of *multiplicative polynomial law* $d : R \to S$ that is

$$d(ab) = d(a)d(b),\ \forall a, b \in R \otimes_F B,\ \forall B.$$

For the general definition of Cayley Hamilton algebra for any characteristic or even for any commutative ring A we can follow Chenevier [4]:

Definition 3.5

1. Given an algebra R over a commutative ring A and $n \in \mathbb{N}$, an *n-norm* is a multiplicative polynomial law $N : R \to A$ homogeneous of degree n (see Definition 3.4).
2. An algebra R over a commutative ring A with an n-norm $N : R \to A$ is a Cayley Hamilton algebra if each $a \in R$ satisfies its *characteristic polynomial* $\chi_a(x) := N(x - a)$, that is $\chi_a(a) = 0$.

Remark 3.6 It is not hard to prove that, if A contains $\mathbb{Q}$, then Definition 3.5 is equivalent to Definition 2.5 by taking, as trace $t(a)$ of an element a, minus the coefficient of x^{n-1} in $\chi_a(x) := N(x - a) = x^n - t(a)x^{n-1} + \dots$.

Although this is certainly the correct definition in general, there are several technical problems with this definition. The first is to develop the theory in such a way that an analogue of Theorem 3.2 holds. This in fact can be done but it is particularly difficult and to this is devoted the book with De Concini [5].

The second and main question we want to address is, under which conditions an n-Cayley Hamilton algebra R can be embedded in an algebra of $n \times n$ matrices over a commutative A algebra B so that the Norm is the restriction to R of the determinant.

In fact this question can be reformulated as follows. One constructs a *universal map* j_R of R in $n \times n$ matrices and then asks if j_R is injective.

The construction of j_R is in two steps. First let $R = A\langle x_i\rangle_{i\in I}$ be a free algebra over a commutative ring A then:

Definition 3.7 Let $A[\xi_{h,k}^{(i)}]$ be the polynomial algebra over A in the variables $\xi_{h,k}^{(i)}$, $i \in I$, $h, k = 1, \dots, n$ and set $j_R(x_i) := \xi_i := (\xi_{h,k}^{(i)})$ the *generic matrix* with entries $\xi_{h,k}^{(i)}$.

The subalgebra $A\langle\xi_i\rangle$ of $M_n(A[\xi_{h,k}^{(i)}])$, $i \in I$, $h, k = 1, \dots, n$ is called the algebra of *generic matrices.*

Next consider the subalgebra $T_n\langle\xi_i\rangle$ of $A[\xi_{h,k}^{(i)}]$ generated by all the coefficients of the characteristic polynomial of all elements of $A\langle\xi_i\rangle$ and finally the algebra $T_n\langle\xi_i\rangle \cdot A\langle\xi_i\rangle$. A first fact is that $T_n\langle\xi_i\rangle \cdot A\langle\xi_i\rangle$ is closed under the determinant, this follows from Amitsur's Formula (24) and hence it is a n-Cayley–Hamilton algebra. In fact if $A = \mathbb{Q}$ it coincides with the algebra $F_{T,n}\langle X\rangle$ of Formula (13). If A is any field or the integers, its main property is that it behaves as a *free n-Cayley–Hamilton algebra*. In general this is explained as follows.

One may present a general algebra R as a quotient $R = A\langle x_i\rangle/I$.

If R is an n-Cayley–Hamilton algebra one can prove, see Theorem 3.18, that in fact R is also a quotient of the algebra $T_n\langle\xi_i\rangle \cdot A\langle\xi_i\rangle$ and the quotient is compatible with the two norms, where the first is the determinant.

Let thus $R = T_n\langle\xi_i\rangle \cdot A\langle\xi_i\rangle/I$. The ideal I generates in $M_n(A[\xi^{(i)}_{h,k}])$, $i \in I$, $h, k = 1, \dots, n$ an ideal which is, as any ideal in a matrix algebra, of the form $M_n(J)$, with J an ideal of $A[\xi^{(i)}_{h,k}]$.

Then the universal map is given by $j_R : R \to M_n(A[\xi^{(i)}_{h,k}]/J)$. By the universal property this is independent of the presentation of R.

Question Under which conditions one has that $j_R : R \to M_n(A[\xi^{(i)}_{h,k}]/J)$ is injective?

This is equivalent to ask if $M_n(J) \cap T_n\langle\xi_i\rangle = I$. The main Theorem is that this is true when R is an n-Cayley–Hamilton algebra over the rational numbers $\mathbb{Q}$, Procesi [13].

The reason why this Theorem holds over $\mathbb{Q}$ is strictly related to invariant theory and the fact that the group $GL(n)$ of invertible $n \times n$ matrices in characteristic 0 is linearly reductive.

The group $GL(n)$ of invertible $n \times n$ matrices acts by conjugation on I-tuples of matrices and thus on the polynomial ring in the entries $\xi^{(k)}_{i,j}$ of the generic matrices ξ_k. The action is via the formula

$$\xi_k \mapsto X^{-1}\xi_k X, \ \xi^{(k)}_{i,j} \mapsto \sum_{a,b} y_{i,a}\xi^{(k)}_{a,b}x_{b,j}, \ X^{-1} = (y_{i.j}).$$

First fact is a generalization of Theorem 3.2, that is (Theorem 1.10 of [5]):

Theorem 3.8 *$T_n\langle\xi_i\rangle \cdot A\langle\xi_i\rangle$ is the ring of polynomial maps φ from I-tuples of $n \times n$ matrices to $n \times n$ matrices which are* equivariant *under the action of conjugation that is*

$$X\varphi(X^{-1}\xi_1 X, \dots, X^{-1}\xi_j X, \dots)X^{-1} = \varphi(\xi_1, \dots, \xi_j, \dots) \tag{15}$$

In this setting the commutative algebra $T_n\langle\xi_i\rangle$ is the ring of polynomial invariants under conjugation.

This Theorem proved by Procesi in [11] but also implicit in Sibirskii [23] in characteristic 0, and it follows in general from the work of Donkin [6].

The next point is when presenting $R = T_n\langle\xi_i\rangle \cdot A\langle\xi_i\rangle/I$, the ideal I generates in $M_n(A[\xi^{(i)}_{h,k}])$, $i \in I$, $h, k = 1, \dots, n$ an ideal $M_n(J)$, with J an ideal of $A[\xi^{(i)}_{h,k}]$ which is stable under the conjugation action of $GL(n)$.

Then the universal map, given by $j_R : R \to M_n(A[\xi^{(i)}_{h,k}]/J)^{GL(n)}$ maps R into the $GL(n)$ invariants.

Theorem 3.9 (Strong Embedding) *We have a commutative diagram in which, if R is a $\mathbb{Q}$ n-CH algebra, the first horizontal arrows are isomorphisms the second injective and the vertical maps surjective:*

$$\begin{array}{ccccc} T_n\langle \xi_i\rangle \cdot A\langle \xi_i\rangle & \xrightarrow[\cong]{j} & M_n[A[\xi_{h,k}^{(i)}]]^{GL(n)} & \longrightarrow & M_n[A[\xi_{h,k}^{(i)}]] \\ \downarrow & & \downarrow & & \downarrow \\ R = T_n\langle \xi_i\rangle \cdot A\langle \xi_i\rangle / I & \xrightarrow[\cong]{j_R} & M_n[A[\xi_{h,k}^{(i)}]/J]^{GL(n)} & \longrightarrow & M_n[A[\xi_{h,k}^{(i)}]/J] \end{array} \tag{16}$$

Notice that the previous commutative diagram exists in general, and j is always an isomorphism. The fact that in characteristic 0, j_R is an isomorphism depends upon the fact that $GL(n)$, in characteristic 0, is linearly reductive, and then the proof, see [13] or [1] Theorem 14.2.1, of this Theorem is based on the so called Reynold's identities.

The main open question is thus to understand for general R the map j_R. For this a useful remark is that an n Cayley–Hamilton algebra satisfies all polynomial identities of $n \times n$ matrices since it is a quotient of the free CH algebra. The norm algebra of R is $N_R := T_n\langle \xi_i\rangle / T_n\langle \xi_i\rangle \cap I$. Then one may apply the known results on PI algebras in particular Artin's characterization of Azumaya algebras [3]. In particular if for all maximal ideals $\mathfrak{m}$ of N_R we have that $R/\mathfrak{m}R$ is a central simple algebra of rank n^2 over $N_R/\mathfrak{m}$ then R is a rank n^2 Azumaya algebra, cf. [22] and for such algebras j_R is an isomorphism, see [1] §10.4.

Determinants Given an algebra R let us denote, for S any algebra, by $\mathcal{M}_n(R, S)$ the set of multiplicative polynomial maps homogeneous of degree n from R to S. This is a functor on algebras S and, by Roby's Theory it is representable.

This is done by constructing the divided powers $\Gamma_n(R)$ (over the base A) together with the map $i_R : r \mapsto r^{[n]}$ of R to $\Gamma_n(R)$.

One proves that, if R is an algebra, then $\Gamma_n(R)$ is also an algebra and i_R is a universal multiplicative polynomial map, homogeneous of degree n. That is any multiplicative polynomial map, homogeneous of degree n from R to an algebra S factors through i_R and a homomorphism of algebras $\Gamma_n(R) \to S$ giving an isomorphism of functors.

$$\mathcal{M}_n(R, S) \simeq \hom_{\mathcal{R}}(\Gamma_n(R), S), \quad S \in \mathcal{R}. \tag{17}$$

The divided powers $\Gamma_n(R)$ are constructed by generators and relations.

In fact in most applications there is a more concrete description of $\Gamma_n(R)$. For instance if R is a free (or just projective) A module one describes the divided power as symmetric tensors:

$$\Gamma_n(R) \simeq (R^{\otimes n})^{S_n}, \quad R^{\otimes n} = R \otimes_A R \otimes_A R \ldots \otimes_A R, \; i_R : r \mapsto r^{[n]} = r^{\otimes n}. \tag{18}$$

The algebra structure on $\Gamma_n(R)$ is induced by the tensor product of algebras. When we restrict, in Formula (17), to S commutative, this functor, on commutative algebras, is also representable and studied in [21], Roby. It is represented by the abelianization $\pi : \Gamma_n(R) \to \Gamma_n(R)_{ab}$ that is $\Gamma_n(R)$ modulo the ideal generated by the commutators $[a, b]$.

$$\mathcal{M}_n(R, A) \simeq \hom_C(\Gamma_n(R)_{ab}, A), \quad A \in C. \tag{19}$$

The Approach of Zieplies Vaccarino This is based on a very remarkable fact, the ring of invariants $T_n(\xi_1, \dots, \xi_n)$ of m-tuples of $n \times n$ matrices can be defined making no reference to matrices!

In fact if A is a field or the integers $T_n(\xi_1, \dots, \xi_n)$ is isomorphic to

$$\Gamma_n(A\langle x_1, \dots, x_m\rangle)_{ab}, \ A\langle x_1, \dots, x_m\rangle \quad \text{the free algebra.}$$

According to Vaccarino [26] the relationship between the two rings arises by composing the map $j : A\langle x_i\rangle_{i\in I} \to A\langle \xi_i\rangle_{i\in I}$ to generic matrices with the multiplicative map $\det : A\langle \xi_i\rangle_{i\in I} \to T_n\langle \xi_i\rangle$.

From Roby one has then a factorization:

$$\begin{array}{ccc} A\langle x_i\rangle & \xrightarrow{\ j\ } & A\langle \xi_i\rangle \\ {\scriptstyle i}\downarrow & & \downarrow{\scriptstyle \det} \\ \Gamma_n(A\langle x_i\rangle)_{ab} & \xrightarrow{\ D\ } & T_n\langle \xi_i\rangle \end{array} \tag{20}$$

We have a fundamental result, Theorem 20.24 of [5].

Theorem 3.10 *If $R = A\langle X\rangle$ is a free algebra, in some variables $X = \{x_i\}_{i\in I}$ with either $A = F$ a field or $A = \mathbb{Z}$ the integers.*

Then $D : \Gamma_n(A\langle x_i\rangle)_{ab} \to T_n\langle \xi_i\rangle$ is an isomorphism.

This Theorem is proved by Zieplies [27] and Vaccarino [25] and [26], when $A = \mathbb{Q}$. The proof is based on a Theorem of Procesi and Razmyslov (see [11, 12, 16–18]). The general case is fully treated in [5]. It is based on the characteristic free results of Donkin [6] and Zubkov [30] on the invariants of matrices ad a careful combinatorial study of $\Gamma_n(A\langle X\rangle)$ (we call it a *symbolic approach Sect. 3.12*) inspired by the work of Zieplies [28].

In fact since $A\langle X\rangle$ is a free A module, its divided power is more conveniently described as the symmetric tensors:

$$\Gamma_n(A\langle X\rangle) \simeq (A\langle X\rangle^{\otimes n})^{S_n}.$$

Since, using the basis of monomials, the space $A\langle X\rangle^{\otimes n}$ is a permutation representation of S_n, one has a combinatorial description of $(A\langle X\rangle^{\otimes n})^{S_n}$. The abelian quotient, isomorphic to the ring of invariants of matrices, does not have a combinatorial description and it is a rather hard object to study.

Example 3.11 If $X = \{x\}$ is a single variable we have that $A\langle X\rangle = A[x]$ is the commutative ring of polynomials. We identify $A[x]^{\otimes n} = A[x_1, \dots, x_n]$ by setting $x_i := 1^{\otimes i-1} \otimes x \otimes 1^{n-i}$, so $\Gamma_n(A[x])$ is the algebra of symmetric polynomials in n-variables, commutative.

This algebra is the polynomial ring in the *elementary symmetric functions* e_i given by:

$$(1+x)^{\otimes n} = \prod_{i=1}^{n} 1^{\otimes i-1} \otimes (1+x) \otimes 1^{n-i} = \prod_{i=1}^{n} (1+x_i) = 1+e_1+e_2+\ldots+e_n. \quad (21)$$

The map D maps $(1+x)^{\otimes n}$ to $\det(1+\xi)$, so the elementary symmetric function e_i maps to $\sigma_i(\xi)$ with $t^n + \sum_{i=1}^{n}(-1)^i \sigma_i(\xi) t^{n-i}$, the characteristic polynomial $\det(t-\xi)$ of a generic matrix $\xi = (\xi_{i,j})$, cf. Proposition 2.1.

3.12 Symbolic Approach

The proof of Theorem 3.10 given in [5] is based on several steps, which involve in particular the development of a symbolic method.

Formula (2) can be developed in the framework of a purely symbolic calculus. We fix an alphabet X and want to define an *integral form* of the free trace algebra $\mathbb{Q}\langle x_i\rangle_{i\in I}[t(M)]$. For a monomial M we set:

$$\sum_{i=0}^{m} \sigma_i(M) y^i := \exp\Big(\sum_{j=1}^{\infty} (-1)^{j+1} \frac{t(M^j)}{j} y^j\Big). \quad (22)$$

Recall that two monomials NP, PN are called cyclically equivalent and $t(M)$ depends only on monomials up to cyclic equivalence.

There is a canonical choice of a representative in cyclic equivalence class of monomials: the one minimal, in the lexicographic order, in its class of cyclic equivalence.

Recall that a monomial M of positive length, is called *primitive* if it is not a power N^k, $k > 1$. In particular a *Lyndon word*, is a primitive monomial minimal, in the lexicographic order, in its class of cyclic equivalence.

One then defines the *integral form* of the free algebra $\mathbb{Q}\langle x_i\rangle_{i\in I}[t(M)]$, where M runs over all monomials up to cyclic equivalence, to be the polynomial algebra

$$\mathbb{Z}\langle x_i\rangle_{i\in I}[\sigma_j(M)]_{j\in\mathbb{N}}, \ M \quad \text{runs over Lyndon words} \quad (23)$$

in the variables $\sigma_j(M)$, with M a Lyndon word.

This algebra can then be extended by base change to $A\langle x_i\rangle_{i\in I}[\sigma_j(M)]$, for A any commutative ring. In particular one has

$$\mathbb{Q}\langle x_i\rangle_{i\in I}[t(M)] = \mathbb{Q}\otimes\mathbb{Z}\langle x_i\rangle_{i\in I}[\sigma_j(M)]_{j\in\mathbb{N}} = \mathbb{Q}\langle x_i\rangle_{i\in I}[\sigma_j(M)]_{j\in\mathbb{N}}.$$

Where in the L.H.S. M runs over all monomials while in the R.H.S. only primitive monomials, both up to cyclic equivalence.

Then $A\langle x_i\rangle_{i\in I}[\sigma_j(M)]$ can be evaluated, for each positive integer n, as equivariant polynomial maps from I-tuples of $n\times n$ matrices (over any commutative A algebra) to matrices. The Theorem of Donkin states that this map is surjective for A a field or the integers, and the Theorem of Zubkov, Theorems 1.14, 16.3 and 18.4 of [5], gives a description of the kernel.

3.12.1 The Approach of Zieplies and Vaccarino

In the approach of Zieplies and Vaccarino the algebra $A\langle x_i\rangle_{i\in I}[\sigma_j(M)]$ appears in a different and useful way which then explains better the relations of Zubkov. It is constructed as the abelian quotient of a limit of divided powers of the free algebra.

After this is done one will finish with Theorem 3.17, which over the integers or any field of any characteristic is proved in [5] Theorem 20.13 Formula (111) and Theorem 20.24.

First remark that the construction $R\mapsto\Gamma_n(R)$ is functorial in R so given $r\in R$ the map $A[x]\to R,\ x\to r$ induces a map $e_i\to\tau_i(r)$, with $\tau_i(r)$ defined by $(1+r)^{\otimes n} = 1+\tau_1(r)+\tau_2(r)+\ldots+\tau_n(r)$.

One can prove, Theorem 20.13 of [5] that:[1]

Proposition 3.13 *The elements $\tau_i(M)$, $i=1,\ldots,n$ as M runs over the primitive monomials generate $\Gamma_n(A\langle X\rangle)$.*

Denote by $\mathbb{S}_n\langle X\rangle$ the abelian quotient of $\Gamma_n(A\langle X\rangle)$ and by $\sigma_i(M)$ the class of $\tau_i(M)$ in $\mathbb{S}_n\langle X\rangle$. We shall presently show why this notation is compatible with that of Formula (22), Proposition 3.14.

One can prove, Proposition 20.20 of [5], that if $M = NP$ one has $\sigma_i(NP) = \sigma_i(PN)$.

As for the theory of symmetric functions one can pass to the limit as $n\to\infty$ of the algebras $\Gamma_n(A\langle X\rangle)$ and their abelian quotients.

If $\epsilon : A\langle X\rangle\to A$ is the evaluation of X in 0, we have the map

$$\pi_n : A\langle X\rangle^{\otimes n+1}\to A\langle X\rangle^{\otimes n},\ \pi_n(a_1\otimes\ldots\otimes a_n\otimes a_{n+1}) = a_1\otimes\ldots\otimes a_n\otimes\epsilon(a_{n+1}).$$

[1] In [5] the symbol τ_i is replaced by σ_i while our σ_i is $\bar{\sigma}_i$.

This induces a map, still called $\pi_n : \Gamma_{n+1}(A\langle X\rangle) \to \Gamma_n(A\langle X\rangle)$. We have

$$\pi_n(\tau_i(M)) = \begin{cases} \tau_i(M) \text{ if } i \le n \\ 0 \text{ if } i = n+1 \end{cases} .$$

One can then define a limit algebra $\Gamma_\infty(A\langle X\rangle)$ generated by the elements $\tau_i(M)$, $i = 1, \dots, \infty$ as M runs over the primitive monomials and its abelian quotient $\mathbb{S}\langle X\rangle$.

Theorem 20.22 of [5] states that in $\mathbb{S}\langle X\rangle$ one has $\sigma_i(NP) = \sigma_i(PN)$ for all monomials N, P and finally that $\mathbb{S}\langle X\rangle = A[\sigma_i(M)]$, is the free polynomial ring in the variables $\sigma_i(M)$ as M varies among the Lyndon words.

Proposition 3.14 *In this way we recover the algebra of Formula* (23) *(for any A, but in fact it is enough to do it for $A = \mathbb{Z}$).*

$$\mathbb{S}\langle X\rangle \simeq A\langle x_i\rangle_{i\in I}[\sigma_j(M)]_{j\in\mathbb{N}}$$

Let $T_A(X)$, respectively $T_{A,+}(X)$, denote the monoid of endomorphisms of $A\langle X\rangle$ given by mapping each variable $x_i \in X$ to some element $f_i \in A\langle X\rangle$, respectively $f_i \in A\langle X\rangle_+$. The second condition imposes that the endomorphism preserves the ideal kernel of ϵ of elements with no constant term. Each endomorphism $\varphi \in T_A(X)$ induces an endomorphism of each $\Gamma_n(A\langle X\rangle)$ and on the quotient $\mathbb{S}_n$.

If $\varphi \in T_{A,+}(X)$ the endomorphisms induced in $\Gamma_n(A\langle X\rangle)$ are compatible with the maps π_n and hence φ induces an endomorphism on $\Gamma_\infty(A\langle X\rangle)$ and $\mathbb{S}\langle X\rangle$.

Definition 3.15 A T-ideal I of $A\langle X\rangle$ or of $\Gamma_n(A\langle X\rangle)$, or $\mathbb{S}_n\langle X\rangle$, $n = 1, \dots, \infty$ is a multigraded ideal closed under all endomorphisms induced by $T_A(X)$.

For $\Gamma_\infty(A\langle X\rangle)$ or $\mathbb{S}\langle X\rangle$ the condition is to be closed under all endomorphisms induced by $T_{A,+}(X)$.

Remark 3.16 The condition of I to be multigraded can be replaced by the condition that, for every commutative A algebra B, the ideal $I \otimes_A B$ is closed under all endomorphisms induced by $T_B(X)$, as in polynomial laws.

For each $i = 1, 2, \dots$ we have the maps

$$f \mapsto \tau_i(f),\ A\langle X\rangle_+ \to \Gamma_\infty(A\langle X\rangle); \quad f \mapsto \sigma_i(f),\ A\langle X\rangle_+ \to \mathbb{S}.$$

They are both polynomial laws, of A-modules, homogeneous of degree i which commute with the action of the endomorphisms $T_{A,+}(X)$.

There is an explicit Formula which allows us to compute these laws. It is due to Amitsur, [2], (who stated it for matrix invariants), see also [19] or Theorem 4.15 of

[5] or Theorem 3.3.8. of [1]. Given non commutative variables x_i and commutative parameters u_i:

$$\sigma_n(\sum_i u_i x_i) = \sum_{\substack{(p_1 < \ldots < p_k) \subset W_0, \\ j_1, \ldots, j_k \in \mathbb{N}, \ \sum j_i \ell(p_i) = n}} (-1)^{n - \sum j_i} u^{\sum_{i=1}^k j_i \nu(p_i)} \sigma_{j_1}(p_1) \ldots \sigma_{j_k}(p_k) \tag{24}$$

with W_0 the set of Lyndon words ordered by the degree lexicographic order, and, for a word p, $\nu(p)$ is the vector $(a_1, \ldots, a_n)$ with a_i counting how many times the variable x_i appears in p. Finally $u^{(a_1,\ldots,a_n)} := \prod_{i=1}^n u_i^{a_i}$.

In particular one can collect, in Formula (24) the terms of the same degree in the variables x_i and have an explicit expression of the *polarized* forms of $\sigma_n(x)$:

$$\sigma_n(\sum_i u_i x_i) = \sum_{(a_1,\ldots,a_n) \mid \sum_i a_i = n} \prod_{i=1}^n u_i^{a_i} \sigma_{n;a_1,\ldots,a_n}(x_1, \ldots, x_n). \tag{25}$$

Example:

$$\sigma_{3;1,1,1}(a, b, c)$$

$$= \sigma_1(a)\sigma_1(b)\sigma_1(c) - \sigma_1(a)\sigma_1(bc) - \sigma_1(b)\sigma_1(ac) - \sigma_1(c)\sigma_1(ab) + \sigma_1(abc) + \sigma_1(acb) \tag{26}$$

$$\sigma_{3;2,1}(a, b) = -\sigma_1(a)\sigma_1(ab) + \sigma_1(b)\sigma_2(a), \quad \sigma_{3;3}(a) = \sigma_3(a).$$

Substituting for a variable x a linear combination $\sum_j u_j M_j$ of monomials and applying Formula (24) to $\sigma_n(\sum_j u_j M_j)$ one obtains an element of $\mathbb{S}\langle X\rangle$ provided one has a further law. In fact a primitive word computed in monomials need no more be primitive so we also need the expression of the elements $\sigma_i(x^j)$ in terms of the $\sigma_k(x)$, $k \le i \cdot j$. We then use Formula (3).

One has from [5] Theorem 20.13 Formula (111) and Theorem 20.24:

Theorem 3.17 *The Kernels of the maps $\Gamma_\infty(A\langle X\rangle) \to \Gamma_n(A\langle X\rangle)$, respectively $\mathbb{S} \to \mathbb{S}_n$ are the T-ideals generated by all the elements $\tau_i(f)$, $i > n$, respectively the T-ideal generated by all the elements $\sigma_i(f)$, $i > n$, $f \in A\langle X\rangle_+$.*

In other words, in the case $\mathbb{S}_n\langle X\rangle$, the Kernel of π_n is the ideal generated by all the polarized forms $\sigma_{m;a_1,\ldots,a_n}(p_1, \ldots, p_m)$, $m > n$ with $p_1, \ldots, p_m$ monomials of positive degree.

The Theorem of Zubkov then states that the ring of invariants of matrices has the same generators and relations as $\mathbb{S}_n\langle X\rangle$ hence the isomorphism of Theorem 3.10.

The following example shows for $n = 2$ an explicit deduction of the multiplicative nature of the determinant from these relations.

$$\sigma_{3;1,1,1}(a, b, ba) =$$

$$+\sigma_1(a)\sigma_1(b)\sigma_1(ab) - \sigma_1(a)\sigma_1(ab^2) - \sigma_1(b)\sigma_1(a^2b) + \sigma_1(a^2b^2) - 2\sigma_2(ab)$$

$$\sigma_{4;2,2}(a, b) =$$

$$-\sigma_1(a)\sigma_1(b)\sigma_1(ab) + \sigma_1(a)\sigma_1(ab^2) + \sigma_1(b)\sigma_1(a^2b) - \sigma_1(a^2b^2) + \sigma_2(ab) + \sigma_2(b)\sigma_2(a)$$

$$\sigma_{3;1,1,1}(a, b, ba) + \sigma_{4;2,2}(a, b) = \sigma_2(ab) - \sigma_2(b)\sigma_2(a).$$

One can in fact take the basic identity for invariants of $n \times n$ matrices.

$$\det(ab) = \det(a)\det(b) \iff \sigma_n(ab) - \sigma_n(a)\sigma_n(b) = 0. \tag{27}$$

Then consider the polynomial ring $A[\sigma_i(p)],\ p \in W_0,\ i \leq n$. Using Formula (24) one defines for each $f = \sum_i u_i M_i \in A\langle X\rangle$ the element $\sigma_k(f),\ k \leq n$ as follows.

If one substitutes each x_i with M_i in Formula (24) one has a formal expression containing symbols $\sigma_i(M),\ i \leq k$ where M may be an arbitrary monomial (including 1). Then M is cyclically equivalent to some power N^j with $N \in W_0$ a Lyndon word. One then sets, using Formula (3):

$$\sigma_i(N^j) = P_{i,j}(\sigma_1(N), \ldots, \sigma_n(N), 0, \ldots, 0),\ \sigma_i(1) := \binom{n}{i}. \tag{28}$$

Given $f = \sum_i u_i M_i,\ g = \sum_i v_i M_i \in A\langle X\rangle$ one may consider

$$\sigma_n(fg) - \sigma_n(f)\sigma_n(g) = \sum_{\underline{h},\underline{k}} u^{\underline{h}} v^{\underline{k}} \varphi_{\underline{h},\underline{k}},\ \varphi_{\underline{h},\underline{k}} \in A[\sigma_i(p)]. \tag{29}$$

Evaluating the variables $\xi_i \in X$ in the generic $n \times n$ matrices one has a homomorphism $\rho : A\langle X\rangle \to A[\xi_i]$ to the algebra of generic matrices which extends to a homomorphism, $\rho : A[\sigma_i(p)] \to A[\xi_{i,j}^k]^{PGL(n)}$, of the symbolic algebra to the ring of invariants of matrices. By the Theorem of Donkin this is surjective. Moreover clearly the identity given by (28) holds for the corresponding matrix invariants. As for (29) we have that $\sigma_n(\rho(fg)) - \sigma_n(\rho(f))\sigma_n(\rho(g)) = \det(\rho(f)\rho(g)) - \det(\rho(f))\det(\rho(g)) = 0$ so all the elements $\varphi_{\underline{h},\underline{k}}$ map to 0.

Theorem 3.18 *The Kernel of ρ is the ideal K of $A[\sigma_i(p)]$ generated by the elements $\varphi_{\underline{h},\underline{k}}$ of Formula* (29)*, when computed using Formula* (28) *and* (24)*.*

Proof The previous relations express the identity $\sigma_n(fg) = \sigma_n(f)\sigma_n(g)$. Consider the algebra $A[\sigma_i(p)]/K$, and the map $A\langle X\rangle \to A[\sigma_i(p)]/K$ mapping $f \in A\langle X\rangle$ to the class $\bar{\sigma}_n(f)$ of $\sigma_n(f)$ modulo K.

By construction this is a multiplicative map homogeneous of degree n so it factors through a map $A\langle X\rangle \to \Gamma_n(A\langle X\rangle)_{ab} \xrightarrow{\bar{\rho}} A[\sigma_i(p)]/K$. On the other hand $\Gamma_n(A\langle X\rangle)_{ab}$ is generated by the elements $\sigma_i(p)$ and the generators of K are 0 in $\Gamma_n(A\langle X\rangle)_{ab}$ hence $\bar{\rho}$ is an isomorphism and so the claim follows from Theorem 3.10. □

References

1. Aljadeff, E., Giambruno, A., Procesi, C., Regev, A.: Rings with Polynomial Identities and Finite Dimensional Representations of Algebras, vol. 1, pp. 1–12. A.M.S. Colloquium Publications, Providence (2020)
2. Amitsur, S.A.: On the characteristic polynomial of a sum of matrices. Linear Multilinear Algebra **8**(3), 177–182 (1979/1980)
3. Artin, M.: On Azumaya algebras and finite-dimensional representations of rings. J. Algebra **11**, 532–563 (1969)
4. Chenevier, G.: The p-adic analytic space of pseudocharacters of a profinite group and pseudorepresentations over arbitrary rings. In: Diamond F., Kassaei P., Kim M. (eds.) Automorphic Forms and Galois Representations. London Math. Soc. Lecture Note Ser., 414, vol. 1, pp. 221–285. Cambridge Univ. Press, Cambridge (2014)
5. De Concini, C., Procesi, C.: The Invariant Theory of Matrices, vol. 69, 151 pp. A.M.S. University Lecture Series v., Providence (2017)
6. Donkin, S.: Invariants of several matrices. Invent. Math. **110**(2), 389–401 (1992)
7. Frobenius, G.: Über die Primfactoren der Gruppendeterminante. ETH-Bibliothek Zürich, Göttingen, [s.a.]. Shelf Mark: Rar 1524 Persistent Link https://doi.org/10.3931/e-rara-18882
8. Lew, J.S.: The generalized Cayley-Hamilton theorem in n dimensions. Z. Angew. Math. Phys. **17**, 650–653 (1966)
9. Mazur, B.: Deforming Galois representations. In: Galois Groups over $\mathbb{Q}$, vol. 16, pp. 385–437. MSRI Publications, Springer, Berlin (1989)
10. Nyssen, L.: Pseudo-représentations. Math. Ann. **306**(2), 257–283 (1996)
11. Procesi, C.: The invariant theory of $n \times n$ matrices. Adv. Math. **19**, 306–381 (1976)
12. Procesi, C.: Trace identities and standard diagrams. In: Ring Theory. Lecture Notes in Pure and Appl. Mathematics, vol. 51, pp. 191–218. M. Dekker, New York (1979)
13. Procesi, C.: A formal inverse to the Cayley Hamilton theorem. J. Algebra **107**, 63–74 (1987)
14. Procesi, C.: Deformations of Representations. Lecture Notes in Pure and Appl. Math, vol. 198, pp. 247–276. Dekker, New York (1998)
15. Procesi, C.: Lie Groups, An Approach Through Invariants and Representations, pp. xxiv+596. Springer Universitext, New York (2007)
16. Razmyslov, Yu.P.: Identities with trace in full matrix algebras over a field of characteristic zero. Izv. Akad. Nauk SSSR Ser. Mat. **38**, 723–756 (1974) (Russian)
17. Razmyslov, Yu.P.: Identities of algebras and their representations. In: Proceedings of the International Conference on Algebra, Part 2, Novosibirsk, pp. 173–192 (1989). Contemp. Math., vol. 131, Part 2. Amer. Math. Soc., Providence (1992)
18. Razmyslov, Yu.P.: Trace identities of matrix algebras via a field of characteristic zero. Math. USSR Izvestia (translation) **8**, 727–760 (1974)
19. Reutenauer, C., Schützenberger, M.-P.: A formula for the determinant of a sum of matrices. Lett. Math. Phys. **13**(4), 299–302 (1987)
20. Roby, N.: Lois polynomes et lois formelles en théorie des modules. Ann. Sci. École Norm. Supér. Troisième Série **80**, 213–348 (1963)

21. Roby, N.: Lois polynômes multiplicatives universelles. C R Hebd. Séances Acad. Sci. Sér. A et B **290**(19), A869–A871 (1980)
22. Rouquier, R.: Caractérisation des caractères et pseudocaractères. J. Algebra **180**(1996), 571–586
23. Sibirskii, K.S.: Algebraic invariants for a set of matrices. Sib. Mat. Zh. **9**, 152–164 (1968)
24. Taylor, R.: Galois representations associated to Siegel modular forms of low weight. Duke Math. J. **63**(2), 281–332 (1991)
25. Vaccarino, F.: Generalised symmetric functions and invariants of matrices. Math. Z. **260**, 509–526 (2008)
26. Vaccarino, F.: Homogeneous multiplicative polynomial laws are determinants. J. Pure Appl. Algebra **213**(7), 1283–1289 (2009)
27. Ziplies, D.: Generators for the divided powers algebra of an algebra and trace identities. Beiträge zur Algebra Geom. **24**, 9–27 (1987)
28. Ziplies, D.: Abelianizing the divided powers algebra of an algebra. J. Algebra **122**(2), 261–274 (1989)
29. Wiles, A.: On ordinary λ-adic representations associated to modular forms. Invent. Math. **94**(3), 529–573 (1988)
30. Zubkov, A.N.: On a generalization of the Razmyslov-Procesi theorem. Algebra Logika Sibirskiĭ Fond Algebry i Logiki. Algebra Logika **35**, 433–457 (1996)

Growth of Differential Identities

Carla Rizzo

Abstract In this paper we study the growth of the differential identities of some algebras with derivations, i.e., associative algebras where a Lie algebra L (and its universal enveloping algebra $U(L)$) acts on them by derivations. In particular, we study in detail the differential identities and the cocharacter sequences of some algebras whose sequence of differential codimensions has polynomial growth. Moreover, we shall give a complete description of the differential identities of the algebra UT_2 of 2×2 upper triangular matrices endowed with all possible action of a Lie algebra by derivations. Finally, we present the structure of the differential identities of the infinite dimensional Grassmann G with respect to the action of a finite dimensional Lie algebra L of inner derivations.

Keywords Polynomial identity · Differential identity · Codimension · Cocharacter

1 Introduction

Let A be an associative algebra over a field F of characteristic zero and assume that a Lie algebra L acts on it by derivations. Such an action can be naturally extended to the action of the universal enveloping algebra $U(L)$ of L and in this case we say that A is an algebra with derivations or an L-algebra. In this context it is natural to define the differential identities of A, i.e., the polynomials in non-commutative variables $x^h = h(x)$, $h \in U(L)$, vanishing in A.

An effective way of measuring the differential identities satisfied by a given L-algebra A is provided by its sequence of differential codimensions $c_n^L(A)$, $n = 1, 2, \ldots$. The nth term of such sequence measures the dimension of the space of multilinear differential polynomials in n variables of the relatively free

C. Rizzo (✉)
Università degli Studi di Palermo, Palermo, Italy
e-mail: carla.rizzo@unipa.it

O. M. Di Vincenzo, A. Giambruno (eds.), *Polynomial Identities in Algebras*,
Springer INdAM Series 44, https://doi.org/10.1007/978-3-030-63111-6_20

algebra with derivations of countable rank of A. Since in characteristic zero, by the multilinearization process, every differential identity is equivalent to a system of multilinear ones, the sequence of differential codimensions of A gives a quantitative measure of the differential identities satisfied by the given L-algebra. Maybe the most important feature of this sequence proved by Gordienko in [6] is that in case A is a finite dimensional L-algebra, $c_n^L(A)$ is exponentially bounded. Moreover, he determined the exponential rate of growth of the sequence of differential codimensions, i.e., he proved that for any finite dimensional L-algebra A, the limit $\lim_{n\to\infty} \sqrt[n]{c_n^L(A)}$ exists and is a non-negative integer. Such integer, denoted $\exp^L(A)$, is called the differential PI-exponent of the algebra A and it provides a scale allowing us to measure the rate of growth of the identities of any finite dimensional L-algebra. As a consequence of this result it follows that the differential codimensions of a finite dimensional L-algebra A are either polynomially bounded or grow exponentially. Hence no intermediate growth is allowed.

When studying the polynomial identities of an L-algebra A, one is lead to consider $\mathrm{var}^L(A)$, the L-variety of algebras with derivations generated by A, that is the class of L-algebras satisfying all differential identities satisfied by A. Thus we define the growth of $\mathcal{V} = \mathrm{var}^L(A)$ to be the growth of the sequence $c_n^L(\mathcal{V}) = c_n^L(A), n = 1, 2, \ldots$ and we say that a variety $\mathcal{V}$ has almost polynomial growth if $\mathcal{V}$ has exponential growth but every proper subvariety has polynomial growth. Since the ordinary polynomial identities and corresponding codimensions are obtained by leting L act on A trivially (or L is the trivial Lie algebra), the algebra UT_2 of 2×2 upper triangular matrices regarded as L-algebra where L acts trivially on it generates an L-variety of almost polynomial growth (see [4, 8]). Clearly another example of algebras generating an L-variety of almost polynomial growth is the infinite dimensional Grassmann algebra G where L acts trivially on it (see [8, 13]). Notice that in the ordinary case Kemer in [8] proved that UT_2 and G are the only algebras generating varieties of almost polynomial growth.

Recently in [4] the authors introduced another algebra with derivations generating a L-variety of almost polynomial growth. They considered UT_2^ε to be the algebra UT_2 with the action of the 1-dimensional Lie algebra spanned by the inner derivation ε induced by $2^{-1}(e_{11} - e_{22})$, where the e_{ij}'s are the usual matrix units. Also they proved that when the Lie algebra $\mathrm{Der}(UT_2)$ of all derivations acts on UT_2, the variety with derivations generated by UT_2 has no almost polynomial growth.

Notice that if δ is the inner derivation of UT_2 induced by $2^{-1}e_{12}$, then $\mathrm{Der}(UT_2)$ is a 2-dimensional metabelian Lie algebra with basis $\{\varepsilon, \delta\}$. Here we shall study the differential identities of UT_2^δ, i.e., the algebra UT_2 with the action of the 1-dimensional Lie algebra spanned by δ. In particular we shall prove that UT_2^δ does not generate an L-variety of almost polynomial growth. Moreover, in order to complete the description of the differential identities of UT_2, we shall study the T_L-ideal of the differential identities of UT_2 with the action of an arbitrary 1-dimensional Lie subalgebra of $\mathrm{Der}(UT_2)$.

Furthermore, we shall study the differential identities of some particular L-algebras whose sequence of differential codimensions has polynomial growth. In particular we shall exhibit an example of a commutative algebra with derivations that generates a L-variety of linear growth.

Finally, we shall give an example of an infinite dimensional L-algebra of exponential growth. We shall present the structure of the differential identities of $\widetilde{G}$, i.e., the infinite dimensional Grassmann algebra with the action of a finite dimensional abelian Lie algebra and we shall show that, unlike the ordinary case, $\widetilde{G}$ does not generate an L-variety of almost polynomial growth.

2 L-Algebras and Differential Identities

Throughout this paper F will denote a field of characteristic zero. Let A be an associative algebra over F. Recall that a derivation of A is a linear map $\partial : A \to A$ such that

$$\partial(ab) = \partial(a)b + a\partial(b), \qquad \text{for all } a, b \in A.$$

In particular an inner derivation induced by $a \in A$ is the derivation $\operatorname{ad} a : A \to A$ of A defined by $(\operatorname{ad} a)(b) = [a, b] = ab - ba$, for all $b \in A$. The set of all derivations of A is a Lie algebra denoted by $\operatorname{Der}(A)$, and the set $\operatorname{ad}(A)$ of all inner derivations of A is a Lie subalgebra of $\operatorname{Der}(A)$.

Let L be a Lie algebra over F acting on A by derivations. If $U(L)$ is its universal enveloping algebra, the L-action on A can be naturally extended to an $U(L)$-action. In this case we say that A is an algebra with derivations or an L-algebra.

Let L be a Lie algebra. Given a basis $\mathcal{B} = \{h_i \mid i \in I\}$ of the universal enveloping algebra $U(L)$ of L, we let $F\langle X|L\rangle$ be the free associative algebra over F with free formal generators $x_j^{h_i}$, $i \in I$, $j \in \mathbb{N}$. We write $x_i = x_i^1$, $1 \in U(L)$, and then we set $X = \{x_1, x_2, \dots\}$. We let $U(L)$ act on $F\langle X|L\rangle$ by setting

$$\gamma(x_{j_1}^{h_{i_1}} x_{j_2}^{h_{i_2}} \dots x_{j_n}^{h_{i_n}}) = x_{j_1}^{\gamma h_{i_1}} x_{j_2}^{h_{i_2}} \dots x_{j_n}^{h_{i_n}} + \dots + x_{j_1}^{h_{i_1}} x_{j_2}^{h_{i_2}} \dots x_{j_n}^{\gamma h_{i_n}},$$

where $\gamma \in L$ and $x_{j_1}^{h_{i_1}} x_{j_2}^{h_{i_2}} \dots x_{j_n}^{h_{i_n}} \in F\langle X|L\rangle$. The algebra $F\langle X|L\rangle$ is called the free associative algebra with derivations on the countable set X and its elements are called differential polynomials (see [4, 7, 9]).

Given an L-algebra A, a polynomial $f(x_1, \dots, x_n) \in F\langle X|L\rangle$ is a polynomial identity with derivation of A, or a differential identity of A, if $f(a_1, \dots, a_n) = 0$ for all $a_i \in A$, and, in this case, we write $f \equiv 0$.

Let $\operatorname{Id}^L(A) = \{f \in F\langle X|L\rangle \mid f \equiv 0 \text{ on } A\}$ be the set of all differential identities of A. It is readily seen that $\operatorname{Id}^L(A)$ is a T_L-ideal of $F\langle X|L\rangle$, i.e., an ideal invariant under the endomorphisms of $F\langle X|L\rangle$. In characteristic zero every differential identity is equivalent to a system of multilinear differential identities. Hence $\operatorname{Id}^L(A)$ is completely determined by its multilinear polynomial.

Let

$$P_n^L = \operatorname{span}\{x_{\sigma(1)}^{h_1} \cdots x_{\sigma(n)}^{h_n} \mid \sigma \in S_n, h_i \in \mathcal{B}\}$$

be the space of multilinear differential polynomials in the variables $x_1, \ldots, x_n$, $n \geq 1$. We act on P_n^L via the symmetric group S_n as follows: for $\sigma \in S_n$, $\sigma(x_i^h) = x_{\sigma(i)}^h$. For every L-algebra A, the vector space $P_n^L \cap \mathrm{Id}^L(A)$ is invariant under this action. Hence the space $P_n^L(A) = P_n^L/(P_n^L \cap \mathrm{Id}^L(A))$ has a structure of left S_n-module. The non-negative integer $c_n^L(A) = \dim P_n^L(A)$ is called nth differential codimension of A and the character $\chi_n^L(A)$ of $P_n^L(A)$ is called nth differential cocharacter of A. Since char $F = 0$, we can write

$$\chi_n^L(A) = \sum_{\lambda \vdash n} m_\lambda^L \chi_\lambda,$$

where λ is a partition of n, χ_λ is the irreducible S_n-character associated to λ and $m_\lambda^L \geq 0$ is the corresponding multiplicity.

Let L be a Lie algebra and H be a Lie subalgebra of L. If A is an L-algebra, then by restricting the action, A can be regarded as a H-algebra. In this case we say that A is an L-algebra where L acts on it as the Lie algebra H and we identify the T_L-ideal $\mathrm{Id}^L(A)$ and the T_H-ideal $\mathrm{Id}^H(A)$, i.e., in $\mathrm{Id}^L(A)$ we omit the differential identities $x^\gamma \equiv 0$, for all $\gamma \in L \backslash H$.

Notice that any algebra A can be regarded as L-algebra by letting L act on A trivially, i.e., L acts on A as the trivial Lie algebra. Hence the theory of differential identities generalizes the ordinary theory of polynomial identities.

We denote by P_n the space of multilinear ordinary polynomials in $x_1, \ldots, x_n$ and by $\mathrm{Id}(A)$ the T-ideal of the free algebra $F\langle X\rangle$ of polynomial identities of A. We also write $c_n(A)$ for the nth codimension of A and $\chi_n(A)$ for the nth cocharacter of A. Since the field F is of characteristic zero, we have $\chi_n(A) = \sum_{\lambda \vdash n} m_\lambda \chi_\lambda$, where $m_\lambda \geq 0$ is the multiplicity of χ_λ in the given decomposition.

Since $U(L)$ is an algebra with unit, we can identify in a natural way P_n with a subspace of P_n^L. Hence $P_n \subseteq P_n^L$ and $P_n \cap \mathrm{Id}(A) = P_n \cap \mathrm{Id}^L(A)$. As a consequence we have the following relations.

Remark 1 For all $n \geq 1$,

1. $c_n(A) \leq c_n^L(A)$;
2. $m_\lambda \leq m_\lambda^L$, for any $\lambda \vdash n$.

Recall that if A is an L-algebra then the variety of algebras with derivations generated by A is denoted by $\mathrm{var}^L(A)$ and is called L-variety. The growth of $\mathcal{V} = \mathrm{var}^L(A)$ is the growth of the sequence $c_n^L(\mathcal{V}) = c_n^L(A)$, $n = 1, 2, \ldots$.

We say that the L-variety $\mathcal{V}$ has polynomial growth if $c_n^L(\mathcal{V})$ is polynomially bounded and $\mathcal{V}$ has almost polynomial growth if $c_n^L(\mathcal{V})$ is not polynomially bounded but every proper L-subvariety of $\mathcal{V}$ has polynomial growth.

3 On Algebras with Derivations of Polynomial Growth

In this section we study some algebras with derivations whose sequence of differential codimension has linear growth.

Let us first consider the algebra $C = F(e_{11} + e_{22}) \oplus Fe_{12}$ where the e_{ij}'s are the usual matrix units. The Lie algebra $\mathrm{Der}(C)$ of derivations of C is a 1-dimensional Lie algebra generated by ε where

$$\varepsilon(\alpha(e_{11} + e_{22}) + \beta e_{12}) = \beta e_{12},$$

for all $\alpha, \beta \in F$.

Let C^ε denote the L-algebra C where L acts on it as the Lie algebra $\mathrm{Der}(C)$. Thus we have the following.

Theorem 1

1. $\mathrm{Id}^L(C^\varepsilon) = \langle [x, y],\ x^\varepsilon y^\varepsilon,\ x^{\varepsilon^2} - x^\varepsilon \rangle_{T_L}$.
2. $c_n^L(C^\varepsilon) = n + 1$.
3. $\chi_n^L(C^\varepsilon) = 2\chi_{(n)} + \chi_{(n-1,1)}$.

Proof Let $Q = \langle [x, y], x^\varepsilon y^\varepsilon,\ x^{\varepsilon^2} - x^\varepsilon \rangle_{T_L}$. It is easily checked that $Q \subseteq \mathrm{Id}^L(C^\varepsilon)$. Since $x^\varepsilon w y^\varepsilon \in Q$, where w is a (eventually trivial) monomial of $F\langle X|L\rangle$, we may write any multilinear polynomial f, modulo Q, as a linear combination of the polynomials

$$x_1 \dots x_n,\ x_k^\varepsilon x_{i_1} \dots x_{i_{n-1}}, \quad i_1 < \dots < i_{n-1}.$$

We next show that these polynomials are linearly independent modulo $\mathrm{Id}^L(C^\varepsilon)$. Suppose that

$$\alpha x_1 \dots x_n + \sum_{k=1}^{n} \beta_k x_{i_1} \dots x_{i_{n-1}} x_k^\varepsilon \equiv 0 \ (\mathrm{mod}\, P_n^L \cap \mathrm{Id}^L(C^\varepsilon)).$$

By making the evaluation $x_j = e_{11} + e_{22}$, for all $j = 1, \dots, n$, we get $\alpha = 0$. Also for fixed k, the evaluation $x_k = e_{12}$ and $x_j = e_{11} + e_{22}$ for $j \neq k$ gives $\beta_k = 0$. Thus the above polynomials are linearly independent modulo $P_n^L \cap \mathrm{Id}^L(C^\varepsilon)$. Since $P_n^L \cap Q \subseteq P_n^L \cap \mathrm{Id}^L(C^\varepsilon)$, this proves that $\mathrm{Id}^L(C^\varepsilon) = Q$ and the above polynomials are a basis of P_n^L modulo $P_n^L \cap \mathrm{Id}^L(C^\varepsilon)$. Hence $c_n^L(C^\varepsilon) = n + 1$.

We now determine the decomposition of the nth differential cocharacter of this algebra. Suppose that $\chi_n^L(C^\varepsilon) = \sum_{\lambda \vdash n} m_\lambda \chi_\lambda$. Let us consider the standard tableau

$$T_{(n)} = \boxed{1\,|\,2\,|\dots|\,n}$$

and the monomials

$$f_{(n)} = x^n, \qquad f^{\varepsilon}_{(n)} = x^{\varepsilon} x^{n-1} \tag{1}$$

obtained from the essential idempotents corresponding to the tableau $T_{(n)}$ by identifying all the elements in the row. Clearly $f_{(n)}$ and $f^{\varepsilon}_{(n)}$ are not identities of C^{ε}. Moreover, they are linearly independent modulo $\mathrm{Id}^L(C^{\varepsilon})$. In fact, suppose that $\alpha f_{(n)} + \beta f^{\varepsilon}_{(n)} \equiv 0(\mathrm{mod}\ \mathrm{Id}^L(C^{\varepsilon}))$. By making the evaluation $x = e_{11} + e_{22}$ we get $\alpha = 0$. Moreover, if we evaluate $x = e_{11} + e_{22} + e_{12}$, we obtain $\beta = 0$. Thus it follows that $m_{(n)} \geq 2$.

Since $\deg \chi_{(n)} = 1$ and $\deg \chi_{(n-1,1)} = n - 1$, if we find a differential polynomial corresponding to the partition $(n-1, 1)$ which is not a differential identity of C^{ε}, we may conclude that $\chi^L_n(C^{\varepsilon}) = 2\chi_{(n)} + \chi_{(n-1,1)}$.

Let us consider the polynomial

$$f_{(n-1,1)} = (x^{\varepsilon} y - y^{\varepsilon} x)x^{n-2}$$

obtained from the essential idempotent corresponding to the standard tableau

$$T_{(n-1,1)} = \begin{array}{|c|c|c|c|}\hline 1 & 3 & \ldots & n \\ \hline 2 \\ \cline{1-1}\end{array}$$

by identifying all the elements in each row of the tableau. Evaluating $x = e_{11} + e_{22}$ and $y = e_{12}$ we get $f_{(n-1,1)} = -e_{12} \neq 0$ and $f_{(n-1,1)}$ is not a differential identity of C^{ε}. Thus the claim is proved. □

Let us now consider the algebra $M_1 = Fe_{22} \oplus Fe_{12}$ and let ε and δ be derivations of M_1 such that

$$\varepsilon(\alpha e_{22} + \beta e_{12}) = \beta e_{12}, \qquad \delta(\alpha e_{22} + \beta e_{12}) = \alpha e_{12}, \tag{2}$$

for all $\alpha, \beta \in F$.

Lemma 1 *$Der(M_1)$ is a 2-dimensional metabelian Lie algebra spanned by ε and δ defined in* (2).

Proof Let us consider the Lie algebra D spanned by ε and δ. Since $[\varepsilon, \delta] = \delta$, D is a 2-dimensional metabelian Lie algebra and $D \subseteq \mathrm{Der}(M_1)$.

Now consider $\gamma \in \mathrm{Der}(M_1)$. Notice that $\gamma(e_{22}e_{12}) = \gamma(e_{22})e_{12} + e_{22}\gamma(e_{12}) = e_{22}\gamma(e_{12})$. Since $\gamma(e_{22}e_{12}) = 0$, it follows that

$$\gamma(e_{12}) = \alpha e_{12},$$

for some $\alpha \in F$. On the other hand, $\gamma(e_{12}) = \gamma(e_{12}e_{22}) = \alpha e_{12} + e_{12}\gamma(e_{22})$. Thus it follows that $e_{12}\gamma(e_{22}) = 0$. Hence

$$\gamma(e_{22}) = \beta e_{12},$$

for some $\beta \in F$. Thus we have that $\gamma = \alpha\varepsilon + \beta\delta \in D$ and the claim is proved. □

Similarly, if we consider the algebra $M_2 = Fe_{11} \oplus Fe_{12}$ and we assume that ε and δ are derivation of M_2 such that

$$\varepsilon(\alpha e_{11} + \beta e_{12}) = \beta e_{12}, \qquad \delta(\alpha e_{11} + \beta e_{12}) = \alpha e_{12}, \tag{3}$$

for all $\alpha, \beta \in F$, then we have the following.

Lemma 2 *$Der(M_2)$ is a 2-dimensional metabelian Lie algebra spanned by ε and δ defined in* (3).

Let L be any Lie algebra. We shall denote by M_1 and M_2 the L-algebras M_1 and M_2 where L acts trivially on them. Since $x^\gamma \equiv 0$ for all $\gamma \in L$, in this case we are dealing with ordinary identities. Thus we have the following result.

Theorem 2 ([3, Lemma 3])

1. $Id^L(M_1) = \langle x[y,z]\rangle_{T_L}$ *and* $Id^L(M_2) = \langle [x,y]z\rangle_{T_L}$.
2. $c_n^L(M_1) = c_n^L(M_2) = n$.
3. $\chi_n^L(M_1) = \chi_n^L(M_2) = \chi_{(n)} + \chi_{(n-1,1)}$.

Denote by M_1^ε and M_2^ε the L-algebras M_1 and M_2 where L acts on them as the 1-dimensional Lie algebra spanned by the derivation ε defined in (2) and (3), respectively.

Theorem 3

1. $Id^L(M_1^\varepsilon) = \langle xy^\varepsilon,\ x^\varepsilon y - y^\varepsilon x - [x,y],\ x^{\varepsilon^2} - x^\varepsilon\rangle_{T_L}$ *and* $Id^L(M_2^\varepsilon) = \langle x^\varepsilon y,\ xy^\varepsilon - yx^\varepsilon - [x,y],\ x^{\varepsilon^2} - x^\varepsilon\rangle_{T_L}$.
2. $c_n^L(M_1^\varepsilon) = c_n^L(M_2^\varepsilon) = n+1$.
3. $\chi_n^L(M_1^\varepsilon) = \chi_n^L(M_2^\varepsilon) = 2\chi_{(n)} + \chi_{(n-1,1)}$.

Proof If Q is the T_L-ideal generated by the polynomials $xy^\varepsilon,\ x^\varepsilon y - y^\varepsilon x - [x,y],\ x^{\varepsilon^2} - x^\varepsilon$, then it easy to check that $Q \subseteq \mathrm{Id}^L(M_1^\varepsilon)$.

Since $x^\varepsilon y^\varepsilon,\ x[y,z] \in Q$, the polynomials

$$x_j x_{i_1} \dots x_{i_n-1},\ x_1^\varepsilon x_2 \dots x_n, \quad i_1 < \dots < i_{n-1},$$

span P_n^L modulo $P_n^L \cap Q$ and we claim that they are linearly independent modulo $\mathrm{Id}^L(M_1^\varepsilon)$. In fact, let $f \in P_n^L \cap \mathrm{Id}^L(M_1^\varepsilon)$ be a linear combination of these polynomials, i.e.,

$$f = \sum_{j=1}^{n} \alpha_j x_j x_{i_1} \dots x_{i_n-1} + \beta x_1^\varepsilon x_2 \dots x_n \equiv 0 \ (\mathrm{mod}\ P_n^L \cap \mathrm{Id}^L(M_1^\varepsilon)).$$

For fixed $j \neq 1$, from the substitutions $x_j = e_{12}$ and $x_k = e_{22}$ for $k \neq j$ we get $\alpha_j = 0$, $j \neq 1$. By making the evaluation $x_k = e_{22}$ for all $k = 1, \ldots, n$, we obtain $\alpha_1 = 0$. Finally by evaluating $x_1 = e_{12}$ and $x_k = e_{22}$ for $k \neq 1$, we get $\beta = 0$. Thus the above polynomials are linearly independent modulo $P_n^L \cap \mathrm{Id}^L(M_1^\varepsilon)$. Since $P_n^L \cap Q \subseteq P_n^L \cap \mathrm{Id}^L(M_1^\varepsilon)$, this proves that $\mathrm{Id}^L(M_1^\varepsilon) = Q$ and the above polynomials are a basis of P_n^L modulo $P_n^L \cap \mathrm{Id}^L(M_1^\varepsilon)$. Clearly $c_n^L(M_1^\varepsilon) = n + 1$.

We now determine the decomposition of the nth differential cocharacter $\chi_n^L(M_1^\varepsilon)$ of this algebra. Suppose that $\chi_n^L(M_1^\varepsilon) = \sum_{\lambda \vdash n} m_\lambda \chi_\lambda$. We consider the tableau $T_{(n)}$ defined in Theorem 1 and let $f_{(n)}$ and $f_{(n)}^\varepsilon$ be the corresponding polynomials defined in (1). It is clear that $f_{(n)}$ and $f_{(n)}^\varepsilon$ are not identities of M_1^ε. Moreover, they are linearly independent modulo $\mathrm{Id}^L(M_1^\varepsilon)$. In fact, suppose that $\alpha f_{(n)} + \beta f_{(n)}^\varepsilon \equiv 0(\mathrm{mod}\ \mathrm{Id}^L(M_1^\varepsilon))$. By making the evaluation $x = e_{22}$ we get $\alpha = 0$. Moreover, if we evaluate $x = e_{22} + e_{12}$, we obtain $\beta = 0$. Thus it follows that $m_{(n)} \geq 2$. By Remark 1 and Theorem 2 we have $m_{(n-1,1)} \geq 1$. Thus, since $\deg \chi_{(n)} = 1$ and $\deg \chi_{(n-1,1)} = n - 1$, it follows that $\chi_n^L(M_1^\varepsilon) = 2\chi_{(n)} + \chi_{(n-1,1)}$.

A similar proof holds for the algebra M_2^ε. □

Let M_1^δ and M_2^δ be the L-algebras M_1 and M_2 where L acts on them as the 1-dimensional Lie algebra spanned by the derivation δ defined in (2) and (3), respectively. The proof on the next theorem is similar to the above proof and is omitted.

Theorem 4

1. *$Id^L(M_1^\delta) = \langle x[y,z],\ xy^\delta,\ x^\delta y - y^\delta x,\ x^{\delta^2}\rangle_{T_L}$ and $Id^L(M_2^\delta) = \langle [x,y]z,\ x^\delta y,\ xy^\delta - yx^\delta,\ x^{\delta^2}\rangle_{T_L}$.*
2. *$c_n^L(M_1^\delta) = c_n^L(M_2^\delta) = n + 1$.*
3. *$\chi_n^L(M_1^\delta) = \chi_n^L(M_2^\delta) = 2\chi_{(n)} + \chi_{(n-1,1)}$.*

Let now L be a 2-dimensional metabelian Lie algebra. Let denote by M_1^D the L-algebra M_1 where L acts on it as the Lie algebra $\mathrm{Der}(M_1)$ and M_2^D the L-algebra M_2 where L acts on it as the Lie algebra $\mathrm{Der}(M_2)$.

Remark 2

1. $x^\delta y - y^\delta x \in \langle xy^\varepsilon,\ x^\varepsilon y - y^\varepsilon x - [x,y],\ x^{\varepsilon\delta} - x^\delta\rangle_{T_L}$.
2. $xy^\delta - yx^\delta \in \langle x^\varepsilon y,\ xy^\varepsilon - yx^\varepsilon - [x,y],\ x^{\varepsilon\delta} - x^\delta\rangle_{T_L}$.

Proof First notice that $[x,y]^\delta \in \langle xy^\varepsilon,\ [x,y]^\varepsilon - [x,y]\rangle_{T_L}$. Thus, since $[x,y]^\varepsilon \equiv [x,y](\mathrm{mod}\ \langle xy^\varepsilon,\ x^\varepsilon y - y^\varepsilon x - [x,y]\rangle_{T_L})$, it follows that

$$[x,y]^\delta \in \langle xy^\varepsilon,\ x^\varepsilon y - y^\varepsilon x - [x,y],\ x^{\varepsilon\delta} - x^\delta\rangle_{T_L}.$$

Moreover, since $xy^\delta \in \langle xy^\varepsilon,\ x^{\varepsilon\delta} - x^\delta\rangle_{T_L}$, we get

$$x^\delta y - y^\delta x \in \langle xy^\varepsilon,\ x^\varepsilon y - y^\varepsilon x - [x,y],\ x^{\varepsilon\delta} - x^\delta\rangle_{T_L}.$$

A similar proof holds for the other statement. □

We do not present the proof of next theorem since it can be easily deduced by using the strategy of proof given in Theorem 3.

Theorem 5

1. $Id^L(M_1^D) = \langle xy^\varepsilon,\ x^\varepsilon y - y^\varepsilon x - [x,y],\ x^{\varepsilon^2} - x^\varepsilon,\ x^{\delta\varepsilon},\ x^{\varepsilon\delta} - x^\delta\rangle_{T_L}$ *and* $Id^L(M_2^D) = \langle x^\varepsilon y,\ xy^\varepsilon - yx^\varepsilon - [x,y],\ x^{\varepsilon^2} - x^\varepsilon,\ x^{\delta\varepsilon},\ x^{\varepsilon\delta} - x^\delta\rangle_{T_L}$.
2. $c_n^L(M_1^D) = c_n^L(M_2^D) = n + 2$.
3. $\chi_n^L(M_1^D) = \chi_n^L(M_2^D) = 3\chi_{(n)} + \chi_{(n-1,1)}$.

4 The Algebra of 2 × 2 Upper Triangular Matrices and Its Differential Identities

In this section we study the growth of differential identities of the algebra UT_2 of 2×2 upper triangular matrices over F.

Let L be any Lie algebra over F and denote by UT_2 the L-algebra UT_2 where L acts trivially on it. Since $x^\gamma \equiv 0$, for all $\gamma \in L$, is a differential identity of UT_2, we are dealing with ordinary identities. Thus by Malcev [11], Kemer [8] and by the proof of Lemma 3.5 in [1], we have the following results.

Theorem 6

1. $\mathrm{Id}^L(UT_2) = \langle [x_1, x_2][x_3, x_4]\rangle_{T_L}$.
2. $c_n^L(UT_2) = 2^{n-1}(n-2) + 2$.
3. *If* $\chi_n^L(UT_2) = \sum_{\lambda\vdash n} m_\lambda \chi_\lambda$ *is the nth differential cocharacter of* UT_2, *then*

$$m_\lambda = \begin{cases} 1, & \text{if } \lambda = (n) \\ q+1, & \text{if } \lambda = (p+q,p) \text{ or } \lambda = (p+q,p,1) \\ 0 & \text{in all other cases} \end{cases}.$$

Theorem 7 $\mathrm{var}^L(UT_2)$ *has almost polynomial growth.*

Let now ε be the inner derivation of UT_2 induced by $2^{-1}(e_{11} - e_{22})$, i.e.,

$$\varepsilon(a) = 2^{-1}[e_{11} - e_{22}, a], \qquad \text{for all } a \in UT_2, \tag{4}$$

where the e_{ij}'s are the usual matrix units. We shall denote by UT_2^ε the L-algebra UT_2 where L acts on it as the 1-dimensional Lie algebra spanned by ε. In [4] the authors proved the following.

Theorem 8 ([4, Theorems 5 and 12])

1. $\mathrm{Id}^L(UT_2^\varepsilon) = \langle [x,y]^\varepsilon - [x,y],\ x^\varepsilon y^\varepsilon,\ x^{\varepsilon^2} - x^\varepsilon\rangle_{T_L}$.

2. $c_n^L(UT_2^\varepsilon) = 2^{n-1}n + 1$.
3. *If* $\chi_n^L(UT_2^\varepsilon) = \sum_{\lambda \vdash n} m_\lambda^\varepsilon \chi_\lambda$ *is the nth differential cocharacter of* UT_2^ε, *then*

$$m_\lambda^\varepsilon = \begin{cases} n+1, & \text{if } \lambda = (n) \\ 2(q+1), & \text{if } \lambda = (p+q, p) \\ q+1, & \text{if } \lambda = (p+q, p, 1) \\ 0 & \text{in all other cases} \end{cases}.$$

Theorem 9 ([4, Theorem 15]) $\mathrm{var}^L(UT_2^\varepsilon)$ *has almost polynomial growth.*

Let now δ be the inner derivation of UT_2 induced by $2^{-1}e_{12}$, i.e.,

$$\delta(a) = 2^{-1}[e_{12}, a], \qquad \text{for all } a \in UT_2. \tag{5}$$

Denote by UT_2^δ the L-algebra UT_2 where L acts as the 1-dimensional Lie algebra spanned by δ. The following remarks are easily verified.

Remark 3 $[x, y][z, w] \equiv 0$, $[x, y]^\delta \equiv 0$, $x^\delta y^\delta \equiv 0$, $x^\delta[y, z] \equiv 0$ and $x^{\delta^2} \equiv 0$ are differential identities of UT_2^δ.

Remark 4 $x^\delta y[z, w],\ [x, y]zw^\delta,\ x^\delta yz^\delta \in \langle x^\delta y^\delta,\ x^\delta[y, z],\ [x, y]^\delta\rangle_{T_L}$.

Remark 5 For any permutations $\sigma \in S_t$, we have

$$[x_{\sigma(1)}^\delta, x_{\sigma(2)}, \ldots, x_{\sigma(t)}] \equiv [x_1^\delta, x_2, \ldots, x_t] \pmod{\langle x^\delta[y, z],\ [x, y]^\delta\rangle_{T_L}}.$$

Proof Let u_1, u_2, u_3 be monomials. We consider $w = u_1x_ix_ju_2y^\delta u_3$. Since $x_ix_j = x_jx_i + [x_i, x_j]$, it follows that $w \equiv u_1x_jx_iu_2y^\delta u_3 \pmod{\langle x^\delta[y, z],\ [x, y]^\delta\rangle_{T_L}}$. In the same way we can show that $u_1y^\delta u_2z_iz_ju_3 \equiv u_1y^\delta u_2z_jz_iu_3 \pmod{\langle x^\delta[w, z]\rangle_{T_L}}$. Hence in every monomial

$$x_{i_1} \ldots x_{i_t} y^\delta z_{j_1} \ldots z_{j_p}$$

we can reorder the variables to the left and to the right of y^δ. Since $[x, y]^\delta = [x^\delta, y] - [y^\delta, x]$, we can reorder all the variables in any commutator $[x_{i_1}^\delta, x_{i_2}, \ldots, x_{i_t}]$ as claimed. □

Lemma 3 *The T_L-ideal of identities of UT_2^δ is generated by the following polynomials*

$$[x, y][z, w],\ [x, y]^\delta,\ x^\delta[y, z],\ x^\delta y^\delta,\ x^{\delta^2}.$$

Proof Let $Q = \langle [x, y][z, w],\ [x, y]^\delta,\ x^\delta[y, z],\ x^\delta y^\delta,\ x^{\delta^2}\rangle_{T_L}$. By Remark 3, $Q \subseteq \mathrm{Id}^L(UT_2^\delta)$.

By the Poincaré-Birkhoff-Witt Theorem (see [12]) every differential multilinear polynomial in $x_1, \ldots, x_n$ can be written as a linear combination of products of the type

$$x_{i_1}^{\alpha_1} \ldots x_{i_k}^{\alpha_k} w_1 \ldots w_m \tag{6}$$

where $\alpha_1, \ldots, \alpha_k \in U(L)$, $w_1 \ldots, w_m$ are left normed commutators in the $x_i^{\alpha_j}$s, $\alpha_j \in U(L)$, and $i_1 < \cdots < i_k$. Since $[x_1^{\alpha_1}, x_2^{\alpha_2}][x_3^{\alpha_3}, x_4^{\alpha_4}] \in Q$, with $\alpha_1, \alpha_2, \alpha_3, \alpha_4 \in \{1, \delta\}$, then, modulo $\langle [x_1^{\alpha_1}, x_2^{\alpha_2}][x_3^{\alpha_3}, x_4^{\alpha_4}], x^{\delta^2} \rangle_{T_L}$, in (6) we have $\alpha_j \in \{1, \delta\}$ and $m \leq 1$, so, only at most one commutator can appear in (6). Thus by Remark 4 every multilinear monomial in P_n^L can be written, modulo Q, as linear combination of the elements of the type

$$x_1 \ldots x_n, \quad x_{h_1} \ldots x_{h_{n-1}} x_j^{\delta}, \quad x_{i_1} \ldots x_{i_k} [x_{j_1}^{\gamma}, x_{j_2}, \ldots, x_{j_m}],$$

where $h_1 < \cdots < h_{n-1}, i_1 < \cdots < i_k, m + k = n, m \geq 2, \gamma \in \{1, \delta\}$.

Let us now consider the left normed commutators $[x_{j_1}^{\gamma}, x_{j_2}, \ldots, x_{j_m}]$ and suppose first that $\gamma = 1$. Since $[x_1, x_2][x_3, x_4] \in Q$, then it is already known that (see for example [5, Theorem 4.1.5])

$$[x_{j_1}, x_{j_2}, \ldots, x_{j_m}] \equiv [x_k, x_{h_1}, \ldots, x_{h_{m-1}}] \pmod{Q},$$

where $k > h_1 < \cdots < h_{m-1}$.

Suppose now $\gamma = \delta$, then by Remark 5 we get

$$[x_{j_1}^{\delta}, x_{j_2}, \ldots, x_{j_m}] \equiv [x_1^{\delta}, x_2, \ldots, x_t] \pmod{\langle x^{\delta}[y, z], [x, y]^{\delta} \rangle_{T_L}}.$$

It follows that P_n^L is spanned, modulo $P_n^L \cap Q$, by the polynomials

$$\begin{aligned} &x_1 \ldots x_n, && x_{i_1} \ldots x_{i_m} [x_k, x_{j_1}, \ldots, x_{j_{n-m-1}}], \\ &x_{h_1} \ldots x_{h_{n-1}} x_r^{\delta}, && x_{i_1} \ldots x_{i_m} [x_{l_1}^{\delta}, x_{l_2}, \ldots, x_{l_{n-m}}], \end{aligned} \tag{7}$$

where $i_1 < \cdots < i_m, k > j_1 < \cdots < j_{n-m-1}, h_1 < \cdots < h_{n-1}, l_1 < \cdots < l_{n-m}$, $m \neq n-1, n$.

Next we show that these polynomials are linearly independent modulo $\mathrm{Id}^L(UT_2^{\delta})$. Let $I = \{i_1, \ldots, i_m\}$ be a subset of $\{1, \ldots, n\}$ and $k \in \{1, \ldots, n\} \setminus I$ such that $k > \min(\{1, \ldots, n\} \setminus I)$, then set $X_{I,k} = x_{i_1} \ldots x_{i_m} [x_k, x_{j_1}, \ldots, x_{j_{n-m-1}}]$. Also for $I' = \{i_1, \ldots, i_m\} \subseteq \{1, \ldots, n\}$, $0 \leq |I'| < n-1$, set $X_{I'}^{\delta} = x_{i_1} \ldots x_{i_m} [x_{l_1}^{\delta}, x_{l_2}, \ldots, x_{l_{n-m}}]$ and suppose that

$$f = \sum_{I,J} \alpha_{I,k} X_{I,k} + \sum_{I'} \alpha_{I'}^{\delta} X_{I'}^{\delta} + \sum_{k=1}^{n} \alpha_r^{\delta} x_{h_1} \ldots x_{h_{n-1}} x_r^{\delta}$$
$$+ \beta x_1 \ldots x_n \equiv 0 \pmod{P_n^L \cap \mathrm{Id}^L(UT_2^{\delta})}.$$

In order to show that all coefficients $\alpha_{I,k}$, $\alpha^{\delta}_{I'}$, α^{δ}_{r}, β are zero we will make some evaluations. If we evaluate $x_1 = \cdots = x_n = e_{11} + e_{22}$ we get $\beta = 0$. For a fixed r, by setting $x_{h_1} = \cdots = x_{h_{n-1}} = e_{11} + e_{22}$ and $x_r = e_{22}$ we get $\alpha^{\delta}_r = 0$. Also, for a fixed $I' = \{i_1, \ldots, i_m\}$, by making the evaluations $x_{i_1} = \cdots = x_{i_m} = e_{11} + e_{22}$, $x_{l_1} = \cdots = x_{l_{n-m}} = e_{22}$ we obtain $\alpha^{\delta}_{I'} = 0$. Finally, for fixed $I = \{i_1, \ldots, i_m\}$ and $J = \{j_1, \ldots, j_{n-m-1}\}$, from the substitutions $x_{i_1} = \cdots = x_{i_m} = e_{11} + e_{22}$, $x_k = e_{12}$, $x_{j_1} = \cdots = x_{j_{n-m-1}} = e_{22}$, it follows that $\alpha_{I,k} = 0$.

We have proved that $\mathrm{Id}^L(UT_2^{\delta}) = Q$ and the elements in (7) are a basis of P_n^L modulo $P_n^L \cap \mathrm{Id}^L(UT_2^{\delta})$. □

We now compute the nth differential cocharacter of UT_2^{δ}. Write

$$\chi_n^L(UT_2^{\delta}) = \sum_{\lambda \vdash n} m^{\delta}_{\lambda}\chi_{\lambda}. \tag{8}$$

In the following lemmas we compute the non-zero multiplicities of such cocharacter.

Lemma 4 *In* (8) $m^{\delta}_{(n)} \geq n + 1$.

Proof We consider the following tableau:

$$T_{(n)} = \boxed{1\,|\,2\,|\ldots|\,n}\,.$$

We associate to $T_{(n)}$ the monomials

$$a(x) = x^n, \tag{9}$$

$$a_k^{(\delta)}(x) = x^{k-1}x^{\delta}x^{n-k}, \tag{10}$$

for all $k = 1, \ldots, n$. These monomials are obtained from the essential idempotents corresponding to the tableau $T_{(n)}$ by identifying all the elements in the row. It is easily checked that $a(x)$, $a_k^{(\delta)}(x)$, $k = 1, \ldots, n$, do not vanish in UT_2^{δ}.

Next we shall prove that the $n + 1$ monomials $a(x)$, $a_k^{(\delta)}(x)$, $k = 1, \ldots, n$, are linearly independent modulo $\mathrm{Id}^L(UT_2^{\delta})$. In fact, suppose that

$$\alpha a(x) + \sum_{k=1}^{n} \alpha_k^{\delta} a_k^{(\delta)}(x) \equiv 0 \pmod{\mathrm{Id}^L(UT_2^{\delta})}.$$

By setting $x = e_{11}+e_{22}$ it follows that $\alpha = 0$. Moreover, if we substitute $x = \beta e_{11}+ e_{22}$ where $\beta \in F$, $\beta \neq 0$, we get $\sum_{k=1}^{n}(1-\beta)\beta^{k-1}\alpha_k^{\delta} = 0$. Since $|F| = \infty$, we can choose $\beta_1, \ldots, \beta_n \in F$, where $\beta_i \neq 0$ and $\beta_i \neq \beta_j$, for all $1 \leq i \neq j \leq n$. Then we get the following homogeneous linear system of n equations in the n variables α_k^{δ}, $k = 1, \ldots, n$,

$$\sum_{k=1}^{n} \beta_i^{k-1}\alpha_k^{\delta} = 0, \quad i = 1, \ldots, n. \tag{11}$$

Since the matrix associated to the system (11) is a Vandermonde matrix, it follows that $\alpha_k^\delta = 0$, for all $k = 1, \ldots, n$. Thus the monomials $a(x)$, $a_k^{(\delta)}(x)$, $k = 1, \ldots, n$, are linearly independent modulo $\mathrm{Id}^L(UT_2^\delta)$. This says that $m_{(n)}^\delta \geq n + 1$. □

Lemma 5 *Let $p \geq 1$ and $q \geq 0$. If $\lambda = (p + q, p)$ then in* (8) *we have $m_\lambda^\delta \geq 2(q + 1)$.*

Proof For every $i = 0, \ldots, q$ we define $T_\lambda^{(i)}$ to be the tableau

$i+1$	$i+2$	$\ldots$	$i+p-1$	$i+p$	1	$\ldots$	i	$i+2p+1$	$\ldots$	n
$i+p+2$	$i+p+3$	$\ldots$	$i+2p$	$i+p+1$						

.

We associate to $T_\lambda^{(i)}$ the polynomials

$$b_i^{(p,q)}(x, y) = x^i \underbrace{\bar{x} \ldots \tilde{x}}_{p-1}[x, y] \underbrace{\bar{y} \ldots \tilde{y}}_{p-1} x^{q-i}, \tag{12}$$

$$b_i^{(p,q,\delta)}(x, y) = x^i \underbrace{\bar{x} \ldots \tilde{x}}_{p-1}(x^\delta y - y^\delta x) \underbrace{\bar{y} \ldots \tilde{y}}_{p-1} x^{q-i}, \tag{13}$$

where the symbols $-$ or $\sim$ means alternation on the corresponding variables. The polynomials $b_i^{(p,q)}$, $b_i^{(p,q,\delta)}$ are obtained from the essential idempotents corresponding to the tableau $T_\lambda^{(i)}$ by identifying all the elements in each row of the tableau. It is clear that $b_i^{(p,q)}$, $b_i^{(p,q,\delta)}$, $i = 0, \ldots, q$, are not differential identities of UT_2^δ. We shall prove that the above $2(q + 1)$ polynomials are linearly independent modulo $\mathrm{Id}^L(UT_2^\delta)$. Suppose that

$$\sum_{i=0}^{q} \alpha_i b_i^{(p,q)} + \sum_{i=0}^{q} \alpha_i^\delta b_i^{(p,q,\delta)} \equiv 0 \pmod{\mathrm{Id}^L(UT_2^\delta)}.$$

If we set $x = \beta e_{11} + e_{22}$, with $\beta \in F$, $\beta \neq 0$, and $y = e_{11}$, we obtain

$$\sum_{i=0}^{q} (-1)^{p-1} \beta^i \alpha_i^\delta = 0.$$

Since $|F| = \infty$, we can take $\beta_1, \ldots, \beta_{q+1} \in F$, where $\beta_j \neq 0$, $\beta_j \neq \beta_k$, for all $1 \leq j \neq k \leq q + 1$. Then we obtain the following homogeneous linear system of $q + 1$ equations in the $q + 1$ variables α_i^δ, $i = 0, \ldots, q$,

$$\sum_{i=0}^{q} \beta_j^i \alpha_i^\delta = 0, \quad j = 1, \ldots, q + 1. \tag{14}$$

Since the matrix of this system is a Vandermonde matrix, it follows that $\alpha_i^\delta = 0$, for all $i = 0, \ldots, q$. Hence we may assume that the following identity holds

$$\sum_{i=0}^{q} \alpha_i b_i^{(p,q)} \equiv 0 \pmod{\mathrm{Id}^L(UT_2^\delta)}.$$

If we evaluate $x = \beta e_{11} + e_{12} + e_{22}$, where $\beta \in F$, $\beta \neq 0$, and $y = e_{11}$, then we get

$$\sum_{i=0}^{q} (-1)^{p-1} \beta^i \alpha_i = 0. \tag{15}$$

Since $|F| = \infty$, we choose $\beta_1, \ldots, \beta_{q+1} \in F$, where $\beta_j \neq 0$, $\beta_j \neq \beta_k$, for all $1 \leq j \neq k \leq q+1$. Then from (15) we obtain a homogeneous linear system of $q+1$ equations in the $q+1$ variables α_i, $i = 0, \ldots, q$, equivalent to the linear system (14). Therefore $\alpha_i = 0$, for all $i = 0, \ldots, q$. Hence the polynomials $b_i^{(p,q)}$, $b_i^{(p,q,\delta)}$, $i = 0, \ldots, q$, are linearly independent modulo $\mathrm{Id}^L(UT_2^\delta)$ and, so, $m_\lambda^\delta \geq 2(q+1)$. □

As an immediate consequence of Remark 1 and Theorem 6 we have the following.

Lemma 6 *Let $p \geq 1$ and $q \geq 0$. If $\lambda = (p+q, p, 1)$, then in* (8) *we have $m_\lambda^\delta \geq q+1$.*

We are now in a position to prove the following theorem about the L-algebra UT_2^δ.

Theorem 10

1. $\mathrm{Id}^L(UT_2^\delta) = \langle [x,y][z,w],\ [x,y]^\delta,\ x^\delta[y,z],\ x^\delta y^\delta,\ x^{\delta^2} \rangle_{T_L}$.
2. $c_n^L(UT_2^\delta) = 2^{n-1}n + 1$.
3. If $\chi_n^L(UT_2^\delta) = \sum_{\lambda \vdash n} m_\lambda^\delta \chi_\lambda$ is the nth differential cocharacter of UT_2^δ, then

$$m_\lambda^\delta = \begin{cases} n+1, & \text{if } \lambda = (n) \\ 2(q+1), & \text{if } \lambda = (p+q, p) \\ q+1, & \text{if } \lambda = (p+q, p, 1) \\ 0 & \text{in all other cases} \end{cases}. \tag{16}$$

Proof By Lemma 3 the T_L-ideal of differential identities of UT_2^δ is generated by the polynomials $[x,y][z,w]$, $[x,y]^\delta$, $x^\delta[y,z]$, $x^\delta y^\delta$, x^{δ^2} and the elements in (7) are a basis of P_n^L modulo $P_n^L \cap \mathrm{Id}^L(UT_2^\delta)$. Thus by counting these elements we get that $c_n^L(UT_2^\delta) = 2^{n-1}n + 1$.

Finally, as a consequence of Lemmas 4, 5, 6 and by following verbatim the proof of [4, Theorem 12] we get the decomposition into irreducible characters of $\chi_n^L(UT_2^\delta)$. □

Notice that $\mathrm{var}^L(UT_2^\delta)$ has exponential growth, nevertheless it has no almost polynomial growth. In fact, the algebra UT_2 (ordinary case) is an algebra with $F\delta$-action where δ acts trivially on UT_2, i.e., $x^\delta \equiv 0$ is differential identity of UT_2. Then it follows that $UT_2 \in \mathrm{var}^L(UT_2^\delta)$, but $\mathrm{var}^L(UT_2)$ growths exponentially. Thus we have the following result.

Theorem 11 $\mathrm{var}^L(UT_2^\delta)$ *has no almost polynomial growth.*

Now denote by UT_2^η the L-algebra UT_2 where L acts on it as the 1-dimensional Lie algebra spanned by a non-trivial derivation η of UT_2. Notice that since any derivation of UT_2 is inner (see [2]), it can be easily checked that the algebra $\mathrm{Der}(UT_2)$ of all derivations of UT_2 is the 2-dimensional metabelian Lie algebra with basis $\{\varepsilon, \delta\}$ defined in (4) and (5), respectively. Thus

$$\eta = \alpha\,\varepsilon + \beta\,\delta, \quad \text{for some } \alpha, \beta \in F \text{ not both zero.}$$

Remark 6 $[x,y]^\eta - \alpha[x,y] \equiv 0$, $x^\eta y^\eta \equiv 0$, $x^{\eta^2} - \alpha x^\eta \equiv 0$, $[x,y][z,w] \equiv 0$, $x^\eta[y,z] \equiv 0$ are differential identities of UT_2^η. Moreover, if $\alpha \neq 0$, then $[x,y][z,w],\ x^\eta[y,z] \in \langle [x,y]^\eta - \alpha[x,y],\ x^\eta y^\eta \rangle_{T_L}$.

We do not present the proof of next theorem since it can be deduced by using the strategy of proofs given in [4, Theorems 5 and 12] and Theorem 10.

Theorem 12

1. *If* $\alpha \neq 0$, *then* $\mathrm{Id}^L(UT_2^\eta) = \langle [x,y]^\eta - \alpha[x,y],\ x^\eta y^\eta,\ x^{\eta^2} - \alpha x^\eta \rangle_{T_L}$. *Otherwise,* $\mathrm{Id}^L(UT_2^\eta) = \langle [x,y][z,w],\ x^\eta[y,z],\ [x,y]^\eta,\ x^\eta y^\eta,\ x^{\eta^2} \rangle_{T_L}$.
2. $c_n^L(UT_2^\eta) = 2^{n-1}n + 1$.
3. *If* $\chi_n^L(UT_2^\eta) = \sum_{\lambda \vdash n} m_\lambda^\eta \chi_\lambda$ *is the nth differential cocharacter of* UT_2^η, *then*

$$m_\lambda^\eta = \begin{cases} n+1, & \text{if } \lambda = (n) \\ 2(q+1), & \text{if } \lambda = (p+q, p) \\ q+1, & \text{if } \lambda = (p+q, p, 1) \\ 0 & \text{in all other cases} \end{cases}.$$

Notice that if $\alpha = 0$, $\mathrm{var}^L(UT_2^\eta) = \mathrm{var}^L(UT_2^\delta)$. Thus by Theorem 11 and by following closely the proof of [4, Theorem 15], taking into account the due changes, we get the following.

Theorem 13 *If* $\alpha \neq 0$, *then* $\mathrm{var}^L(UT_2^\eta)$ *has almost polynomial growth. Otherwise it has no almost polynomial growth.*

Finally let us assume that L is a 2-dimensional metabelian Lie algebra and denote by UT_2^D the L-algebra UT_2 where L acts on it as the Lie algebra $\mathrm{Der}(UT_2)$. Giambruno and Rizzo in [4] proved the following result.

Theorem 14 ([4, Theorems 19 and 25])

1. $\mathrm{Id}^L(UT_2^D) = \langle [x,y]^\varepsilon - [x,y],\ x^\varepsilon y^\varepsilon,\ x^{\varepsilon^2} - x^\varepsilon,\ x^{\delta\varepsilon},\ x^{\varepsilon\delta} - x^\delta \rangle_{T_L}$.
2. $c_n^L(UT_2^D) = 2^{n-1}(n+2)$.
3. If $\chi_n^L(UT_2^D) = \sum_{\lambda\vdash n} m_\lambda^D \chi_\lambda$ *is the nth differential cocharacter of* UT_2^D*, then*

$$m_\lambda^D = \begin{cases} 2n+1, & \text{if } \lambda = (n) \\ 3(q+1), & \text{if } \lambda = (p+q, p) \\ q+1, & \text{if } \lambda = (p+q, p, 1) \\ 0 & \text{in all other cases} \end{cases}.$$

Since $x^\delta \equiv 0$ is a differential identity of UT_2^ε, $\mathrm{var}^L(UT_2^\varepsilon) \subseteq \mathrm{var}^L(UT_2^D)$. Then by Theorem 8, we have the following.

Theorem 15 ([4, Theorem 26]) $\mathrm{var}^L(UT_2^D)$ *has no almost polynomial growth.*

5 On Differential Identities of the Grassmann Algebra

In this section we present an example of infinite dimensional algebra with derivations of exponential growth.

Let L be a finite dimensional abelian Lie algebra and G the infinite dimensional Grassmann algebra over F. Recall that G is the algebra generated by 1 and a countable set of elements $e_1, e_2, \ldots$ subjected to the condition $e_i e_j = -e_j e_i$, for all $i, j \geq 1$.

Notice that G can be decomposed in a natural way as the direct sum of the subspaces

$$G_0 = \mathrm{span}_F\{e_{i_1} \ldots e_{i_{2k}} \mid i_1 < \cdots < i_{2k}, k \geq 0\}$$

and

$$G_1 = \mathrm{span}_F\{e_{i_1} \ldots e_{i_{2k+1}} \mid i_1 < \cdots < i_{2k+1}, k \geq 0\},$$

i.e., $G = G_0 \oplus G_1$.

Now consider the algebra G where L acts trivially on it. Since $x^\gamma \equiv 0$, for all $\gamma \in L$, is a differential identity of G, we are dealing with ordinary identities. Thus by Krakowski and Regev [10] we have the following results.

Theorem 16

1. $\mathrm{Id}^L(G) = \langle [x, y, z] \rangle_T$.
2. $c_n^L(G) = 2^{n-1}$.
3. $\chi_n^L(G) = \sum_{j=1}^{n} \chi_{(j,1^{n-j})}$.

Theorem 17 $\mathrm{var}^L(G)$ *has almost polynomial growth.*

Recall that if $g = e_{i_1} \dots e_{i_n} \in G$, the set $\mathrm{Supp}\{g\} = \{e_{i_1}, \dots, e_{i_n}\}$ is called the support of g. Let now $g_1, \dots, g_t \in G_1$ be such that $\mathrm{Supp}\{g_i\} \cap \mathrm{Supp}\{g_j\} = \emptyset$, for all $i, j \in \{1, \dots, t\}$. We set

$$\delta_i = 2^{-1} \operatorname{ad} g_i, \quad i = 1, \dots, t.$$

Then for all $g \in G$ we have

$$\delta_i(g) = \begin{cases} 0, & \text{if } g \in G_0 \\ g_i g, & \text{if } g \in G_1 \end{cases}, \quad i = 1, \dots, t.$$

Since for all $g \in G$, $[\delta_i, \delta_j](g) = 0$, $i, j \in \{1, \dots, t\}$, $L = \mathrm{span}_F\{\delta_1, \dots, \delta_t\}$ is a t-dimensional abelian Lie algebra of inner derivations of G. We shall denote by $\widetilde{G}$ the algebra G with this L-action.

Recall that for a real number x we denote by $\lfloor x \rfloor$ its integer part.

Theorem 18 ([13, Theorems 3 and 9])

1. $\mathrm{Id}^L(\widetilde{G}) = \langle [x, y, z], [x^{\delta_i}, y], x^{\delta_i \delta_j} \rangle_{T_L}$, $i, j = 1, \dots, t$.
2. $c_n^L(\widetilde{G}) = 2^t 2^{n-1} - \sum_{j=1}^{\lfloor t/2 \rfloor} \sum_{i=2j}^{t} \binom{t}{i} \binom{n}{i-2j}$.
3. *If* $\chi_n^L(\widetilde{G}) = \sum_{\lambda \vdash n} m_\lambda^L \chi_\lambda$ *is the nth differential cocharacter of* $\widetilde{G}$, *then*

$$m_\lambda^L = \begin{cases} \sum_{i=0}^{r} \binom{t}{i}, & \text{if } \lambda = (n-r+1, 1^{r-1}) \text{ and } r < t \\ 2^t, & \text{if } \lambda = (n-r+1, 1^{r-1}) \text{ and } r \geq t \\ 0 & \text{in all other cases} \end{cases}.$$

Recall that two functions $\varphi_1(n)$ and $\varphi_2(n)$ are asymptotically equal and we write $\varphi_1(n) \approx \varphi_2(n)$ if $\lim_{n \to \infty} \varphi_1(n)/\varphi_2(n) = 1$. Then the following corollary is an obvious consequence of the previous theorem.

Corollary 1 $c_n^L(\widetilde{G}) \approx 2^t 2^{n-1}$.

Notice that by Corollary 1 $\mathrm{var}^L(\widetilde{G})$ has exponential growth, nevertheless it has no almost polynomial growth. In fact, the Grassmann algebra G (ordinary case) is an algebra with L-action where δ_i, $i = 1, \dots, t$, acts trivially on G, i.e., $x^{\delta_i} \equiv 0$, $i = 1, \dots, t$, are differential identities of G. Then it follows that $G \in \mathrm{var}^L(\widetilde{G})$, but by Theorem 16 $c_n^L(G) = 2^{n-1}$. Thus we have the following result.

Theorem 19 ([13, Theorem 6]) $\mathrm{var}^L(\widetilde{G})$ *has no almost polynomial growth.*

References

1. Benanti, F., Giambruno, A., Sviridova, I.: Asymptotics for the multiplicities in the cocharacters of some PI-algebras. Proc. Amer. Math. Soc. **132**(3), 669–679 (2004)
2. Coelho, S.P., Polcino Milies, C.: Derivations of upper triangular matrix rings. Linear Algebra Appl. **187**, 263–267 (1993)
3. Giambruno, A., La Mattina, D.: PI-algebras with slow codimension growth. J. Algebra **284**(1), 371—391 (2005)
4. Giambruno, A., Rizzo, C.: Differential identities, 2×2 upper triangular matrices and varieties of almost polynomial growth. J. Pure Appl. Algebra **223**(4), 1710–1727 (2019)
5. Giambruno, A., Zaicev, M.: Polynomial Identities and Asymptotic Methods. Mathematical Surveys and Monographs, vol. 122. AMS, Providence (2005)
6. Gordienko, A.S.: Asymptotics of H-identities for associative algebras with an H-invariant radical. J. Algebra **393**, 92–101 (2013)
7. Gordienko, A.S., Kochetov, M.V.: Derivations, gradings, actions of algebraic groups, and codimension growth of polynomial identities. Algebr. Represent. Th. **17**(2), 539–563 (2014)
8. Kemer, A.R.: Varieties of finite rank. In: Proceedings of the 15-th All the Union Algebraic Conference, Krasnoyarsk (in Russian), vol. 2, p. 73 (1979).
9. Kharchenko, V.K.: Differential identities of semiprime rings. Algebra Logic **18**, 86–119 (1979)
10. Krakowski, D., Regev, A.: The polynomial identities of the Grassmann algebra. Trans. Amer. Math. Soc. **181**, 429–438 (1973)
11. Malcev, J.N.: A basis for the identities of the algebra of upper triangular matrices. Algebra i Logika **10**, 393–400 (1971)
12. Procesi, C.: Lie Groups: An Approach Through Invariants and Representations, Universitext. Springer, New York (2006)
13. Rizzo, C.: The Grassmann algebra and its differential identities. Algebr. Represent. Theory **23**(1), 125–134 (2020)

Derived Lengths of Symmetric Poisson Algebras

Salvatore Siciliano

Abstract Let L be a Lie algebra over a field of positive characteristic. We survey the known results about the Lie structure of the symmetric Poisson algebra $S(L)$ and the truncated symmetric Poisson algebra $\mathbf{s}(L)$ of L. In particular, some results about the derived lenghts of $\mathbf{s}(L)$ are discussed.

Keywords Symmetric Poisson algebra · Truncated symmetric Poisson algebra · Poisson identity · Derived length · Metabelian Lie algebra.

1 Introduction

We recall that a Poisson algebra over a field $\mathbb{F}$ is a triple $(A, \cdot, \{,\})$ where A is a commutative associative $\mathbb{F}$-algebra with unity, $(A, \{,\})$ is a Lie algebra, and the two operations are related by the Leibniz rule, that is, for all $a, b, c \in A$ one has

$$\{a \cdot b, c\} = \{a, c\} \cdot b + a \cdot \{b, c\}.$$

Poisson algebras have many applications in algebra, differential geometry and mathematical physics, and attracted a lot of attention over the decades.

Now, for a Lie algebra L over $\mathbb{F}$, we identify the symmetric algebra $S(L)$ of L with the polynomial ring $\mathbb{F}[x_1, x_2, \ldots]$, where $x_1, x_2, \ldots$ is an $\mathbb{F}$-basis of L over $\mathbb{F}$. By linearity and the Leibniz rule, the Lie bracket of L can be uniquely extended to a Poisson bracket of $S(L)$ so that this commutative algebra becomes a Poisson algebra, called the *symmetric Poisson algebra* of L. Moreover, if the ground field has positive characteristic p, then the Poisson bracket of $S(L)$ naturally induces a Poisson bracket on $\mathbf{s}(L) = S(L)/I$, where I is the ideal generated by the elements x^p with $x \in L$. The Poisson algebra $\mathbf{s}(L)$ is called the *truncated symmetric Poisson algebra* of L.

S. Siciliano (✉)
Dipartimento di Matematica e Fisica "Ennio De Giorgi", Università del Salento, Lecce, Italy
e-mail: salvatore.siciliano@unisalento.it

O. M. Di Vincenzo, A. Giambruno (eds.), *Polynomial Identities in Algebras*, Springer INdAM Series 44, https://doi.org/10.1007/978-3-030-63111-6_21

Poisson identities of symmetric Poisson algebras of Lie algebras started to be investigated by Kostant [6], Shestakov [15], and Farkas [3, 4]. In [4] Farkas showed that, in characteristic zero, $S(L)$ satisfies a nontrivial Poisson identity if and only if L contains an abelian subalgebra of finite codimension. Some years later, in [5] Giambruno and Petrogradsky generalized Farkas' result to arbitrary characteristic. Furthermore, in the same paper they established when the truncated symmetric Poisson algebra of a restricted Lie algebra satisfies a nontrivial multilinear Poisson identical relation. In [9], Monteiro Alves and Petrogradsky studied the Lie identities of $S(L)$ and $\mathbf{s}(L)$. In particular, they determined necessary and sufficient conditions on L such that $S(L)$ or $\mathbf{s}(L)$ is Lie nilpotent, studied the Lie nilpotence class of $\mathbf{s}(L)$ and, in odd characteristic, established when $S(L)$ and $\mathbf{s}(L)$ are solvable. Further developments of these topics have been recently carried out by the author in [18]. It should be mentioned that similar problems have been also considered in other settings, such as restricted enveloping algebras (see [2, 13, 16, 17, 19, 20]) and group algebras (see e.g. [8, 10, 14]).

In this note we survey the known results about solvable (truncated) symmetric Poisson algebras and their derived lengths. In Sect. 3 we recall some theorems about the Lie structure of ordinary and restricted enveloping algebras, which originally motivated the present subject. In Sect. 4 we summarize results on the existence of nontrivial Poisson identities in symmetric and truncated symmetric Poisson algebras, and in Sect. 5 we consider Lie nilpotence and solvability of these Poisson algebras. Finally, in Sect. 6 some results on the derived lengths of a truncated symmetric Poisson algebras are collected.

2 Definitions and Notation

We fix some notation and terminology. Let $\mathbb{F}$ be a field. We denote by $\langle S\rangle_{\mathbb{F}}$ the subspace spanned by a subset S of a $\mathbb{F}$-vector space. For a Lie algebra L, we use the symbol $Z(L)$ for denoting the center of L. The terms of the derived series of L are defined by $\delta_0(L) = L$ and $\delta_n(L) = [\delta_{n-1}(L), \delta_{n-1}(L)]$ for $n > 0$. Moreover, we denote by $\gamma_n(L)$ $(n \geq 1)$ the terms of the descending central series of L. The derived subalgebra $\gamma_2(L) = \delta_1(L)$ of L will be also denoted by L'. For every $x \in L$, the adjoint map of x is defined by $\operatorname{ad} x : L \to L, a \mapsto [x, a]$.

Let A be a unital associative algebra over a field $\mathbb{F}$. Then A can be regarded as a Lie algebra via the Lie bracket defined by $[x, y] = xy - yx$, for all $x, y \in A$. Longer Lie products in A are interpreted using the left-normed convention.

We say that A is *Lie nilpotent* when A is nilpotent as a Lie algebra. The algebra A is *bounded Lie Engel* if there exists a positive integer n such that A satisfies the identity $[x, y, \dots, y] = 0$, where y appears n times in the expression. The nth *upper Lie power* of A is the ideal defined inductively by $A^{(1)} = A$ and $A^{(i)} = [A^{(i-1)}, A]A$. We say that A is *strongly Lie nilpotent* if $A^{(i)} = 0$, for some positive integer i.

Moreover, we say that A is *Lie solvable* if A is solvable as a Lie algebra. The upper derived series of A is defined by setting $\tilde{\delta}_0(A) = A$ and $\tilde{\delta}_n(A) = [\tilde{\delta}_{n-1}(A), \tilde{\delta}_{n-1}(A)] \cdot A$ for every $n > 0$. We say that A is *strongly solvable* if $\tilde{\delta}_n(A) = 0$ for some n. Obviously, strong Lie nilpotence (strong solvability, respectively) implies Lie nilpotence (solvability, respectively), but the converse is in general not true.

Let X be a set. By the *free Poisson algebra on* X we mean a Poisson algebra $F(X)$ together with a map $i : X \to F(X)$ such that for every map $f : X \to B$ into a Poisson algebra B there exists a unique Poisson algebra homomorphism $\theta : F(X) \to B$ such that the following diagram commutes

$$\begin{array}{ccc} X & \xrightarrow{i} & F(X) \\ f\downarrow & \swarrow\theta & \\ B & & \end{array}$$

It was shown by Shestakov in [15] that if $L(X)$ is the free Lie algebra on X, then $S(L(X))$ is a free Poisson algebra on X.

One says that a Poisson algebra P *satisfies a nontrivial Poisson identity* if there exists a nonzero element in the free Poisson algebra of countable rank which vanishes under any substitution in P.

3 The Lie Structure of Enveloping Algebras

As the study of the Lie structure of symmetric and truncated symmetric Poisson algebras is also motivated by similar problems for ordinary and restricted enveloping algebras, in this section we illustrate the picture for these algebras.

Let L be a restricted Lie algebra over a field of characteristic $p > 0$ and denote by $u(L)$ the restricted enveloping algebra of L. We recall that a subset S of L is said to be *p-nilpotent* if there exists a positive integer n such that $x^{[p]^n} = 0$ for every $x \in S$. The characterization of restricted enveloping algebras has been obtained by Passman in [11] and, independently, by Petrogradsky in [12]. Their result can be stated as follows:

Theorem 1 ([11, 12]) *Let L be a restricted Lie algebra over a field of characteristic $p > 0$. Then the restricted enveloping algebra $u(L)$ satisfies a polynomial identity if and only if L has restricted subalgebras $B \subseteq A$ such that:*

(i) $\dim L/A < \infty$ *and* $\dim B < \infty$*;*
(ii) *A/B is abelian and B is central in A;*
(iii) *B is p-nilpotent.*

The conditions under which $u(L)$ is Lie nilpotent, bounded Lie Engel, or Lie solvable in odd characteristic were determined by Riley and Shalev in the following theorems.

Theorem 2 ([13]) *Let L be a restricted Lie algebra over a field of characteristic $p > 0$. The following statements are equivalent:*

1. *$u(L)$ is Lie nilpotent;*
2. *$u(L)$ is strongly Lie nilpotent;*
3. *L is nilpotent and L' is finite-dimensional and p-nilpotent.*

Theorem 3 ([13]) *Let L be a restricted Lie algebra over a field of characteristic $p > 0$. Then $u(L)$ is bounded Lie Engel if and only if L is nilpotent, L' is p-nilpotent, and L contains a restricted ideal I such that L/I and I' are finite-dimensional.*

Theorem 4 *Let L be a restricted Lie algebra over a field of characteristic $p > 2$. The following statements are equivalent:*

1. *$u(L)$ is Lie solvable;*
2. *$u(L)$ is strongly Lie solvable;*
3. *L' is finite-dimensional and p-nilpotent.*

The equivalence of (1) and (3) in Theorem 4 is shown in [13] whereas it is shown in [16] that (2) and (3) are equivalent for all $p > 0$. On the other hand, the characterization of Lie solvable restricted enveloping algebras in characteristic 2 has been settled quite recently. We recall that a restricted Lie algebra is said to be *strongly abelian* if it is abelian and its power mapping is trivial. In a joint paper with Usefi, the following theorem was proved:

Theorem 5 ([20]) *Let L be a restricted Lie algebra over a field $\mathbb{F}$ of characteristic 2. Let $\bar{\mathbb{F}}$ be the algebraic closure of $\mathbb{F}$ and set $\mathfrak{L} = L \otimes_{\mathbb{F}} \bar{\mathbb{F}}$. Then $u(L)$ is Lie solvable if and only if $\mathfrak{L}$ has a finite-dimensional 2-nilpotent restricted ideal I such that $\bar{\mathfrak{L}} = \mathfrak{L}/I$ satisfies one of the following conditions:*

(i) *$\bar{\mathfrak{L}}$ has an abelian restricted ideal of codimension at most 1;*
(ii) *$\bar{\mathfrak{L}}$ is nilpotent of class 2 and $\dim \bar{\mathfrak{L}}/Z(\bar{\mathfrak{L}}) = 3$;*
(iii) *$\bar{\mathfrak{L}} = \langle x_1, x_2, y\rangle_{\bar{\mathbb{F}}} \oplus Z(\bar{\mathfrak{L}})$, where $[x_1, y] = x_1$, $[x_2, y] = x_2$, and $[x_1, x_2] \in Z(\bar{\mathfrak{L}})$;*
(iv) *$\bar{\mathfrak{L}} = \langle x, y\rangle_{\bar{\mathbb{F}}} \oplus H \oplus Z(\bar{\mathfrak{L}})$, where H is a strongly abelian finite-dimensional restricted subalgebra of $\bar{\mathfrak{L}}$ such that $[x, y] = x$, $[y, h] = h$, and $[x, h] \in Z(\bar{\mathfrak{L}})$ for every $h \in H$;*
(v) *$\bar{\mathfrak{L}} = \langle x, y\rangle_{\bar{\mathbb{F}}} \oplus H \oplus Z(\bar{\mathfrak{L}})$, where H is a finite-dimensional abelian subalgebra of $\bar{\mathfrak{L}}$ such that $[x, y] = x$, $[y, h] = h$, $[x, h] \in Z(\bar{\mathfrak{L}})$, and $[x, h]^{[2]} = h^{[2]}$, for every $h \in H$.*

Note that the cases (ii)–(v) can occur only when L' is finite-dimensional. In other words, if $u(L)$ is Lie solvable and L' is infinite-dimensional, then L has a restricted ideal of codimension at most 1 whose derived subalgebra is finite-dimensional and 2-nilpotent.

Now, let L be an ordinary Lie algebra over an arbitrary field $\mathbb{F}$ and denote by $U(L)$ the universal enveloping algebra of L. Latyšev in [7] proved that over a field of characteristic zero, $U(L)$ satisfies a polynomial identity if and only if L

is abelian. Subsequently, Bahturin in [1] extended Latyshev's result to the positive characteristic:

Theorem 6 ([1]) *Let L be a Lie algebra over a field of characteristic $p > 0$. Then $U(L)$ satisfies a polynomial identity if and only if the following conditions are satisfied:*

1. *L has an abelian ideal of finite codimension;*
2. *all inner derivations adx, $x \in L$, are algebraic of bounded degree.*

In [13] Riley and Shalev showed that if L is defined over a field of characteristic different from 2, then $U(L)$ is Lie solvable only when L is abelian. This is no longer true in characteristic 2. However, one can apply Theorem 5 with respect to the restricted Lie algebra $\hat{L}$ consisting of all the primitive elements of the Hopf algebra $U(L)$. In this way, by using the fact that $u(\hat{L}) \cong U(L)$, necessary and sufficient conditions on L such that $U(L)$ is Lie solvable can be obtained, thereby completing the classification also in the ordinary case. Indeed, in [20] the following result was proved:

Theorem 7 ([20]) *Let L be a Lie algebra over a field $\mathbb{F}$ of characteristic 2. Then $U(L)$ is Lie solvable if and only if one of the following conditions is satisfied:*

(i) *L contains an abelian ideal of codimension 1 and, for every $x \in L$, one has $(adx)^2 = \lambda adx$ for some $\lambda \in \mathbb{F}$;*
(ii) *L is nilpotent of class 2 and $\dim_{\mathbb{F}} L/Z(L) = 3$;*
(iii) *$L = \langle x_1, x_2, y\rangle_{\mathbb{F}} \oplus Z(L)$, where $[x_1, y] = x_1$, $[x_2, y] = x_2$, and $[x_1, x_2] \in Z(L)$.*

4 Symmetric Poisson Algebras Satisfying a Poisson Identity

In characteristic zero, the characterization of symmetric Poisson algebras satisfying a nontrivial Poisson identity is given by the following result of Farkas:

Theorem 8 ([4]) *Let L be a Lie algebra over a field of characteristic zero. Then $S(L)$ satisfies a nontrivial Poisson identity if and only if L contains an abelian subalgebra of finite codimension.*

Afterwards, in [5] Giambruno and Petrogradsky extended Farkas' result to Lie algebras defined over arbitrary fields:

Theorem 9 ([5]) *Let L be a Lie algebra over an arbitrary field. Then $S(L)$ satisfies a nontrivial multilinear Poisson identity if and only if L contains an abelian subalgebra of finite codimension.*

Furthermore, Giambruno and Petrogradsky established when the truncated symmetric Poisson algebra of a Lie algebra satisfies a nontrivial multilinear Poisson identical relation. In fact, they proved the following theorem for restricted Lie algebras, but their proof holds for arbitrary Lie algebras as well.

Theorem 10 ([5]) *Let L be a restricted Lie algebra over a field of characteristic $p > 0$. Then $\mathbf{s}(L)$ satisfies a nontrivial multilinear Poisson identity if and only if there exists a restricted ideal H of L such that*

1. $\dim L/H < \infty$;
2. $\dim H' < \infty$;
3. *H is nilpotent of class 2.*

5 Lie Nilpotence and Solvability of $S(L)$ and $\mathbf{s}(L)$

In this section we focus on Lie identities of symmetric Poisson algebras. Let P be a Poisson algebra. One says that P is *Lie nilpotent* if P is nilpotent as a Lie algebra. In this case, the minimal n such that $\gamma_{n+1}(L) = 0$ is called the *Lie nilpotence class* of L. The upper Lie power series of P is the chain of Poisson ideals of P defined by $P^{(1)} = P$ and $P^{(n)} = \{P^{(n-1)}, P^{(n-1)}\} \cdot P$ for every $n > 1$. The Poisson algebra P is said to be *strongly Lie nilpotent* of class c if $P^{(c+1)} = 0$ and $P^{(c)} \neq 0$.

Moreover, we say that P is *solvable* if P is solvable as a Lie algebra. In this case, the minimal n such that $\delta_n(P) = 0$ is called the *derived length* of P and denoted by $\mathrm{dl}_{Lie}(P)$. In particular, P is said to be *metabelian* if $\delta_2(P) = 0$. The upper derived series of P is defined by setting $\tilde{\delta}_0(P) = P$ and $\tilde{\delta}_n(P) = \{\tilde{\delta}_{n-1}(P), \tilde{\delta}_{n-1}(P)\} \cdot P$ for every $n > 0$. Note that $\tilde{\delta}_n(P)$ is a Poisson ideal of P for every n. The Poisson algebra P is said to be *strongly solvable* if $\tilde{\delta}_n(P) = 0$ for some n. In this case, the minimal n with such a property is called the *strong derived length* of P and denoted by $\mathrm{dl}^{Lie}(P)$.

The Lie properties of symmetric and truncated symmetric Poisson algebras have been studied by Monteiro Alves and Petrogradsky in [9]. In the following theorem, a characterization of Lie nilpotence of $\mathbf{s}(L)$ is obtained:

Theorem 11 ([9]) *Let L be a Lie algebra over a field of characteristic $p > 0$. The following conditions are equivalent:*

(1) *$\mathbf{s}(L)$ is strongly Lie nilpotent;*
(2) *$\mathbf{s}(L)$ is Lie nilpotent;*
(3) *L is nilpotent and* $\dim L' < \infty$.

In the next result, an explicit formula for the strong Lie nilpotence class of $\mathbf{s}(L)$ is provided.

Theorem 12 ([9]) *Let L be a Lie algebra over a field of characteristic $p > 3$. The following numbers are equal:*

(1) *the strong Lie nilpotence class of $\mathbf{s}(L)$;*
(2) *the Lie nilpotence class of $\mathbf{s}(L)$;*

(3)

$$1 + (p-1)\sum_{n\geq 1} n \cdot \dim(\gamma_{n+1}(L)/\gamma_{n+2}(L)).$$

In cases $p = 2, 3$, the numbers (1) and (3) remain equal.

In characteristics $p = 2, 3$, the possible equality of (2) and (3) in the previous theorem remains unclear, and (3) only yields an upper bound for the Lie nilpotence class in these cases.

As shown in the next result, a symmetric Poisson algebra $S(L)$ is strongly solvable only when L is abelian, and the same conclusion holds for solvability provided that the ground field has characteristic different from 2.

Theorem 13 ([9]) *Let L be a Lie algebra over a field $\mathbb{F}$, and $S(L)$ its symmetric Poisson algebra. Then the following statements hold.*

(1) *$S(L)$ is strongly solvable if and only if L is abelian.*
(2) *If $\mathbb{F}$ has characteristic different from 2, then $S(L)$ is solvable if and only if L is abelian.*

Finally, in [9] the following characterization of solvable truncated symmetric Poisson algebras over fields of odd characteristic is obtained:

Theorem 14 ([9]) *Let L be a Lie algebra over a field of characteristic $p \geq 3$. Consider its truncated symmetric Poisson algebra $\mathbf{s}(L)$. The following conditions are equivalent:*

(1) *$\mathbf{s}(L)$ is strongly solvable;*
(2) *$\mathbf{s}(L)$ is solvable;*
(3) *L is solvable and* $\dim L' < \infty$.

In the case $p = 2$, conditions (1) and (3) remain equivalent.

In characteristic 2, solvability and strong solvability are not equivalent properties for $S(L)$ or $\mathbf{s}(L)$ (see [9, Lemmas 11.1 and 11.2]). In this respect, Monteiro Alves and Petrogradsky conjectured in [9, §5.3] that a symmetric Poisson algebra $S(L)$ over a field $\mathbb{F}$ of characteristic 2 is solvable if and only if $L = \langle x\rangle_{\mathbb{F}} \oplus A$, where A is an abelian ideal of L on which $\operatorname{ad} x$ acts algebraically. However, this conjecture was disproved by the author in [18]. In fact, one has

Proposition 1 ([18]) *Let L be a Lie algebra over a field of characteristic 2. If L is nilpotent of class 2 and* $\dim L/Z(L) = 3$, *then $S(L)$ is solvable of derived length 3.*

For an explicit counterexample to the aforementioned conjecture in [9, §5.3], let L be the relatively free nilpotent Lie algebra of class 2 on three generators over a field of characteristic 2. Then L does not contain any abelian ideal of codimension 1 in L. On the other hand, by Proposition 1, $S(L)$ is solvable as $Z(L)$ has codimension 3 in L.

As for restricted enveloping algebras, solvability of $S(L)$ and $\mathbf{s}(L)$ in characteristic 2 turns out to be a rather complicated problem, whose solution will appear in a forthcoming paper.

6 Derived Lengths of $\mathbf{s}(L)$

In this section we deal with the derived lengths of truncated symmetric Poisson algebras. While Theorem 12 provides a rather satisfactory method to determine the Lie nilpotence classes of $\mathbf{s}(L)$, the computation of the derived lengths is a more difficult task. A lower and a upper bound for the strong derived length is given by the following result. As usual, we will denote by $\lceil t \rceil$ the upper integral part of the real number t.

Proposition 2 ([18]) *Let L be a Lie algebra over a field $\mathbb{F}$ of characteristic $p > 0$. If $\mathbf{s}(L)$ is strongly solvable, then*

$$\lceil \log_2(2 + (p-1)d) \rceil \leq dl^{Lie}(\mathbf{s}(L)) \leq 1 + \sum_{i \geq 1} \lceil \log_2(1 + (p-1)d_i) \rceil,$$

where $d = \dim L'$ and $d_i = \dim \delta_i(L)/\delta_{i+1}(L)$ for every $i \geq 1$.

When L is metabelian, the previous result almost determines the strong derived length of $\mathbf{s}(L)$, as in this case the difference between the upper and the lower bound is at most 1. In particular, if L is nilpotent of class 2, then we obtain the exact value of $\mathrm{dl}^{Lie}(\mathbf{s}(L))$:

Corollary 1 *Let L be a nilpotent Lie algebra of class 2 over a field $\mathbb{F}$ of characteristic $p > 0$. If $\mathbf{s}(L)$ is strongly solvable, then*

$$dl^{Lie}(\mathbf{s}(L)) = \lceil \log_2(2 + (p-1) \dim L') \rceil.$$

If P is a strongly solvable Poisson algebra, then we clearly have $\mathrm{dl}_{Lie}(P) \leq \mathrm{dl}^{Lie}(P)$. For truncated symmetric Poisson algebras, the derived lengths can be actually different. For instance, let $n > 6$ and consider the Lie algebra L over a field $\mathbb{F}$ of characteristic 2 having an $\mathbb{F}$-basis $x, y_1, \ldots, y_n, z_1, \ldots, z_n$ such that $[x, y_i] = z_i$ and other commutators are zero. By [9, Lemma 11.2] and Corollary 1 we have

$$\mathrm{dl}^{Lie}(\mathbf{s}(L)) = \lceil \log_2(2 + (p-1) \dim L') \rceil > 3 = \mathrm{dl}_{Lie}(\mathbf{s}(L)).$$

Lie algebras whose truncated symmetric Poisson algebras is metabelian are described in the following theorem:

Theorem 15 ([18]) *Let L be a nonabelian Lie algebra over a field $\mathbb{F}$ of characteristic $p > 0$. Then $\mathbf{s}(L)$ is metabelian if and only if one of the following conditions is satisfied:*

(1) $p = 3$, *L is nilpotent and* $\dim L' = 1$;
(2) $p = 2$ *and* $\dim L' = 1$;
(3) $p = 2$, *L is nilpotent of class 2 and* $\dim L' = 2$.

The next result provides a lower bound for the derived length of $\mathbf{s}(L)$ for a nonabelian Lie algebra L.

Theorem 16 ([18]) *Let L be a nonabelian Lie algebra over a field of characteristic $p > 0$. If $\mathbf{s}(L)$ is solvable, then $\mathrm{dl}_{Lie}(\mathbf{s}(L)) \geq \lceil \log_2(p+1) \rceil$. Moreover, if equality holds and $p > 2$, then L is nilpotent.*

Let L be a nonabelian Lie algebra over a field of characteristic $p > 0$. We say that $\mathbf{s}(L)$ has *minimal* derived length if the lower bound in Theorem 16 is attained. We have the following characterization:

Theorem 17 ([18]) *Let L be a nonabelian Lie algebra over a field of characteristic $p > 0$. Then $\mathbf{s}(L)$ has minimal derived length if and only if one of the following conditions is satisfied:*

(1) $p > 2$, *L is nilpotent and* $\dim L' = 1$;
(2) $p = 2$ *and* $\dim L' = 1$;
(3) $p = 2$, *L is nilpotent of class 2 and* $\dim L' = 2$.

Plainly, the lower bound in Theorem 17 also represents the smallest possible value for the strong derived length of a nonabelian truncated symmetric Poisson algebra over a field of characteristic $p > 0$. We say that $\mathbf{s}(L)$ has *minimal* strong derived length if this bound is attained. A combination Theorem 17, Proposition 2 and Corollary 1 yields

Corollary 2 *Let L be a nonabelian Lie algebra over a field of characteristic $p > 0$. Then $\mathbf{s}(L)$ has minimal derived length if and only it has minimal strong derived length.*

References

1. Bahturin, Yu.A.: Identities in the universal envelopes of Lie algebras. J. Austral. Math. Soc. **18**, 19–27 (1974)
2. Catino, F., Siciliano, S., Spinelli, E.: Restricted enveloping algebras with minimal Lie derived length. Algebr. Represent. Theory **13**, 653–660 (2010)
3. Farkas, D.R.: Poisson polynomial identities. Comm. Algebra **26**, 401–416 (1998)
4. Farkas, D.R.: Poisson polynomial identities. II. Arch. Math. (Basel) **72**, 252–260 (1999)
5. Giambruno, A., Petrogradsky, V.: Poisson identities of enveloping algebras. Arch. Math. (Basel) **87**, 505–515 (2006)

6. Kostant, B.: A Lie algebra generalization of the Amitsur-Levitski theorem. Adv. Math. **40**, 155–175 (1981)
7. Latyshev, V.N.: Two remarks on PI-algebras. Sibirsk. Mat. Zh. **4**, 1120–1121 (1963)
8. Levin, F., Rosenberger, G.: On Lie metabelian group rings. Results Math. **26**, 83–88 (1994)
9. Monteiro Alves, I.Z., Petrogradsky, V.M.: Lie structure of truncated symmetric Poisson algebras. J. Algebra **488**, 244–281 (2017)
10. Passi, I.B.S., Passman, D.S., Sehgal, S.K.: Lie solvable group rings. Canad. J. Math. **25**, 748–757 (1973)
11. Passman, D.S.: Enveloping algebras satisfying a polynomial identity. J. Algebra **134**, 469–490 (1990)
12. Petrogradsky, V.M.: Existence of identities in the restricted enveloping algebra. Math. Zametki **49**, 84–93 (1991)
13. Riley, D.M., Shalev, A.: The Lie structure of enveloping algebras. J. Algebra **162**, 46–61 (1993)
14. Shalev, A.: The derived length of Lie soluble group rings I. J. Pure Appl. Algebra **78**, 291–300 (1992)
15. Shestakov, I.P.: Quantization of Poisson superalgebras and speciality of Jordan Poisson superalgebras. Algebra Logika **32**, 571–584 (1993); Translation Algebra and Logic **32**, 309–317 (1994)
16. Siciliano, S.: Lie derived lengths of restricted universal enveloping algebras. Publ. Math. Debrecen **68**, 503–513 (2006)
17. Siciliano, S.: On Lie solvable restricted enveloping algebras. J. Algebra **314**, 226–234 (2007)
18. Siciliano, S.: Solvable symmetric Poisson algebras and their derived lengths. J. Algebra **543**, 98–110 (2020)
19. Siciliano, S., Spinelli, E.: Lie metabelian restricted universal enveloping algebras. Arch. Math. (Basel) **84**, 398–405 (2005)
20. Siciliano, S., Usefi, H.: Lie solvable enveloping algebras of characteristic two. J. Algebra **382**, 314–331 (2013)

Group and Polynomial Identities in Group Rings

Ernesto Spinelli

To Antonio Giambruno on his 70th birthday.

Abstract In the 1980s Brian Hartley conjectured that if the unit group, $\mathcal{U}(FG)$, of a torsion group ring FG satisfies a group identity, then FG satisfies a polynomial identity. The aim of this survey is to review the most relevant results that arose from the proof of this conjecture and discuss some recent developments and open questions concerning $*$-group identities for $\mathcal{U}(FG)$ and group identities for the subgroup of its unitary units.

Keywords Group Algebras · Unit Group · Group Identities · Polynomial Identities · Involution

1 Introduction

Throughout let F be a field of characteristic $p \geq 0$ and G a group. Write $\mathcal{U}(FG)$ for the unit group of the group ring FG. We say that a subset $S \subseteq \mathcal{U}(FG)$ satisfies a *group identity* if there exists a non-trivial reduced word $w(x_1, \ldots, x_n)$ in the free group on countably many generators, $\langle x_1, x_2, \ldots \rangle$, such that $w(g_1, \ldots, g_n) = 1$ for all $g_1, \ldots, g_n \in S$.

In the attempt to connect the algebraic structure of FG with the group structure of its unit group, Brian Hartley made the following conjecture.

The author Ernesto Spinelli is member of INDAM-GNSAGA.

E. Spinelli (✉)
Dipartimento di Matematica "G. Castelnuovo", Università degli Studi di Roma "La Sapienza", Roma, Italy
e-mail: spinelli@mat.uniroma1.it

O. M. Di Vincenzo, A. Giambruno (eds.), *Polynomial Identities in Algebras*, Springer INdAM Series 44, https://doi.org/10.1007/978-3-030-63111-6_22

Conjecture 1.1 *Let G be a torsion group. If $\mathcal{U}(FG)$ satisfies a group identity, then FG satisfies a polynomial identity.*

We recall that a subset $V \subseteq FG$ satisfies a *polynomial identity* if there exists a non-zero element $f(x_1, \ldots, x_m)$ in the free algebra on non-commuting indeterminates $F\{x_1, x_2, \ldots\}$ such that $f(a_1, \ldots, a_m) = 0$ for all $a_i \in V$.

When F is infinite, after a first result of Gonçalves and Mandel [13] dealing with the special case of semigroup identities (that is, identities of the form $x_{i_1} \cdots x_{i_k} = x_{j_1} \cdots x_{j_l}$), Giambruno, Jespers and Valenti [6] confirmed Hartley's Conjecture under the assumption that G does not contain elements of p-power order if $p > 0$. Some years later Giambruno, Sehgal and Valenti [7] were able to remove the hypothesis on G.

Finally, modifying the original proof of [7], Liu [22] positively answered the question for fields of any size.

Many years before, Isaacs and Passman (see Corollaries 5.3.8 and 5.3.10 of [24]) characterized group rings satisfying a polynomial identity. Recalling that a group G is called p-abelian if its commutator subgroup, G', is a finite p-group, and that 0-abelian means abelian, their result was as follows.

Theorem 1.2 *The group ring FG satisfies a polynomial identity if, and only if, G has a p-abelian subgroup of finite index.*

From this theorem one deduces that if FG satisfies a polynomial identity, then $\mathcal{U}(FG)$ does not necessarily satisfy a group identity. In fact, one of the easiest consequences of the solution of Conjecture 1.1 is that, in characteristic 0, $\mathcal{U}(FG)$ satisfies a group identity if, and only if, G is abelian (see Corollary 1.2.21 of [17]). Thus, if you take any finite non-abelian group G, then $\mathbb{C}G$ satisfies a polynomial identity, but $\mathcal{U}(\mathbb{C}G)$ does not satisfy a group identity.

Anyway, the positive solution of Hartley's Conjecture was the crucial step leading to the establishment of necessary and sufficient conditions for $\mathcal{U}(FG)$ to satisfy a group identity. This was first done by Passman [25] for infinite fields and then by Liu and Passman [23] for arbitrary fields.

Theorem 1.3 *Let $p > 0$ and G a torsion group. If G' is a p-group, then the following are equivalent:*

(i) *$\mathcal{U}(FG)$ satisfies a group identity;*
(ii) *$\mathcal{U}(FG)$ satisfies the group identity $(x_1^{-1}x_2^{-1}x_1x_2)^{p^r} = 1$ for some positive integer r;*
(ii) *FG satisfies a polynomial identity and G' has bounded exponent.*

Theorem 1.4 *Let $p > 0$ and G a torsion group. If G' is not a p-group, then the following are equivalent:*

(i) *$\mathcal{U}(FG)$ satisfies a group identity;*
(ii) *$\mathcal{U}(FG)$ has bounded exponent;*
(ii) *FG satisfies a polynomial identity, G has bounded exponent and F is finite.*

It turns out that the solution for finite fields is different if G' is not a p-group, but, in any case, if the unit group of a torsion group ring satisfies a group identity, then it satisfies an identity of a particularly nice form.

Once that the torsion case was settled, it was natural to investigate what happens when the group G contains elements of infinite order. Here the situation is much more complicated because of the difficulty in handling the torsion-free part of the group. Indeed, for any such result, a restriction will occur for the sufficiency, pending a positive answer to the following celebrated conjecture by Kaplansky.

Conjecture 1.5 *Let G be a torsion-free group. Then $\mathcal{U}(FG)$ contains only trivial units, that is, units of the form λg, where $0 \neq \lambda \in F$ and $g \in G$.*

In this setting Hartley's Conjecture is not expected to hold in general. In fact, if G is isomorphic to the direct product of infinitely many copies of a non-abelian torsion-free nilpotent group, for any field F, $\mathcal{U}(FG)$ has only trivial units, and hence is nilpotent, but, according to Theorem 1.2, FG does not satisfy a polynomial identity. But it is true if one restricts the assumptions on G, as deduced from Theorem 5.5 of [9].

Theorem 1.6 *Let $p > 0$ and G a group with an element of infinite order and infinitely many p-elements. If $\mathcal{U}(FG)$ satisfies a group identity, then FG satisfies a polynomial identity.*

In [9] Giambruno, Sehgal and Valenti proved more, classifying group rings of non-torsion groups whose group of units satisfies a group identity. In more detail, they proved that under this assumption the torsion elements of G form a subgroup, T. For the converse, a suitable restriction on G/T was required, namely that it is a *unique product group*, that is, for every pair of non-empty sets H_1 and H_2 of G/T there exists an element $g \in G/T$ that can be uniquely written as $g = h_1h_2$ with $h_i \in H_i$, in order to force the units of $F(G/T)$ to be trivial. For the complete result (which is quite technical and split in several cases), we refer to the original paper or to Chapter 1 of [17]. We confine ourselves to report here the only part which does not require any restriction on the torsion-free part of G, as summarized in Theorem 1.5.16 of [17].

Theorem 1.7 *Let $p > 0$ and G a group with an element of infinite order and let the p-elements of G have unbounded exponent. Then $\mathcal{U}(FG)$ satisfies a group identity if, and only if, FG satisfies a polynomial identity and G' is a p-group of bounded exponent.*

All these results allowed researchers to solve problems open for decades, such as, for instance, the classification of group rings whose unit group is solvable, concluded by A. Bovdi [3] after a series of papers over many years beginning with Bateman [2]. For these and other results of the same type we refer to the monograph [17]. The aim of this note is to present some recent developments concerning group identities for symmetric and unitary units (with respect to an involution) of FG and discuss some open questions which naturally arise in these frameworks. To this end, we recall that for what concerns symmetric units one can find a comprehensive outline of the known results for the classical involution in [17]. For this reason, in the sequel we avoid reporting in detail what is already collected in that book.

2 $*$-Group Identities for $\mathcal{U}(FG)$

Assume that the group G is endowed with an involution $*$. The F-linear extension of $*$ to FG is an involution of FG, also denoted by $*$. An element $\alpha \in FG$ is said to be *symmetric* (with respect to $*$) if $\alpha^* = \alpha$. Write FG^+ for the set of symmetric elements and $\mathcal{U}^+(FG)$ for the set of symmetric units of FG. One of the problems of main interest is to understand if group identities satisfied by $\mathcal{U}^+(FG)$ can be lifted to $\mathcal{U}(FG)$ or, if this is not the case, how they influence the structure of FG.

Since the 1990s a lot of attention has been devoted to the *classical involution* of FG, that is, the one induced from the map $g \mapsto g^{-1}$ on G. In the last decade Giambruno, Polcino Milies and Sehgal considered involutions as above other than the classical one, and recently they framed the problem in a different setting, inspired by a classical result of Amitsur. Specifically, let R be an F-algebra with F-linear involution $*$. We say that R satisfies a $*$-polynomial identity if there exists a non-zero element $f(x_1, x_1^*, \ldots, x_m, x_m^*)$ in the free algebra with involution $F\{x_1, x_1^*, x_2, x_2^*, \ldots\}$ such that $f(a_1, a_1^*, \ldots, a_m, a_m^*) = 0$ for all $a_i \in R$. Obviously, if the symmetric elements of R satisfy the polynomial identity $f(x_1, \ldots, x_m)$, then R satisfies the $*$-polynomial identity $f(x_1+x_1^*, \ldots, x_m+x_m^*)$. Amitsur [1] proved that also the converse is true: indeed, if R satisfies a $*$-polynomial identity, then R satisfies a polynomial identity and, consequently, also R^+ does the same.

Following this direction, in [11] they considered $*$-group identities for $\mathcal{U}(FG)$. We say that $\mathcal{U}(FG)$ satisfies a $*$-group identity if there exists a non-trivial word $w(x_1, x_1^*, \ldots, x_n, x_n^*)$ in the free group with involution $\langle x_1, x_1^*, x_2, x_2^*, \ldots\rangle$ such that $w(a_1, a_1^*, \ldots, a_n, a_n^*) = 1$ for all $a_i \in \mathcal{U}(FG)$. It is clear that if $\mathcal{U}^+(FG)$ satisfies the group identity $w(x_1, \ldots, x_n)$, then $\mathcal{U}(FG)$ satisfies the $*$-group identity $w(x_1x_1^*, \ldots, x_nx_n^*)$. In the same paper the torsion case was settled proving a quite surprising result for the formulation of which we need the notion of an *SLC-group*. We recall that a non-abelian group G is said to be an LC-group (for *lack of commutativity*) if, for any pair of commuting elements $g, h \in G$, at least one among g, h and gh is central. According to Proposition III.3.6 of [15], G is an LC-group with a unique non-identity commutator if, and only if, $G/\zeta(G) \cong C_2 \times C_2$, where $\zeta(G)$ is the center of G. An LC-group G with involution $*$ is called a *special LC-group*, or *SLC-group*, if it has a unique non-identity commutator z, and for all $g \in G$ we have $g^* = g$ if $g \in \zeta(G)$, and otherwise $g^* = gz$. In group ring theory SLC-groups have a special role, since they occur in the characterization of group rings whose symmetric elements commute, as proved by Jespers and Ruiz Marin in [16]: in more detail, they stated that if $p \neq 2$ and G is a non-abelian group with involution linearly extended to FG, then FG^+ is commutative if, and only if, G is an SLC-group. In particular, if $*$ is the classical involution, G is a Hamiltonian 2-group.

The main result of [11], as formulated in the survey paper [18], is the following.

Theorem 2.1 *Let F be an infinite field of characteristic $p \neq 2$ and G a torsion group with involution linearly extended to FG. Then the following are equivalent:*

(i) *$\mathcal{U}(FG)$ satisfies a $*$-group identity;*
(ii) *$\mathcal{U}^+(FG)$ satisfies a group identity;*
(iii) *one of the following occurs:*
   (a) *$\mathcal{U}(FG)$ satisfies a group identity,*
   (b) *$p = 0$ and G is an SLC-group, or*
   (c) *$p > 2$, FG satisfies a polynomial identity, and G contains a $*$-invariant normal p-subgroup N of bounded exponent such that G/N is an SLC-group.*

For the sake of completeness, we recall that, under the same assumptions, Giambruno, Polcino Milies and Sehgal [10] had already provided necessary and sufficient condition so that $\mathcal{U}^+(FG)$ satisfies a group identity, and the same result was previously established by Giambruno, Sehgal and Valenti [8] for the classical involution.

According to the above statements, $*$-group identities on $\mathcal{U}(FG)$ do not force group identities on $\mathcal{U}(FG)$, but Hartley's Conjecture remains true under the weaker assumption that $\mathcal{U}(FG)$ satisfies a $*$-group identity.

Very recently in [12] the non-torsion case was investigated. Here the situation is much more involved and an analogue of Theorem 2.1 was proven with some restrictions upon G. For the rest of the section, let us denote by T the set of torsion elements of G, and by P that of p-elements, respectively. For the semiprime case the result is the following.

Theorem 2.2 *Let F be an infinite field of characteristic $p \neq 2$ and G a group with involution $*$ linearly extended to FG. Assume that G contains no 2-elements and T is a subgroup of G. If FG is semiprime and $\mathcal{U}(FG)$ satisfies a $*$-group identity, then*

(a) *T is an abelian p'-subgroup such that every idempotent of FT is central in FG (and, consequently, every subgroup of T is normal in G), and*
(b) *G/T satisfies a $*$-group identity.*

Conversely, if (a) holds and G/T is a unique product group satisfying a group identity, then $\mathcal{U}(FG)$ satisfies a group identity.

Assume now that F, G and $\mathcal{U}(FG)$ are as in Theorem 2.2, but FG is not necessarily semiprime. As stressed at the beginning of Section 4 of [12], P is a (normal) subgroup of G. The solution for the general case is split in two parts, just according to the structure of P, as shown in the following

Theorem 2.3 *Let F be an infinite field of characteristic $p \neq 2$ and G a group with involution $*$ linearly extended to FG. Assume that G contains no 2-elements, T is a subgroup of G and FG is not semiprime.*

(1) *If $\mathcal{U}(FG)$ satisfies a $*$-group identity and P is of bounded exponent, then*

   (a) *T/P is abelian and every idempotent of $F(T/P)$ is central in $F(G/P)$, and*
   (b) *G/T satisfies a $*$-group identity.*

   Conversely, if (a) holds and G/T is a unique product group satisfying a group identity, then $\mathcal{U}(FG)$ satisfies a group identity.

(2) *If $\mathcal{U}(FG)$ satisfies a $*$-group identity and P is of unbounded exponent, then*

   (a') *FG satisfies a polynomial identity, and*
   (b') *G' has bounded p-power exponent.*

   Conversely, if G satisfies (a') and (b'), then $\mathcal{U}(FG)$ satisfies a group identity.

Of particular interest is the following

Corollary 2.4 *Let F and G be as in Theorem 2.3. If P is of unbounded exponent, then $\mathcal{U}(FG)$ satisfies a $*$-group identity if, and only if, $\mathcal{U}(FG)$ satisfies a group identity.*

A first question naturally arises from the above mentioned results.

Question 2.5 *Let F be an infinite field of characteristic $p \neq 2$ and G a group with involution $*$ (and, if it helps, with no 2-elements) linearly extended to FG. Assume that $\mathcal{U}(FG)$ satisfies a $*$-group identity (or, if it helps, that $\mathcal{U}^+(FG)$ satisfies a group identity). Is it true that T is a subgroup of G?*

Sehgal and Valenti in [26] gave a positive answer in the case in which the symmetric units of FG with respect to the classical involution satisfy a group identity, and characterized when this happens (under the same restrictions on G/T discussed before Theorem 1.7). Among other things (for the complete result we refer to the original paper or to Chapter 2 of [17]), they proved that

- the statement of Theorem 1.6 remains true under the weaker assumption that $\mathcal{U}^+(FG)$ satisfies a group identity, and
- if FG is as in Theorem 1.7, $\mathcal{U}^+(FG)$ satisfies a group identity if, and only if, $\mathcal{U}(FG)$ satisfies a group identity.

In particular, unlike in [12], they allow the presence of 2-elements in G. This generates a second question.

Question 2.6 *How to modify the above Theorems if G contains 2-elements, even under the assumption that T is a subgroup of G (and, if it helps, that $\mathcal{U}^+(FG)$ satisfies a group identity).*

Let us briefly discuss the possible obstacles just in the semiprime case. First of all, we must be careful with Remark 3.3 of [12], as it will not work if g has even order. This becomes an issue in Lemma 3.8 of [12], as in the final part of the proof, we cannot be sure that $1 + \pi$ is not a zero divisor. But in any case, that result cannot extend to the case where H is an SLC-group. Indeed, assuming that F is algebraically closed, we see that FH must include non-commutative matrix rings among its Wedderburn components, and therefore it has idempotents that are not even central in FH, let alone in FG. And such a case can arise. If H is any finite SLC-group, then let $G = H \times \langle x \rangle$, where x is a symmetric element of infinite order. Then G is an SLC-group, and hence $\mathcal{U}^+(FG)$ is commutative. Why is this important? Let us come back to the classical involution. In Theorem 4 of [26] it was established that, if $p = 0$ and $\mathcal{U}^+(FG)$ satisfies a group identity, then T is either abelian or a Hamiltonian 2-group and every idempotent of FT is central in FG. According to the previously discussed characterization of group rings whose symmetric elements commute, one could expect to generalize the result of [26] just replacing Hamiltonian 2-groups with SLC-groups, but, as seen above, this cannot be the case.

3 Group Identities for Unitary Units of FG

It is natural to ask the same questions discussed for symmetric units of FG for the subgroup of its *unitary units*

$$Un(FG) := \{\alpha \mid \alpha \in FG,\ \alpha\alpha^* = 1\}.$$

But the picture is not as clear here, and just a few general results have been proved.

Assume for the rest of this section that FG is equipped with the classical involution. In [14] Gonçalves and Passman studied group rings whose unitary units contain no non-abelian free subgroup (in the paper they call a group satisfying this property *2-related*). In more detail, suppose that G is finite. If F is an absolute field, that is, algebraic over a finite field, then $\mathcal{U}(FG)$ is locally finite and, hence, $Un(FG)$ contains no non-abelian free subgroup. Therefore it has sense to ask the question when F is non-absolute (in other words, when either $p = 0$ or $p > 0$ and F has an element trascendental over its prime subfield). The main result they proved is the following

Theorem 3.1 *Let F be a non-absolute field of characteristic $p \neq 2$ and G a finite group. Then $Un(FG)$ contains no non-abelian free subgroup if, and only if,*

(a) *G has a normal Sylow p-subgroup P (by convention, $P = \{1\}$ if $p = 0$), and*
(b) *either $\overline{G} := G/P$ is abelian or it has an abelian subgroup $\overline{A}$ of index 2. Furthermore, if the latter occurs, then either $\overline{G} = \overline{A} \rtimes \langle \overline{y} \rangle$ is dihedral, or $\overline{A}$ is an elementary abelian 2-group.*

Obviously, if $Un(FG)$ contains a non-abelian free subgroup, it cannot satisfy a group identity. This means that if one wants to classify group rings whose unitary units satisfy a group identity, one has to concentrate on the class of groups appearing in Theorem 3.1. This inspired the work of Giambruno and Polcino Milies [5], where they reached this objective when this identity is 2-*free*, that is, it does not vanish on elements of order 2 (for instance nilpotency and the bounded Engel condition), provided F a field of characteristic 0 and G a torsion group, as shown in the following

Theorem 3.2 *Let $p = 0$ and T the set of torsion elements of G. If $Un(FG)$ satisfies a group identity which is 2-free, then T is a subgroup of G and one of the following conditions holds:*

(a) *T is abelian,*
(b) *$A := \langle g \mid g \in T,\ o(g) \neq 2 \rangle$ is a normal abelian subgroup of G and $(T \setminus A)^2 = \{1\}$, or*
(c) *T contains an elementary abelian 2-subgroup B of index 2.*

Conversely, if G is a torsion group satisfying one of the above conditions, then $Un(FG)$ satisfies a group identity.

Going through the details of [5], we notice that one of the main issues is the relation between the existence of free groups in $Un(FG)$ and the nilpotency of the Lie algebra of skew elements, $FG^- := \{x \mid x \in FG,\ x^* = -x\}$, of FG (that should not be surprising if one looks at the general linear group, as stressed in the Introduction of [5]). More generally, the Lie structure of FG^- (which has been extensively investigated in the last decades: for an overview we refer to [20]) seems to deeply influence the structure of $Un(FG)$ and Lee, Sehgal and Spinelli used Lie methods as a main tool to explore the conditions under which the subgroup of unitary units of FG satisfies certain group identities ([19] and [21]).

In particular, in [21] they studied the question of when $Un(FG)$ is both bounded Engel and solvable (as a natural extension of what done in [19]). Of course, every nilpotent group satisfies these properties, but even the bounded Engel property and solvability together are not enough to guarantee nilpotence (see Section 4 of [27] for examples): according to a classical result of Gruenberg, under these hypotheses one can only conclude that it is locally nilpotent. However, Fisher, Parmenter and Sehgal [4] showed that if FG is not modular (recall that FG is said to be modular if $p > 0$ and G has an element of order p), then whenever $\mathcal{U}(FG)$ is both bounded Engel and solvable, it is also nilpotent. Inspired by this result, Lee,

Sehgal and Spinelli asked if it is sufficient to assume that the unitary units are both bounded Engel and solvable, in order to prove that the entire unit group is nilpotent. They showed that, under certain restrictions upon the field, this is the case.

Theorem 3.3 *Let F be an infinite field of characteristic $p > 2$ and G a group such that FG is modular. Then the following are equivalent:*

(i) *$Un(FG)$ is bounded Engel and solvable;*
(ii) *$\mathcal{U}(FG)$ is nilpotent;*
(iii) *G is nilpotent and p-abelian.*

When FG is not modular, one has

Theorem 3.4 *Let $p \neq 2$ and G a torsion group such that FG is not modular and G has no elements of order 2. Then $Un(FG)$ is bounded Engel and solvable if, and only if, G is abelian.*

However, restricting the field suitably, we obtain

Theorem 3.5 *Let F be an algebraically closed field of characteristic $p \neq 2$ and G a group such that FG is not modular. Then the following are equivalent:*

(i) *$Un(FG)$ is bounded Engel and solvable;*
(ii) *$\mathcal{U}(FG)$ is nilpotent;*
(iii) *G is nilpotent and the torsion elements of G are central.*

The assumption on the field in the above statement is not imposed frivolously; indeed, in [19], it was pointed out that if F is the field of 5 elements and G is the dihedral group of order 8, then $Un(FG)$ is nilpotent; however, $\mathcal{U}(FG)$ is neither bounded Engel nor solvable (see Theorems 5.2.1 and 6.2.2 of [17]). Thus Theorem 3.4 fails for an arbitrary field if we allow 2-elements.

In particular, we see that if F is algebraically closed and $p \neq 2$, then $\mathcal{U}(FG)$ is nilpotent whenever $Un(FG)$ is nilpotent. This is quite different from the situation for the symmetric units, where there are counterexamples (recall that when G is isomorphic to a Hamiltonian 2-group, $\mathcal{U}^+(FG)$ is commutative, but, according to Proposition 4.2.6 of [17], $\mathcal{U}(FG)$ is not nilpotent). Anyway, the state of the art for what concerns unitary units is still too fragmentary and we are very far from a knowledge comparable with that of symmetric units. A classification of when $Un(FG)$ satisfies a group identity would be a very appreciable result, but, at the moment, the tools to attack it still seem unclear. Maybe, as in the ordinary case, a decisive step could be to give an answer to the following question, which is an analogue of Hartley's Conjecture,

Question 3.6 *Let G be a torsion group (and, if it helps, F an infinite or algebraically closed field of characteristic $p \neq 2$). Is it true that if $Un(FG)$ satisfies a group identity, then FG satisfies a polynomial identity?*

References

1. Amitsur, S.A.: Identities in rings with involution. Isr. J. Math. **7**, 63–68 (1969)
2. Bateman, J.M.: On the solvability of unit groups of group algebras. Trans. Am. Math. Soc. **157**, 73–86 (1971)
3. Bovdi, A.: Group algebras with a solvable group of units. Commun. Algebra **33**, 3725–3738 (2005)
4. Fisher, J.L., Parmenter, M.M., Sehgal, S.K.: Group rings with solvable n-Engel unit groups. Proc. Am. Math. Soc. **59**, 195–200 (1976)
5. Giambruno, A., Polcino Milies, C.: Unitary units and skew elements in group algebras. Manuscripta Math. **111**, 195–209 (2003)
6. Giambruno, A., Jespers, E., Valenti, A.: Group identities on units of rings. Arch. Math. **63**, 291–296 (1994)
7. Giambruno, A., Sehgal, S.K., Valenti, A.: Group algebras whose units satisfy a group identity. Proc. Am. Math. Soc. **125**, 629–634 (1997)
8. Giambruno, A., Sehgal, S.K., Valenti, A.: Symmetric units and group identities. Manuscripta Math. **96**, 443–461 (1998)
9. Giambruno, A., Sehgal, S.K., Valenti, A.: Group identities on of group algebras. J. Algebra **226**, 488–504 (2000)
10. Giambruno, A., Polcino Milies, C., Sehgal, S.K.: Group identities on symmetric units. J. Algebra **322**, 2801–2815 (2009)
11. Giambruno, A., Polcino Milies, C., Sehgal, S.K.: Star-group identities and group of units. Arch. Math. **95**, 501–508 (2010)
12. Giambruno, A., Polcino Milies, C., Sehgal, S.K.: Star-group identities on units of group algebras: the non-torsion case. Forum Math. **30**, 213–225 (2018)
13. Gonçalves, J.Z., Mandel, A.: Semigroup identities on units of group algebras. Arch. Math. **57**, 539–545 (1991)
14. Gonçalves, J.Z., Passman, D.S.: Unitary units in group algebras. Israel J. Math. **125**, 131–155 (2001)
15. Goodaire, E.J., Jespers, E., Polcino Milies, C.: Alternative Loop Rings. North-Holland, Amsterdam (1996)
16. Jespers, E., Ruiz Marin, M.: On symmetric elements and symmetric units in group rings. Commun. Algebra **34**, 727–736 (2006)
17. Lee, G.T.: Group identities on units and symmetric units of group rings. In: Algebra and Applications, vol. 12. Springer, London (2010)
18. Lee, G.T.: A survey on $*$-group identities on units of group rings. Commun. Algebra **40**, 1–28 (2012)
19. Lee, G.T., Sehgal, S.K., Spinelli, E.: Group rings whose unitary units are nilpotent. J. Algebra **410**, 343–354 (2014)
20. Lee, G.T., Sehgal, S.K., Spinelli, E.: Lie identities on skew elements in group algebras. In: Lie Algebras and Related Topics. Contemporary Mathematics, vol. 652, pp. 103–121. American Mathematical Society, Providence (2015)
21. Lee, G.T., Sehgal, S.K., Spinelli, E.: Bounded Engel and solvable unitary units in group rings. J. Algebra **501**, 225–232 (2018)
22. Liu, C.H.: Group algebras with units satisfying a group identity. Proc. Am. Math. Soc. **127**, 327–336 (1999)
23. Liu, C.H., Passman, D.S.: Group algebras with units satisfying a group identity II. Proc. Am. Math. Soc. **127**, 337–341 (1999)
24. Passman, D.S.: The Algebraic Structure of Group Rings. Wiley, New York (1977)

25. Passman, D.S.: Group algebras whose units satisfy a group identity II. Proc. Am. Math. Soc. **125**, 657–662 (1997)
26. Sehgal, S.K., Valenti, A.: Group algebras with symmetric units satisfying a group identity. Manuscripta Math. **119**, 243–254 (2006)
27. Traustason, G.: Engel groups. In: Groups St Andrews 2009 in Bath, vol. 2, pp. 520–550. London Mathematical Society Lecture Note Series 388. Cambridge University Press, Cambridge (2011)